T0205959

ELECTRONS AND PHONONS IN SEMICONDUCTOR MULTILAYERS

Second Edition

Advances in nanotechnology have generated semiconductor structures that are only a few molecular layers thick, and this has important consequences for the physics of electrons and phonons in such structures. This book describes in detail how confinement of electrons and phonons in quantum wells and wires affects the physical properties of the semiconductor.

This second edition contains four new chapters on spin relaxation, based on recent theoretical research; the hexagonal wurtzite lattice; nitride structures, whose novel properties stem from their spontaneous electric polarization; and on terahertz sources, which includes an account of the controversies that surrounded the concepts of Bloch oscillations and Wannier–Stark states.

The book is unique in describing the microscopic theory of optical phonons, the radical change in their nature due to confinement and how they interact with electrons. It will interest graduate students and researchers working in semiconductor physics.

B. K. RIDLEY is Professor Emeritus in the School of Computer Science and Electronic Engineering at the University of Essex, Colchester. He is a Fellow of the Royal Society, and was awarded the Paul Dirac Prize and Medal in 2001 by the Institute of Physics.

ELECTRONS AND PHONONS IN SEMICONDUCTOR MULTILAYERS

Second Edition

B. K. RIDLEY

University of Essex

CAMBRIDGE
UNIVERSITY PRESS

University Printing House, Cambridge CB2 8BS, United Kingdom

Cambridge University Press is part of the University of Cambridge.

It furthers the University's mission by disseminating knowledge in the pursuit of education, learning and research at the highest international levels of excellence.

www.cambridge.org
Information on this title: www.cambridge.org/9781107424579

First published 2009
First paperback edition 2014

A catalogue record for this publication is available from the British Library

Library of Congress Cataloguing in Publication data
Ridley, B. K.
Electrons and phonons in semiconductor multilayers / B.K. Ridley. –
2nd ed.
p. cm.
Includes bibliographical references and index.
ISBN 978-0-521-51627-3 (hardback)
1. Semiconductors–Surfaces. 2. Layer structure (Solids)
3. Electron-phonon interactions. 4. Quantum wells.
I. Title.
QC611.6.S9R54 2009
537.6′226–dc22 2009000778

ISBN 978-0-521-51627-3 Hardback
ISBN 978-1-107-42457-9 Paperback

Contents

Preface		*page* xi
Introduction		1
1	Simple Models of the Electron–Phonon Interaction	9
1.1	General Remarks	9
1.2	Early Models of Optical-Phonon Confinement	10
1.2.1	The Dielectric-Continuum (DC) Model	11
1.2.2	The Hydrodynamic (HD) Model	16
1.2.3	The Reformulated-Mode (RM) Model	18
1.2.4	Hybrid Modes	21
1.3	The Interaction of Electrons with Bulk Phonons	22
1.3.1	The Scattering Rate	22
1.3.2	The Coupling Coefficients	24
1.3.3	The Overlap Integral in 2D	27
1.3.4	The 2D Rates	29
1.3.5	The 1D Rates	34
1.4	The Interaction with Model Confined Phonons	35
2	Quantum Confinement of Carriers	42
2.1	The Effective-Mass Equation	42
2.1.1	Introduction	42
2.1.2	The Envelope-Function Equation	44
2.1.3	The Local Approximation	46
2.1.4	The Effective-Mass Approximation	48
2.2	The Confinement of Electrons	49
2.3	The Confinement of Holes	53
2.4	Angular Dependence of Matrix Elements	62
2.5	Non-Parabolicity	64
2.6	Band-Mixing	66

3 Quasi-Continuum Theory of Lattice Vibrations 67
3.1 Introduction 67
3.2 Linear-Chain Models 69
3.2.1 Bulk Solutions 69
3.2.2 Interface between Nearly Matched Media 71
3.2.3 Interface between Mismatched Media 75
3.2.4 Free Surface 75
3.2.5 Summary 76
3.3 The Envelope Function 76
3.4 Non-Local Operators 78
3.5 Acoustic and Optical Modes 80
3.6 Boundary Conditions 83
3.7 Interface Model 85
3.8 Summary 91
Appendix: The Local Approximation 94
4 Bulk Vibrational Modes in an Isotropic Continuum 97
4.1 Elasticity Theory 97
4.2 Polar Material 104
4.3 Polar Optical Waves 105
4.4 Energy Density 107
4.5 Two-Mode Alloys 114
5 Optical Modes in a Quantum Well 119
5.1 Non-Polar Material 119
5.2 Polar Material 122
5.3 Barrier Modes: Optical-Phonon Tunnelling 127
5.4 The Effect of Dispersion 137
5.5 Quantization of Hybrid Modes 137
6 Superlattice Modes 141
6.1 Superlattice Hybrids 141
6.2 Superlattice Dispersion 144
6.3 General Features 148
6.4 Interface Polaritons in a Superlattice 154
6.5 The Role of LO and TO Dispersion 155
6.6 Acoustic Phonons 157
7 Optical Modes in Various Structures 160
7.1 Introduction 160
7.2 Monolayers 160
7.2.1 Single Monolayer 162
7.2.2 Double Monolayer 166
7.3 Metal–Semiconductor Structures 170

7.4 Slab Modes 173
7.5 Quantum Wires 176
7.6 Quantum Dots 181
8 Electron–Optical Phonon Interaction in a Quantum Well 182
8.1 Introduction 182
8.2 Scattering Rate 183
8.3 Scattering Potentials for Hybrids 184
8.4 Matrix Elements for an Indefinitely Deep Well 185
8.5 Scattering Rates for Hybrids 187
8.6 Threshold Rates 189
8.7 Scattering by Barrier LO Modes 192
8.8 Scattering by Interface Polaritons 194
8.9 Summary of Threshold Rates in an Indefinitely Deep Well 197
8.9.1 Intrasubband Rates 197
8.9.2 Intersubband Rates 198
8.10 Comparison with Simple Models 199
8.11 The Interaction in a Superlattice 202
8.12 The Interaction in an Alloy 205
8.13 Phonon Resonances 206
8.14 Quantum Wire 208
8.15 The Sum-Rule 209
Appendix: Scalar and Vector Potentials 212
9 Other Scattering Mechanisms 217
9.1 Charged-Impurity Scattering 217
9.1.1 Introduction 217
9.1.2 The Coulomb Scattering Rate 220
9.1.3 Scattering by Single Charges 221
9.1.4 Scattering by Fluctuations in a Donor Array 223
9.1.5 An Example 225
9.2 Interface-Roughness Scattering 227
9.3 Alloy Scattering 230
9.4 Electron–Electron Scattering 231
9.4.1 Basic Formulae for the 2D Case 231
9.4.2 Discussion 234
9.4.3 Electron–Hole Scattering 236
9.5 Phonon Scattering 236
9.5.1 Phonon–Phonon Processes 236
9.5.2 Charged-Impurity Scattering 239
9.5.3 Alloy Fluctuations and Neutral Impurities 240
9.5.4 Interface-Roughness Scattering 241

10 Quantum Screening 244
10.1 Introduction 244
10.2 The Density Matrix 245
10.3 The Dielectric Function 248
10.4 The 3D Dielectric Function 250
10.5 The Quasi-2D Dielectric Function 252
10.6 The Quasi-1D Dielectric Function 259
10.7 Lattice Screening 265
10.8 Image Charges 266
10.9 The Electron-Plasma/Coupled-Mode Interaction 268
10.10 Discussion 272
11 The Electron Distribution Function 275
11.1 The Boltzmann Equation 275
11.2 Net Scattering Rate by Bulk Polar-Optical Phonons 276
11.3 Optical Excitation 278
11.4 Transport 281
11.4.1 The 3D Case 284
11.4.2 The 2D Case 286
11.4.3 The 1D Case 288
11.4.4 Discussion 289
11.5 Acoustic-Phonon Scattering 290
11.5.1 The 3D Case 291
11.5.2 The 2D Case 293
11.5.3 The 1D Case 294
11.5.4 Piezoelectric Scattering 296
11.6 Discussion 296
11.7 Acoustic-Phonon Scattering in a Degenerate Gas 300
11.7.1 Introduction 300
11.7.2 Energy- and Momentum-Relaxation Rates 300
11.7.3 Low-Temperature Approximation 304
11.7.4 The Electron Temperature 306
11.7.5 The High-Temperature Approximation 306
12 Spin Relaxation 311
12.1 Introduction 311
12.2 The Elliot–Yafet process 313
12.3 The D'yakonov–Perel Process 317
12.3.1 The DP Mechanism in a Quantum Well 322
12.3.2 Quantum Wires 324
12.4 The Rashba Mechanism 326
12.5 The Bir–Aranov–Pikus Mechanism 326

12.6 Hyperfine Coupling 329
Appendix 1 332
Appendix 2 333
Appendix 3 335
13 Electrons and Phonons in the Wurtzite Lattice 336
13.1 The Wurtzite Lattice 336
13.2 Energy Band Structure 338
13.3 Eigenfunctions 340
13.4 Optical Phonons 343
13.5 Spontaneous Polarization 346
Appendix 1 Symmetry 347
14 Nitride Heterostructures 349
14.1 Single Heterostructures 349
14.2 Piezoelectric Polarization 351
14.3 Polarization Model of Passivated HFET with Field Plate 354
14.4 The Polarization Superlattice 358
14.4.1 Strain 358
14.4.2 Deformation Potentials 359
14.4.3 Fields 359
14.5 The AlN/GaN Superlattice 360
14.6 The Quantum-Cascade Laser 366
Appendix Airy Functions 368
15 Terahertz Sources 369
15.1 Introduction 369
15.2 Bloch Oscillations 370
15.3 Negative-Mass NDR 373
15.3.1 The Esaki–Tsu Approach 375
15.3.2 Lucky Drift 376
15.3.3 The Hydrodynamic Model 377
15.4 Ballistic Transport 378
15.4.1 Optical-Phonon-Determined Transit-Time Oscillations 379
15.4.2 Transit-Time Oscillations in a Short Diode 379
15.4.3 Negative-Mass NDR 380
15.4.4 Bloch Oscillations 383
15.5 Femtosecond Generators 387
15.5.1 Optical Non-Linear Rectification. 387
15.5.2 Surge Current 388
15.5.3 Dember Diffusion 388
15.5.4 Coherent Phonons 389
15.5.5 Photoconductive Switch 389

15.6 CW Generators 389
15.6.1 Photomixing 389
15.6.2 Quantum-Cascade Lasers 390
Appendix 392
Appendix 1 The Polar-Optical Momentum-Relaxation Time in a 2D Degenerate Gas 393
Appendix 2 Electron/Polar Optical Phonon Scattering Rates in a Spherical Cosine Band 395
References 397
Index 406

Preface

A Second Edition has given me several opportunities, which I have grasped with some eagerness. The first involved a matter of house-keeping – the correction of typos and an equation (Eq. (11.31)) in the First Edition. The second has allowed me to add the theory of the elasticity of optical modes, which became formulated too late to be included in the First Edition, but which resolved some uncertainties regarding mechanical boundary conditions. Given the current interest in hot phonons in high-power devices, I have updated the section on phonon lifetime. But the greatest opportunity was to expand the book to include some of the topics that grew in significance during the decade following the writing of the First Edition. Four new chapters focus on spin relaxation, the III–V nitrides, and the generation of terahertz radiation. In the burgeoning technology of spintronics, the rate of decay of the spin-polarization of the electron gas is a crucial parameter, and I have reviewed the mechanisms for this in both bulk and low-dimensional material. The advance of growth techniques and the technological need for higher-powered devices and for visible LEDs have thrust AlN, GaN, and InN and their alloys into the forefront of semiconductor physics. New properties associated with the hexagonal lattice have presented a challenge, and these are described in the chapter on electrons and phonons in the wurtzite lattice, and their role in heterostructures and multilayers is reviewed in a further chapter. Considerable effort is being made to close the gap in the electromagnetic spectrum, roughly between 300GHz and 30THz, in the emission spectrum of sources of radiation. The final chapter focuses on some of the physical mechanisms that are used or proposed to fill this gap. Terahertz radiation has applications in astrophysics, study of the atmosphere, biology, medicine, security screening, illicit material detection, non-destructive analysis, communications, and ultra-fast spectroscopy. The development of a compact, coherent continuous-wave, solid-state source is therefore of considerable technological importance.

It was timely, therefore, that John Fowler of CUP suggested I tackle a second edition of my book. His efforts and those of his CUP colleagues, Dawn Preston, Lindsay Barnes and Anne Rix, in gently guiding an author, only partly computer literate, through the mysteries of electronic publishing, were much appreciated. I was also much delighted by Ann Ridley's cover design of the wurtzite lattice that has captured some of its quantum mystery.

In preparing this new edition I have been generously helped by friends, who also happen to be colleagues. My warm thanks go to Angela Dyson, Ceyhun Bulutay, Martin Vaughan, and Alan Brannick for much practical assistance, and to Nic Zakhlenuik for his attempts (not always successful) to curb what passes for intuition in favour of formal proof. But my timeless gratitude is for my wife, Sylvia, who has borne the frequent fate over the last few months of book-widowhood with unfailing good nature.

Thorpe-le-Soken BKR

Introduction

It is the intellect's ambition to seem
no longer to belong to an individual.
Human, All Too Human, *F. Nietzsche*

If one tells the truth, one is sure, sooner or later, to be found out.
(Phrases and Philosophies for the Use of the Young, *O. Wilde*)

This book has grown out of my own research interests in semiconductor multilayers, which date from 1980. It therefore runs the risk of being far too limited in scope, of prime interest only to the author, his colleagues and his research students. I hope that this is not the case, and of course I believe that it will be found useful by a large number of people in the field; otherwise I would not have written it. Nevertheless, knowledgeable readers will remark on the lack of such fashionable topics as the quantum-Hall effect, Coulomb blockade, quantized resistance, quantum tunnelling and any physical process that can be studied only in the millikelvin regime of temperature. This has more to do with my own ignorance than any lack of feeling that these phenomena are important. My research interests have not lain there. My priorities have always been to try to understand what goes on in practical devices, and as these work more or less at room temperature, the tendency has been for my interest to cool as the temperature drops. The essential entities in semiconductor multilayers are electrons and phonons, and it has seemed to me fundamental to the study and exploitation of these systems that the effect of confinement on these particles and their interactions be fully understood. This book is an attempt to discuss what understanding has been achieved and to discover where it is weak or missing. Inevitably it emphasizes concepts over qualitative description, and experimentalists may find the paucity of experimental detail regrettable. I hope not, though I would appreciate their point, but the book is long enough as it is, and there are excellent review articles in the literature.

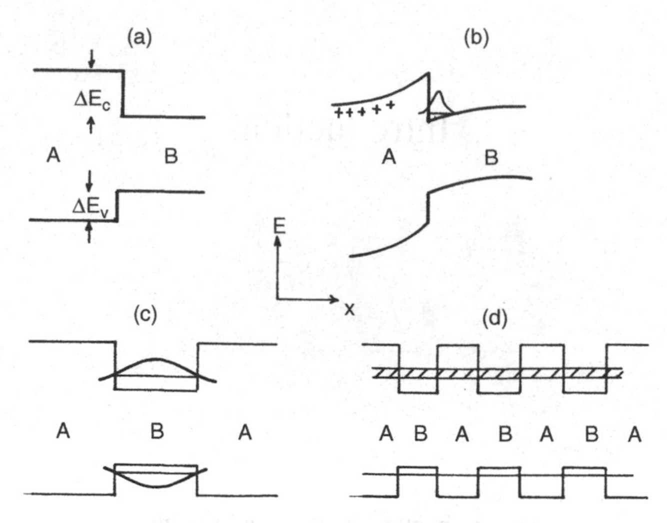

Fig. 1 Types of multilayered structures: (a) Single heterojunction, (b) Modulation-doped heterojunction, (c) Quantum well, (d) Superlattice.

At the risk of being boring, let me remind the reader what a semiconductor multilayer is about. There are several kinds of layered structure, of which those shown in Fig. 1 are the most common. They are interesting only insofar as they have a dimension that is smaller than the coherence length of an electron, in which case the electron becomes quantum confined between two potential steps. The free motion of the electron is then confined to a plane and many of its properties stem from this two-dimensional (2D) space. Electrons are not 2D objects and never could be, but for brevity they are usually referred to as 2D electrons when they are in the sort of layered structure shown in Fig. 1. The most striking effects are the quantization of energy into subbands, as depicted in Fig. 2 and Fig. 3, and the consequent transformation of the density-of-states function as shown in Fig. 4 for 3D, 2D, 1D and 0D electrons. Scattering events must now be classified into intrasubband, intersubband and capture processes, as indicated in Fig. 3. All of this is qualitatively well understood. The real problem here concerns the description of the confined-electron wavefunction (Fig. 5), which involves solving the Schrödinger equation in an inhomogeneous system. This problem is as basic as one can get in a multilayer system and calls for a comprehensive pseudopotential band structure computation. But for me, and anyone interested in further describing scattering events, this approach is too computer-intensive and inflexible, though in some cases there may be little alternative. An attractive (because simple) approximation is to take as known and unchanged the Bloch functions in each bulk medium and connect them at the interface, satisfying the

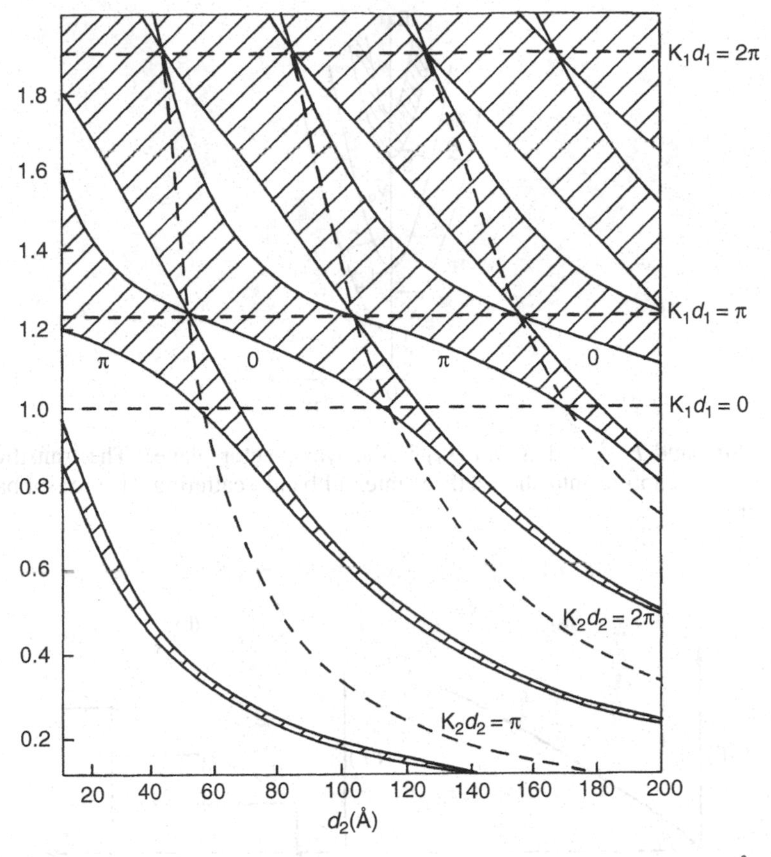

Fig. 2 Subband structures in a GaAs/$Al_{0.3}Ga_{0.7}As$ superlattice with 100 Å ($=d_1$) barriers. The energy is in units of $V_0 = 0.2$eV and d_2 is the well-width.

usual condition of current continuity. An even more attractive approximation is to forget about the cell-periodic part of each Bloch function and have a rule for joining the envelope functions. This is the effective-mass approximation. It is widely used and the energy levels it predicts are close to what is observed in a number of practical cases. Why it works so well is by no means obvious. A discussion of this basic issue will be found in Chapter 2.

Identical problems are found in connection with the confinement of optical phonons (Fig. 6). Adjacent media with different elastic and dielectric properties obviously affect the propagation of elastic waves, but whereas the appropriate boundary conditions for long-wavelength acoustic modes are well known, those for optical modes are not. The interaction between electrons and optical phonons is arguably the most important in semiconductor physics so it is truly important to know how optical modes are confined in a layer. Once again one can resort to

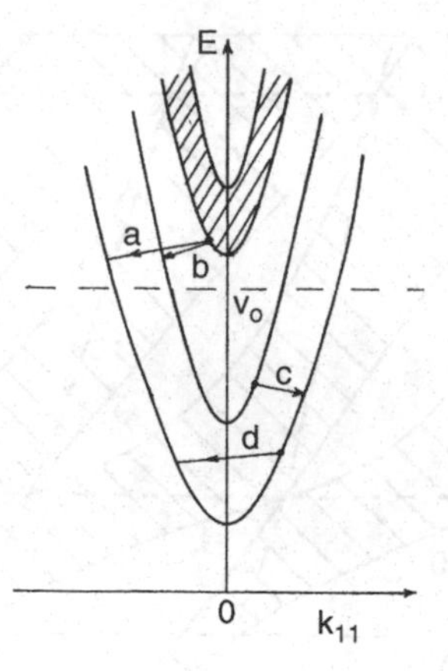

Fig. 3 Subband E–k_{11}, diagram (k_{11} is the wavevector plane). The transitions denote: a, b, capture into the well; c, intersubband scattering; d, intrasubband scattering.

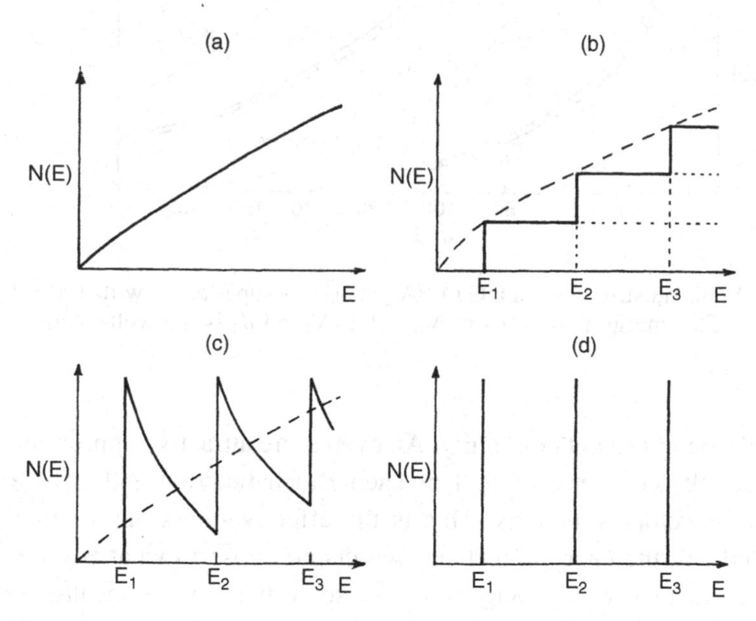

Fig. 4 Density states function of energy: (a) 3D; (b) 2D; (c) 10; (d) 0D.

extensive numerical computation of the relevant lattice electrodynamics, but, for the same reason as in the case of electrons, what is needed is a reliable envelope-function theory. This has proved to be difficult to come by, as the discussion in Chapter 3 amply illustrates.

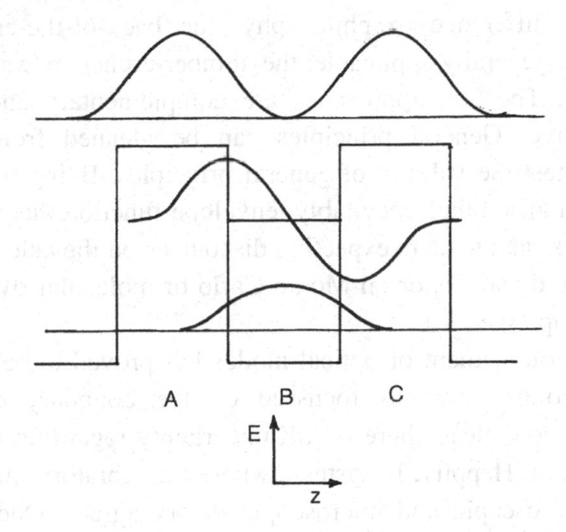

Fig. 5 Ground and excited state wavefunctions for an electron in a superlattice.

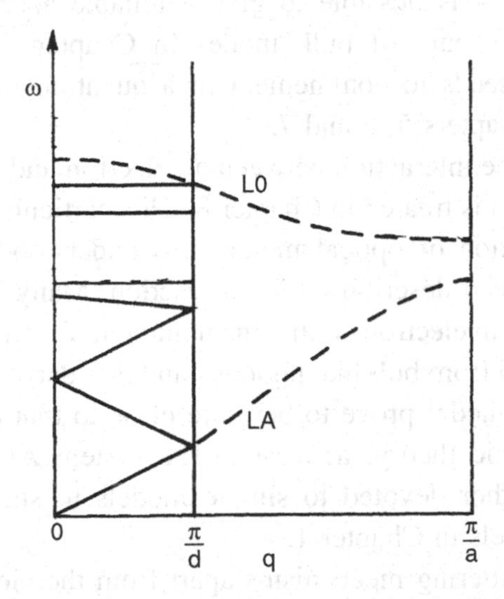

Fig. 6 Phonon dispersion influenced by mechanical/electrical mismatch at a heterojunction. Acoustic modes exhibit folding, optical modes exhibit confinement.

The whole question of envelope-function theory versus numerical computation – effectively continuum theory versus discrete theory – is extremely interesting. Though there is no earthly reason why it should do so, it tends to invoke a partisan response in the practitioners of the two approaches and may occasionally reveal a distinct difference in philosophy: the back-of-the-envelopers want general principles, generally applicable; the number-crunchers want to get at least one system right. The two approaches are complementary and, in principle, mutually supportive. General principles can be gleaned from computation; computation can test the validity of general principles. Being only dimly computer literate I am afraid that, inevitably, envelope-function theory and its usage permeate this book, and readers expecting discourses on the calculations of band structure or lattice dynamics or on Monte Carlo or molecular dynamics simulations will be disappointed.

Modelling the confinement of optical modes has proved to be unusually controversial. The controversy has focussed on the boundary conditions, and although much is now clear, there is still uncertainty regarding the elasticity of optical displacement. Happily, in systems where the vibratory mismatch is large the results of macroscopic and microscopic theory agree. (Oddly enough, the situation in perhaps the first system studied – the free-standing slab of NaCl – is still somewhat unclear.) In these systems the different polarizations of optical modes hybridize and it is possible to give a reliable account of confinement. Beginning with an account of bulk modes in Chapter 4, the description of vibratory modes proceeds to confinement in a quantum well, superlattice and other structures in Chapters 5, 6 and 7.

A central issue is the interaction between an electron and an optical mode in a quantum well, and this is treated in Chapter 8. This particular topic has a history. Before the hybridization of optical modes was understood there were several simple models advanced describing this interaction. Many important aspects of these models arose from electron confinement and remain relevant. Moreover, the scattering rate derived from bulklike phonons and that derived from the so-called dielectric-continuum model prove to be quite close to that derived from hybrid theory and microscopic theory, at least in the system AlAs/GaAs. It seemed worthwhile to an author devoted to simple models to start the book with an account of these models in Chapter 1.

There are other scattering mechanisms apart from the electron–phonon interaction. Charged-impurity and interface-roughness scattering often determine the mobility of electrons, and, of course, there is always alloy scattering in alloys. These processes are described in Chapter 9. Electron–electron scattering is also described there. It shares many grave problems with charged-impurity scattering that are difficult to solve, and it cannot be said with confidence that the rapid thermalization observed in optical experiments is fully explained by electron–electron scattering. A somewhat different topic, also in the same chapter, is the phenomenon of phonon scattering. Optical phonons are produced in abundance by hot electrons and may themselves become hot and take on any drifted motion of the electrons. How far they do so depends upon their lifetime and how rapidly they scatter. Hot-phonon effects are frequently observed when the density of electrons is large.

A substantial population of electrons has a direct effect on all scattering mechanisms through its screening action. Screening is a complex phenomenon, particularly so when more than one subband is involved. The whole topic occupies Chapter 10, and it could easily occupy a complete book.

I have a similar sentiment regarding Chapter 11, which gets down to the statistics of scattering in a population of electrons. Two-body scattering events are all very well to study on their own, but experiments generally look at populations, not individuals – average quantities, not instances – so it is necessary to look at the distribution function when the system is prodded optically or electrically, and to look at how energy and momentum are relaxed and how that depends on dimensionality. In this, for simplicity, I have sometimes committed the heresy of ignoring any phonon confinement. This is not serious for acoustic phonons.

In writing this book I have been keenly aware that references to all the work that has been done in this field are far from being comprehensive. I have told myself that this is not a review article, after all, but I feel that a suitable acknowledgement may not have been made in every case. I hope I am wrong.

Where I am most certainly not wrong is the feeling that much of this book owes its existence to my friends and colleagues at Essex, Cornell and elsewhere. Mohamed Babiker and Nick Constantinou have contributed enormously, and my account of the confinement of optical phonons and their interaction with electrons has been informed significantly by our collaboration over the years. There have been similarly important inputs from Collin Bennett and erstwhile collaborators Martyn Chamberlain, Rita Gupta and Frances Riddoch. The Platonic forms of theory are often grossly distorted in reality and I am grateful to my experimentalist colleagues Pam Bishop, Mike Daniels, Naci Balkan and Anthony Vickers for their attempts, not always successful, to anchor my feet to the ground. In a similar vein I am invigorated by the thought that semiconductors are actually useful, which my annual wintering in Lester Eastman's department at Cornell has reinforced in a delightfully stimulating way, especially by my interaction with graduate engineers like Luke Lester, Sean O'Keefe, Glen Martin, Matt Seaford and Trung LeTran. And, of course, I am grateful to Brad Foreman, whose graduate work at Cornell on quasi-continuum theory illuminated the whole field of optical-phonon confinement, and Mike Burt at British Telecom, whose analysis of the effective-mass approximation stimulated us all. My admiration of analytic theory generated in the former Soviet republics is almost as fervent as my admiration of its literature. It was, therefore, lucky for me to benefit from a

Royal Society Fellowship held by Nicolai Zakhleniuk from Kiev, taken up at Essex, and the insights gained regarding the electron distribution function were exceptionally germane.

Books are not written in a vacuum. This one was written at home in between, as is no doubt usual, myriad other things from professional duties and obligations to gardening. During its production my wife suffered all kinds of chores with remarkable good nature (which only occasionally degenerated into frightening hysteria) and I am indebted to her for her extraordinary efforts on my behalf. Nor did my offspring escape. In an extended encounter with what is known as an equation editor Aaron Ridley developed a remarkable talent, and in certain calculations of power-loss rates Melissa Ridley discovered numerical skills hitherto dormant. Both took to unfamiliar tasks with great good nature, and their contributions were much appreciated.

1

Simple Models of the Electron–Phonon Interaction

Teach us delight in simple things.
The Children's Song, *R. Kipling*

1.1 General Remarks

Evidently, the advent of mesoscopic layered semiconductor structures generated a need for a simple analytic description of the confinement of electrons and phonons within a layer and of how that confinement affected their mutual interaction. The difficulties encountered in the creation of a reliable description of excitations of one sort or another in layered material are familiar in many branches of physics. They are to do with boundary conditions. The usual treatment of electrons, phonons, plasmons, excitons, etc., in homogeneous bulk crystals simply breaks down when there is an interface separating materials with different properties. Attempts to fit bulk solutions across such an interface using simple, physically plausible connection rules are not always valid. How useful these rules are can be assessed only by an approach that obtains solutions of the relevant equations of motion in the presence of an interface, and there are two types of such an approach. One is to compute the microscopic band structure and lattice dynamics numerically; the other is to use a macroscopic model of long-wavelength excitations spanning the interface. The latter is particularly appropriate for generating physical concepts of general applicability. Examples are the quasi-continuum approach of Kunin (1982) for elastic waves, the envelope-function method of Burt (1988) for electrons and the wavevector-space model of Chen and Nelson (1993) for electromagnetic waves and excitons. Some of this will be discussed in later chapters in connection with the boundary conditions that are useful for electrons and phonons.

Effective-mass theory has proved to be remarkably good for describing the situation for electrons and holes (although a rigorous justification for its use has not been available until recently), but obtaining a theory of equivalent simplicity for optical phonons has been more problematic. Historically, boundary problems connected with phonons were ignored in calculations of the electron–phonon interaction, and only confinement effects associated with electrons were taken into account. A review of these calculations is given later. It turns out that in some cases the estimate of the scattering rate obtained by using a bulk phonon spectrum is a reasonable approximation. An account of this work is in any case a good introduction to the electron–phonon interaction, where differences from the bulk interaction are solely due to electron confinement.

Nevertheless, the folding of the acoustic-mode spectrum and the confinement of optical modes are readily observable in spectroscopy. In many cases the effect of folding can be ignored, as far as the interaction with electrons is concerned, but the same cannot be said about optical-mode confinement. Early models of optical-mode confinement are described later, after which we return to the polar interaction between electrons and optical phonons to describe the effect of phonon confinement using these early models. In doing so we will assume that the confinement of electrons is adequately described by particle-in-a-box, effective-mass theory, with boundary conditions entailing the continuity of wave amplitude and of $m^{*-1}d\psi/dz$ (Friedman, 1956; BenDaniel and Duke, 1966).

1.2 Early Models of Optical-Phonon Confinement

The effects of confinement (Fig. 1.1) are clearly seen in a number of studies of Raman scattering from zone-centre modes in the GaAs/$Ga_x\,Al_{1-x}$ system, and this work has been reviewed by Klein (1986), Cardona (1989, 1990) and Menendez (1989). The typical range of wavevectors, **k**, observed by Raman scattering is of order 10^4 to $10^5\,cm^{-1}$ and it is found that k is quantized in correspondence with the observed quantization of phonon frequency. In the GaAs/AlAs system, $k = n\pi/a_0(m+\gamma)$, where n is an integer greater than zero, a_0 is the thickness of a monolayer of GaAs in the direction perpendicular to the planes (usually [100], in which case $a_0 \approx 2.8\,Å$), m is the number of monolayers in the GaAs layer, and γ is a correction to the expected relationship $k = n\pi/a_0 m$, which would apply if the interfaces between GaAs and AlAs were infinitely rigid. It is found that $\gamma \approx 1$ as if the effective interfaces coincided with the Al ions immediately adjacent to the GaAs. Observed frequencies falling in between modes of this scheme are usually interpreted as interface modes of the type first described for an ionic slab by Fuchs and Kliewer (1965). Raman experiments thus confirm the existence of optical-mode confinement and the presence of other frequencies, plausibly identified as interface modes (Fig. 1.2).

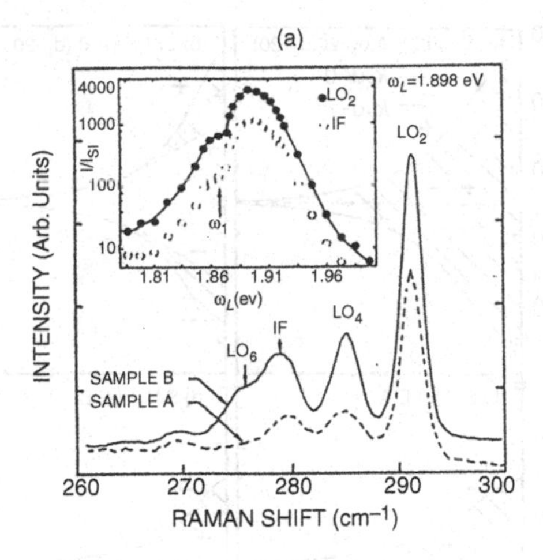

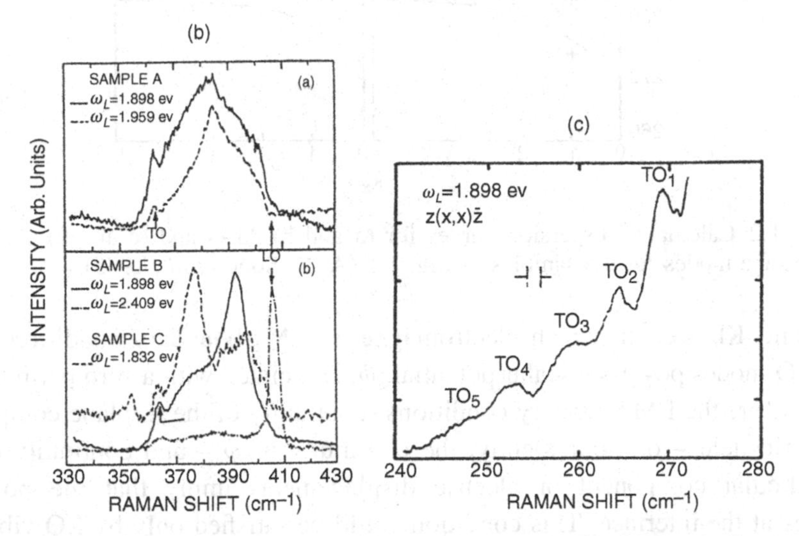

Fig. 1.1 Raman spectra for a GaAs(a)/AlAs(b) superlattice in the configuration at 10K. The scattered light is in resonance with the frequency corresponding to the first electron-heavy-hole transition ($\hbar\omega = 1.898$ eV): (a) even order GaAs LO confined modes and interface modes (IFs); (b) AlAs interface modes and off-resonance LO and TO modes; (c) GaAs TO modes. Sample A, $a = 20$, $b = 20$; sample B, $a = 20$, $b = 60$; sample C, $a = 60$, $b = 20$, Å (A. K. Sood *et al.*, 1985).

1.2.1 The Dielectric-Continuum (DC) Model

The search for a continuum model of phonon confinement was motivated by the need to describe the electron–phonon interaction in layered materials, particularly that involving longitudinally polarized (LO) modes, its being the strongest in bulk material at room temperature. The first approaches (Lassnig, 1984; Riddoch and Ridley, 1985; and Sawaki, 1986) used the dielectric-continuum (DC) model of

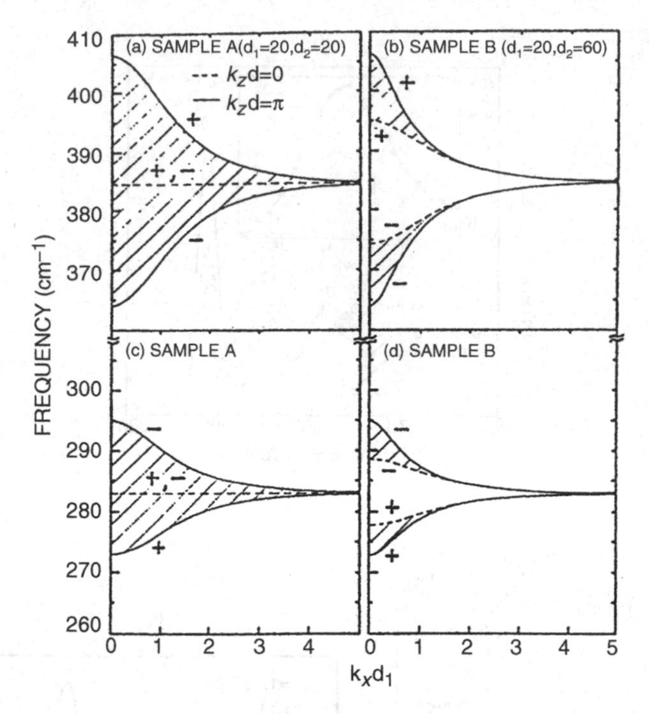

Fig. 1.2 Calculated dispersion curves for (a and b) AlAs and (c and d) GaAs interface modes for the samples of Fig. 1.1 (A. K. Sood *et al.*, 1985).

Fuchs and Kliewer, in which electromagnetic (EM) boundary conditions were used. LO modes possess a scalar potential, ϕ, associated with a zero permittivity, and therefore the EM boundary conditions (continuity of the in-plane component of electric field – or, equivalently, the continuity of ϕ – and continuity of the perpendicular component of electric displacement) imply that the potential vanishes at the interface. This condition could be satisfied only by LO vibration patterns with antinodes of u_z, the relative ionic displacement in the direction perpendicular to the layers, at the interfaces. The allowed modes could be classified as symmetric and antisymmetric, referring to the potential relative to the midplane of the layer (Fig. 1.3), viz.:

$$\text{symmetric}\qquad \begin{aligned} \phi \propto \mathbf{u}_r &= \mathbf{k}_{||} A_L e^{i\mathbf{k}_{||}\cdot\mathbf{r}} \cos k_L z \\ u_z &= 1 k_L A_L e^{i\mathbf{k}_{||}\cdot\mathbf{r}} \sin k_L z \\ k_L a &= (2n-1)\pi \end{aligned} \tag{1.1a}$$

$$\text{antisymmetric}\qquad \begin{aligned} \phi \propto \mathbf{u_r} &= i k_{||} A_L e^{i\mathbf{k}_{||}\cdot\mathbf{r}} \sin k_L z \\ u_z &= k_L A_L e^{i\mathbf{k}_{||}\cdot\mathbf{r}} \cos k_L z \\ k_L a &= 2n\pi \end{aligned} \tag{1.1b}$$

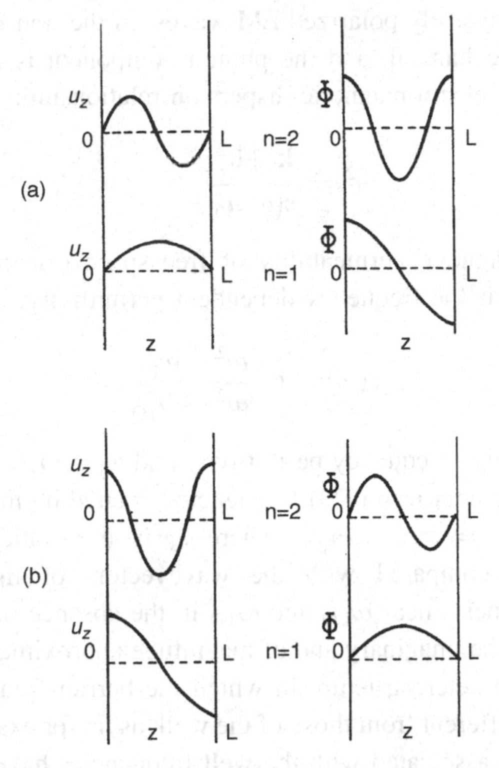

Fig. 1.3 LO mode patterns predicted by (a) the hydrodynamic model; (b) the dielectric-continuum model, u_z is the optical displacement along the confinement direction and φ is the scalar potential (proportional to the optical displacement in the plane).

We adopt here the common convention of orienting Cartesian axes so that the z axis is normal to the plane of the layer (corresponding to the direction of crystal growth), with **r** a position vector in the plane, and $\mathbf{k_L}$, $\mathbf{k}_{\|}$ the corresponding wavevector components. The layer lies in the range $-a/2 \leq z \leq a/2$. The factor A_L is a normalizing constant proportional to $(k_{\|}^2 + k_L^2)^{-1/2}$. (Note that these displacements satisfy the condition for a longitudinally polarized mode, namely $\nabla.\mathbf{u} = 0$.) The well-known dependence of the potential ϕ on the wavevector, characteristic of the Fröhlich interaction, $\phi \propto k^{-1}$, stimulated the observation that the strength of the interaction with electrons would be reduced by confinement in thin layers as a consequence of the quantization of $\mathbf{k_L}$.

The DC model also describes interface polaritons, which in the so-called unretarded limit (infinitely large velocity of light) are the quanta of Fuchs–Kliewer surface modes. Interface polaritons, often referred to as interface phonons (a practice that can be confusing given that mechanical interface modes may also exist), are transversely polarized EM waves. In the non-retarded limit, their energy is totally mechanical, and the photon component is zero. Nevertheless, they obey the usual electromagnetic dispersion relationship:

$$\omega^2 = \frac{(\mathbf{k}_{||}^2 + \mathbf{k}_z^2)}{\varepsilon(\omega)\mu_0} \tag{1.2}$$

where μ_0 is the magnetic permeability of free space (assuming non-magnetic material), and $\varepsilon(\omega)$ is the frequency-dependent permittivity:

$$\varepsilon(\omega) = \varepsilon_\infty \frac{\omega^2 - \omega_{\mathrm{LO}}^2}{\omega^2 - \omega_{\mathrm{TO}}^2} \tag{1.3}$$

in which ε_∞ is the high-frequency permittivity and ω_{LO}, ω_{TO} are the zone-centre LO and transversely polarized (TO) frequencies, related by the Lyddane–Sachs–Teller formula $\omega_{\mathrm{LO}}^2 = (\varepsilon_s/\varepsilon_\infty)\omega_{\mathrm{TO}}^2$, where ε_s is the static permittivity. For wavevectors large compared with the wavevectors of light, which would propagate at frequencies near ω_{LO} and ω_{TO} in the absence of ionic motion, the wavevector k_z must be imaginary and of magnitude approximately equal to $\mathbf{k}_{||}$. In a symmetric double heterostructure in which the barrier-frequencies of the LO and TO bands are different from those of the well, as in, for example, AlAs/GaAs, the interface modes associated with the well frequencies have the form

$$\text{symmetric} \quad \begin{aligned} \mathbf{u}_r &= \mathbf{k}_{||} A_{\mathrm{I}} e^{i\mathbf{k}_{||}\cdot\mathbf{r}} \cosh k_{||} z \\ u_z &= -ik_{||} A_{\mathrm{I}} e^{i\mathbf{k}_{||}\cdot\mathbf{r}} \sinh k_{||} z \\ \coth(k_{||} a/2) &= -r \end{aligned} \tag{1.4a}$$

$$\text{antisymmetric} \quad \begin{aligned} \mathbf{u}_{\mathbf{r}} &= i\mathbf{k}_{||} A_{\mathrm{I}} e^{i\mathbf{k}_{||}\cdot\mathbf{r}} \sinh k_{||} z \\ u_z &= k_{||} A_{\mathrm{I}} e^{i\mathbf{k}_{||}\cdot\mathbf{r}} \cosh k_{||} z \\ \tanh(k_{||} a/2) &= -r \end{aligned} \tag{1.4b}$$

where r is the permittivity ratio $\varepsilon(\omega)/\varepsilon_B(\omega)$, where $\varepsilon_B(\omega)$ is the permittivity of the barrier layers. The antisymmetric mode has the higher frequency (Fig. 1.4). Associated with any ionic displacement is an electric field $\mathbf{E}$ given by

$$\mathbf{E} = \frac{-s\rho_0 \mathbf{u}}{\varepsilon_\infty} \tag{1.5a}$$

$$s = \frac{\omega^2 - \omega_{\mathrm{TO}}^2}{\omega_{\mathrm{LO}}^2 - \omega_{\mathrm{TO}}^2} \tag{1.5b}$$

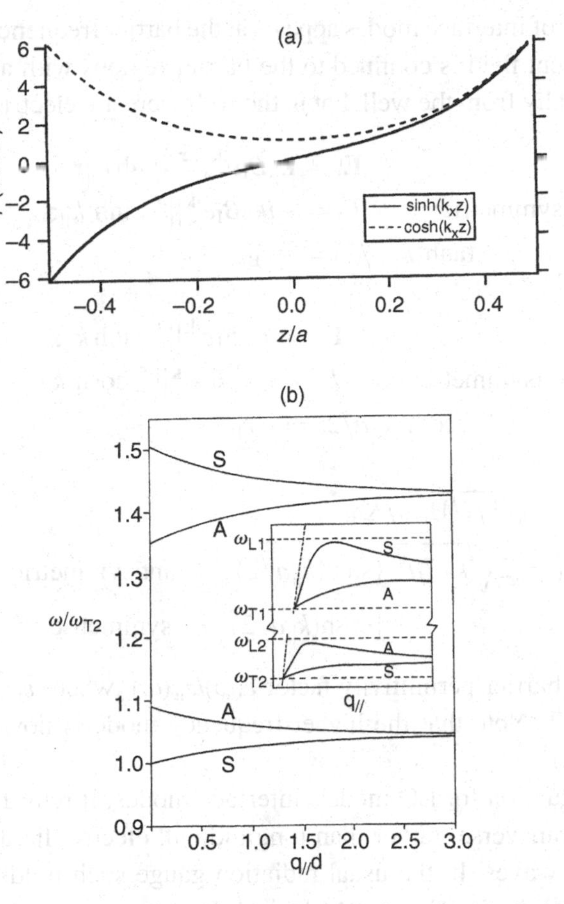

Fig. 1.4 (a) Scalar potential profiles of the two interface polaritons; (b) dispersion of GaAs and AlAs interface polaritons in a GaAs quantum well, width d. The inset shows the dispersion near $q_{11} = 0$ where the two modes coincide at the light line (O. Al-Dossary *et al.*, 1992).

where s is the field factor and ρ_0 is the ion charge density, so that when $\omega = \omega_{\mathrm{LO}}$, then $s = 1$ and $\mathbf{E} = \mathbf{E}_{\mathrm{LO}}$, the field of the LO mode. In the barrier region the fields fall away exponentially according to exp $(-z)$, for $z > a/2$. The amplitude has the following proportionality (arising out of energy normalization):

$$A_{\mathrm{I}} = \frac{\sqrt{\hbar/NM_r\omega}}{K_{\mathrm{I}}}$$
$$K_{\mathrm{I}} = \frac{\sqrt{(2k_{||}\ \sinh k_{||}a)}}{a} \tag{1.6}$$

where N is the number of unit cells and M_r is the reduced mass.

A similar pair of interface modes appears at the barrier frequencies. In this case the ionic displacement field is confined to the barrier regions with amplitudes that fall away exponentially from the well, but in the well there are electric fields of the form

$$\begin{aligned} &\mathbf{E}_r = \mathbf{k}_{||} B_{\mathrm{I}} e^{i\mathbf{k}_{||}\cdot\mathbf{r}} \cosh k_{||} z \\ \text{symmetric} \quad &E_z = -i k_{||} B_{\mathrm{I}} e^{i\mathbf{k}_{||}\cdot\mathbf{r}} \sinh k_{||} z \\ &\tanh(k_{||} a/2) = -r_{\mathrm{B}} \end{aligned} \tag{1.7a}$$

$$\begin{aligned} &\mathbf{E}_r = i\mathbf{k}_{||} B_{\mathrm{I}} e^{i\mathbf{k}_{||}\cdot\mathbf{r}} \sinh k_{||} z \\ \text{antisymmetric} \quad &E_z = -k_{||} B_{\mathrm{I}} e^{i\mathbf{k}_{||}\cdot\mathbf{r}} \cosh k_{||} z \\ &\coth(k_{||} a/2) = -r_{\mathrm{B}} \end{aligned} \tag{1.7b}$$

$$\begin{aligned} &B_{\mathrm{I}} = \sqrt{\hbar/NM_r\omega}/K_{\mathrm{IB}} \\ &K_{\mathrm{IB}} = 2\sqrt{k_{||}/a)}\ \{\sinh(k_{||} a/2) \quad \text{antisymmetric} \\ &\qquad\qquad\qquad\ \{\cosh(k_{||} a/2) \quad \text{symmetric} \end{aligned} \tag{1.7c}$$

where r_{B} is the barrier permittivity factor $\varepsilon(\omega)/\varepsilon_w(\omega)$, where $\varepsilon_w\ (\omega)$ is the permittivity of the well. Note that the higher-frequency mode is now symmetric in the well (Fig. 1.4(b)).

Unlike the situation for LO modes, interface modes, if retardation is taken into account, have transverse, rather than longitudinal, electric fields, consonant with their being EM waves. In the usual radiation gauge such fields are described by a vector potential **A** viz: $\mathbf{E} = -\partial\mathbf{A}/\partial t$; and the interaction with an electron is of the form (e/m) **A.p** rather than $e\phi$. Light interacts with electrons only weakly, and one might therefore expect that interface modes would contribute little to the scattering rate, but because these modes reside far from the light line this turns out not to be the case. Because the non-retarded waves are essentially mechanical in character, the electric field of an interface mode can be regarded as deriving from a scalar-potential reduced in magnitude from that of an LO mode only by the field factor s, Eq. (1.5b), and this turns out to be a valid assumption, as we will show in Chapter 8. As a consequence, any expected diminution of the scattering rate as a result of LO confinement is going to be countered to some degree by the action of interface modes.

1.2.2 The Hydrodynamic (HD) Model

The DC model thus provides a simple picture of the confinement of LO modes and of unretarded interface polaritons, and it allows the scattering rate to be determined in a straightforward way (Wendler *et al.*, 1987, 1988; Mori and Ando, 1989). Unfortunately, the DC model predicts the wrong mode patterns as seen in Raman scattering experiments. That this is so is evident from theory. The objection from theory is that the model takes no account of mechanical boundary conditions, and because the mechanical energy of optical modes is far greater than EM energy, this neglect cannot be easily justified. It is, moreover, obvious that the DC model could not be applied to a non-polar system such as Ge/Si, so it clearly does not provide an account of confinement that is generally applicable. The objection from experiment is that the mode patterns it predicts are simply not those observed. In spite of this it turns out that the DC model provides a reasonable estimate of scattering rates in some cases. This is discussed fully in Chapter 8.

An alternative to the DC model was presented by Babiker (1986), who replaced EM boundary conditions by mechanical boundary conditions used for acoustic waves in a fluid, namely the continuity of displacement and pressure. Quite apart from the question of connection rules, this hydrodynamic (HD) model contained the important step of introducing dispersion, which was necessary in order to describe pressure. In systems with large mismatch of frequency, the HD conditions reduce to $u_z = 0$ at the interface and one obtains (Fig 1.3)

$$\text{symmetric}\quad \begin{aligned} \phi \propto \mathbf{u}_r &= \mathbf{k}_{\|} A_L e^{i\mathbf{k}_{\|}\cdot\mathbf{r}} \cos k_L z \\ u_z &= i k_L A_L e^{i\mathbf{k}_{\|}\cdot\mathbf{r}} \sin k_L z \\ k_L a &= 2n\pi \end{aligned} \tag{1.8a}$$

$$\text{antisymmetric}\quad \begin{aligned} \phi \propto \mathbf{u}_r &= i\mathbf{k}_{\|} A_L e^{i\mathbf{k}_{\|}\cdot\mathbf{r}} \sin k_L z \\ u_z &= k_L A_L e^{i\mathbf{k}_{\|}\cdot\mathbf{r}} \cos k_L z \\ k_L a &= (2n-1)\pi \end{aligned} \tag{1.8b}$$

These modes have parities opposite to those of DC modes of the same order number. One consequence of this was that the scattering rates of electrons differed markedly from the rates predicted by the DC model (Ridley, 1989).

Another consequence affected resonant Raman scattering. Scattering at a frequency corresponding to an electron interband transition, besides being usefully strong, is mediated in a quantum well by both the deformation-potential interaction of the form **D.u**, where **D** is the optical deformation-potential constant, and the polar interaction of form $e\phi$. The two interactions have different polarization selection rules, which for backscattering configurations are $[z(x, y) - z]$ for the deformation interaction and $[z(x, x) - z]$ for the polar interaction, and they can thus be distinguished. For transitions involving electron states of the same parity, which are those principally involved in an interband process, the interaction must have even parity. The deformation interaction effectively probes u_z, and the polar interaction probes ϕ. We have defined the symmetry of the LO modes with respect to ϕ; that means that the polar interaction reveals symmetric modes, whereas the deformation interaction reveals antisymmetric modes. Because of dispersion each order of mode has a characteristic frequency, with the lowest-order mode ($n = 1$) having the highest frequency. If the DC model were correct for the AlAs/GaAs system, the $n = 1$ mode would show up in the configuration $[z(x, x) - z]$. In fact, the $n = 1$ mode shows up in the configuration $[z(x, y) - z]$ (Zucker *et al.*, 1984; Sood *et al.*, 1985), as predicted by the HD model.

1.2.3 The Reformulated-Mode (RM) Model

Microscopic calculations based on the bond-charge (Yip and Chang, 1984), shell (Richter and Strauch, 1987) or rigid ion model (Molinari *et al.*, 1986; Ren *et al.*, 1987, 1988; and Bechstedt and Gerecke, 1989) provided overwhelming confirmation that the ionic displacement vanished near the interfaces of the AlAs/GaAs system. Nevertheless, the mode patterns generated were found to be more complex than either the HD or the DC models predicted (Fig. 1.5). Certainly it was clear that the HD model could not satisfy EM boundary conditions for finite $\mathbf{k}_{\|}$, and it was also clear that the neglect of shear stresses in formulating the mechanical boundary conditions made the HD mode inapplicable to the problem of TO confinement. In spite of this, the HD modes, having the right parity, were used by a number of authors to estimate scattering rates (Trallero-Giner and Comas, 1988; Ridley, 1989; and Guillemot and Clerot, 1991).

One of the curious features discovered by microscopic calculations was the disappearance with increasing $\mathbf{k}_{\|}$ of the LO1 mode near $k = 0$, with the result that the highest-frequency mode became LO2. This latter mode was symmetric as predicted by the HD model, but its form was not simply sinusoidal in that the potential vanished at the interface as entailed in the DC model. In short, the actual mode patterns appeared to be some mixture of the HD and DC models. Huang and Zhu (1988) described the potential associated with these modes as follows (Fig. 1.6):

$$\text{symmetric} \qquad \phi \propto \cos\left[\frac{n\pi z}{(m+1)a_0}\right] - (-1)^{n/2} \qquad n = 2, 4, 6, \ldots \tag{1.9a}$$

$$\phi \propto \sin\left[\frac{\mu_0 \pi z}{(m+1)a_0}\right] + \frac{C_n z}{(m+1)a_0}$$

$$\text{antisymmetric} \qquad \tan\left(\frac{\mu_n \pi}{2}\right) = \frac{\mu_n \pi}{2} \qquad C_n = -2\sinh\left(\frac{\mu_n \pi}{2}\right) \tag{1.9b}$$

where $\mu_3 = 2.86$, $\mu_5 = 4.91$, $\mu_7 = 6.95$, etc.; m is the number of monolayers; and account has been taken of the penetration of the vibration as far as the first Al atom in the barrier. These reformulated modes represent the situation quite well, provided dispersion is neglected. When dispersion is included, as it is in the microscopic models, it is clear that there is significant mixing of LO and interface modes. In the absence of dispersion, however, a modified DC continuum model could be envisaged – the reformulated-mode (RM) model, which described bulklike modes according to Eq. (1.9) and interface modes according to the DC model – even though adoption of the DC model for interface modes ignores mechanical boundary conditions once more. Estimates of the scattering rates for inter- and intrasubband transitions using the RM model gave results close to the results using the DC model

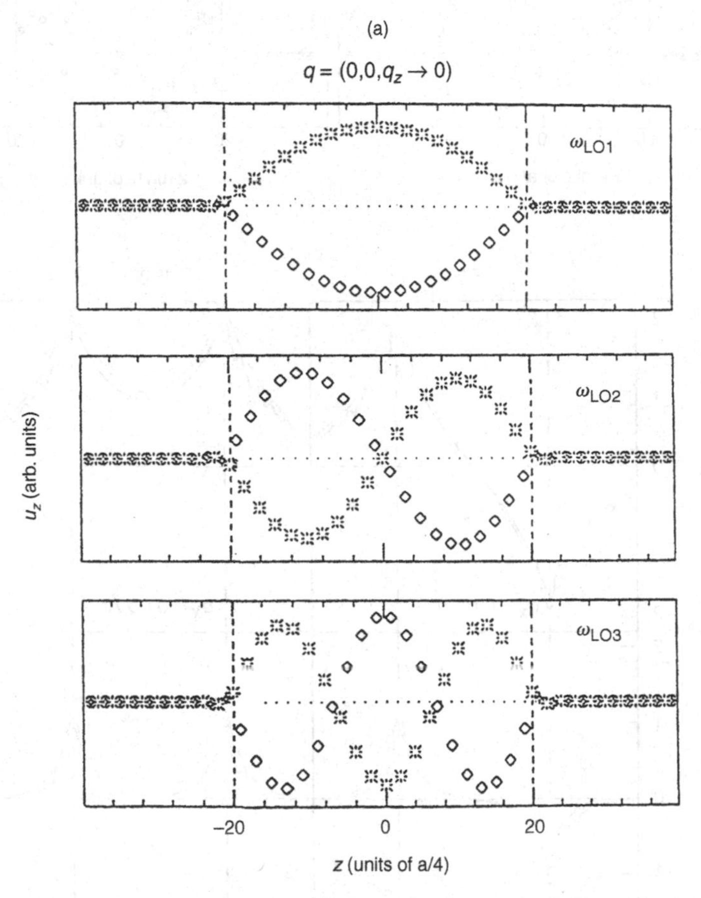

Fig. 1.5 Calculated optical displacement, u_z, for low-order LO modes in a GaAs quantum well: (a) $q = (0,0, q_z \to 0)$; (b) $q_z \neq 0$; (c) IP-like modes. For q_x (in-plane wavevector) $= 0$, the patterns agree with the HD model except that LO1 is missing when q_z differs from zero. For $q_x \neq 0$, the patterns become more complex (Rucker *et al.*, 1992).

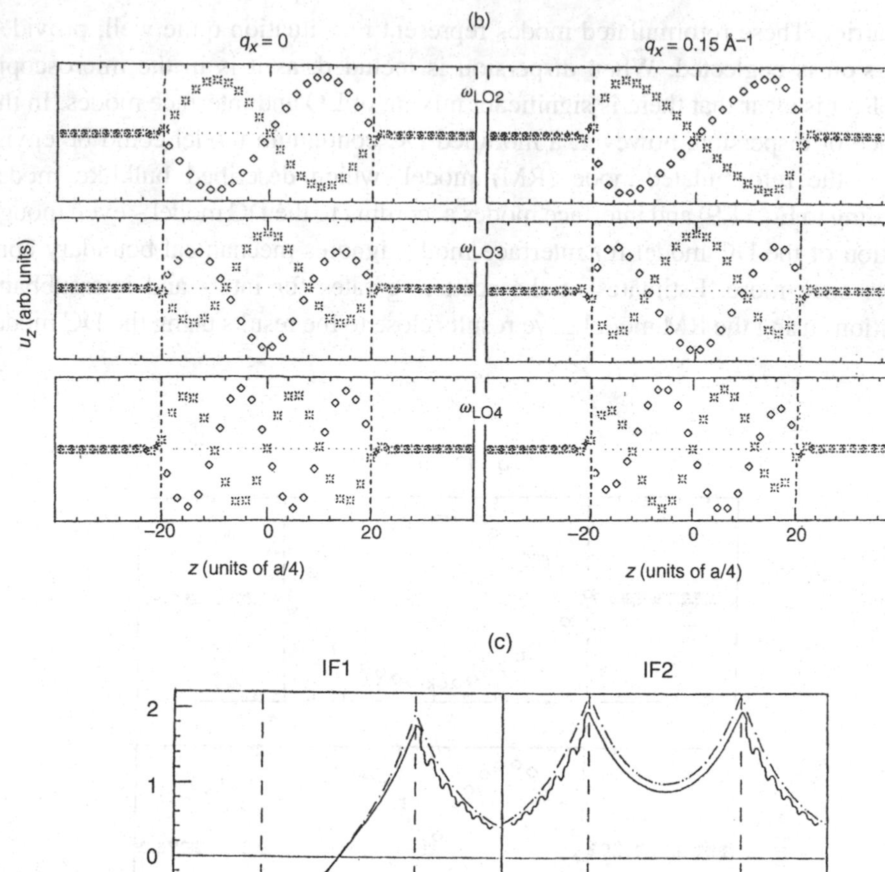
(b)
$q_x = 0$
$q_x = 0.15$ Å^{-1}
ω_{LO2}
ω_{LO3}
ω_{LO4}
u_z (arb. units)
−20
0
20
z (units of a/4)
(c)
IF1
IF2
$q_x = 0.05$ Å^{-1}
$q_x = 0.15$ Å^{-1}
potential (arb. units)
−60 −40 −20 0 20 40 60
z (Å)

Fig. 1.5 (*cont.*)

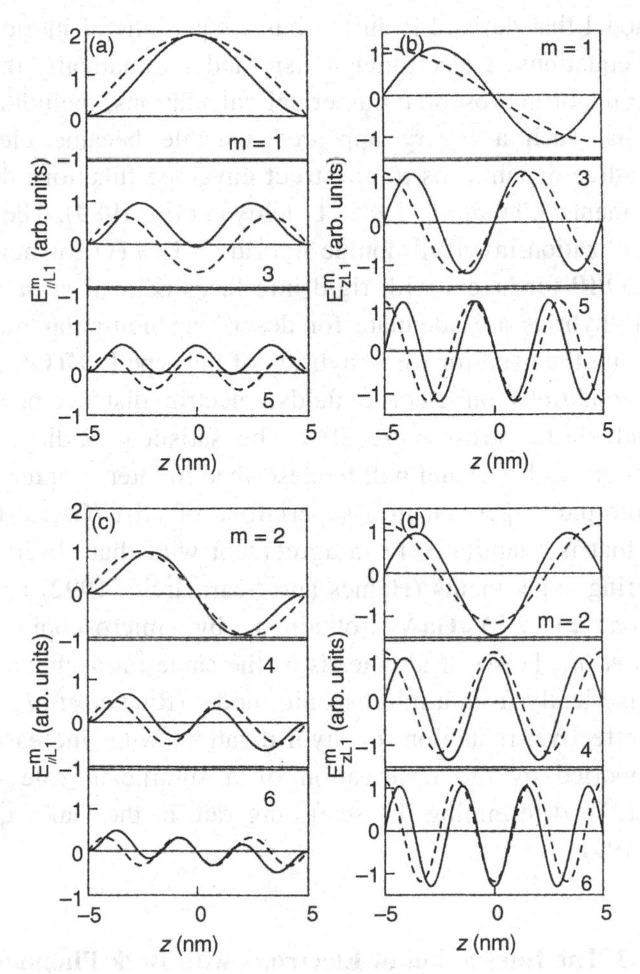

Fig. 1.6 Reformulated LO modes (solid lines) compared with DC model modes (dashed lines): (a) and (b) symmetric; (c) and (d) antisymmetric; (a) and (c) in-plane field; (b) and (d) axial field (Huang and Zhu, 1988; Haupt and Wendler, 1991).

(Rudin and Reinecke, 1990, 1991; Haupt and Wendler, 1991; and Weber and Ryan, 1992), but it has been pointed out by Haupt and Wendler that in this dispersionless model, modes with the same parity described by Eq. (1.9) are not mutually orthogonal and must be made so. Once this is done, identical rates to those of the DC model are obtained (Nash, 1992).

1.2.4 Hybrid Modes

Although the RM model was a significant step forward it obviously could not claim that it had a satisfactory theoretical foundation. What was needed was a continuum model that derived from the basic equations of motion of the ions and from the equations of electromagnetism and automatically reproduced the principal features of microscopic numerical calculations, including the mixing of modes. That such a theory appeared feasible became clear from the attempts by lattice dynamicists to construct envelope functions describing the ionic displacements (Chu *et al.*, 1988; Tsuchiya *et al.*, 1989). The first theories of mode hybridization involved double hybrids – LO/TO for non-polar slabs, LO/TO and LO/IP for layers with rigid interfaces (Zianni *et al.*, 1992; Nash, 1992). Double hybrids are adequate for describing non-polar modes, but not polar modes, and the need for triple hybrids of the type LO/TO/IP was evident if boundary conditions on electric fields, electric displacement, ionic displacement and elastic stress were all to be satisfied (Ridley, 1992, 1993; Trallero-Giner *et al.*, 1992) and will be described in later chapters. This theory predicts anisotropic dispersion in a superlattice of AlAs/GaAs (Constantinou *et al.*, 1993) that is quantitatively in agreement with that observed in micro-Raman scattering experiments (Haines and Scarmarcio, 1992) and reproduces the dispersion in AlAs/GaAs obtained by microscopic calculations (Chamberlain *et al.*, 1994). It also leads to the same intrasubband rate as that calculated numerically from a microscopic model (Rucker *et al.*, 1991), and it predicts an effective reduction of hybridization with increasing polarity, which is supported by the observation of a significant role of the AlAs interface mode in determining the scattering rate in the GaAs quantum well (Tsen *et al.*, 1991).

1.3 The Interaction of Electrons with Bulk Phonons

1.3.1 The Scattering Rate

The early calculations of the electron–phonon scattering in multilayered semiconductors ignored the folding of the acoustic modes and the confinement of the optical modes and simply used the bulklike phonon spectrum and confined electrons. The assumption regarding the acoustic modes is not a bad one, but the same cannot be said about the assumption regarding the optical modes, though it has been claimed that there exists a sum rule whereby the use of any complete set of modes will give the same result (Herbert, 1973; Mori and Ando, 1989; and Register, 1992). We will discuss this sum rule further in Chapter 8. Whatever the theoretical status of such a sum rule, it is found that in the GaAs/AlAs system, which is the only one comprehensively researched at the present time, the overall scattering rates obtained by using a bulk spectrum are very roughly the same as those calculated taking phonon confinement into account (Rucker *et al.*, 1991).

More precisely, the actual rate varies between that obtained using GaAs bulk modes for large well-widths to that obtained using AlAs bulk modes for narrow wells. It is therefore useful to examine the rates using a bulk spectrum in order to provide a reference for the purpose of assessing the effect of phonon confinement. We give an account of the calculation of the scattering rates of confined electrons caused by their interaction with bulk phonons. We assume that the interaction between an electron and a phonon is weak enough for their respective states to be relatively long-lived and therefore well defined, so that the scattering rate is given by first-order perturbation theory embodied in the Fermi golden rule:

$$W(\mathbf{k}) = \frac{2\pi}{\hbar}\int |M(\mathbf{k}',\mathbf{k})|^2\delta(E_f - E_i)dN_f \tag{1.10}$$

where $M(\mathbf{k}', \mathbf{k})$ is the matrix element connecting electron states $\mathbf{k} >$ and $\mathbf{k}' >$; E_f, E_i are the final and initial energies and N_f is the number of final states. For the electron–phonon interaction in which only one-phonon processes are allowed, with the assumption that the final state is unoccupied, the rate can be written as follows:

$$W(\mathbf{k}) = \int W(\mathbf{k}',\mathbf{k})\,\delta(E(\mathbf{k}') - E(\mathbf{k}) \pm \hbar\omega(q))\mathbf{dk}' \tag{1.11}$$

where

$$W(\mathbf{k}',\mathbf{k}) = \sum_{\mathbf{q}} \frac{C^2(q)|I(\mathbf{k}',\mathbf{k},\mathbf{q})|^2(n(\omega) + 1/2 \pm 1/2)}{8\pi^2\rho\omega(q)} \tag{1.12}$$

Here, $C(q)$ is the coupling coefficient, $n(\omega)$ is the phonon occupation number, ρ is the mass density in the case of the interaction with acoustic phonons and the reduced-mass density for optical phonons, $\omega(q)$ is the phonon angular frequency and the upper sign is for emission, the lower for absorption. $I(\mathbf{k}', \mathbf{k}, \mathbf{q})$ is an overlap integral of the form

$$I(\mathbf{k}',\mathbf{k},\mathbf{q}) = I(\mathbf{k}',\mathbf{k})G(\mathbf{k}',\mathbf{k},\mathbf{q}) \tag{1.13}$$

where, with $u_{\mathbf{k}}(\mathbf{r})$ as the cell-periodic part of the electron wavefunction

$$I(\mathbf{k}',\mathbf{k}) = \int u^*_{\mathbf{k}'}(\mathbf{r})u_k(\mathbf{r})d\mathbf{r}_0 \tag{1.14}$$

and the integral is over the unit cell. This integral will be discussed at length in Section 2.4 and no more needs to be said here. $G(\mathbf{k}',\mathbf{k},\mathbf{q})$ is an overlap integral involving the envelope functions of the electron and phonon, viz.:

$$G(\mathbf{k}',\mathbf{k},\mathbf{q}) = \int \psi^*(\mathbf{k}',\mathbf{r})\varphi(\mathbf{q},\mathbf{r})\psi(\mathbf{k},\mathbf{r})d\mathbf{r} \tag{1.15}$$

where $\psi(\mathbf{k}, \mathbf{r})$ and $\varphi(\mathbf{q}, \mathbf{r})$ are electron and phonon envelope functions, and the integral is over the normalizing volume.

Confinement of electron and phonon states affects scattering in a number of ways. Most obvious is the quantization of electron states, introducing subbands with a modified density-of-states function (Fig. 1.7). Since all scattering processes are proportional to the density of final states available, the effect of confinement is direct. Moreover, the appearance of subbands calls for a classification of scattering into intra- and intersubband processes (Fig. 1.8). Quantization of phonon wavevectors also has to be taken into account; in practice this means replacing integrals by sums. Confinement also affects the envelope functions of electrons and phonons, of which the major consequence is that crystal momentum in the confinement direction is not, in general, strictly conserved. The parameter that quantifies this effect is $G(\mathbf{k}', \mathbf{k}, \mathbf{q})$, which becomes, in general, a complicated function of the wavevectors involved. However, it is usually safe to assume that the cell-periodic parts of the electron wavefunction are unaffected by confinement, and hence the cell overlap-integral $I(\mathbf{k}', \mathbf{k})$ is still obtainable from the bulk expression (Section 2.4), though the dependence on scattering angle will be affected by any restrictions on the wavevector component.

Concerning notation, where both electrons and phonons are involved, the convention of denoting electron wavevectors by $\mathbf{k}$ and phonon wavevectors by $\mathbf{q}$ has been adopted.

In what immediately follows we intend to focus on the scattering rate as distinct from the rates for momentum-relaxation and energy-relaxation. These latter are needed for transport and hot-electron theory, topics that will be treated in Chapter 11. The scattering rate itself determines the time the electron spends in a given state and, of course, underpins the physically important energy- and momentum-relaxation processes.

1.3.2 The Coupling Coefficients

Electrons and phonons react via the lattice distortion induced by vibration and, in polar material, via the long-range electric fields that accompany ionic displacements. The interaction via lattice distortion occurs in both polar and non-polar materials and is quantified by a deformation potential, a vector $\mathbf{D}_0$ in the case of optical waves, a tensor Ξ_{ij} in the case of acoustic waves. The symmetry of the relevant electron states is important in determining the non-zero elements of these deformation potentials. Thus the deformation-potential interaction with optical modes is important for electrons only for intervalley processes and intravalley processes within an L-valley, but it is always important for holes, and with (table) respect to this scattering mechanism we will restrict our attention to holes.

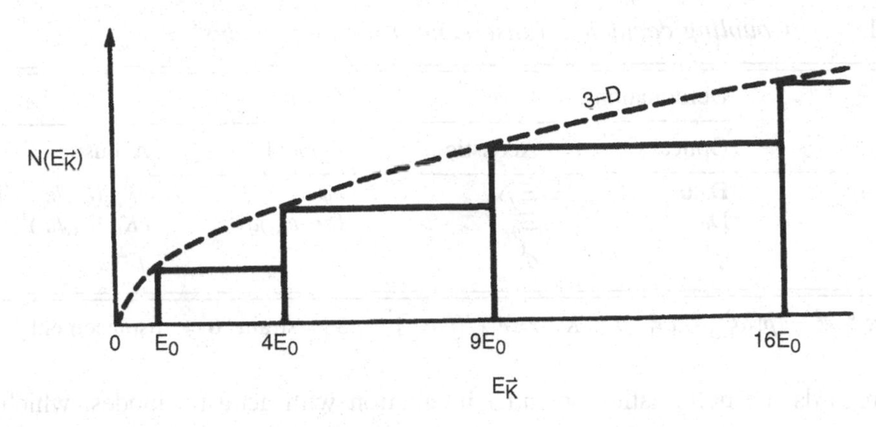

Fig. 1.7 2D Density-of-states function.

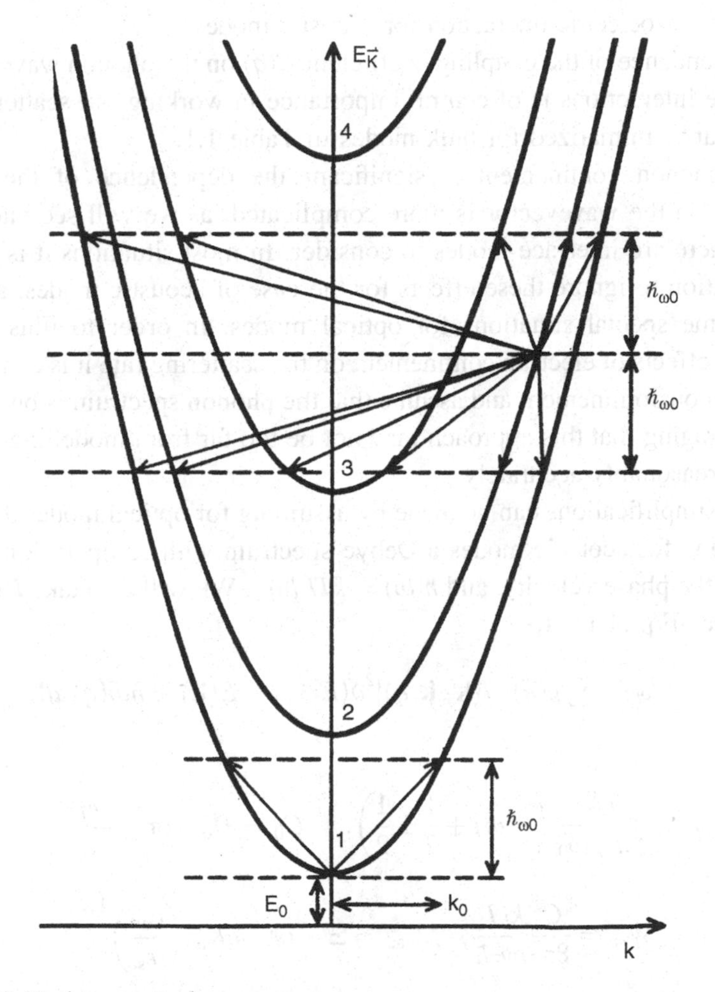

Fig. 1.8 Various scattering processes.

Table 1.1. *Coupling coefficient (unscreened) for bulk modes*[a]

	Deformation		Polar	
	Optical	Acoustic	Optical	Acoustic
Energy	$\mathbf{D}_0.\mathbf{u}$	$\Xi_{ij}S_{ij}$	$e\varphi$	$eK_{av}(c_{av}/\varepsilon_s)^{1/2}u$
$C(q)$	D_0	Ξ_q	$(ee_i/\varepsilon_\infty)q^{-1}$	$eK_{av}(c_{av}/\varepsilon_s)^{1/2}$
$f(q)$	q^0	q^0	q^{-2}	q^{-2}

Note: [a] $e_i^2 = \rho(\omega_{LO}^2 - \omega_{TO}^2)\varepsilon_\infty, K_{av} = e_{14}/(\varepsilon_s c_{av})^{1/2}$, S = strain, $\mathbf{u}$ = displacement

As regards the deformation-potential interaction with acoustic modes, which is always present, we will for our purposes here ignore its tensor nature. As regards the polar interactions we take them to be standard Fröhlich interaction for optical modes and piezoelectric interaction for acoustic modes.

The dependence of the coupling coefficient $C(q)$ on the phonon wavevector for each of the interactions is of central importance in working out scattering rates, and these are summarized for bulk modes in Table 1.1.

When phonon confinement is significant the dependence of the coupling coefficient on the wavevector is more complicated, as we will see later, and in addition there are interface modes to consider. In most situations it is not a bad approximation to ignore these effects for the case of acoustic modes, and this is true in some special situations for optical modes. In order to illustrate most simply the effects of electron confinement on the scattering rate it is convenient to ignore phonon confinement and assume that the phonon spectrum is bulklike, and it is encouraging that this approach may not be too far from modelling some real situations reasonably accurately.

Further simplifications can be made by assuming for optical modes an Einstein spectrum and for acoustic modes a Debye spectrum with equipartition ($\omega = vq$, where v is the phase velocity, and $n(\omega) = kT/\hbar\omega$). We will also take $\boldsymbol{I}(\mathbf{k}', \mathbf{k}) = 1$. This reduces Eq. (1.11) to

$$W(\mathbf{k}) = W_{\substack{opt \\ ac}} \int \sum_q f(q)|G(\mathbf{k}', \mathbf{k}, q)|^2 \delta(E(\mathbf{k}') - E(\mathbf{k}) \pm \hbar\omega(q))d\mathbf{k}' \tag{1.16}$$

where

$$W_{opt} = \frac{C_2^0}{8\pi^2\rho\omega_0}\left(n(\omega) + \frac{1}{2} \pm \frac{1}{2}\right), \quad C_0 = D_0 \quad \text{or} \quad \frac{ee_i}{\varepsilon_\infty} \tag{1.17a}$$

$$W_{ac} = \frac{C_{ac}^2 k_b T}{8\pi^2 \rho v^2 \hbar}, \quad C_{ac} = \Xi \quad \text{or} \quad eK_{av}\left(\frac{c_{av}}{\varepsilon_s}\right)^{1/2} \tag{1.17b}$$

$$f(q) = q^n \tag{1.17c}$$

and $n = 0$ for deformation-potential scattering, $n = -2$ for polar scattering (see Table 1.1).

1.3.3 The Overlap Integral in 2D

We need to know the wavevector dependence of the overlap integral:

$$G(\mathbf{k}',\mathbf{k},\mathbf{q}) = \int \psi^*(\mathbf{k}',\mathbf{r})e^{i\mathbf{q}.\mathbf{r}}\psi(\mathbf{k},\mathbf{r})d\mathbf{r} \tag{1.18}$$

Let us assume that the electron is confined totally in the region $0 \leq z \leq a$; then

$$\psi(\mathbf{k},\mathbf{r}) = \left(\frac{2}{aA}\right)^{1/2} e^{i\mathbf{k}_{\parallel}\cdot\mathbf{r}_{\parallel}} \sin k_z z, \quad k_z = \frac{n\pi}{a} \tag{1.19}$$

where $\mathbf{k}_{\parallel}$ and $\mathbf{r}_{\parallel}$ are vectors in the plane perpendicular to the z axis whose area is A. This situation corresponds to a quasi-2D electron gas. Integration over the plane gives

$$G(\mathbf{k}',\mathbf{k},\mathbf{q}) = \delta_{\mathbf{k}'_{\parallel},\mathbf{k}_{\parallel}+\mathbf{q}} G(k'_z, k_z, q_z) \tag{1.20}$$

$$G(k'_z, k_z, q_z) = \int_0^a \psi(k', z)e^{iq_z z}\psi(k, z)\, dz \tag{1.21}$$

Crystal momentum in the plane is conserved for wavevectors small enough for umklapp processes to be impossible, as is the usual case. But as regards momentum in the confining direction we have

$$\begin{aligned} G(k'_z, k_z, q_z) = \frac{1}{2}\Bigg[& \frac{\sin\{(q_z + k'_z - k_z)a/2\}}{(q_z + k'_z - k_z)a/2} e^{i(q_z+k'_z-k_z)a/2} \\ & + \frac{\sin\{(q_z + k'_z + k_z)a/2)\}}{(q_z - k'_z + k_z)a/2} e^{i(q_z+k'_z-k_z)a/2} \\ & - \frac{\sin\{(q_z + k'_z + k_z)a/2)\}}{(q_z + k'_z + k_z)a/2} e^{i(q_z+k'_z+k_z)a/2} \\ & - \frac{\sin\{(q_z - k'_z - k_z)a/2)\}}{(q_z - k'_z - k_z)a/2} e^{i(q_z-k'_z-k_z)a/2} \Bigg] \end{aligned} \tag{1.22}$$

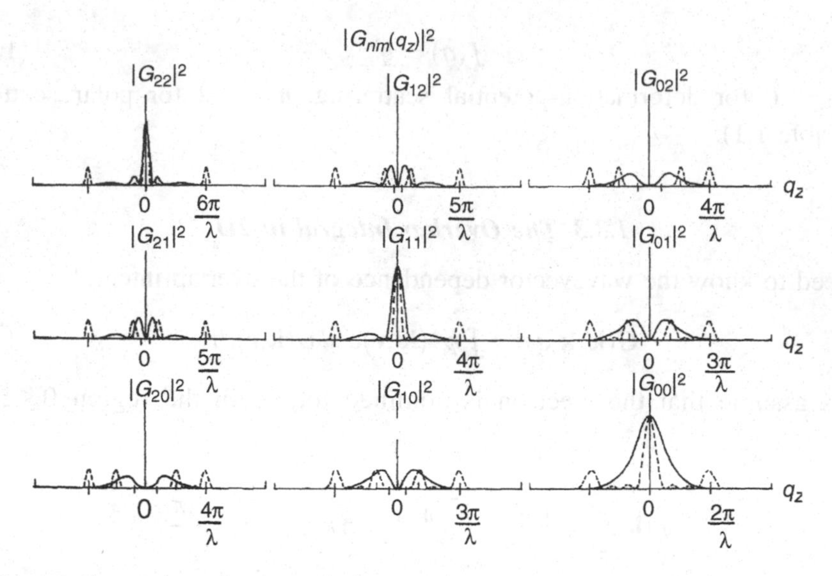

Fig. 1.9 Squared overlap integral for a single heterojunction (full lines) and for a square well (dashed lines). In $G_{nm}(q_z)^2$ n and m are subband suffices (Polonowski and Tomizawa, 1985).

with no restriction on q_z for given initial and final subband wavevectors, although maxima in $G(\mathbf{k}'_z, \mathbf{k}_z, \mathbf{q}_z)$ occur (Fig. 1.9) for the four momentum conserving values:

$$q_z = \pm(k'_z \pm k_z) \tag{1.23}$$

Regarding the acoustic-phonon energy as negligible and taking the Einstein approximation for all the optical modes allow us to decouple the sum over q_z from energy conservation for a given intra- or intersubband transition. Note that the sum over $q_{\|}$ has been reduced to one term by in-plane momentum conservation. Converting the sum over q_z to an integral we have to evaluate

$$\sum_{qz} f(q)\left|G(k'_z, k_z, q_z)\right|^2 = \int_{-\infty}^{\infty} f(q)\left|G(k'_z, k_z, q_z)\right|^2 dq_z \frac{a}{2\pi} \tag{1.24}$$

and we have exploited the form of $G(\mathbf{k'_z}, \mathbf{k_z}, \mathbf{q_z})$ to allow us to extend the limits of integration to $\pm\infty$ without too much error.

For the deformation-potential interaction we take $f(q) = 1$ and obtain

$$\int_{-\infty}^{\infty} \left|G(k'_z, k_z, q_z)\right|^2 dq_z \frac{a}{2\pi} = 1 + \frac{1}{2}\delta_{n,m} \tag{1.25}$$

where $k_z = n\pi/a$ and $k_z' = m\pi/a$. This result is exactly the one that would be obtained by assuming strict momentum conservation in the confinement direction. On the other hand, for the polar interaction we have $f(q) = (q_{\|}^2 + q_z^2)^{-1}$, and

$$
\begin{aligned}
L^2(q_{\|}) &= \int_{-\infty}^{\infty} \frac{|G(k_z', k_z, q_z)|^2}{q_{\|}^2 + q_z^2} dq_z \frac{a}{2\pi} \\
&= \frac{1}{2}\left[\frac{1+\delta_{n,m}}{q_{\|}^2 + (k_z' - k_z)^2} + \frac{1}{q_{\|}^2 + (k_z' - k_z)^2}\right. \\
&\quad - 2\frac{q_{\|}}{a}\left(1 - e^{-q_{\|}a}\cos(k_z' - k_z)a\right) \\
&\quad \left.\times\left(\frac{1}{q_{\|}^2 + (k_z' - k_z)^2} - \frac{1}{q_{\|}^2 + (k_z' - k_z)^2}\right)^2\right]
\end{aligned}
\tag{1.26}
$$

The first two terms are what would be obtained with strict momentum conservation, which becomes a better approximation the larger the confinement length. The momentum conservation approximation (MCA) becomes more accurate when the phonon in-plane wavevector is small compared with $k'_z - k_z$, as can only happen for intersubband transitions.

1.3.4 The 2D Rates

Obtaining the scattering rates associated with the non-polar interaction is straightforward. For an inter- or intrasubband transition we replace the elementary volume in $\mathbf{k}'$ space as follows:

$$
d\mathbf{k}' = k_{\|}' dk_{\|}' d\theta \frac{2\pi}{a} = \frac{m^*}{\hbar^2} dE(k_{\|}') d\theta \frac{2\pi}{a}, \quad 0 \le \theta \le 2\pi \tag{1.27}
$$

Thus

$$
W_{nmopt}(E) = \frac{D_0^2 m^*}{2\rho\omega_0\hbar^2 a}\left(\delta_{nm} + \frac{1}{2}\right)\left(n(\omega) + \frac{1}{2} \pm \frac{1}{2}\right) \tag{1.28}
$$

$$
W_{nmac}(E) = \frac{\Xi^2 k_B T m^*}{\rho v^2 \hbar^3 a}\left(\delta_{nm} + \frac{1}{2}\right) \tag{1.29}
$$

Both of these rates are proportional to the single-spin 2D density of states, $m^*/2\pi\hbar^2 a$, where m^* is the effective mass of the electron (Fig. 1.10).

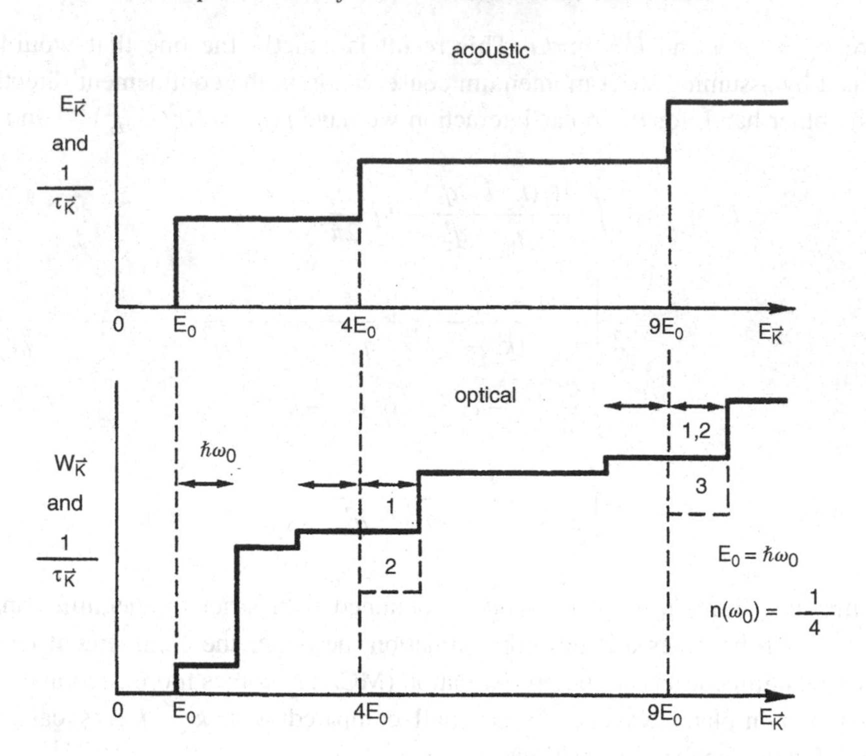

Fig. 1.10 Scattering rates (arbitrary units) for non-polar phonons as a function of electron energy. (τ_{K}^{-1} is the momentum-relaxation rate. The numbers refer to the subbands occupied initially.)

The polar rates are more awkward and must be obtained numerically. Analytical results can be obtained by adopting the MCA (Ridley, 1982). If this is done, we obtain for the polar optical rate

$$W_{nmopt}(E) = \frac{e^2\omega_0}{8a}\left(\frac{1}{\varepsilon_\infty} - \frac{1}{\varepsilon_s}\right)\left(n(\omega) + \frac{1}{2} \pm \frac{1}{2}\right)$$
$$\times \left[\frac{1+\delta_{mn}}{\left[(m-n)^4 E_0^2 + 2(m-n)^2 E_0(2E(k) \mp \hbar\omega^*) + (\hbar\omega^*)^2\right]^{1/2}}\right.$$
$$\left. + \frac{1}{\left[(m+n)^4 E_0^2 + 2(m+n)^2 E_0(2E(k) \mp \hbar\omega^*) + (\hbar\omega^*)^2\right]^{1/2}}\right] \tag{1.30}$$

where

$$\hbar\omega^* = \hbar\omega_0 \mp +(m^2 - n^2)E_0 \tag{1.31a}$$

$$E_0 = \frac{\hbar^2\pi^2}{2m^* a^2} \tag{1.31b}$$

$$E(k) = \frac{\hbar^2 k^2}{2m^*} \tag{1.31c}$$

and for the piezoelectric rate (adding absorption and emission rates):

$$W_{nmac}(E) = \frac{e^2 K_{av}^2 k_B T}{4\varepsilon_s \hbar a} \times \left[\frac{1 + \delta_{m,n}}{\left[(m-n)^4 E_0^4 + 2(m-n)^2 E_0(2E(k) + (m^2 - n^2)E_0) + (m^2 - n^2)^2 E_0^2\right]^{1/2}} + \frac{1}{\left[(m+n)^4 E_0^4 + 2(m+n)^2 E_0(2E(k) + (m^2 - n^2)E_0) + (m^2 - n^2)^2 E_0^2\right]^{1/2}} \right] \tag{1.32}$$

Unfortunately, in this approximation the piezoelectric rate diverges for intrasubband transitions, a catastrophe that comes about basically because of the neglect of screening. In Fig. 1.11 the MCA results for polar optical-phonon scattering are compared with those obtained numerically (Riddoch and Ridley, 1983). Marked deviations from the bulk scattering rate occur only for small well-widths and for small energies, and the MCA results are grossly unreliable for intrasubband transitions at small well-widths.

An analytic result for the polar optical-mode scattering rate can be obtained for the important special case of threshold emission, i.e. when the electron has just enough energy to emit a phonon and make an intrasubband transition. We return to Eq. (1.16), but instead of integrating over z to obtain Eq. (1.22) we introduce a form factor $F(q_{\|})$ thus:

$$W_{11opt}(E) = \frac{e^2\omega}{8\pi^2}\left(\frac{1}{\varepsilon_\infty} - \frac{1}{\varepsilon_s}\right) \int \frac{a}{2q_{\|}} F_{nm}(q_{\|}, q_z)\delta_{\mathbf{k}'_{\|},\mathbf{k}_{\|}+\mathbf{q}_{\|}}\delta(E(k') - E(k) + \hbar\omega)d\mathbf{k}' \tag{1.33}$$

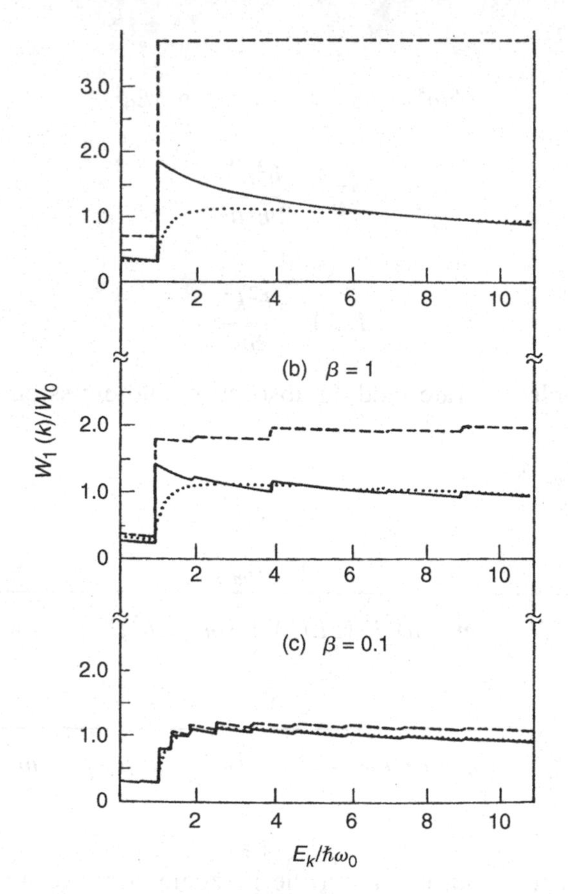

Fig. 1.11 Scattering rate for polar optical phonons ($\beta = \hbar\omega_0/E_1$, E_1 = energy of lowest subband). Dashed line: MCA result; dotted line: bulk result.

where

$$F_{nm}(q_{\|}, q_z) = \frac{q_{\|}}{\pi} \int \psi^*(k'_z, k)\psi_i(k_z, z)\psi_f(k'_z, z')\psi_i^*(k, z') \times \frac{e^{iq_z(z-z')}}{q_{\|}^2 + q_z^2} dz'\, dz\, dq_z \tag{1.34}$$

The integration over q_z can be carried out immediately, giving

$$F_{nm}(q_{\|}) = \int \psi^*(k_z', z)\psi(k_z, z)\psi(k_z', z')\psi^*(k, z')e^{-q_{\|}|z-z'|}dz'\, dz \tag{1.35}$$

Now, at the threshold $q_{\|}$ is fixed by momentum and energy conservation at the value given by

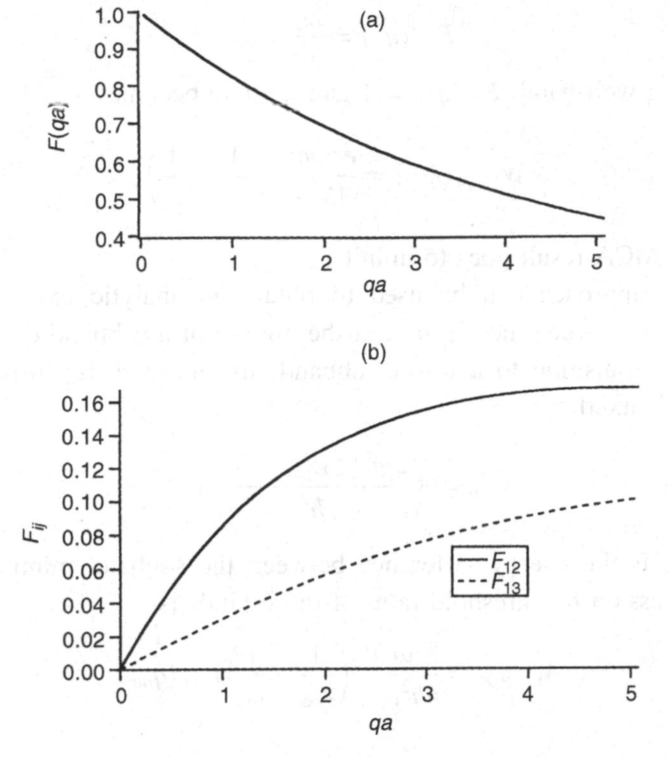

Fig. 1.12 Form factors: (a) for transitions within the lowest subband; (b) for transitions between the lowest subband and subbands 2 and 3.

$$q_{\|} = \left(\frac{2m^*\omega}{\hbar}\right)^{1/2} \equiv q_0 \tag{1.36}$$

and we obtain

$$W_{11opt}(\hbar\omega) = \frac{e^2\omega m^*}{4\hbar^2 q_0}\left(\frac{1}{\varepsilon_\infty} - \frac{1}{\varepsilon_s}\right)F_{11}(q_0) \tag{1.37}$$

where, for fully confined electron states

$$F_{11}(q_{\|}) = \frac{1}{2}\frac{\eta(\eta^2+\pi^2)(3\eta^2+2\pi^2) - \pi^4(1-e^{-2\eta})}{[\eta(\eta^2+\pi^2)]^2} \tag{1.38}$$

Here, $\eta = q_{\|}a/2$. Fig. 1.12 shows $F(q_{\|})$. Note that $F_{nm}(q_{\|})$ is essentially a normalized form of $L^2(q_{\|})$, Eq. (1.26). Thus

$$F_{nm}(q_{\|}) = \frac{2q}{a}L^2(q_{\|}) \tag{1.39}$$

For vanishing well-width $F_{11}(q_{\|}) = 1$ and the rate becomes

$$W_{11opt}(\hbar\omega) = \frac{e^2\omega m^*}{4\hbar^2 q_{\|}}\left(\frac{1}{\varepsilon_\infty} - \frac{1}{\varepsilon_s}\right) \tag{1.40}$$

whereas the MCA result goes to infinity.

The same approach can be used to obtain an analytic expression for an intersubband rate when the electron at the bottom of a subband emits a phonon and makes a transition to a lower subband. In this case the in-plane phonon wavevector is fixed at

$$q_{nm} = \left(\frac{2m^*(\Delta E_{nm} - \hbar\omega)}{\hbar^2}\right)^{1/2} \tag{1.41}$$

where ΔE_{nm} is the energy difference between the subband minima. Thus the general expression for threshold rates of these kinds is

$$W_{nmopt} = \frac{e^2\omega m^*}{4\hbar^2 q_{nm}}\left(\frac{1}{\varepsilon_\infty} - \frac{1}{\varepsilon_s}\right)F_{nm}(q_{nm}) \tag{1.42}$$

where

$$\begin{aligned} F_{nm}(q_{nm}) = 2\eta\Bigg[&\frac{1+\delta_{n,m}}{4\eta^2 + (m-n)^2\pi^2} + \frac{1}{4\eta^2 + (m+n)^2\pi^2} \\ &- 4\eta(1 - e^{-2\eta}\cos(m-n)\pi) \\ &\times\left(\frac{1}{4\eta^2 + (m-n)^2\pi^2} - \frac{1}{4\eta^2 + (m+n)^2\pi^2}\right)^2\Bigg] \end{aligned} \tag{1.43}$$

and $\eta = q_{nm}a/2$ (Fig. 1.12(b)).

1.3.5 The 1D Rates

A similar analysis can be carried out for rectangular wires with electron confinement in the y and z directions. For non-polar scattering the rates are

$$\begin{aligned} W_{n_y n_z, m_y m_z \text{opt}}(E) = \frac{(2+\delta_{n_y m_y})(2+\delta_{n_z m_z})\pi D_0^2}{4\rho\omega_0} &\times [n(\omega)N(E(k) + \hbar\omega^*) \\ &+ n(\omega+1)N(E(k) - \hbar\omega^*)] \end{aligned} \tag{1.44}$$

$$W_{n_yn_z,m_ym_{zac}}(E) = \frac{(2+\delta_{n_ym_y})(3+\delta_{n_zm_z})\pi\Xi^2 k_B T N(E(k))}{2\rho v^2\hbar} \tag{1.45}$$

where $N(E(k))$ is the 1D density of states and

$$\hbar\omega^* = \hbar\omega_0 \mp [(m_y^2 - n_y^2)E_{0y} + (m_z^2 - n_z^2)E_{0z}] \tag{1.46}$$

where the upper sign denotes emission. The polar optical rate is shown in Fig. 1.13 (Riddoch and Ridley, 1984). In these cases the divergencies appearing in the density of states at the subband edges produce striking departures of the rates from those for bulk material. Fig. 1.14 gives a comparison of 1D and 2D rates.

1.4 The Interaction with Model Confined Phonons

In most cases the effect of confinement on the acoustic-phonon spectrum will be small and the rates derived previously ought to be reasonably accurate. This is also the case for optical modes interacting via the deformation-potential under the conditions discussed. Here the scattering rate is simply proportional to the density of final states, with no dependence on phonon wavevector. The situation is quite different for the polar interaction. Here the interaction strength is proportional to q^{-2}. In bulk material q is limited only by momentum and energy conservation, but confinement does not allow q to become arbitrarily small. In the 2D case with complete confinement of the optical modes within a layer of width 'a', $q^2 = q_{\|}^2 + (n\pi/a)^2$, and the freedom allowed by momentum and energy conservation applies only to the in-plane component $q_{\|}$. As a result the interaction is forced to occur with modes whose wavevectors may be appreciably larger than those of the comparable interaction in bulk material, and consequently the scattering rate is reduced. This effect was noticed for the case of a polar slab by Riddoch and Ridley (1985) and for a quantum well by Sawaki (1986).

Restricting our attention to the polar interaction with optical modes, we note that the part of the scattering rate immediately affected by confinement is the overlap integral, $G(\mathbf{k}'\ \mathbf{k},\ \mathbf{q})$. In the DC and HD models, the 2D overlap terms for the LO modes are

$$G(k_z', k_z, q_z) = \int_0^a \psi(k_z', z)\begin{Bmatrix}\sin q_z z\\ \cos q_z z\end{Bmatrix}\psi(k_z, z)dz, \qquad q_z = \frac{n\pi}{a} \tag{1.47}$$

where the sine is for the DC model and the cosine is for the HD model. In the case of complete electron confinement, transverse momentum is conserved in the HD model and $G(\mathbf{k}_z',\ \mathbf{k}_z,\ \mathbf{q}_z)$ vanishes unless

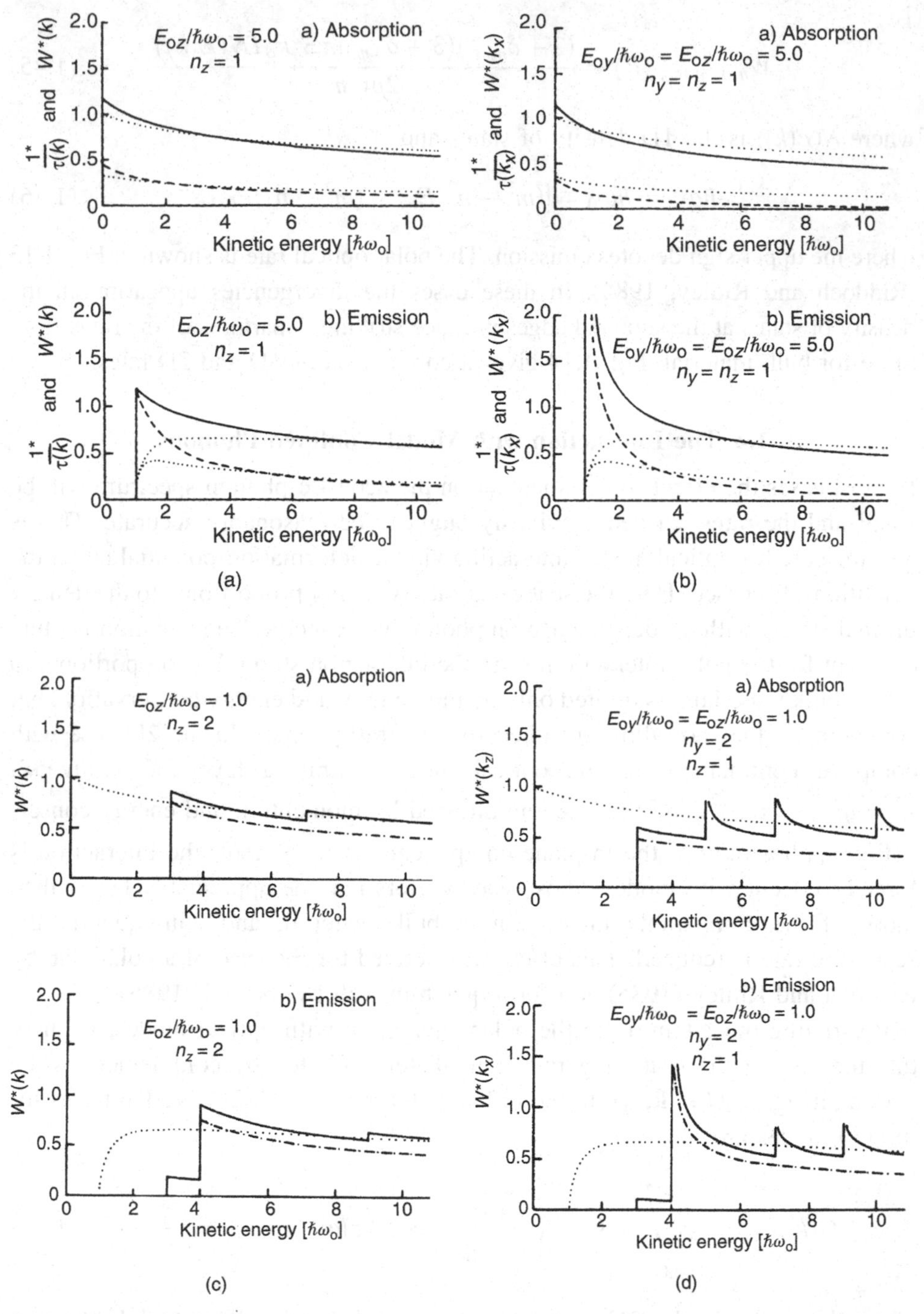

Fig. 1.13 Polar mode scattering rates in 2D and 1D. Intrasubband: (a) 2D; (b) 1D; intersubband: (c) 2D, (d) 1D. Dashed lines: momentum-relaxation rates; dotted lines: bulk rates; dash–dot lines in (c) and (d): intrasubband contribution (Riddoch and Ridley, 1984).

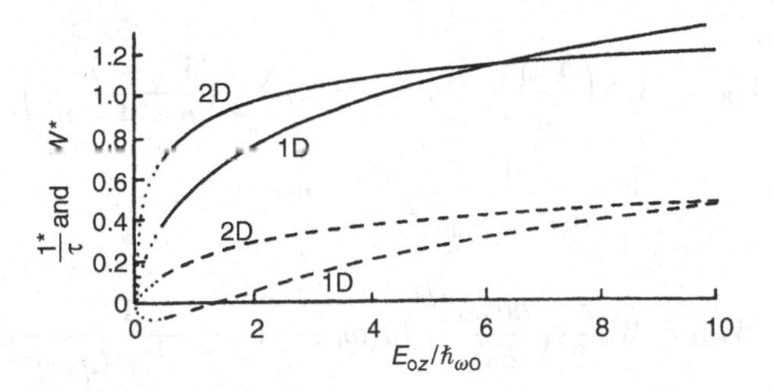

Fig. 1.14 Comparison of 2D and 1D rates. Dashed lines depict momentum-relaxation rates; note that they can go negative in 1D as a consequence of the bias of polar scattering toward forward scattering. Dotted lines indicate where intersubband scattering can occur. For both cases $E(\boldsymbol{K})/\ \hbar\omega_0 = 0.3$ (Riddoch and Ridley, 1984).

$$q_z = \begin{cases} 2k_z & k_z' = k_z \\ |k_z' \pm k_z| & k_z' \neq k_z \end{cases} \tag{1.48}$$

in which case it equals 1/2. For the DC model

$$G(1,1,n) = \frac{-2n}{\pi}(-1)^{(n+1)/2}\left[\frac{1}{n^2} + \frac{1}{4-n^2}\right] \quad n = 1, 3\ldots \ \text{intrasubband}\ (1 \to 1) \tag{1.49a}$$

$$G(2,1,n) = \frac{2n}{\pi}(-1)^{(n/2)}\left[\frac{1}{n^2-9} - \frac{1}{n^2-1}\right] \quad n = 2, 4\ldots \ \text{intrasubband}\ (2 \to 1) \tag{1.49b}$$

The rate is given in either case by

$$W_{ij}(k) = \frac{e^2\omega_0 m^*}{\hbar^2 a}\left(\frac{1}{\varepsilon_\infty} - \frac{1}{\varepsilon_s}\right)\left(n(\omega_0) + \frac{1}{2} \pm \frac{1}{2}\right) \sum_n |G(i,j,n)|^2 \left[q_z^4 + 2q_z^2(2k^2 \mp q_0^{*2}) + q_0^{*4}\right]^{-1/2} \tag{1.50}$$

where $q_0^{*2} = 2m^*\omega_0^*/\hbar$ and $G(1, 1, n) = 1/2\delta_{n,2}, G(2,1,n) = 1/2(\delta_{n,1} + \delta_{n,3})$ for the HD model. Here, k is the in-plane wavevector of the electron. In particular, for the intrasubband threshold emission rates, $q_0^2 = 2m^*\omega_0/\hbar = k^2$, and

$$W_{\mathrm{DC}} = W_0 8\left(\frac{\hbar\omega_0}{E_0}\right)^{1/2}(n(\omega_0)+1)\sum_n\left(\frac{1}{n}+\frac{n}{4-n^2}\right)^2 \times\frac{1}{(n\pi)^2+(q_0a)^2},\qquad n=1,3,\ldots \tag{1.51}$$

$$W_{\mathrm{HD}} = W_0\frac{\pi^2}{2}\left(\frac{\hbar\omega_0}{E_0}\right)^{1/2}(n(\omega_0)+1)\frac{1}{(2\pi)^2+(q_0a)^2} \tag{1.52}$$

For the intersubband rate from the bottom of band 2 to band 1 we get

$$W_{\mathrm{DC}} = W_0 8\left(\frac{\hbar\omega_0}{E_0}\right)^{1/2}(n(\omega_0)+1)\sum_n\left(\frac{1}{n^2-9}-\frac{1}{n^2-1}\right)^2 \times\frac{n^2}{(n\pi)^2+(q_0^*a)^2},\qquad n=2,4,\ldots \tag{1.53}$$

$$W_{\mathrm{HD}} = W_0\frac{\pi^2}{2}\left(\frac{\hbar\omega_0}{E_0}\right)^{1/2}(n(\omega_0)+1)\left[\frac{1}{(4\pi)^2-(q_0a)^2}+\frac{1}{(12\pi)^2-(q_0a)^2}\right] \tag{1.54}$$

$$W_0 = \frac{e^2}{4\pi\hbar}\left(\frac{2m^*\omega_0}{\hbar}\right)^{1/2}\left(\frac{1}{\varepsilon_\infty}-\frac{1}{\varepsilon_s}\right) \tag{1.55}$$

(N.B.: Since $\hbar\omega_0 = \Delta E_{12} - \hbar\omega_0$, we have $(q_0^*a)^2 = 3\pi^2 - (q_0a)^2$, hence the denominators in Eq. (1.54).) Fig. 1.15 compares these rates as a function of well-width with the rates obtained with a bulk spectrum. Both models predict a reduction in rates with increasing confinement. The DC model gives much higher intrasubband rates than the HD model, and the reverse is true for intersubband rates. This reflects the opposite symmetries of the modes in the two models.

Scattering by polar-optical phonons is often the most dominant process at room temperature (and indeed down to about 40 K) in III–V compound semiconductors, and so a reduction in this rate through confinement is attractive for many technological applications. Unfortunately this reduction is countered by an increase arising from the polar-optical interface modes whose influence strengthens with decreasing well-width.

As mentioned in Section 1.2, each material composing the interface contributes two interface modes, one of which has a symmetric potential in the well, the other an antisymmetric potential. Symmetric modes are responsible for intrasubband

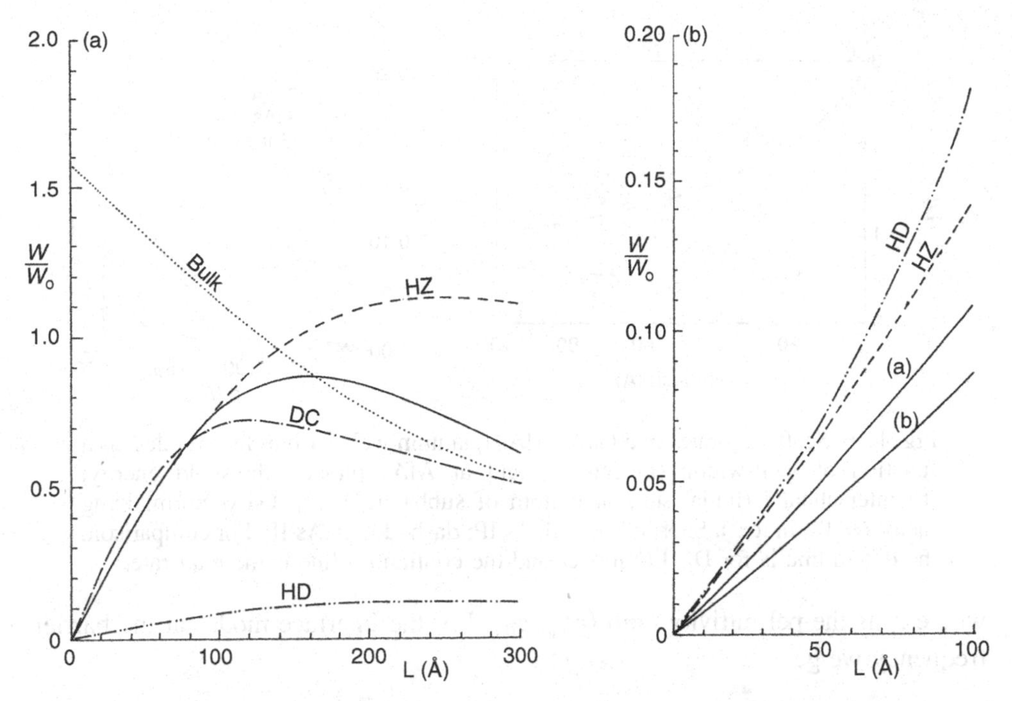

Fig. 1.15 Threshold scattering rates in GaAs for bulk (dotted), HD (broken), DC (dot–dash), HZ (reformulate modes) (dash) and hybrid (continuous) models as a function of well-width: (a) intrasubband, (b) intersubband. ($k_LL = 2.8\pi$ (a), 3π (b))

transitions, asymmetric modes for intersubband transitions. The overlap integrals now have hyperbolic sines and cosines, and there is no question of momentum conservation in the confinement direction. Threshold rates can be readily obtained. For the interface modes at the frequency of the material in the well we obtain

$$W_{\text{well}}(1 \to 1) = W_0 2^5 \pi^6 \left(\frac{\hbar\omega_0}{E_0}\right)^{1/2} \times \frac{r_0^2 \tanh(q_0 a/2)}{(q_0 a)^3((2\pi)^2 + (q_0 a)^2)^2(r_0 + \coth(q_0 a/2))^2} \tag{1.56}$$

$$W_{\text{well}}(2 \to 1) = W_0 2^7 \pi^6 \left(\frac{\hbar\omega_0}{E_0}\right)^{1/2} \times \frac{r_0^2 q_0 a \coth(q_0 a/2)}{(\pi^2 + (q_0 a)^2)^2((3\pi)^2 + (q_0 a)^2)^2(r_0 + \tanh(q_0 a/2))^2} \tag{1.57}$$

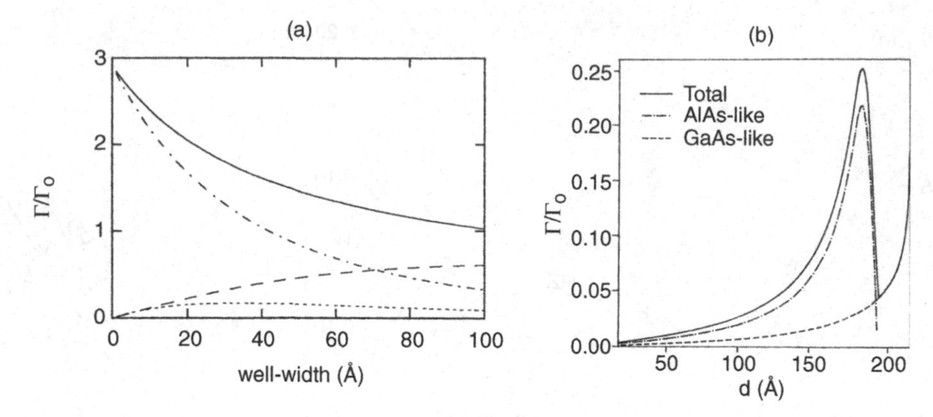

Fig. 1.16 Scattering rates in a GaAs/AlAs quantum well for interface modes, as a function of well-width: (a) intrasubband at AlAs phonon threshold energy; (b) intersubband (initial state at bottom of subband 2); Γ_0, GaAs normalizing factor (= W_0 in text). Short dashes, GaAs IP; dash–dot, AlAs IP. For comparison, the dotted line is for DC LO modes and the continuous line is the total rate.

where r_0 is the permittivity ratio $(\varepsilon_{\infty_w}/\varepsilon_B)$. For the interface modes at the barrier frequency we get

$$W_{\text{Barrier}}(1 \to 1) = W_0 2^5 \pi^6 \left(\frac{\hbar\omega_0}{E_0}\right)^{1/2} \frac{r_0^2 \tanh^2(q_0 a/2)}{(q_0 a)^3((2\pi)^2 + (q_0 a)^2)^2 (r_0 + \tanh(q_0 a/2))^2} \tag{1.58}$$

$$W_{\text{Barrier}}(2 \to 1) = W_0 2^7 \pi^6 \left(\frac{\hbar\omega_0}{E_0}\right)^{1/2} \frac{r_0^2 q_0 a \coth^2(q_0 a/2)}{(\pi^2 + (q_0 a)^2)^2((3\pi)^2 + (q_0 a)^2)^2 (r_0 + \coth(q_0 a/2))^2} \tag{1.59}$$

In Eqs. (1.56) and (1.57) all parameters (e.g., ω_0, q_0, r_0, E_0) refer to values for the well, and in Eqs. (1.58) and (1.59) all except Eo refer to values for the barrier. The reader will note that both rates associated with the well modes vanish in the limit of zero well-width (Fig. 1.16). In the same limit the intrasubband rate for barrier modes is finite viz.

$$W_{\text{Barrier}}(1 \to 1) = W_0 \frac{\pi}{2} \tag{1.60}$$

which is equal to the rate for bulk phonons in the same limit (cf. Eq. (1.40)). The intersubband rate vanishes. Thus, it is the barrier interface mode producing a symmetric potential in the well that maintains high intrasubband scattering rates in narrow wells and vitiates the effect of the LO modes.

This conclusion, however, is not obviously valid in general because it neglects the phenomenon of optical-mode hybridization. If hybridization of LO, TO, and interface-polariton (IP) modes occurs, the components do not act independently, and we have seen that confinement can have a marked effect. In subsequent chapters we will explore this important effect of hybridization on scattering and evaluate what may be termed the electron–hybridon interaction. We will find, however, that in spite of the unphysical nature of the DC model, it reproduces the scattering rates of the hybrid model remarkably well.

2

Quantum Confinement of Carriers

Thou hast set them bounds
which they shall not pass.
Psalms ciii

2.1 The Effective-Mass Equation

2.1.1 Introduction

It is well known that an electron wave incident at a potential discontinuity has to satisfy the twin conditions of continuity of probability and continuity of current. This translates into the continuity of wavefunction and of its gradient. These rules must apply in all cases, and in particular to heterostructures. The trouble here is that the wavefunction of an electron in a solid is very complicated, in general, and the Schrodinger equation of which it is a solution is usually far too complicated to solve accurately. We know that in a periodic structure it has the form of a Bloch wave, viz.:

$$\psi(\mathbf{r}) = V^{-1/2}\mathbf{u}_{\mathbf{k}}(\mathbf{r})e^{i\mathbf{k.r}} \tag{2.1}$$

where V is the normalization volume, and $\mathbf{u}_{\mathbf{k}}(\mathbf{r})$ has the periodicity of the lattice together with a symmetry for $\mathbf{k} = \mathbf{k}_0$ characteristic of the position $\mathbf{k}_0$ in the Brillouin zone. In a layered structure that is lattice-matched there is still periodicity, but with discontinuities in electron affinity there will be reflections setting up standing waves with perhaps more than one conduction band involved.

In general, we can adopt one of two approaches. The first is to use the techniques of band-structure calculations and calculate the eigenvalues and eigenstates of the whole heterostructure. This is fundamentally sound in principle but cumbersome in practice. Each new heterostructure requires time-consuming numerical computation, and although this is worth doing for particularly important structures, the method does not provide criteria for predicting the properties of the virtually infinite number of possible structures that can be conceived. The second approach is to use the effective-mass approximation, which, though imprecise, is well established conceptually in semiconductor physics and is easily applicable in analytic form to all manner of heterostructures. We will adopt this second approach here.

Actually, it is not obvious that effective-mass theory can apply in the presence of abrupt steps in the potential that an electron encounters at a heterojunction. The well-known effective-mass equation describing the energy levels of a shallow-level donor impurity is

$$\left[\left(\frac{-\hbar^2}{2m^*}\right)\nabla^2 + V(\mathbf{r})\right]F(\mathbf{r}) = (E - E_n(0))\,F(\mathbf{r}) \tag{2.2}$$

where m^* is the effective mass, $F(\mathbf{r})$ is an envelope function, and $\mathbf{E}_n(0)$ is the nth conduction-band edge, and its validity rests on the assumption that $V(\mathbf{r})$ is a potential energy that varies little over a unit cell – clearly not the case for a heterojunction. But it can be true on either side. For example, if there are no space charge effects, $V(\mathbf{r}) = 0$. Thus we can take the solutions on either side of the interface and connect them according to some rule. The question is, what rule?

We know the rule for the total wavefunction, of course. Amplitude and gradient across the interface must be continuous. But we have condensed all the microscopic properties into the effective-mass tensor via **k.p** theory; the result for a simple band is

$$\frac{1}{m^*_{ij}} = \frac{1}{m_0}\left[\delta_{ij} + \frac{1}{m_0}\sum_{m\neq n}\frac{(p^i_{nm}p^j_{nm} + p^j_{nm}p^i_{nm})}{E_n(0) - E(0)}\right] \tag{2.3}$$

where i, j are x, y, and z; the $p_{nm}{}^i$ are momentum matrix elements between bands n and m, and m_0 is the free electron mass. The connection between the envelope function $F(\mathbf{r})$ and the total wave function in the effective-mass approximation is derived from an expansion over the first Brillouin zone

$$\psi_n(\mathbf{r}) = \sum_{\mathbf{k}} a(\mathbf{k})u_{n\mathbf{k}}(\mathbf{r})e^{i\mathbf{k}.\mathbf{r}} \tag{2.4}$$

where, to first order in **k.p** theory

$$u_{n\mathbf{k}}(\mathbf{r}) = u_{no}(\mathbf{r}) + \frac{\hbar}{m_o}\sum_{m\neq n}\frac{\mathbf{k}.\mathbf{p}_{mn}}{E_n(0) - E_m(0)}u_{mo}(\mathbf{r}) \tag{2.5}$$

and

$$\psi_n(\mathbf{r}) = F(\mathbf{r})u_{no} - \frac{i\hbar}{m_o}\sum_{m\neq n}\frac{\nabla F(\mathbf{r}).\mathbf{p}_{mn}}{E_n(0) - E_m(0)}u_{mo} \tag{2.6}$$

The lowest order $F(\mathbf{r})$ is seen to be the comparatively slowly varying envelope of $\mathbf{u}_{no}$.

As long as m^* is a constant in space, the effective-mass equation in Eq. (2.2) is acceptable, but this is not the case in the presence of a heterojunction. In reality the interface will occupy a number of atomic sites and thus the effective mass must be regarded as varying with distance. This means that the effective-mass equation must be modified to maintain hermiticity. In the case where motion in the plane of the interface is unfettered, say the (x, y) plane, the modified equation is

$$\left[-\frac{\hbar}{2}\nabla.\left(\frac{1}{m^*(z)}\nabla\right)+V(\mathbf{r})\right]F(\mathbf{r}) = (E - E_n(0,z))\,F(\mathbf{r}) \tag{2.7}$$

Integrating this equation over an infinitesimally thin interface region leads to the necessity for the continuity of

$$F(\mathbf{r}) \quad \text{and} \quad \frac{1}{m^*}\frac{dF(\mathbf{r})}{dz} \tag{2.8}$$

This makes sense in that the current across the interface is therefore continuous. But reference to Eq. (2.6) indicates that for $\psi(\mathbf{r})$ to be continuous if $F(\mathbf{r})$ is, then the second term has to be negligible and the difference between the microscopic components of the Bloch functions also has to be negligible. These conditions are, unfortunately, not compatible with the continuity of the gradient of ψ since the gradient of $F(\mathbf{r})$ is discontinuous if the effective mass is. This suggests that the effective-mass approach is unworkable, but in fact the boundary conditions of Eq. (2.8) give results in agreement with experiment.

The reason why Eq. (2.8) works so well can be elucidated by starting off with a more general expansion of the wavefunction, viz.:

$$\psi(\mathbf{r}) = \sum_n F_n(\mathbf{r})u_n(\mathbf{r}) \tag{2.9}$$

where $u_n(\mathbf{r})$ is one of a complete set of periodic functions and $F_n(\mathbf{r})$ is a slowly varying envelope function with Fourier components confined to the first Brillouin zone. Unlike the conventional approach, the expansion is for the whole structure, not separate expansions for the semiconductors involved. Once $\psi(\mathbf{r})$ is given for the whole structure, the $F_n(\mathbf{r})$ are given by an inversion formula, and are unique. In this approach boundary conditions are entirely circumvented and clearly $F_n(\mathbf{r})$ and its gradient are continuous everywhere.

2.1.2 The Envelope-Function Equation

An equation for the $F_n(\mathbf{r})$ can be obtained as follows (Burt, 1988). The Schrödinger equation is

$$-\frac{\hbar^2}{2m_0}\nabla^2\psi + V\psi = E\,\psi \tag{2.10}$$

The kinetic energy is therefore

$$T\psi = \frac{\hbar}{2m_0}\sum_n \left[(\nabla^2 F_n)u_n + 2(\nabla F_n)\cdot\nabla u_n + F_n\nabla^2 u_n\right] \tag{2.11}$$

(where the explicit dependencies on **r** have been dropped for brevity). This can be expressed in terms of a slowly varying function with Fourier components restricted to the first Brillouin zone by multiplying on the left by u_n^* and integrating over a unit cell. The result is

$$T\psi \to \sum_n \left[\frac{\hbar^2}{2m_0}\nabla^2 F_n + \sum_m -\frac{i\hbar}{m_0}P_{nm}\cdot\nabla F_m + \sum_m T_{nm}F_m\right]u_n \tag{2.12}$$

where

$$\mathbf{p}_{nm} = \int u_n^* \mathbf{p} u_m d\mathbf{r}_0, \qquad T_{nm} = \int u_u^* T u_m d\mathbf{r}_0 \tag{2.13}$$

are constants, and the u_n are normalized over the volume of the unit cell.

The potential energy is more awkward because of the step at the interface, which introduces Fourier components outside the first zone. The way forward is to expand everything in plane waves, viz.:

$$V\psi = \sum_n \sum_{\mathbf{kk}'} \sum_{\mathbf{GG}'} V_{\mathbf{G}}(\mathbf{k})F_n(\mathbf{k}')u_{n_\mathbf{G}}e^{i(\mathbf{k}+\mathbf{k}'+G+G').\mathbf{r}} \tag{2.14}$$

where G and G' are reciprocal-lattice vectors. Putting $G \to G + G'$ and $\mathbf{k} + \mathbf{k}' = \mathbf{k}_1 + \mathbf{G}_1$ (since $\mathbf{k} + \mathbf{k}'$ may introduce a wavevector lying outside the first zone), one obtains

$$V\psi = \sum_n \sum_m \sum_{\mathbf{kk}'} \sum_{\mathbf{GG}'} \left(V_{\mathbf{G}-\mathbf{G}'}(\mathbf{k})F_n(\mathbf{k}')u_{n\mathbf{G}'}e^{i\mathbf{k}_1.\mathbf{r}}u_{m,\mathbf{G}+\mathbf{G}_1}^*\right)u_n \tag{2.15}$$

In this derivation use is made of the following:

$$u_n = \sum_{\mathbf{G}} u_{n\mathbf{G}}e^{i\mathbf{G}.\mathbf{r}}, \qquad e^{i\mathbf{G}.\mathbf{r}} = \sum_n (u^{-1})_{Gn}u_n = \sum_n u_{n\mathbf{G}}^* u_n \tag{2.16}$$

The envelope-function equation is thus

$$\frac{-\hbar^2}{2m_0}\nabla^2 F_n(\mathbf{r}) + \sum_m -\frac{i\hbar}{m_0}\mathbf{P}_{nm}\nabla F_m(\mathbf{r}) + \sum_m \int H_{nm}(\mathbf{r},\mathbf{r}')F_m(\mathbf{r}')d\mathbf{r}' = EF_n(\mathbf{r}) \tag{2.17}$$

where

$$H_{nm}(\mathbf{r},\mathbf{r}') = T_{nm}\Delta(\mathbf{r}-\mathbf{r}') + V_{nm}(\mathbf{r},\mathbf{r}') \tag{2.18}$$

$$\Delta(\mathbf{r}-\mathbf{r}') = \frac{1}{\Omega}\sum_{\mathbf{k}} e^{i\mathbf{k}.(\mathbf{r}-\mathbf{r}')} \tag{2.19}$$

$$V_{nm}(\mathbf{r},\mathbf{r}') = \sum_{\mathbf{kk}'}\sum_{\mathbf{GG}'} \overset{*}{u}_{n,\mathbf{G}+\mathbf{G}_1} V_{\mathbf{G}-\mathbf{G}'}(\mathbf{k}) u_{m\mathbf{G}'} e^{i(\mathbf{k}_1.r-\mathbf{k}'.r')} \tag{2.20}$$

and $\mathbf{k}$, $\mathbf{k}'$ are wavevectors within the first zone.

2.1.3 The Local Approximation

Equation (2.17) for the envelope function has a wide range of validity since nothing has been assumed about the potential, but it is complicated by the non-local nature of the latter, which arises if $\mathbf{k}+\mathbf{k}'$ lies outside the first zone. If $\mathbf{k}+\mathbf{k}'$ does not lie outside the first zone, then $\mathbf{G}_1 = 0$, $\mathbf{k}_1 = \mathbf{k}+\mathbf{k}'$, and

$$V_{nm}^{loc}(\mathbf{r},\mathbf{r}') = \sum_{\mathbf{kk}'}\sum_{\mathbf{GG}'} \overset{*}{u}_{n\mathbf{G}} V_{\mathbf{G}-\mathbf{G}'}(\mathbf{k}) u_{m'\mathbf{G}'} e^{i\mathbf{k},\mathbf{r}} \Delta(\mathbf{r}-\mathbf{r}') \tag{2.21}$$

which reduces Eq. (2.17) to an ordinary differential equation (or rather a set of coupled differential equations), viz.:

$$\frac{-\hbar^2}{2m_0}\nabla^2 F_n(\mathbf{r}) + \sum_m \frac{-ik}{m_0}\mathbf{P}_{nm}\cdot\nabla F_m(\mathbf{r}) + \sum_m H_{nm}(\mathbf{r})F_m(\mathbf{r}) = EF_n(\mathbf{r}) \tag{2.22}$$

where

$$H_{nm}(\mathbf{r}) = T_{nm}(\mathbf{r}) + V_{nm}^{loc}(\mathbf{r}) \tag{2.23}$$

$$V_{nm}^{loc}(\mathbf{r}) = \sum_{\mathbf{k}}\sum_{\mathbf{GG}'} \overset{*}{u}_{n\mathbf{G}}.V_{\mathbf{G}-\mathbf{G}'}(\mathbf{k}) u_{m\mathbf{G}'} e^{i\mathbf{k}.\mathbf{r}} \tag{2.24}$$

It can be shown that for local microscopic potentials and slowly varying envelope functions the non-local contributions are negligible, and so for most purposes the envelope-function equation in the local approximation, Eq. (2.22) is adequate. “Slowly varying” in this context is not the usual condition of restricting k to be near zero, but the much less restrictive condition entailing insignificant spectral weight outside the Brillouin zone. (See Foreman, 1995, for a justification of this point from the standpoint of quasi-continuum theory. Quasi-continuum theory will be described in the next chapter.)

Writing Eq. (2.24) in terms of the local potential $V(\mathbf{r})$ we obtain

$$\begin{aligned} V_{nm}^{loc}(\mathbf{r}) &= \sum_{\mathbf{GG}'} \overset{*}{u}_{n\mathbf{G}} u_{m\mathbf{G}'} \sum_{\mathbf{k}} \int V(\mathbf{r}')e^{-i(\mathbf{k}+\mathbf{G}-\mathbf{G}').\mathbf{r}'} e^{i\mathbf{k}.\mathbf{r}} d\mathbf{r}'/\Omega \\ &= \sum_{\mathbf{GG}'} \overset{*}{u}_{n\mathbf{G}} u_{m\mathbf{G}'} \int V(\mathbf{r}')e^{-i(\mathbf{G}-\mathbf{G}').\mathbf{r}'} \Delta(\mathbf{r}-\mathbf{r}')d\mathbf{r}' \end{aligned} \tag{2.25}$$

Consider an interface at which $V(\mathbf{r})$ changes abruptly, situated at $(x, y, 0)$. Far from the interface, say at (x, y, z), the integrand in Eq. (2.25) will be appreciable only when $r' = \mathbf{r}$. In the bulk of the material $V(\mathbf{r})$ is periodic:

$$V(\mathbf{r}) = \sum_{\mathbf{G}} V_{\mathbf{G}}^{+} e^{i\mathbf{G.r.}} \quad \text{and} \quad V_{nm}^{loc}(\mathbf{r}) = \begin{cases} V^{+} & z > 0 \\ V^{-} & z < 0 \end{cases} \tag{2.26}$$

where

$$V^{\pm} = \sum_{\mathbf{GG'}} \overset{*}{u}_{n\mathbf{G}} \cdot V_{\mathbf{G}-\mathbf{G'}}^{\pm} u_{m\mathbf{G'}} = \int \overset{*}{u}_{n}(\mathbf{r}) V^{\pm}(\mathbf{r}) u_{m}(\mathbf{r}) \frac{dr_0}{\Omega_0} \tag{2.27}$$

i.e. $V^{\pm}$ are the matrix elements of the local potential with respect to the periodic functions u_n and u_m, independent of position within the bulk of each material.

If we take the interface to be $z = 0$, then the local potential is of the form

$$V(z) = \theta^{+}(z)V^{+}(z) + \theta^{-}(z)V^{-}(z) \tag{2.28}$$

where $\theta^{+}(z)$ is the usual step function:

$$\theta^{+}(z) = \begin{cases} 0 & z < 0 \\ 1 & z > 0 \end{cases} \tag{2.29}$$

and

$$\theta^{-}(z) = \theta^{+}(-z) \tag{2.30}$$

the evaluation of $V_{nm'}^{loc}(z)$ in Eq. (2.25) can be carried out using Eq. (2.28) to give

$$V_{nm'}^{loc}(z) = \theta^{+}(z)V^{+} + \theta^{-}(z)V^{-} + \text{oscillatory terms} \tag{2.31}$$

The oscillatory terms that arise are well known to Fourier analysis and are referred to as the Gibbs phenomenon. They can be approximately obtained by noting that in bulk material $V_{nm'}^{loc}(z)$ is defined only at each unit cell and it is the same in each unit cell. It can thus be represented by the continuous function

$$V_{nm'}^{loc}(z) = a\sum_{l=0}^{\infty} V^{+}\delta_B\left(z - \left(l + \frac{l}{2}\right)a\right) + a\sum_{l=0}^{-\infty} V^{-}\delta_B\left(z + \left(l + \frac{l}{2}\right)a\right) \tag{2.32}$$

where a is the unit-cell dimension, and

$$\delta_B(z) = \frac{1}{2\pi} \int_{-\pi/a}^{\pi/a} e^{ikz} dk = \frac{\sin(\pi z/a)}{\pi z} \tag{2.33}$$

This gives a spatial variation of the cell-averaged potential that applies everywhere, including the interface. The envelope-function equation in the local approximation then contains variables whose z-dependence is explicit.

2.1.4 The Effective-Mass Approximation

A common situation is when the states of interest have an energy close to that at the zone-centre of one of the bands, and the zone-centre eigenfunctions are similar in the different materials forming the interface. In this case the non-diagonal elements of $H_{nm}(\mathbf{r})$ in Eq. (2.22) will be small and we can make the approximation

$$\sum_m H_{nm}(\mathbf{r})F_m(\mathbf{r}) = E_n(\mathbf{r})F_n(\mathbf{r}) \tag{2.34}$$

and the envelope-function equation becomes

$$-\frac{\hbar^2}{2m_0}\nabla^2 F_n(\mathbf{r}) - \frac{i\hbar}{m_0}\sum_m \mathbf{P}_{nm}\cdot\nabla F_m(\mathbf{r}) = (E - E_n(\mathbf{r}))F_n(\mathbf{r}) \tag{2.35}$$

In the bulk, E_n is the zone-centre energy of the nth band. If the energy is close to the zone-centre energy of the sth band, $E_S(z)$, $F_S(\mathbf{r})$ will be bigger than any other envelope function and then, to leading order

$$F_m(\mathbf{r}) = -\frac{i\hbar}{m_0}\frac{\mathbf{P}_{ms}\cdot\nabla F_s(\mathbf{r})}{E - E_m(\mathbf{r})}, \qquad m \neq s \tag{2.36}$$

Substitution in Eq. (2.35) with $n = s$ gives the required effective-mass equation

$$-\frac{\hbar^2}{2}\nabla\cdot\left[\frac{1}{m^*(\mathbf{r})}\nabla F_s(\mathbf{r})\right] + E_s(\mathbf{r})F_s(\mathbf{r}) = EF_s(\mathbf{r}) \tag{2.37}$$

where

$$\frac{1}{m^*(\mathbf{r})} = \frac{1}{m_0} + \frac{2}{m_0^2}\sum_{m\neq s}\frac{|\mathbf{P}_{ms}|^2}{E - E_m(\mathbf{r})} \tag{2.38}$$

A further approximation is to ignore the oscillatory terms in Eq. (2.31) and assume that the matrix element of the potential changes abruptly in crossing the interface (the abrupt-interface approximation). Integration across this abrupt interface leads to the necessity for $F_s(z)$ and $(1/m^*(z))\, dF_s(z)/dz$ to be continuous. More accurate profiles can be obtained by integrating the continuous function. In this case $F_s(z)$ and $dF_s(z)/dz$ are, of course, continuous, as Fig. 2.1 shows. Figure 2.1 also shows that there is no contradiction and that the sharp changes in slope at the interfaces are well approximated by the effective-mass boundary conditions.

When interband coupling cannot be neglected, as for example in narrow-gap semiconductors, the Hamiltonian component in Eq. (2.34) cannot, in general, be regarded as a simple step-function at the interface. Instead, pure interface contributions arise that have a delta-function character, and these, in principle, can

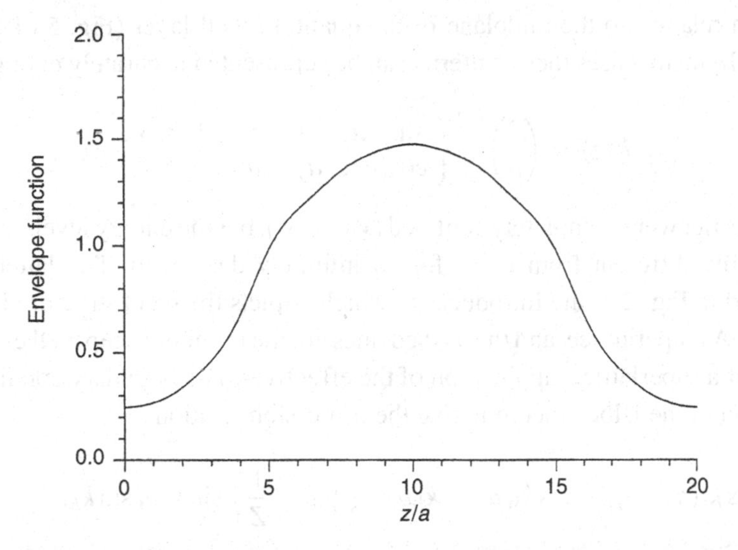

Fig. 2.1 Electron envelope function for a 10/10 superlattice ground state using crystal pseudopotentials. Note that the envelope function is continuous on a microscopic scale but displays a discontinuous slope when viewed mesoscopically (Burt, 1992).

dominate the boundary conditions. In this case the continuity of $(1/m^*(z))dF(z)/dz$ is no longer guaranteed (Foreman, 1995).

A similar analysis can be made for the valence band, with degenerate bands involved, and we will use the results of such an analysis in Section 2.3.

For further reading on the topic of the effective-mass equation the reader is referred to the articles by Altarelli (1986), Bastard and Brum (1986) and Burt (1988, 1992).

2.2 The Confinement of Electrons

Application of the effective-mass boundary conditions to the case of a single quantum well where only simple, zone-centre conduction bands are involved yields the dispersion equation

$$Z = \frac{m_W^* k_B}{m_B^* k_W} = \begin{cases} +i\tan(k_W a/2) & \text{asymmetric} \\ -i\cot(k_W a/2) & \text{symmetric} \end{cases}$$
$$\frac{\hbar^2 k_W^2}{2m_W^*} = E, \qquad \frac{\hbar^2 k_B^2}{2m_B^*} = E - V_0 \tag{2.39}$$

where m_W^*, m_B^* are the effective masses in the well and barrier, k_w k_B are the wave-vector components along the normal to the interface, a is the well-width and V_0 is the discontinuity in energy of the conduction bands. The two solutions are associated with asymmetric and antisymmetric envelope-function patterns, with symmetry defined in relation to the midplane of the quantum-well layer (Fig. 5 of the Introduction). In many cases these patterns can be represented accurately enough by

$$F(z) = \left(\frac{2}{a}\right)^{1/2} \begin{cases} \sin(n\pi z/a) & n = 1, 3, 5... \\ \cos(n\pi z/a) & n = 2, 4, 6... \end{cases} \tag{2.40}$$

as if the carrier were completely confined ($V_0 \to \infty$), but the energy levels are usually considerably different from those for an infinitely deep well. The latter point is underlined in Fig. 2 in the Introduction, which depicts the band structure for GaAs/$Al_{0.3}Ga_{0.7}As$ superlattice, and the dashed lines are the infinitely deep well energies. In the case of a superlattice, application of the effective-mass boundary conditions and application of the Bloch theorem give the dispersion relation

$$\cos k_z(a+b) = \cos k_W a \cos k_B b - \frac{1}{2}\left(Z + \frac{1}{Z}\right) \sin k_W a \sin k_B b \tag{2.41}$$

where k_z is the wavevector for propagation along the superlattice axis, and a and b are the well- and barrier-widths, respectively. Basically, the single energy levels of the single quantum well are broadened into minibands. For states with $k_z = 0$ or $\pi/(a+b)$ the wave patterns are purely symmetric or purely antisymmetric within a well or a barrier, but for intermediate values of k_z the symmetry is mixed.

The electron wavefunction in a well of a superlattice can be represented by

$$F(z) = A(S_1 \cos k_w z + S_2 \sin k_w z), \quad |z| \leq \frac{a}{2} \tag{2.42}$$

where

$$\begin{aligned} S_1 &= (e^{ik_z d} + e^{ik_B b}) \sin\left(\frac{k_w a}{2}\right) + iZ^{-1}(e^{ik_z d} - e^{ik_B b}) \cos\left(\frac{k_w a}{2}\right) \\ S_2 &= (e^{ik_z d} - e^{ik_B b}) \cos\left(\frac{k_w a}{2}\right) - iZ^{-1}(e^{ik_z d} + e^{ik_B b}) \sin\left(\frac{k_w a}{2}\right) \end{aligned} \tag{2.43}$$

and in a barrier by

$$F(z) = A(T_1 \cos k_B z - T_2 \sin k_B z), \quad |z| \leq \frac{b}{2} \tag{2.44}$$

where

$$\begin{aligned} T_1 = {} & \left[e^{ik_z d}(\sin k_W a + iZ^{-1} \cos k_W a) - iZ^{-1}\right] \cos\left(\frac{k_B b}{2}\right) \\ & - iZ^{-1}\left[e^{ik_z d} \cos k_W a - iZ^{-1} \sin k_W a) + 1\right] \sin\left(\frac{k_B b}{2}\right) \\ T_2 = {} & \left[e^{ik_z d}(\sin k_W a - iZ^{-1} \cos k_W a) + iZ^{-1}\right] \sin\left(\frac{k_B b}{2}\right) \\ & - iZ^{-1}\left[e^{ik_z d} \cos k_W a - iZ^{-1} \sin k_W a) - 1\right] \cos\left(\frac{k_B b}{2}\right) \end{aligned} \tag{2.45}$$

Normalization gives

$$A = 2^{1/2}\Big[\Big(|S_1|^2 + |S_2|^2\Big)a + \Big(|T_1|^2 + |T_2|^2\Big)b + \Big(|S_1|^2 - |S_2|^2\Big)\frac{(\sin k_w a)}{k_w} + \Big(|T_1|^2 - |T_2|^2\Big)\frac{(\sin k_B b)}{k_B}\Big] \tag{2.46}$$

When the potential barrier is large, $|Z| \to \infty$, $\exp(ik_B b) \to 0$, and the dispersion relation reduces to

$$\sin k_W a \approx \frac{-2(\cos(k_z(a+b)))}{(Z \sin k_z b)} \approx 0 \tag{2.47}$$

The wavefunction in the well reduces to a single-quantum-well function multiplied by the phase factor $\exp(ik_Z d)$.

Motion parallel to the layers is unrestricted in quasi-2D systems and the full envelope function and its energy for a single quantum well are

$$F(\mathbf{r}) \approx \left(\frac{2}{V}\right)^{1/2} e^{i\mathbf{k}.\mathbf{r}} \begin{cases} \sin(n\pi z/a) \; n \text{ odd} \\ \cos(n\pi z/a) \text{ even} \end{cases} \tag{2.48a}$$

$$E = E_n + \frac{\hbar^2 k^2}{2m_W^*} \tag{2.48b}$$

where V is the normalizing volume, **k** and **r** are vectors in the plane and E_n is the subband energy determined from Eq. (2.39).

Effective-mass theory has also been applied to quantum wires, with elliptical cross-sections of various eccentricity, to quantum dots and to cases where there is a magnetic field.

Another type of confinement operates in the technologically important system of the high-electron-mobility transistor (HEMT). Here, electrons donated by impurities in the barrier fall over the potential cliff at the heterojunction and are confined near the interface by the electrostatic attraction of the positive charges of the impurities. The ground-state wavefunction is often approximated by the Fang–Howard expression (Fig. 2.2):

$$\psi(z) = \left(\frac{b^3}{2}\right)^{1/2} z e^{-bz/2} \tag{2.49}$$

where b is a parameter chosen to minimize the energy

$$E = \int_0^\infty eV(z)\,\psi(z)^2\,dz + \int_0^\infty \psi(z)\frac{\hbar^2}{2m^*}\frac{d^2\psi(z)}{dz^2}\,dz \tag{2.50}$$

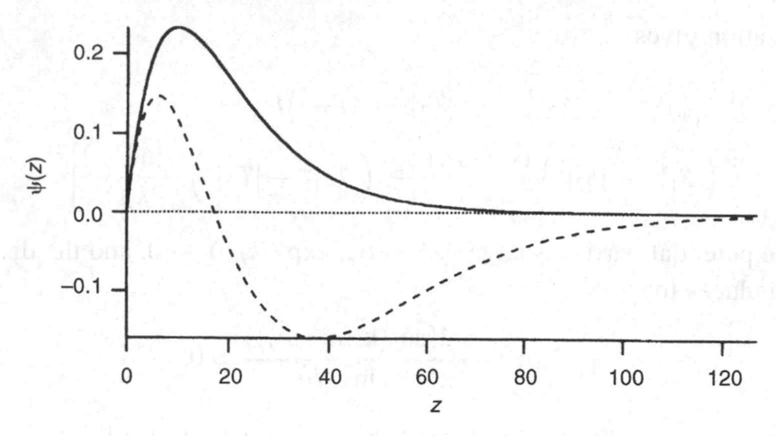

Fig. 2.2 Fang–Howard wavefunctions for the ground and excited states in a single heterojunction system.

and $V(z)$ is the electrostatic potential that comprises components associated with the charged donors, the electrons themselves and image charges induced in the barrier. In III-V systems the latter are often ignored, and we will do that here. If the surface density of donors is N_s, and this is taken to be equal to the carrier density, the field that an electron experiences is given by

$$F(z) = -\frac{eN_s}{2\varepsilon}\left(1 - \int_0^z \psi(z')^2 dz'\right) \quad (2.51)$$

and so

$$V(z) = \frac{eN_s}{2\varepsilon}\int_0^z \left(1 - \int_0^{z'} \psi(z'')^2 dz''\right) \quad (2.52)$$

Thus

$$E = \frac{33e^2N_s}{32\varepsilon b} + \frac{\hbar^2 b^2}{8m^*} \quad (2.53)$$

and the energy is minimized when

$$b = \left(\frac{33e^2N_s}{8\varepsilon\hbar^2}\right)^{1/3} \quad (2.54)$$

More elaborate variational wavefunctions have been used (Ando, 1982; Takada and Uemura, 1977) that give somewhat better agreement with wavefunctions and energies generated numerically by solving the coupled Schrödinger and Poisson equations, but the extra accuracy is often regarded as dispensable.

Transistors with high sheet-charge densities are becoming increasingly common, and there is therefore a need to take into account the population of the second subband. The simplest form for the wavefunction in the second subband that is orthogonal to that for the ground state is

$$\psi(z) = \left(\frac{3b^3}{2}\right)^{1/2} z\left(1 - \frac{b}{3}z\right)e^{-bz/2} \tag{2.55a}$$

Once again, b can be chosen to minimize the total energy, taking into account the population of both subbands. A more accurate form requiring two variational parameters is (Fig. 2.2)

$$\psi(z) = \left(\frac{3b_2^5}{2(b_1^2 - b_1 b_2 + b_2^2)}\right)^{1/2} z\left(1 - \frac{b_1 + b_2}{6}z\right)e^{-b_2 z/2} \tag{2.55b}$$

The scheme can obviously be extended to the case of three or more subbands. More often than not, wavefunctions, energies and populations are generated numerically (Fig. 2.3) from which analytic forms for the wavefunctions can be obtained for use in scattering problems.

The quantum states can sometimes be engineered by inserting a monolayer; that can be done successfully even when there is considerable lattice mismatch. Examples are a layer of AlAs or of InAs in a GaAs quantum well. The former layer introduces a repulsive potential, whereas the latter introduces an attractive potential. These δ-function-like potentials modify the electron wavefunction basically by introducing a change in slope. In addition they modify the optical vibrations, a topic we will return to in Section 7.2. Alloy grading is also used to engineer the quantum confinement of electrons.

2.3 The Confinement of Holes

In bulk material the band structure for holes is complicated by the triple degeneracy (not counting spin) associated with p-orbitals. Spin-orbit coupling reduces the degeneracy by 1. In direct-gap III-V compounds there are, therefore, four zone-centre bands of prime importance: the conduction band, whose unit-cell wavefunction has s-orbital symmetry, $|\ iS\downarrow\rangle$ and $|\ iS\uparrow\rangle$; a doubly degenerate valence band plus split-off band characterized by p-orbitals of the form $|(X - iY)\downarrow/\sqrt{2}\rangle$, $|(X + iY)\downarrow/\sqrt{2}\rangle$ and $|Z\uparrow\rangle, |Z\downarrow\rangle$, $|(X - iY)\uparrow/\sqrt{2}\rangle$, and $|(X - iY)\downarrow/\sqrt{2}\rangle$. Away from the zone-centre these bands couple via **k.p** interaction, which leads to an 8×8 matrix Hamiltonian. The coupling is sufficient to describe the curvatures and hence the effective masses of the conduction, light-hole and split-off bands to a good approximation, but the curvature of the heavy-hole band is determined by the interaction with more

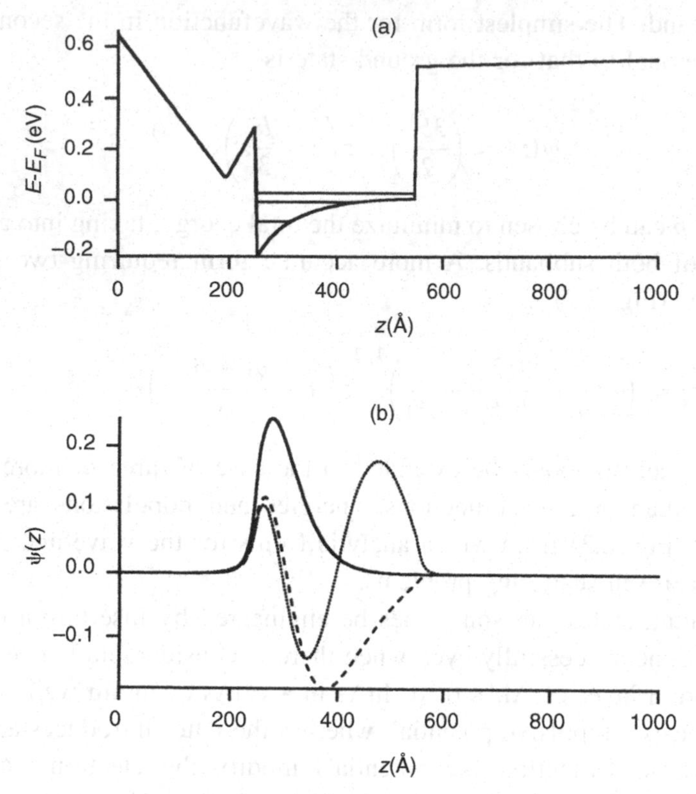

Fig. 2.3 Numerical results for a AlInAs/GaInAs/InP MODFET with δ-doping ($5\times10^{12}\,\text{cm}^{-2}$) and a 50-Å spacer: energy profile and energies of the first three subbands relative to the Fermi level and corresponding wave-functions.

remote bands. Once the effective mass of the conduction band is determined, the 8×8 matrix can be resolved approximately into a 2×2 matrix for the conduction band and a 6×6 matrix for the valence bands. The latter gives rise to the Schrödinger equation:

$$\begin{vmatrix} P+Q-E & L & M & 0 & iL\sqrt{2} & -i\sqrt{2} \\ L^* & P-Q-E & 0 & M & -i\sqrt{2}Q & i\sqrt{3/2}L \\ M^* & 0 & P-Q-E & -L & i\sqrt{3/2}L^* & -i\sqrt{2}Q \\ 0 & M^* & -L^* & P+Q-E & -i\sqrt{2}M^* & -iL^*/\sqrt{2} \\ -iL^*/\sqrt{2} & i\sqrt{2}Q & i\sqrt{3/2}L & i\sqrt{2}M & P-\Delta_0-E & 0 \\ i\sqrt{2}M^* & -i\sqrt{3/2}L^* & i\sqrt{2}Q & iL/\sqrt{2} & 0 & P-\Delta_0-E \end{vmatrix} \begin{vmatrix} |\frac{3}{2},\frac{3}{2}\rangle \\ |\frac{3}{2},\frac{1}{2}\rangle \\ |\frac{3}{2},-\frac{1}{2}\rangle \\ |\frac{3}{2},-\frac{3}{2}\rangle \\ |\frac{1}{2},\frac{1}{2}\rangle \\ |\frac{1}{2},-\frac{1}{2}\rangle \end{vmatrix} = 0 \tag{2.56}$$

where E is the hole energy and Δ_0 is the spin–orbit splitting. The other parameters are

$$P = \frac{\hbar^2}{2m_0}(\gamma_1[k_x^2 + k_y^2] + k_z\gamma_1 k_z), \quad Q = \frac{\hbar^2}{2m_0}(\gamma_2[k_x^2 + k_y^2] + k_z\gamma_2 k_z)$$
$$L = -\frac{i\sqrt{3}}{m_0}\hbar^2(k_x - ik_y)k_z\gamma_3, \quad M = \frac{\hbar^2}{2m_0}(\sqrt{3}(k_x^2 - k_y^2)\gamma_2 - i2\sqrt{3}k_x k_y \gamma_3) \tag{2.57}$$

γ_i are the Luttinger parameters, which arise from the **k.p** interactions with other bands, viz.:

$$\gamma_1 = -\frac{1}{3}\left(1 + \frac{2}{m_0}\sum_n \frac{p_1^2 n}{E_0 - E_n}\right)$$
$$\gamma_2 = -\frac{1}{6}\left(1 + \frac{2}{m_0}\sum_n \frac{p_{1nx}^2 - p_{1ny}^2}{E_0 - E_n}\right) \tag{2.58}$$
$$\gamma_3 = -\frac{1}{3m_0}\sum_n \frac{\left(p_{1nx}p_{n2y} - p_{1ny}p_{n2x}\right)}{E_0 - E_n}$$

where the p_{nma} are momentum matrix elements, i.e. $\langle n \mid pa \mid m \rangle$. Their magnitudes are determined by experiments, e.g. measurements of magneto-resistance and cyclotron resonance. The eigenfunctions are depicted by the total angular momentum quantum number $j = l \pm s$ and its projection, m_j, viz: $\mid J, m_j\rangle$. Terms linear in **k** that arise through the lack of inversion symmetry in III-V compounds are small and have been neglected.

The eigenfunctions can be depicted thus:

$$\left|\frac{3}{2},\frac{3}{2}\right\rangle = \frac{1}{\sqrt{2}}(X + iY)\uparrow \qquad \left|\frac{3}{2},-\frac{3}{2}\right\rangle = \frac{1}{\sqrt{2}}(X - iY)\downarrow$$
$$\left|\frac{3}{2},\frac{3}{2}\right\rangle = \frac{1}{\sqrt{6}}[(X + iY)\downarrow -2Z\uparrow] \qquad \left|\frac{3}{2},-\frac{1}{2}\right\rangle = \frac{1}{\sqrt{6}}[(X - iY)\uparrow +2Z\downarrow]$$
$$\left|\frac{1}{2},\frac{1}{2}\right\rangle = \frac{1}{\sqrt{3}}[(X + iY)\downarrow +Z\uparrow] \qquad \left|\frac{1}{2},-\frac{1}{2}\right\rangle = \frac{1}{\sqrt{3}}[-(X - iY)\uparrow +Z\downarrow] \tag{2.59}$$

If the spin–orbit splitting is large, a further simplification can be made by decoupling the split-off band and reducing the Hamiltonian to a 4 × 4 matrix describing light and heavy holes. This can be block diagonalized by using the unitary transformation UHU^+, where

$$
U = \begin{vmatrix} \frac{1}{\sqrt{2}}e^{-i\phi} & 0 & 0 & -\frac{1}{\sqrt{2}}e^{i\phi} \\ 0 & \frac{1}{\sqrt{2}}e^{-i\eta} & -\frac{1}{\sqrt{2}}e^{i\eta} & 0 \\ 0 & \frac{1}{\sqrt{2}}e^{-i\eta} & \frac{1}{\sqrt{2}}e^{i\eta} & 0 \\ \frac{1}{\sqrt{2}}e^{-i\phi} & 0 & 0 & \frac{1}{\sqrt{2}}e^{i\phi} \end{vmatrix} \tag{2.60}
$$

The result is

$$
\begin{vmatrix} H_{11} & H_{12} & 0 & 0 \\ H_{21} & H_{22} & 0 & 0 \\ 0 & 0 & H_{11} & H_{12} \\ 0 & 0 & H_{21} & H_{22} \end{vmatrix} \begin{vmatrix} \left|\frac{3}{2},\frac{3}{2}\right\rangle \\ \left|\frac{3}{2},\frac{1}{2}\right\rangle \\ \left|\frac{3}{2},-\frac{1}{2}\right\rangle \\ \left|\frac{3}{2},-\frac{3}{2}\right\rangle \end{vmatrix} = 0 \tag{2.61}
$$

where

$$
H_{11} = E_s + \delta - \frac{\hbar^2}{2m_0}\left[\frac{d}{dz}(\gamma_1 - 2\gamma_2)\frac{d}{dz} - (\gamma_1 + \gamma_2)k^2\right] \tag{2.62a}
$$

$$
H_{12} = -\frac{\hbar^2}{2m_0}\left(\sqrt{3}\bar{\gamma}(\theta)k^2 + 2\sqrt{3}k\gamma_3\frac{d}{dz}\right) \tag{2.62b}
$$

$$
H_{22} = E_s - \delta - \frac{\hbar^2}{2m_0}\left[\frac{d}{dz}(\gamma_1 + 2\gamma_2)\frac{d}{dz} - (\gamma_1 - \gamma_2)k^2\right] \tag{2.62c}
$$

$$
H_{21} = -\frac{\hbar^2}{2m_0}\left(\sqrt{3}\bar{\gamma}(\theta)k^2 - 2\sqrt{3}k\frac{d}{dz}\gamma_3\right) \tag{2.62d}
$$

We have included the effect of a biaxial strain so that the important case of a strained layer is accommodated. E_s is the shift in energy due to the hydrostatic component of the stress and δ is the strain-splitting energy, viz.:

$$
E_s = 2a\frac{(c_{11} - c_{12})}{c_{11}}e_s, \qquad \delta = -b\frac{(c_{11} + 2c_{12})}{c_{11}}e_s, \qquad e_s = \frac{\Delta a_0}{a_0} \tag{2.63}
$$

a and b are deformation potentials for the valence band, c_{11} and c_{12} are elastic constants and e_s is the strain related to the difference in the lattice constant a_0 of the two layers. Compressive strains ($\Delta a_0, < 0$) make δ negative since $b < 0$. In our convention we regard hole energy as positive. Thus if $\delta < 0$, the heavy-hole (HH) band corresponds to a rise of electron energy relative to the light-hole (LH) band (Fig. 2.4).

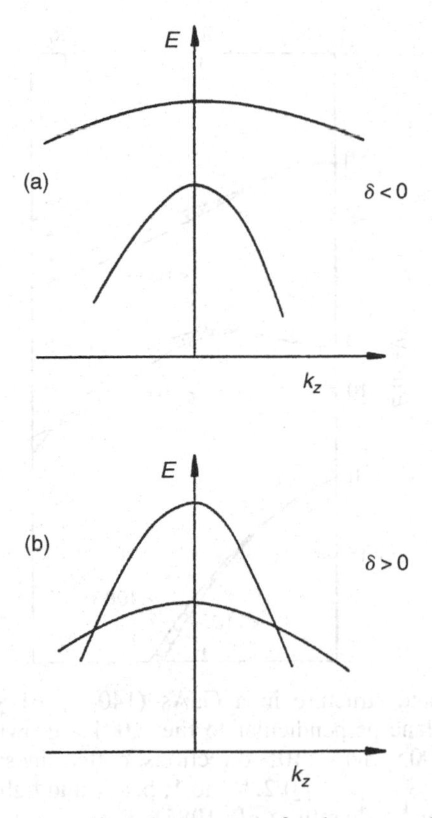

Fig. 2.4 Schematic valence-band structure for heavy and light holes in the presence of (a) compressive ($\delta < 0$) and (b) tensile ($\delta > 0$) biaxial strains.

In Eq. (2.62), $k^2 = k_x{}^2 + k_y{}^2$, with k_z in the confinement direction. The factor $\bar{\gamma}(\theta)$ is given by

$$\bar{\gamma}(\theta) = (\gamma_2^2 \cos^2 2\theta + \gamma_3^2 \sin^2 2\theta)^{1/2} \tag{2.64}$$

and describes the anisotropy in the (x, y) plane. In the bulk case we can replace $\partial/\partial z$ by ik_z, and solving the secular determinant leads to the energy

$$\begin{aligned} E = & E_s + \frac{\hbar^2}{2m_0}\Big\{\gamma_1^2(k_x^2 + k_y^2 + k_z^2) \pm [4\gamma_2^2(k_x^2 + k_y^2 + k_z^2)^2 \\ & + 12(\gamma_3^2 - \gamma_2^2)(k_x^2k_y^2 + k_y^2k_z^2 + k_z^2k_x^2) \\ & + 2\gamma_2(k_x^2 + k_y^2 - 2k_z^2)K^2 + K^4]^{1/2}\Big\} \end{aligned} \tag{2.65}$$

where $K = \sqrt{2m_0\delta/\hbar^2}$ is a wavevector associated with the strain-splitting. The effect of the strain is all contained in E_s and the parameter K^2. Restrictions on k_z through confinement will further determine subband energies. Figure 2.5

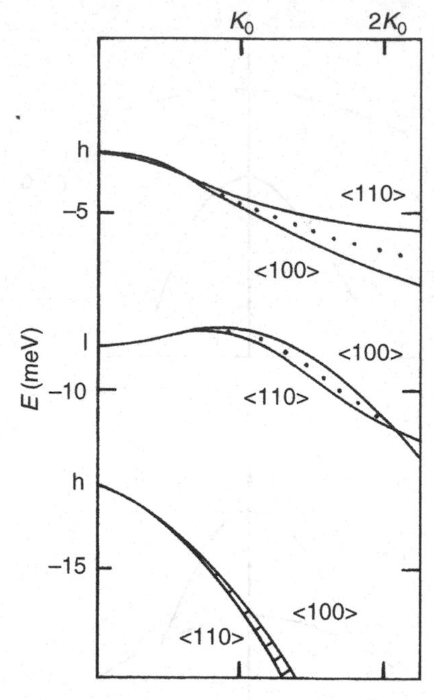

Fig. 2.5 Hole subband structure in a GaAs (140-Å) $Al_{0.21}Ga_{0.79}As$ (200-Å) superlattice in the plane perpendicular to the <001> growth axis. Solid lines: dispersion along <100> and <110> directions, dotted lines: results of the axial approximation, $\gamma_3 \approx \gamma_2 \approx (\gamma_2 + \gamma_2)/2$. h and 1: heavy and light holes; $K_0 = \pi/340$ (Å) $= 9.24 \times 10^5$ cm^{-1} (Altarelli *et al.*, 1985).

illustrates the effect of confinement in the absence of strain. For small in-plane wavevectors the mixing of heavy- and light-hole wavefunctions can be found by standard perturbation techniques (see, for example, Andreani *et al.*, 1987; O'Reilly, 1989; and Foreman, 1994). Integration of Eq. (2.43) across an interface yields boundary conditions that entail the continuity of

$$F_H(z) \quad \text{and} \quad (\gamma_1 - 2\gamma_2)\frac{dF_H(z)}{dz} \tag{2.66a}$$

$$F_L(z) \quad \text{and} \quad (\gamma_1 - 2\gamma_2)\frac{dF_L(z)}{dz} - 2\sqrt{3}\gamma_3 k F_H(z) \tag{2.66b}$$

In the preceding, terms linear in k that arise from lack of inversion symmetry in polar semiconductors, being small, are ignored.

The Hamiltonian of Eq. (2.62) contains off-diagonal elements that are not symmetrized. Thus H_{12} contains the operator $\gamma_3\partial/\partial z$ and H_{21} the operator $(\partial/\partial z)$ γ_3. Many authors used the symmetrized forms $(\gamma_3\partial/\partial z + (\partial/\partial z)\,\gamma_3)/2$ in order to make the individual terms Hermitian, but this is found to give unphysical behaviour (Foreman, 1993), whereas the use of the unsymmetrized Hamiltonian, which Foreman obtains from Burt's general analysis, does not. Thus Eq. (2.62) appears to be the correct Hamiltonian, and it should be noted that only the total Hamiltonian need be Hermitian. That a difference exists between HH and LH boundary conditions is not surprising in view of the different orbital forms. The HH state involves only $|X\rangle$ and $|Y\rangle$ components, i.e. components parallel to the interface, whereas the LH state involves $|X\rangle$, $|Y\rangle$ and $|Z\rangle$.

The mixed wavefunctions are obtained by using perturbation theory, taking the terms in k as the perturbation. Thus we take the envelope function for the HH band to be, for first order in eigenvectors

$$\begin{gathered} F_H = F_H^0 + F_H^1 + F_L^1, \quad E = E_0 + E_1 + E_2, \quad H_{11} = H_{11}^0 + H_{11}^1, \\ H_{12} = H_{12}^2 + H_{12}^1, \quad H_{22} = H_{22}^0 + H_{22}^1, \quad H_{21} = H_{21}^2 + H_{21}^1 \end{gathered} \tag{2.67}$$

where

$$(H_{11}^0 - E_0)F_H^0 = 0 \tag{2.68a}$$

$$(H_{11}^0 - E_0)F_H^1 = (E_1 - H_{11}^1)F_H^0 \tag{2.68b}$$

$$(H_{22}^0 - E_0)F_L^1 = -H_{21}^1 F_H^0 \tag{2.68c}$$

and

$$E_1 = \langle F_H^0 | H_{11}^1 | F_H^0 \rangle = \frac{-\hbar^2}{2m_0}(\gamma_1 + \gamma_2)k^2 \tag{2.69a}$$

$$E_2 = \langle F_H^0 | H_{11}^1 | F_H^1 \rangle + \langle F_H^0 | H_{12}^1 | F_L^1 \rangle \tag{2.69b}$$

where $H_{12}^1 = -(\hbar^2/2m_0)2\sqrt{3}k\gamma_3\partial/\partial z$. Thus, for an infinitely deep well we can take

$$F_H^0 = A\cos k_H z \qquad -\frac{a}{2} \leq z \leq \frac{a}{2} \tag{2.70}$$

whence $F_H^1 = 0$ and

$$F_L^1 = -\frac{\sqrt{3\gamma_3 k}}{\gamma_2 k_H(1 - S_H)} \cdot A \sin k_{H0} z = B \sin k_{H0} z \tag{2.71}$$

where $S_{\mathrm{H}} = m_0\delta/\gamma_2\hbar^2 K_{H_0}^2$ and k_H is the wavevector component in the confinement direction. We see that a symmetric HH state mixes with an antisymmetric LH state. Similarly for an antisymmetric LH state the mixing is with the symmetric HH state, viz.:

$$F_L^0 = C\sin k_L z, \quad F_H^1 = -\frac{\sqrt{3\gamma_3 k}}{\gamma_2 k_L(1 - S_L)} .C\cos k_{L0} z = D\cos k_{L0} z \tag{2.72}$$

where $S_L = m_0\delta/\gamma_2\hbar^2 K_{L0}^2$

The boundary conditions at each interface couple HH and LH states of the same energy. (Note that with strain one or the other wavevectors k_{H0}, k_{L0} may be pure imaginary.) The overall envelope function is of the form

$$F = A\cos k_H z + B\sin k_{H0} z + C\sin k_L z + D\cos k_{L0} z \tag{2.73}$$

and $k_H \to k_{H0} + \varepsilon$, $k_L \to k_{L0} + \varepsilon$, $\varepsilon < k_{H,L}$. For an infinitely deep well $F = 0$ at $z = \pm a/2$, and from this boundary condition the wavevector shift can be obtained.

The energy can then be expressed in terms of an in-plane effective mass m_k^*, thus

$$\begin{aligned} E &= E_s \pm \delta + \frac{\hbar^2}{2m_0}(\gamma_1 \mp 2\gamma_2)(k_{H0}+\varepsilon)^2 + E_1 + E_2 \\ &= E_s \pm \delta + \frac{\hbar^2(\gamma_1 \mp 2\gamma_2)k_{H0}^2}{2m_0} + \frac{\hbar^2 k^2}{2m_k^*} \end{aligned} \tag{2.74}$$

(upper sign for HH).

The in-plane masses are given by

$$\text{HH}\ \frac{m_0}{m_k^*} = \gamma_1 + \gamma_2 - \frac{3\gamma_3^2}{\gamma_2(1-S_H)} + \frac{3\gamma_3^2(\gamma_1 - 2\gamma_2)(\cos k_L a + (-1)^{n+1})}{\gamma_2^2(1-S_H)(1-S_L)k_L a\,\sin k_L a} \tag{2.75a}$$

$$\text{LH}\ \frac{m_0}{m_k^*} = \gamma_1 - \gamma_2 + \frac{3\gamma_3^2}{\gamma_2(1-S_L)} + \frac{3\gamma_3^2(\gamma_1 + 2\gamma_2)(\cos k_L L + (-1)^{n+1})}{\gamma_2^2(1-S_H)(1-S_L)k_H a\,\sin k_H a} \tag{2.75b}$$

where the S factors are the dimensionless ratio of strain-splitting energy to confinement energy, viz.:

$$S_H = \frac{m_0\delta}{\gamma_2\hbar^2 k_H^2}, \quad S_L = \frac{m_0\delta}{\gamma_2\hbar^2 k_L^2} \tag{2.76}$$

and the wavevectors k_H, k_L are those defining equal energies for the two bands for $k = 0$, i.e.

$$(\gamma_1 + 2\gamma_2)k_L^2 - K^2 = (\gamma_1 - 2\gamma_2)k_H^2 + K^2, \quad K^2 = \frac{2m_0\delta}{\hbar^2} \tag{2.77}$$

the well-width is a and n is an integer. With complete confinement the HH state has $k_H = n\pi/a$ and the corresponding value of k_L that enters, Eq. (2.75a) is obtained from Eq. (2.77). If $\delta < 0$, k_L is pure imaginary. For the LH state $k_L = n\pi/a$ and the corresponding value of k_H entering, Eq. (2.75b) is obtained from Eq. (2.77).

For compressive strains ($\delta < 0$), the in-plane masses exhibit the phenomenon of mass reversal: anomalously low for the HH band and anomalously high for the LH

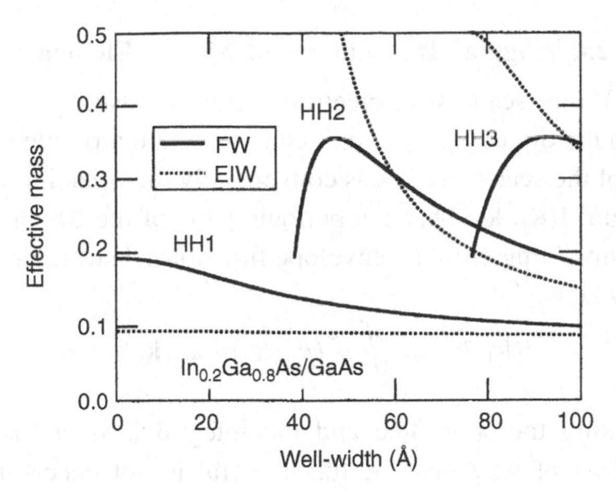

Fig. 2.6. In-plane hole effective mass in the $Ga_{0.8}In_{0.2}As/GaAs$ system as a function of well-width taking into account the finite depth of the well (solid lines). Dotted lines: equivalent infinitely deep well (Foreman, 1994).

band. For tensile strains it is possible for $S_H = S_L = 1$, in which case the preceding equations become invalid. The energy bands under these special circumstances become degenerate at $k = 0$ but have maxima shifted from the zone-centre.

The expressions for the in-plane mass given in Eq. (2.75) illustrate the main features of the effects of strain and confinement, but for wells of finite depth they are quantitatively unreliable. Figure 2.6 depicts this for the GaInAs/GaAs system. The small k approximation that was used is also inadequate to represent the non-parabolicity of the bands and to describe anticrossings, and, of course, the neglect of the effects of the split-off band becomes less justifiable with increasing k. The strain-splitting terms in the 3×3 Hamiltonian are of the form

$$H_S = \begin{vmatrix} \delta & 0 & 0 \\ 0 & -\delta & \sqrt{2\delta} \\ 0 & \sqrt{2\delta} & 0 \end{vmatrix} \tag{2.78}$$

which couples the LH and split-off (SO) bands, so neglecting the effect of the SO band is valid only for strains that are not too large even at small k.

Within the approximations leading to the expressions in Eq. (2.75) for the in-plane masses the bands are isotropic and parabolic. Anisotropy and non-parabolicity appear when terms of order k^4 enter the energy. The interaction between HH1 and HH2 is zero up to terms of order k^6. One would have to extend the perturbation method considerably in order to describe anticrossing effects between these subbands. Hitherto, analytic extension only to order k^4 has been reported (Foreman, 1994).

2.4 Angular Dependence of Matrix Elements

A perturbation V that scatters an electron or hole is usually slowly varying in space relative to the dimension of a unit cell, and the matrix element involved in the description of the scattering rate is conveniently factorized into the product of an overlap integral $\mathbf{I}(\mathbf{k}_1, \mathbf{k}_2)$ between periodic parts of the Bloch functions and a matrix element involving only the envelope functions. Thus the rate is multiplied by $|I(\mathrm{k}_1, \mathrm{k}_2)|^2$, where

$$I(\mathbf{k}_1, \mathbf{k}_2) = \int u^*(\mu_2, \mathbf{k}_2) u(\mu_1, \mathbf{k}_1) d\mathbf{r} \tag{2.79}$$

with μ representing the spin state and the integral is over the unit cell. For arbitrary directions of wavevectors, this integral is not necessarily unity, as a consequence of the change in direction of angular momentum.

A state with total angular momentum J has a component in the z-direction of m where $m = (J, J-1, \ldots, -J)$, i.e. $|J, m\rangle$. A rotation of the axis from xyz to $x'y'z'$ will convert this state to $|J, m'\rangle$ – the old state is connected linearly to the new state via a rotation matrix

$$|J, m\rangle = \sum_{m'} R^J_{m'm}(a, \beta, \gamma) |J, m'\rangle \tag{2.80}$$

where a, β and γ are the angles between the old and the new axes. The rotation matrix derives from the unitary operator for finite rotations, $U(a, \beta, \gamma)$ where

$$U(a, \beta, \gamma) = U_{z'}(\gamma) U_y(\beta) U_z(a) \tag{2.81}$$

corresponding to a rotation through an angle a about the old z axis, then a rotation through an angle β about the new y axis and finally a rotation through an angle γ about the final z axis. A rotation through ϕ about, say, the x axis is given by

$$U_x(\phi) = e^{i\phi J_x} \tag{2.82}$$

The rotation matrix is then given by

$$R^J_{m'm}(a, \beta, \gamma) = e^{im'\gamma} r^J_{m'm}(\beta) e^{ima} \tag{2.83}$$

where

$$r_{m'm}{}^J(\beta) = \left[\frac{(J+m')!(J-m'!)}{(J+m)!(J-m)!}\right]^{1/2} \left[\cos\left(\frac{\beta}{2}\right)\right]^{m'+m} \left[\sin\left(\frac{\beta}{2}\right)\right]^{m'-m} \times P^{m'-m,\, m'+m}_{J-m}(\cos\beta) \tag{2.84a}$$

$$P_n^{ab}(\cos\beta) = \frac{(-1)^n}{2^n n!}(1-\cos\beta)^{-a}(1+\cos\beta)^{-b} \times \left[\frac{d}{d\cos\beta}\right]\left[(1-\cos\beta)^{a+n}(1+\cos\beta)^{b+n}\right] \tag{2.84b}$$

and the P_n^{ab} (cos β) are Jacobi polynomials (see Landau and Lifshitz, 1977). Note that the angles α ($0 \le \alpha \le 2\pi$) and β ($0 \le \beta \le \pi$) are the spherical azimuthal and polar angles of the new z axis and with respect to the xyz axes.

For conduction band electrons in a central zone valley near $\mathbf{k} = 0$ the wavefunction is almost purely s-orbital-like and so $J = 1/2$. In this case

$r^{1/2}_{m'm} =$ (2.85)

$m'\backslash m$	$\frac{1}{2}$	$-\frac{1}{2}$
$\frac{1}{2}$	$\cos\left(\frac{\beta}{2}\right)$	$\sin\left(\frac{\beta}{2}\right)$
$-\frac{1}{2}$	$-\sin\left(\frac{\beta}{2}\right)$	$\cos\left(\frac{\beta}{2}\right)$

This is also applicable to the split-off valence band. For the upper valence bands $J = 3/2$ and

$r^{1/2}_{m'm} =$

$m'\backslash m$	$\frac{3}{2}$	$\frac{1}{2}$	$-\frac{1}{2}$	$-\frac{3}{2}$
$\frac{3}{2}$	$\cos^3\left(\frac{\beta}{2}\right)$	$-\frac{1}{2}\sqrt{3}\cos\left(\frac{\beta}{2}\right)\sin\beta$	$\frac{1}{2}\sqrt{3}\sin\left(\frac{\beta}{2}\right)\sin\beta$	$-\sin^3\left(\frac{\beta}{2}\right)$
$\frac{1}{2}$	$\frac{1}{2}\sqrt{3}\cos\left(\frac{\beta}{2}\right)\sin\beta$	$-\frac{1}{2}\cos\left(\frac{\beta}{2}\right)(1-3\cos\beta)$	$-\frac{1}{2}\sin\left(\frac{\beta}{2}\right)(1-3\cos\beta)$	$-\frac{1}{2}\sqrt{3}\sin\left(\frac{\beta}{2}\right)\sin\beta)$
$-\frac{1}{2}$	$\frac{1}{2}\sqrt{3}\sin\left(\frac{\beta}{2}\right)\sin\beta$	$\frac{1}{2}\sin\left(\frac{\beta}{2}\right)(1+3\cos\beta)$	$-\frac{1}{2}\cos\left(\frac{\beta}{2}\right)(1-3\cos\beta)$	$\frac{1}{2}\sqrt{3}\cos\left(\frac{\beta}{2}\right)\sin\beta$
$-\frac{3}{2}$	$\sin^3\left(\frac{\beta}{2}\right)$	$-\frac{1}{2}\sqrt{3}\sin\left(\frac{\beta}{2}\right)\sin\beta$	$\frac{1}{2}\sqrt{3}\cos\left(\frac{\beta}{2}\right)\sin\beta)$	$\cos^3\left(\frac{\beta}{2}\right)$

(2.86)

We are now in a position to calculate the overlap integral, or, rather, its modulus squared. Thus

$$|I_m(\mathbf{k}_1, \mathbf{k}_2)|^2 = \sum_{m'} \left|R^J_{m'm}(\alpha, \beta, \gamma)\right|^2 \tag{2.87}$$

where β is identified as the angle between $\mathbf{k}_1$ and $\mathbf{k}_2$. For the conduction and split-off bands

$$\left|I_{\pm 1/2}(\mathbf{k}_1, \mathbf{k}_2)\right|^2 = 1 \mp \sin\beta \cos\gamma \tag{2.88}$$

An average over the initial spin states removes all angular dependence and gives unity.

For the LH states $J = 3/2$, $m = 1/2$ and $m' = \pm 1/2$. Averaging over the initial spin state gives

$$|I(\mathbf{k}_1, \mathbf{k}_2)|^2 = \frac{1}{4}(1 + 3\cos^2\beta) \tag{2.89}$$

In the case of the HH states $J = 3/2$, $m = 3/2$ and $m' = \pm 3/2$, and we obtain the same result as for LH states given in Eq. (2.89). For an interband transition between HH and LH bands

$$|I(\mathbf{k}_1, \mathbf{k}_2)|^2 = \frac{3}{4}\sin^2\beta \tag{2.90}$$

When holes are confined the wavevector in the confinement direction, k_z, becomes fixed by the boundary conditions. In the simplest case intraband scattering in the plane occurs with fixed k_z. If φ is the angle between wavevectors in the plane

$$\cos\beta = \cos\theta_1 \cos\theta_2 + \sin\theta_1 \sin\theta_2 \cos\phi \tag{2.91}$$

where θ_1 and θ_2 are the angles that the total wavevectors make with the confinement direction, and thus β is dependent on k_1 and k_2, where k_1 and k_2 are the in-plane wavevectors. In the limit of weak confinement (k_z small) $\cos\beta \approx \cos\phi$, but for strong confinements (k_z large) $\cos\beta \approx 1$ and hence the intrasubband overlap factor is near unity, whereas for HH–LH transitions the overlap factor vanishes.

Introducing band-mixing effects makes the angular dependence in bulk material more complex for the conduction band, LH band and SO band, but not, to the same order, for the HH band. The reader is referred to the papers by Vassel *et al.* (1970), Wiley (1971) and Zawadzki and Szymanska (1971) for details.

2.5 Non-Parabolicity

As seen in a previous subsection, non-parabolicity arises as a consequence of band mixing. For electrons in a central-zone valley the mixing is primarily with the valence band. The conduction band in direct gap III-V compounds has the form

$$E(\mathbf{k}) = \frac{\hbar^2 k^2}{2m_0^*} + \alpha_0 k^4 + \beta_0(k_x^2 k_y^2 + k_y^2 k_z^2 + k_z^2 k_x^2) \pm \gamma_0 \left[k^2(k_x^2 k_y^2 + k_y^2 k_z^2 + k_z^2 k_x^2) - 9k_x^2 k_y^2 k_z^2 \right]^{1/2} \tag{2.92}$$

For confinement along the z direction (the $\langle 001 \rangle$ direction), the subband minima are determined by setting $k_x = k_y = 0$, whence, for weak non-parabolicity

$$E(k_z) = \frac{\hbar^2 k_z^2}{2m_0^*} + \alpha_0 k_z^4 \tag{2.93}$$

or, better

$$k_z^2 \approx \frac{2m_0^*}{\hbar^2}[E(1 + \alpha E)], \qquad \alpha = -\left(\frac{2m}{\hbar^2}\right)\alpha_0 \tag{2.94}$$

(Note that α_0, like β_0 and γ_0, is negative.) This defines a "confinement mass", $m_Z^*(E)$

$$m_z^*(E) = m_0^*(1 + \alpha E) \tag{2.95}$$

A similar expression is defined for the mass in the barrier by replacing E by $(E - V_0)$. If $(E - V_0)$ is large, recourse must be made to the complex band structure of the barrier (Chang, 1982). Non-parabolicity tends to increase the mass in the well and reduce the mass in the barrier. Increasing the mass in the well tends to lower the energy, and reducing the mass in the barrier increases the energy. Either may prevail. Introducing motion in the plane affects the energy-dependence involving k_z and hence the boundary conditions, which, in turn, modifies the energy. This has the effect of modifying the effective mass in the plane, leading to a mass enhancement two to three times larger than that for the confinement mass (Ekenburg, 1989).

It should be noted that some authors (e.g., Welch *et al.*, 1984) use the "momentum mass", i.e. $\hbar k/v$, where v is the group velocity in the matching condition of Eq. (2.8). The basis for this is that the matching condition essentially involves currents, and therefore using an effective mass derivable from crystal momentum and group velocity is more appropriate than using the confinement mass of Eq. (2.95). For weak non-parabolicity, the momentum mass is given by

$$m_p^* = m_0^*(1 + 2\alpha E) \tag{2.96}$$

which shows that it is twice as strongly affected by non-parabolicity as is the confinement mass. The increase of momentum mass with velocity turns out to be of the form familiar in special relativity, viz.

$$m_p^* = \frac{m_0^*}{\sqrt{1 - (v^2/v_0^2)}} \tag{2.97}$$

where v_0 is the limiting velocity given by $\sqrt{2am_0^*}$. But the analysis of Section 2.1 shows that this is not the appropriate mass to use to determine bound states though it more nearly describes the mass parallel to the layers than does the confinement mass.

The presence of non-parabolicity really entails more elaborate matching conditions than those of Eq. (2.8) with Eq. (2.95), but the latter are useful if errors of a few millielectron volts can be tolerated. For further discussion see the papers by Ekenburg (1989) and Burt (1992).

2.6 Band-Mixing

When substantially different conduction bands are involved, such as the Γ and X valleys in the GaAs/AlAs system, the confined state becomes a mixture and we return to the basic envelope-function formalism of Eq. (2.2). The gap between the Γ states in GaAs and AlAs is not known precisely, but it is about 1 eV. With such an energy difference effective-mass theory is inapplicable and it is necessary to use the complex band structure of the bulk materials. Calculations of subband structure use either pseudopotential or tight-binding techniques. The latter have been carried out by Shulman and Chang (1981, 1985) and by Ando and Akera (1989), and pseudopotential calculations have been done by Jaros and coworkers (1984, 1985). These show that the confinement of a Γ electron in GaAs is very largely due to the Γ barrier rather than to the $\Gamma - X$ barrier. In other words, mixing of Γ and X states is small though finite, and this mixing can be directly observed in experiments of resonant tunnelling in which tunnelling paths involve $\Gamma - X$ transitions (Landheer *et al.*, 1989).

In other structures such as GaAs/$Al_xGa_{1-x}As$ (x small), HgTe/CdTe and GaSb/InAs the envelope-function method in the effective-mass approximation works quite well (Ando *et al.*, 1989). Even when the adjacent layers are mismatched, a shift of position variable, following the technique of Pikus and Bir (1959), allows the envelope-function method to be adapted to strained layers (Burt, 1988, 1992).

3

Quasi-Continuum Theory of Lattice Vibrations

. . . a dim and undetermined sense
Of unknown modes . . .
The Prelude, *W. Wordsworth*

3.1 Introduction

We now turn to the problem of describing long-wavelength lattice vibrations in multilayered structures. At a microstructural level this problem is solved by using the techniques of lattice dynamics requiring intense numerical computation. This approach is inconvenient, if not impracticable, at the macroscopic level, where the kinetics and dynamics of large numbers of particles need to be described. In this theoretical regime it is necessary to obtain models that transcend those at the level of individual atoms in order to describe electron and hole scattering, and all the energy and momentum relaxation processes that underlie transport and optical properties of macroscopic structures. The basic problem is to make a bridge between the atomic crystal lattice and the classical continuum. We know from the theory of elasticity and the theory of acoustic waves, which hark back to the nineteenth century, that continuum theory works extremely well for long-wave acoustic waves. The case of optical vibrations is another matter. Here, the essence is one atom vibrating against another in a primitive unit cell and it is by no means obvious that a continuum approach can work in this case. This has been highlighted by controversy concerning the boundary conditions that long-wave optical vibrations obey at each interface of a multilayer structure. On the other hand, no controversy attaches to acoustic waves. The elastic boundary conditions in this case are the familiar ones of continuity of displacement and of normal stress components. These acoustic boundary conditions have sometimes been used for optical waves (Perez–Alvarez *et al.*, 1993; Ridley *et al.*, 1994), but the validity for this has not been clear.

Fortunately, there exists an important category of multilayered structures in which the boundary conditions necessary for the description of how optical modes are confined are reasonably clear and uncontroversial. In this category the adjacent semiconductors are characterized by having widely different optical mode frequencies. This means that an optical mode in A cannot travel in B, and vice versa. The natural boundary condition is then to entail that the amplitude of the vibrations vanishes at the interface or at any rate not very far beyond the interface. The paradigm example is the AlAs/GaAs system, and there are several technologically important systems belonging to this category. In addition to this mechanical boundary condition there are, for polar material, the usual electromagnetic boundary conditions. Satisfying both sets of boundary conditions leads to the hybridization of longitudinally polarized (LO) and transversely polarized (TO) and interface-polariton (IP) modes, as we will see in Chapter 5.

There are, however, equally important systems in which the optical bands overlap or lie very close to one another. A purely binary example is InAs/GaSb, but there are also systems containing two-mode alloys, such as $AlGa_xAs$/GaAs, where this is true. Although electromagnetic boundary conditions still pertain, it is by no means clear what mechanical boundary conditions must be fulfilled. Of course, this question can always be answered by performing intensive numerical computations of the specific lattice dynamics, but we need to establish a general understanding of the physics involved and to develop a continuum theory that can be applied to a wide range of systems. This can be achieved through quasi-continuum theory, the essentials of which we will give later. It is instructive, however, to begin with an application of the familiar linear-chain model to the case where a junction between two materials exists.

Before plunging into detail on this issue, we may find it useful to discover the boundary conditions that emerge naturally from the assumption that the differential equations are, in one dimension, chosen perpendicular to the interfacial plane

$$\rho\omega^2 U = -\frac{\partial}{\partial z}\left(c_a \frac{\partial U}{\partial z}\right)$$
$$\rho_r\omega^2 u = fu + \frac{\partial}{\partial z}\left(c_0 \frac{\partial u}{\partial z}\right) \tag{3.1}$$

for the acoustic and optical waves, respectively. Here, ρ is the mass density, ρ_r is the reduced-mass density, U and u are the acoustic and optical displacements, f is a force constant and c_a, c_0 are elastic constants. At the interface, ρ, ρ_r c_a, f and c_0 change discontinuously, whereas the frequency ω must be the same for waves on either side, and we have taken into account any variation c_a and c_0 by including these parameters in the differentiation. Because of the discontinuities that exist regarding the parameters it is interesting to ask whether either U or u can be discontinuous. If so, there would appear a term in each equation that would involve the differential of a δ-function. There is nothing in either equation that could possibly balance this; therefore, if the equations are to hold, no discontinuity in U or u is possible. The same reasoning leads to the conclusion that the stresses $c_a\, dU/dz$ and $c_0\, du/dz$ cannot be discontinuous. The boundary conditions are therefore

$$\begin{aligned} &U, u \quad \text{continuous} \\ &c_a \frac{\partial U}{\partial z}, c_0 \frac{\partial u}{\partial z} \quad \text{continuous} \end{aligned} \tag{3.2}$$

These may be termed the "classical" boundary conditions, i.e. continuity of mechanical displacement and of mechanical stress.

But, of course, it is by no means obvious that equations that satisfactorily describe long waves in bulk material are valid near an interface. In order to investigate this point we look at two different approaches that have been taken. The first is based on the simple linear-chain model, the second on a rigorous mapping of the atomic model of the world onto the conceptually invaluable world of the continuum.

3.2 Linear-Chain Models

3.2.1 Bulk Solutions

For simplicity we consider the one-dimensional case of a linear chain of alternate cations and anions (Fig. 3.1) and define the displacement of a cation at position n by $u_1(n)$ and that of its anion neighbour by $u_2(n+1)$. The equations of motion for the two species of ion when only nearest-neighbour elastic restoring forces are considered are

$$\begin{aligned} -m_1\omega^2 u_1(n) &= f_1\{u_2(n+1) - u_1(n)\} + f_2\{u_2(n-1) - u_1(n)\} \\ -m_2\omega^2 u_2(n+1) &= f_2\{u_1(n+2) - u_2(n+1)\} + f_1\{u_1(n) - u_2(n+1)\} \end{aligned} \tag{3.3}$$

● n-2 ○ n-1 ● n ○ n+1 ● n+2 ○ n+3

Fig. 3.1 Linear chain.

where f_1 and f_2 are the force constants, ω is the angular frequency and m_1, m_2 are the cation and anion masses. These equations will provide the basis for our discussion.

Plane-wave solutions exist of the form

$$u_1(n) = u_{10} \exp i(kna - \omega t), \quad u_2(n+1) = u_{20} \exp i(k(n+1)a - \omega t) \tag{3.4}$$

where a is the distance separating adjacent ions, provided that the following dispersion relation is satisfied:

$$\omega^4 - \frac{(f_1+f_2)}{\mu}\omega^2 + \frac{4f_1f_2}{m_1m_2}\sin^2 ka = 0 \tag{3.5}$$

where $\mu = m_1m_2/(m_1 + m_2)$ is the reduced mass. The two solutions for ω describe the optical and acoustic branches. For long-wavelength modes ($ka < 1$)

$$\begin{aligned} \omega^2 &\approx \omega_0^2 - \frac{4f_1f_2}{(f_1+f_2)(m_1+m_2)}k^2a^2 \\ &= \omega_0^2 - v_s^2k^2 \quad \text{optical modes} \end{aligned} \tag{3.6}$$

$$\omega^2 \approx \frac{4f_1f_2k^2a^2}{(f_1+f_2)(m_1+m_2)} = v_s^2k^2 \quad \text{acoustic modes} \tag{3.7}$$

where $\omega_0{}^2 = (f_1 + f_2)/\mu$, and we have introduced the velocity of sound v_s, where $v_s{}^2 = 4f_1f_2a^2/((f_1 + f_2)(m_1 + m_2))$. Strictly speaking, the modes described by a one-dimensional chain model must all be longitudinally polarized along the axis of the chain, and thus Eqs. (3.6) and (3.7) describe the dispersion relations for LO and LA modes, with f_1 and f_2 incorporating the effects of the coulomb field. However, the one-dimensionality need not be insisted on, and a linear-chain model can describe transversely polarized waves equally well with a reinterpretation of the force constants. The form of the dispersion relations is just that of continuum theory, which allows us to relate continuum and microscopic parameters; thus for acoustic modes

$$v_s^2 = \frac{c_a}{\rho} = \frac{4f_1f_2a^2}{((f_1+f_2)(m_1+m_2))} \tag{3.8}$$

where c_a is the relevant elastic constant and ρ is the mass density. The latter is $(m_1 + m_2)/2a^3$ and so $c_a = 2f_1f_2/(f_1 + f_2)a$. For optical modes we must use the reduced density, which leads to an optical-mode elastic constant $c_0 = c_a\mu/\ (m_1+m_2)$.

Insertion of the long-wavelength dispersion into Eq. (3.4) leads to the following relations between the ionic displacements (to first order in ka):

$$\begin{aligned} u_2 &= -u_1\left(\frac{m_1}{m_2}\right)(1 - i\gamma ka) \quad \text{optical modes} \\ u_2 &= u_1(1 - \gamma ka) \quad \text{acoustic modes} \end{aligned} \tag{3.9}$$

where $\gamma = (f_1 - f_2)/(f_1 + f_2)$. Optical modes are characterized by the relative displacement, $u = u_1 - u_2$, whereas acoustic modes are characterized by the displacement of the centre of mass, $U = (m_1 u_1 + m_2 u_2)/(m_1 + m_2)$. These are related to the actual ionic displacement according to

$$u_1 = \frac{m_2}{M}\left(1 + \frac{m_1}{M} i\gamma ka\right) u + \left(1 + \frac{m_2}{M} i\gamma ka\right) U \tag{3.10a}$$

$$u_2 = -\frac{m_1}{M}\left(1 - \frac{m_2}{M} i\gamma ka\right) u + \left(1 - \frac{m_1}{M} i\gamma ka\right) U \tag{3.10b}$$

where for long-wavelength optical modes $U = 0$ and for long-wavelength acoustic modes $u = 0$, and $M = m_1 + m_2$. Note that u and U are the displacements that appear in continuum theory. For longitudinally polarized modes $\gamma = 0$, whereas for transversely polarized (TO) modes $\gamma \neq 0$. For simplicity we restrict attention to LO modes.

Equation (3.10) then becomes

$$u_1 = \frac{m_2}{M} u + U \tag{3.11a}$$

$$u_2 = -\frac{m_1}{M} u + U \tag{3.11b}$$

We are now in a position to discuss connection rules at interfaces separating medium A from medium B. Three types of interface will be distinguished, namely (1) nearly matched, (2) grossly mismatched, (3) free surface. We consider each in turn.

3.2.2 Interface between Nearly Matched Media

The topic of media interfaces was first discussed by Akero and Ando (1989) and we follow the spirit of their approach. Let a notional interface exist between atoms n and $n+1$ in a linear chain, and for simplicity we assume that there is "lattice matching" in that the spacing between ions is "a" on both sides. The equations of motion in the vicinity of the interface are

$$-m_{1A}\omega^2 u_{1A}(n) = f^*\{u_{2B}(n+1) - u_{1A}(n)\} + f_A\{u_{2A}(n-1) - u_{1A}(n)\} \tag{3.12a}$$

$$-m_{2B}\omega^2 u_{2B}(n+1) = f_B\{u_{1B}(n+2) - u_{2B}(n+1)\} + f^*\{u_{1A}(n) - u_{2B}(n+1)\} \tag{3.12b}$$

where f^* is the elastic force constant across the interface. In writing Eq. (3.12) we are assuming that a vibratory mode of frequency ω is sustainable in some form in both media. Since we are interested in the relationship between microscopic and continuum models we will concentrate on long-wavelength modes; therefore we make the assumption here that the optical-mode frequencies in the adjacent media are close together, thus defining the media as nearly matched. (This assumption will have to be abandoned for rigid and free interfaces.)

Akero and Ando extrapolate the bulk functions across the interface and derive matching conditions for the extrapolated envelopes. Thus replacing the left-hand sides of Eq. (3.12) using Eq. (3.3) and assuming that $f_1 = f_2$ in both media, we obtain

$$f_{\mathrm{A}}\{u_{2A}(n+1) - u_{1A}(n)\} = f^*\{u_{2B}(n+1) - u_{1A}(n)\} \tag{3.13a}$$

$$f_B\{u_{2B}(n+1) - u_{1B}(n)\} = f^*\{u_{2B}(n+1) - u_{1A}(n)\} \tag{3.13b}$$

Eq. (3.13) provides the basic connection rule, but this needs to be translated into continuum language involving u and U, respectively, for the cases of optical and acoustic vibrations. In order to simplify the discussion we exploit the fact that the force constant of the semiconductors of Groups IV and III-V have approximately the same magnitude, and so we put $f_A = f_B = f^*$, and arrive at the straightforward connection rules

$$\begin{aligned} u_{2A}(n+1) &= u_{2B}(n+1) \\ u_{1A}(n) &= u_{1B}(n) \end{aligned} \tag{3.14}$$

i.e. both anionic and cationic displacements are continuous.

The conclusion that the ionic displacements are continuous follows from the assumptions that the bulk equations, Eq. (3.3), apply near the interface and that the force constants are the same in either material. The force-constant approximation is a reasonable one to take and it is commonly used (and usually referred to as the mass approximation!), but the assumption that components of the equation of motion at the interface can be simply replaced by the corresponding components in the bulk equation and its connection with quasi-continuum theory is more problematical.

For acoustic modes we translate these conditions into conditions applying to the centre-of-mass displacement U using Eq. (3.11) with $u = 0$. We must first decide where the interface is. We have assumed that it is somewhere between ions n and $(n+1)$. Let us assume that it lies a distance pa from ion n, where $0 \leq p \leq 1$, and let U_A and U_B be the acoustic displacement at the interface. Imagining

$U_A(z)$ to continue beyond the interface we can write, for slowly varying displacements

$$u_{2A}(n+1) = U_A + (1-p)a\,dU_A(z)/dz \tag{3.15a}$$

$$u_{1A}(n) = U_A - pa\,dU_A(z)/dz \tag{3.15b}$$

Here we assume that U_A varies little over an ionic spacing so that the Taylor expansion can be terminated at the second term. Putting for brevity $\nabla = ad/dz$, we obtain from Eq. (3.14) the connection rules

$$U_A + (1-p)\nabla U_A = U_B + (1-p)\nabla U_B \tag{3.16a}$$

$$U_A - p\nabla U_A = U_B - p\nabla U_B \tag{3.16b}$$

These can be conveniently expressed in terms of a transfer matrix T_{AB} such that

$$\begin{vmatrix} U_A \\ \nabla U_A \end{vmatrix} = T_{AB} \begin{vmatrix} U_B \\ \nabla U_B \end{vmatrix} \tag{3.17}$$

$$T_{AB} = \begin{vmatrix} t_{11} & t_{12} \\ t_{21} & t_{22} \end{vmatrix} \tag{3.18}$$

where T_{AR} is a 2×2 matrix. In the acoustic case under consideration,

$$T_{AB} = \begin{vmatrix} 1 & 0 \\ 0 & 1 \end{vmatrix} \tag{3.19a}$$

which, given the assumption that $f_A = f_B$, corresponds to the standard classical boundary conditions and is applicable generally to long-wave elastic modes. Note that the precise location of the interface is irrelevant. When $f_A \neq f_B \neq f^*$ the usual hydrodynamic boundary conditions are obtained, viz.:

$$T_{AB} = \begin{vmatrix} 1 & 0 \\ 0 & \frac{f_B}{f_A} \end{vmatrix} \tag{3.19b}$$

only if the interface position is chosen to be

$$p = \frac{f^* - f_B}{f_A - f_B} \cdot \frac{f_A}{f^*} \tag{3.19c}$$

Thus, in general, the position of the interface affects the transfer matrix.

Repeating for optical modes, this time using Eq. (3.11) with $U=0$, we obtain for $f_A=f_B=f^*$

$$\begin{vmatrix} u_A \\ \nabla u_A \end{vmatrix} = \begin{vmatrix} R_1 p + R_2(1-p) & (R_1 - R_2)p(1-p) \\ R_1 - R_2 & R_1(1-p) + R_2 p \end{vmatrix} \begin{vmatrix} u_B \\ \nabla u_B \end{vmatrix} \tag{3.20}$$

where

$$R_1 = \frac{m_{1B}M_A}{m_{1A}M_B}, \qquad R_2 = \frac{m_{2B}M_A}{m_{2A}M_B} \tag{3.21}$$

It is noticeable that the connection rule depends on where, precisely, the interface is considered to be. (Akero and Ando's result is obtained by putting $p=1/2$.) We have taken the interface to lie between ions n and $n+1$. If, instead, we place it between $n-1$ and n, we obtain Eq. (3.20) with R_1 and R_2 interchanged.

For a given interface it is noticeable that the element t_{21} is *not* affected by the position of the interface and it therefore possesses an especial significance. Consider the case $t_{11}=1$, $t_{22}=1$, $t_{12}=0$. The displacement is continuous but the slope is not because t_{21} is not zero. This implies that a delta-function force of magnitude $(c_0/a)t_{21}u\ \delta(z)$ appears on the right-hand side of the differential equation. It follows that if t_{21} is negative, a localized mode can exist with a frequency above that of the bulk mode at the zone centre. The element t_{21}, then, indicates whether a localized mode exists or not as a consequence of the difference in mass at the interface.

In general, on a macroscopic scale, neither u nor ∇u is continuous. The discontinuity of u implies the existence of a further element in the differential equation (provided that a differential equation remains valid). This must have the character of involving the differential of a delta function. In short, the conclusion from this analysis is that the classical mechanical boundary conditions simply do not apply to optical modes.

For a travelling optical wave the energy flux is given by

$$S = \rho_r \omega^2 u^2 v_g \tag{3.22}$$

where v_g is the group velocity given by $v_s^2 k/2\omega$ from Eq. (3.7). The flux can be expressed as follows:

$$S = \frac{\rho_r \omega v_s^2}{4i}(u^* \nabla u - u \nabla u^*) \tag{3.23}$$

The connection rule Eq. (3.18) implies that

$$(u^* \nabla u - u \nabla u^*)_A = \text{Det}\ T_{AB}(u^* \nabla u - u \nabla u^*)_B \tag{3.24}$$

where Det means determinant, and so continuity of flux means that

$$\mathrm{Det}T_{AB} = \frac{(\rho_r v_s^2)_B}{(\rho_r v_s^2)_A} \tag{3.25}$$

With $c_{0A} = c_{0B}$ this is satisfied by Eq. (3.20).

3.2.3 Interface between Mismatched Media

Where the frequencies of long-wavelength optical modes in the adjacent media differ greatly an optical vibration in one medium cannot penetrate significantly into the other. Such a situation occurs, for example, in the GaAs/AlAs system. In this case the interface is effectively defined by the position of the first immobile ion. If this is the anion at $n+1$, i.e. if $u_{2B}(n+1)=0$, then from Eq. (3.13a), assuming $f^* = f_A$, $u_{2A}(n+1)=0$, whence, from Eq. (3.10b), $u_A = 0$ or $U_A = 0$. If the first immobile ion is the cation at n, i.e. if $u_{1B}(n) = 0$, then we reach the same conclusion; the displacement vanishes at the interface, viz.:

$$U_A, u_A = 0 \tag{3.26}$$

Note that there is consequently no mechanical energy flow across the interface.

3.2.4 Free Surface

When there is no restoring force on one side of the interface, ($f^* = 0$), and hence

$$u_2(n+1) - u_1(n) = 0 \tag{3.27}$$

With the interface at $(n + pa)$ we obtain for acoustic modes

$$\nabla U = 0 \tag{3.28}$$

independent of the interface position, and for optical modes

$$u + \left(\frac{m_1}{M} - p\right)\nabla u = 0 \tag{3.29}$$

Choosing $p = m_1/M$ leads to:

$$u = 0 \tag{3.30}$$

The condition for acoustic modes corresponds to the classic case of zero stress. The condition for optical modes, however, is the same as for the highly mismatched case. It would appear that the use of acoustic-like boundary conditions to describe optical modes in a free-standing slab is not justified.

3.2.5 Summary

The linear-chain model reveals a big difference in behaviour of acoustic and optical modes at an interface. This difference will persist and become more complex in three dimensions when long-range electric fields enter along with crystal anisotropy. The usual electromagnetic boundary conditions will be added to the mechanical ones, and, as we will see in later chapters, this produces a profound modification of mode patterns. But at the base of the analysis lies the assumption that we can ignore any short-wavelength components that may be introduced by the mismatch of mechanical properties at an interface. In order to enquire into the validity of this assumption and to put the envelope-function approach on a more rigorous foundation it is necessary to turn to quasi-continuum theory.

3.3 The Envelope Function

The description of optical modes in a polar inhomogeneous medium, i.e. one where an interface separates two different materials, requires a careful analysis of electromagnetic and mechanical properties and how they interact. The description of the electromagnetic field is that of a field in a continuum, whereas the mechanical properties are those of a discrete lattice of ions. In order to study the properties of long-wavelength vibrations of the ions and their interaction with the electromagnetic field it is useful to establish a rigorous connection between the true discrete motion of the ions and a continuum representation of that motion, and this is effected by quasi-continuum theory. This theory grew out of interpolation theory (Whittaker, 1915) and was applied to the problem of noise (Shannon, 1949), to vibrations in elastic media (Krumhansl, 1965; Kunin, 1982) and recently to the problem of polar optical vibration (Foreman, 1995). Here we summarize the main aspects, focussing solely on the mechanical properties and, for simplicity, treating only the one-dimensional situation.

We consider first a monatomic lattice of period a. The transition to the continuum representation can be made by making use of the unique relation between a continuous function of displacement $u(x)$ and the actual displacement $u(n)$:

$$u(x) = a \sum_{n} u(n)\delta_B(x - na) \tag{3.31}$$

where

$$\delta_B(x) = \frac{1}{2\pi} \int_{-\pi/a}^{\pi/a} e^{ikx} dk = \frac{\sin(\pi x/a)}{\pi x} \tag{3.32}$$

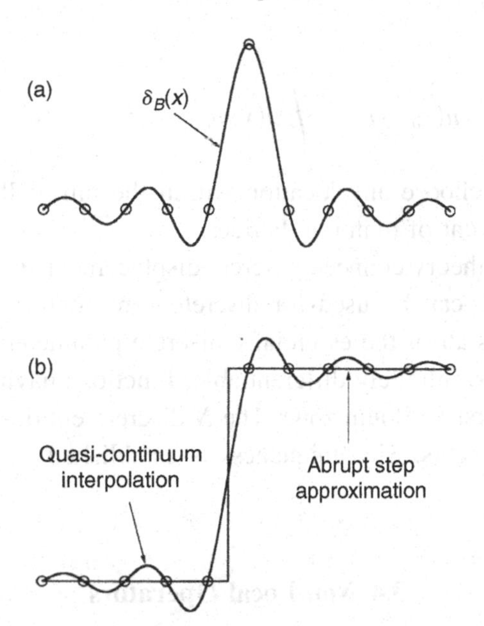

Fig. 3.2 (a) Quasi-continuum function; (b) at an interface.

at each lattice point $\delta_B(na) = (1/a)\ \delta_{n,0}$ (Fig. 3.2). From the continuous function the discrete displacement can be recovered by

$$u(n) = \int u(x)\delta_B(x - na)dx \tag{3.33}$$

The subscript B is to remind us that $\delta_B(x)$ is a Dirac delta function with no Fourier components outside the Brillouin zone. (A similar function was used for electrons in Section 2.1.3.) The continuous function can then be Fourier-analysed:

$$u(x) = \frac{1}{\sqrt{2\pi}} \int_{-\pi/a}^{x/a} u(k)e^{ikx}dk \tag{3.34}$$

and so, from Eq. (3.32)

$$u(n) = \frac{1}{\sqrt{2\pi}} \int_{-\pi/a}^{x/a} u(k)e^{ikna}dk \tag{3.35}$$

The envelope function $u(x)$ is exactly and uniquely related to the original discrete displacement $u(n)$. We can use the function $\delta_B(x)$ to sample $u(x)$ at any location;

thus, with $x_n = na$

$$u(x_n + \varepsilon) = \int u(x)\delta_B(x - x_n - \varepsilon)dx \tag{3.36}$$

which allows us to choose any location within the unit cell as a basis. This is useful for the treatment of diatomic lattices.

Quasi-continuum theory connects discrete displacements with travelling waves. The same procedure can be used for discrete force density, mass density and charge density. Thus all of the essentially discrete parameters of a lattice can be related to continuous, infinitely differentiable, functions having Fourier components limited to the first Brillouin zone. The N discrete entities of the lattice relate to N Fourier components. The uniqueness is established by restricting k to the first Brillouin zone.

3.4 Non-Local Operators

The dynamics of the system in the harmonic approximation is described in terms of the Lagrangian:

$$L = \frac{1}{2}\sum_n m(n)\dot{u}(n)^2 - \frac{1}{2}\sum_{m,n} u(n)\Phi(n, n')u(n') \tag{3.37}$$

where $m(n)$ is the mass of the nth atom and $\Phi(n, n')$ is a force-constant matrix describing the coupling between atoms n and n'. The force-constant matrix is represented in the quasi-continuum by a non-local operator

$$\Phi(x, x') = \sum_{n,n'} \delta_B(x - x_n)\Phi(n, n')\delta_B(x - x_{n'}) \tag{3.38}$$

The kinetic energy is basically local, but a straightforward translation to the quasi-continuum gives the non-linear function $m(x)\dot{u}^2(x)$, which in general has Fourier components outside the first Brillouin zone. This is strictly not allowed in quasi-continuum theory. The way to cope with this problem is to define a non-local mass operator thus:

$$m(n, n') = m(n)\delta_{nn'} \tag{3.39}$$

whence

$$m(x, x') = \sum_n m(n)\delta_B(x - x_n)\delta_B(x' - x_n) \tag{3.40}$$

This solves the problem at the expense of introducing the non-intuitive non-local operator for the mass, analogous to the potential in the electron case, Eq. (2.20).

Thus, in quasi-continuum theory, both kinetic and potential energy become non-local functions.

It can easily be shown that corresponding inverse non-local operators exist and that Φ and m are Hermitian.

The Lagrangian can now be written as follows:

$$\begin{aligned} L &= \frac{1}{2}\sum_{nn'} (\dot{u}(n)m(n,n')u(n') - u(n)\Phi(n,n')u(n')) \\ &= \frac{1}{2}\iint (\dot{u}(x)m(x,x')u(x') - u(x)\Phi(x,x')u(x'))dxdx' \end{aligned} \tag{3.41}$$

This Lagrangian is exactly equivalent to the discrete Lagrangian of Eq. (3.37). Moreover, the functions involved and their derivatives to all orders are continuous everywhere including any interfaces. If these functions were known through an interface, there would be no need of boundary conditions, but to achieve that would be as difficult as making a detailed lattice-dynamic calculation. In order to obtain boundary conditions, it is necessary to convert our non-local expression to some equivalent local form. This implies writing the Lagrangian in the form

$$L = \int L(x)\, dx \tag{3.42}$$

where $L(x)$ is a Lagrangian density.

A non-local/local transformation can be achieved by following the method of Foreman (1995), who extended the treatment of Kunin (1982) to abrupt interfaces. Consider an arbitrary non-local interaction of the form

$$W= \iint u(x)m(x,x')v(x')\, dx\, dx' = \iint u(-k)m(k,k')v(k')\, dk\, dk' \tag{3.43}$$

where $m(x, x')$ is a non-local operator. The method consists in isolating the $k - k'$ dependence of the non-local operator, by transforming to the variables

$$\begin{aligned} y &= \frac{x-x'}{\sqrt{2}}, \quad y' = \frac{x+x'}{\sqrt{2}}, \\ K &= \frac{k+k'}{\sqrt{2}}, \quad K' = \frac{k-k'}{\sqrt{2}} \end{aligned} \tag{3.44}$$

The result is (see Appendix)

$$W = \int W(x)dx \quad W(x) = \sum_{n=0}^{\infty}(-1)^n\left[m_{2n}(x)\frac{\partial^n u}{\partial x^n}\frac{\partial^n v}{\partial x^n} + m_{2n+1}(x)\frac{\partial^n u}{\partial x^n}\frac{\partial^{n+1} v}{\partial x^{n+1}}\right] \tag{3.45}$$

Here

$$m_0(x) = \int m(x,x')\,dx' \tag{3.46}$$

and $m_1(x)$, etc., are more complicated integrals. If the non-local operator is symmetric in x space, all the odd order coefficients vanish. This is true of the mass operator $m(x, x')$.

The conversion of a non-local integral to an infinite sum of local differentials is exact in homogeneous material but only approximately true in quasi-continuum theory where interfaces are involved. This is because the result involves products of quasi-continuum functions that may result in k-values extending beyond the first zone. But for slowly varying envelope functions the local approximation is a good one and we will adopt it here. As in the case for electrons, the local approximation assumes that all physical entities which appear in the theory possess Fourier components that have appreciable magnitudes only within the first Brillouin zone. (The general case involves replacing quasi-continuum theory with metacontinuum theory; see Section 3.8.)

3.5 Acoustic and Optical Modes

It is now time to consider a diatomic lattice and its acoustic and optical vibrations. The atomic positions can be specified by

$$x_{nj} = x_n + x_j \tag{3.47}$$

where, once more, n denotes the unit cell and $j = 1, 2$ denotes the atoms. The Lagrangian is

$$L = \frac{1}{2}\sum_{nj} m_j(n)\dot{u}_j^2(n) - \frac{1}{2}\sum_{nn'}\sum_{jj'} u_j(n)\Phi_{jj'}(n,n')u_{j'}(n') \tag{3.48}$$

where $u_j\,(n)$ is the spatial displacement of ion j in the unit cell n, $m_j\,(n)$ is the mass and $\Phi_{jj'}(n, n')$ is a potential-energy function connecting all ionic displacements. Considering the material to have two ions per unit cell we split the motion into a centre-of-mass and relative components by putting

$$\begin{aligned} u_1 &= U + r_1 u \\ u_2 &= U - r_2 u \end{aligned} \tag{3.49}$$

where U is the displacement of the centre-of-mass, $u = u_1 - u_2$ is the relative displacement and $r_1 = m_1{}^{-1}\mu = M^{-1}m_2$, $r_2 = m_2{}^{-1}\mu = M^{-1}m_1$, where $M = m_1 + m_2$, are mass-ratio operators. Each atomic mass, in general, must now be regarded as intrinsically non-local since the sampling is done at the spatial coordinate that specifies the position of the unit cell. The Lagrangian can be converted into the form

$$L = L_{ac} + L_{op} + L_{inter} \tag{3.50}$$

where, in condensed notation

$$L_{ac} = \frac{1}{2}\langle U|\omega^2 M - \Gamma|U\rangle \tag{3.51}$$

$$L_{op} = \frac{1}{2}\langle u|\omega^2 \mu - \gamma|u\rangle \tag{3.52}$$

$$L_{inter} = -\frac{1}{2}\langle u|\chi|U\rangle \tag{3.53}$$

and we have introduced the frequency ω. For brevity the dependence on (x, x') has been suppressed.

In these equations M is the total-mass operator, μ is the reduced-mass operator, $(m_1 M^{-1} m_2)$ and Γ, γ and χ are modified potential-energy operators related to the interionic potential-energy Φ_{jj} (force constants) as follows:

$$\Gamma = \phi_{11} + \phi_{12} + \phi_{21} + \phi_{22} \tag{3.54}$$

$$\chi^\dagger = [\phi_{11} + \phi_{12}]r_1 - [\phi_{21} + \phi_{22}]r_2 \tag{3.55}$$

$$\chi = r_1^\dagger[\phi_{11} + \phi_{12}] - r_2^\dagger[\phi_{21} + \phi_{22}] \tag{3.56}$$

$$\gamma = r_1^\dagger \phi_{11} r_1 - r_1^\dagger \phi_{12} r_2 - r_2^\dagger \phi_{21} r_1 + r_2^\dagger \phi_{22} r_2 \tag{3.57}$$

The two types of vibration are coupled through the term χ. The corresponding equations of motion are obtained from

$$\frac{\partial}{\partial t}\left(\frac{\partial L}{\partial q}\right) - \frac{\partial L}{\partial q} + \frac{\partial}{\partial x}\left(\frac{\partial L}{\partial q/\partial x}\right) = 0 \tag{3.58}$$

where q is the variable. In simplified notation, they are

$$\begin{aligned}(\Gamma - \omega^2 M)U + \chi^\dagger u = 0\\ (\gamma - \omega^2 \mu)u + \chi U = 0\end{aligned} \tag{3.59}$$

The optical displacement can be eliminated from the acoustic equation by substituting

$$u = g\chi U \tag{3.60}$$

where

$$g = (\omega^2\mu - \gamma)^{-1} \tag{3.61}$$

Provided the acoustic frequency does not coincide with an optical frequency, g is a finite operator. Therefore the acoustic Lagrangian is

$$\begin{aligned} L &= \frac{1}{2}\langle U|\omega^2 M + C|U\rangle \\ C &= -\Gamma + \chi^\dagger g\chi \end{aligned} \tag{3.62}$$

Similarly the acoustic displacement can be eliminated from the optical equation using

$$\begin{aligned} U &= G\chi^\dagger u \\ G &= (\omega^2 M - \Gamma)^{-1} \end{aligned} \tag{3.63}$$

leading to the optical Lagrangian

$$\begin{aligned} L &= \frac{1}{2}\langle u|\omega^2\mu - c|u\rangle \\ c &= \gamma + \chi\, G\chi^\dagger \end{aligned} \tag{3.64}$$

We can now apply the local approximation and obtain

$$\begin{aligned} L_{ac}(x) &= \frac{1}{2}\omega^2 M_0|U|^2 - \frac{1}{2}(\omega^2 M_2 + C_2)\left|\frac{\partial U}{\partial x}\right|^2 + \frac{1}{2}(\omega^2 M_4 + C_4)\left|\frac{\partial^2 U}{\partial x^2}\right|^2 \ldots \\ L_{op}(x) &= \frac{1}{2}(\omega^2\mu_0 - c_0)|u|^2 - \frac{1}{2}(\omega^2\mu_2 + c_2)\left|\frac{\partial u}{\partial x}\right|^2 + \frac{1}{2}(\omega^2\mu_4 + c_4)\left|\frac{\partial^2 u}{\partial x^2}\right|^2 \ldots \end{aligned} \tag{3.65}$$

M_0, μ_0, etc. are now mass-density operators and c_0, etc. are elastic constant operators. All these non-local operators are symmetric in x space, which eliminates the odd coefficients in Eq. (3.45). Furthermore, the term C_0 vanishes as a consequence of the translation invariance of the acoustic vibration. This condition does not apply to optical vibrations.

The resultant equations of motion are

$$\begin{aligned} &\omega^2 M_0 U + \frac{\partial}{\partial x}\left[(\omega^2 M_2 + C_2)\frac{\partial U}{\partial x}\right] + \frac{\partial^2}{\partial x^2}\left[(\omega^2 M_4 + C_4)\left(\frac{\partial^2 U}{\partial x^2}\right)\right] + \ldots = 0 \\ &(\omega^2\mu_0 - c_0)u + \frac{\partial}{\partial x}\left[(\omega^2\mu_2 + c_2)\left(\frac{\partial u}{\partial x}\right)\right] + \frac{\partial^2}{\partial x^2}\left[(\omega^2\mu_4 - c_4)\left(\frac{\partial^2 u}{\partial x^2}\right)\right] \ldots = 0 \end{aligned} \tag{3.66}$$

In the acoustic equation, M_0 is given by

$$M_0(x) = \int M(x,x')dx' \qquad (3.67)$$

where $M_0(x) = m_1(\mathrm{x}) + m_2(x)$, and C_2 is the classic elastic constant. In the optical equation the reduced mass is

$$\mu_0(x) = \int \mu(x,x')\, dx' \qquad (3.68)$$

Also, c_0 is the zone-centre optical force constant and c_2 is the optical elastic constant. Higher-order spatial dispersion is described C_4, c_4, etc.

3.6 Boundary Conditions

We have found equations of motion for the acoustic and optical envelope functions but they are in the form of local differential equations of infinite order. Near an interface, we might expect that envelope functions change rapidly and consequently higher-order derivatives become important, whereas in uniform bulk material their contribution to long wavelengths is negligible. If we are prepared to forgo a detailed description of the variation of the envelope function through the interface we can bypass the difficulty by focussing on the relation between the envelope functions at the boundaries between the bulk and interface regions. These boundaries are, of course, not well-defined but they allow us to relate the comparatively slowly varying functions in the bulk material on either side of the interface. Up to and at these boundaries we can assume that higher-order derivatives are negligible (Fig. 3.3).

We suppose the interface region to lie in the range $-\varepsilon < x < \varepsilon$ and we integrate the equations of motion over this interval. We obtain

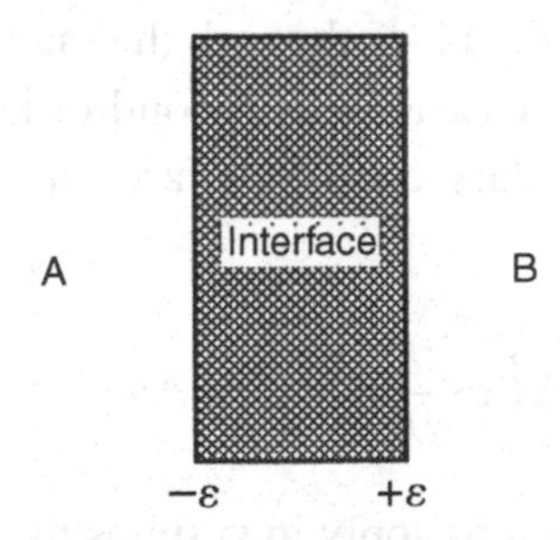

Fig. 3.3 Boundaries between bulk and interface.

$$\omega^2 \int_{-\varepsilon}^{\varepsilon} M_0 U dx + \left[(\omega^2 M_2 + C_2)\frac{\partial U}{\partial x}\right]_{-\varepsilon}^{\varepsilon} + \left(\frac{\partial}{\partial x}\left[(\omega^2 M_4 - c_4)\left(\frac{\partial^2 U}{\partial x^2}\right)\right]\right)_{-\varepsilon}^{\varepsilon} + \cdots = 0$$

$$\int_{-\varepsilon}^{\varepsilon} (\omega^2 \mu_0 - c_0) u dx + \left[(\omega^2 \mu_2 - c_2)\left(\frac{\partial u}{\partial x}\right)\right]_{-\varepsilon}^{\varepsilon} + \left(\frac{\partial}{\partial x}\left[(\omega^2 \mu_4 - c_4)\left(\frac{\partial^2 u}{\partial x^2}\right)\right]\right)_{-\varepsilon}^{\varepsilon} + \cdots = 0 \tag{3.69}$$

The derivatives appear only as their values at $x = \pm\,\varepsilon$, which for long-wavelength modes are small. We can, therefore, discard all terms in the equations of motion higher than second order. Furthermore, any changes in the spatial dispersion coefficients in the interface can be ignored and it is only necessary to specify their values in bulk material. Thus M_2 and μ_2 both vanish, since the mass operators are non-local only in inhomogeneous material, and C_2, c_2 take on their bulk values at $x = \pm\,\varepsilon$.

With regard to the zeroth-order terms each involving an integral, matters are less straightforward. In the acoustic case the frequency is small – we can take it to be of order (a/λ) where λ is the wavelength – and therefore ω^2 contributes $(a/\lambda)^2$. Whatever the coefficient M_0 does in the interface region, its contribution will be at most of order a/λ, and therefore, the zeroth-order acoustic term contributes at most order $(a/\lambda)^3$. The second-order dispersion, however, contributes to order $(a/\lambda)^2$. Thus the interface term in the case of acoustic modes can always be discarded, and the acoustic boundary condition becomes

$$\Delta\left(C_2 \frac{\partial U}{\partial x}\right) = 0 \tag{3.70}$$

where Δ is the difference. A second integration leads to

$$\Delta U = 0 \tag{3.71}$$

These are the classical boundary conditions for an acoustic wave.

Matters are not as straightforward for the optical mode. If μ_0 and c_0 do not deviate significantly from bulk values through the interface, the integral will be of order $(a/\lambda)^3$ as in the acoustic case and the second-order dispersion will dominate. In this special case the boundary conditions are directly analogous to those for acoustic modes, namely

$$\Delta\left(c_2 \frac{\partial u}{\partial x}\right) = 0, \quad \Delta u = 0 \tag{3.72}$$

It is a case that can be expected to apply in systems that share a common cation or anion, such as GaAs/AlAs. The only bonds are GaAs and AlAs and there is no distinct interface layer with different bonds such as occurs in the system InAs/GaSb, for example. The reduced mass runs smoothly from GaAs to AlAs, and the only possibility of interface effects resides in the interface perturbation of the force constants that contribute to $c_0(=\gamma+\chi\Gamma\chi^{\dagger})$, but this is bound to be an extremely weak effect. The same boundary conditions are applicable to the GaAs/$Al_xGa_{1-x}As$ system for the same reason. However, when there is no common cation or anion, as in InAs/GaSb, the interface contribution cannot be ignored. In this case the integral can be of order a/λ tending to outweigh the influence of the second-order dispersion, and the boundary conditions become

$$\int_{-\varepsilon}^{\varepsilon}(c_0-\omega^2\mu_0)udx+\Delta\left(c_2\frac{\partial u}{\partial x}\right)=0, \quad \Delta u=0 \tag{3.73}$$

The main effect will come from the difference in mass of the bulk and interface, since force constants do not vary in a marked way from one semiconductor to another. This difference affects μ_0 directly and c_0 indirectly through the mass dependence of the parameters γ, χ and G. Interface terms introduce a cusp in the envelope function and if strong enough and of the right sign, they allow an interface mode to exist at a frequency different from either bulk frequency.

As far as long-wavelength continuum theory is concerned, the interface parameters can be represented, in general, by

$$\begin{aligned}\mu_0(x)&=\mu_{0b}(x)+\mu_{0i}\delta(x-x_i)\\ c_0(x)&=c_{0b}(x)+c_{0i}\delta(x-x_i)\end{aligned} \tag{3.74}$$

where x_1 is the location of the interface (more of this later). The quantities $\mu_0(x)$ and $c_0(x)$ represent the variation across the interface of the bulk-line functions and $\mu_{0\iota}$, $c_{0\iota}$ are the extra terms introduced by the interface itself. The boundary condition for the optical stress is therefore

$$(c_{0i}-\omega^2\mu_{0i})u(x_i)+\Delta\left(c_{2b}\frac{\partial u}{\partial x}\right)=0 \tag{3.75}$$

(We have put $c_2(x)=c_{2b}$ to emphasize that only bulk values enter here.)

3.7 Interface Model

A rigorous calculation of c_{0i} and μ_{0i} in quasi-continuum theory must take into account the non-local character of the mass and force operators, and this is a somewhat tedious task because of the slow convergence of $\delta_B(x)$. An approximate result can be obtained by (a) ignoring the non-local character of the mass operators (thereby reducing matrix inverses to algebraic inverses), (b) using an abrupt-step representation of the quasi-continuum masses (thereby bypassing the convergence problem), (c) reducing the range of force constants to nearest-neighbour terms and (d) neglecting the variation of force constants between different semiconductors.

As an example of how this scheme works, we consider the InAs/GaSb system. Fig. 3.4(a) depicts the discrete cation and anion masses. In quasi-continuum theory the mass functions exhibit the Gibbs phenomenon, as shown in Figs. 3.4(b) and 3.4(c). Because the cation and anion are separated from one another by half a unit-cell dimension, the reduced mass takes on the value of the interface material in the intermediate region. Thus, although an abrupt-step approximation is clearly a good one for the mass, it is not for the reduced mass, as Fig. 3.4(d) shows. Interface effects are, therefore, expected to enter for optical modes, but not for acoustic modes. In the system there exist two quite different interfaces, one where

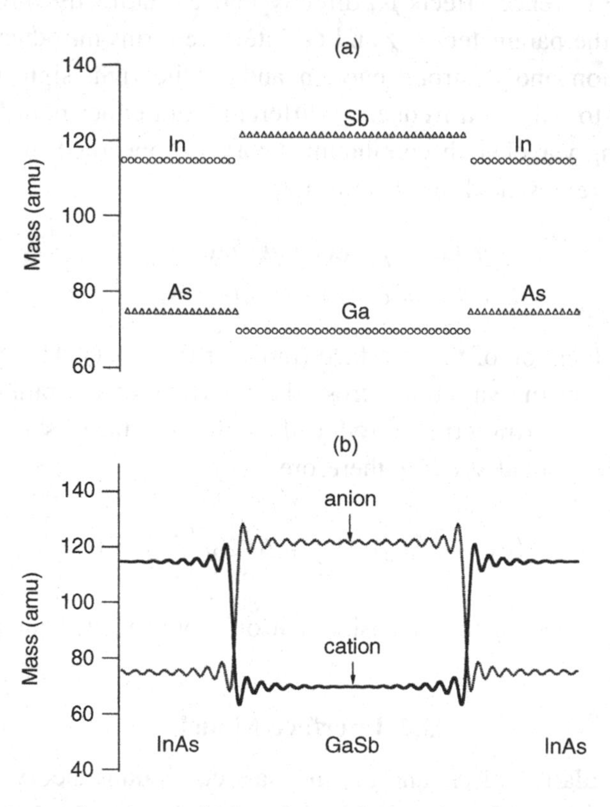

Fig. 3.4 The mass function at the interfaces of the system InAs/GaSb: (a) discrete anion and cation masses; (b) quasi-continuum mass; (c) quasi-continuum reduced mass; (d) continuum reduced mass (Foreman, 1995, private communication).

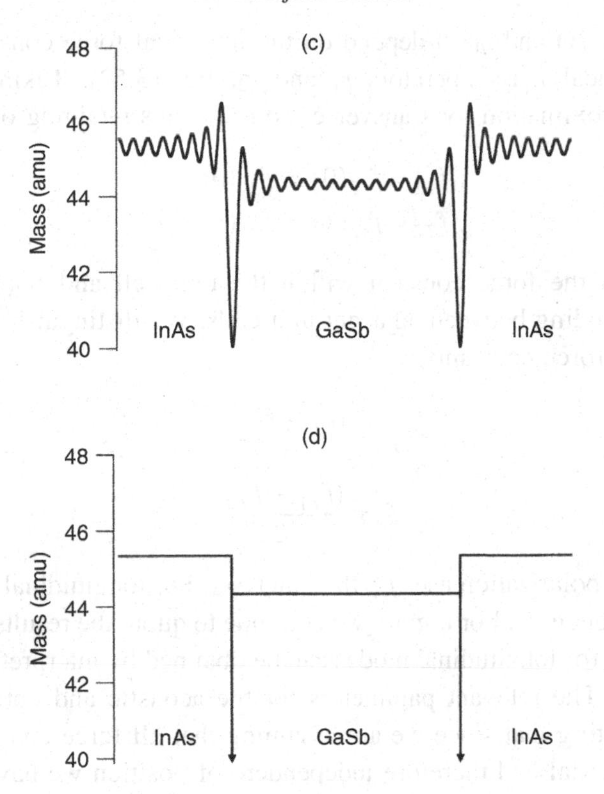

Fig. 3.4 (*cont.*)

Ga abuts As, the other where In abuts Sb. We can, therefore, expect the one to have properties closely related to GaAs, the other to InSb. In the case of the GaAs interface we have in the abrupt-step/local approximation:

$$\mu_0(x) = \frac{1}{a}\left\{\begin{array}{ll} \mu_- & x<\frac{-a}{4} \\ \mu_i & -\frac{1}{a}<x<\frac{a}{4} \\ \mu_+ & x>\frac{a}{4} \end{array}\right\} \tag{3.76}$$

where the origin is taken to be midway between Ga and As, and μ_-, μ_i and μ_+ are, respectively, the reduced masses of InAs, GaAs and GaSb. The spatial extent of the interface is simply described by the parameter $\varepsilon = a/4$. The reduced-mass interface term μ_{0i} is then just the area under the function μ_i, so that

$$\mu_{0i} = \frac{\mu_i}{2}, \tag{3.77}$$

Turning to the force constant c_0, we first remind ourselves that this is made up of two terms, viz.:

$$c_0(x) = \gamma_0(x) + \eta_0(x), \quad \eta_0(x) = (\chi G \chi^\dagger)_0 \tag{3.78}$$

In general, $\gamma_0(x)$ and $\eta_0(x)$ depend on the non-local force constants $\Phi_{ij}(n, n')$ and the non-local mass operators r_1 and r_2, Eq. (3.52). Taking the nearest-neighbour approximation for transverse modes means retaining only

$$\begin{aligned} f_{T1} &= f_{12}(0) = f_{21}(0) \\ f_{T2} &= f_{12}(1) = f_{21}(-1) \end{aligned} \tag{3.79}$$

where $f_{12}(0)$ is the force constant within the unit cell and $f_{12}(1)$ is the force constant for bonding between adjacent unit cells. We distinguish symmetric and antisymmetric force constants

$$\begin{aligned} f_S &= \frac{(f_{T1}+f_{T2})}{2} \\ f_A &= \frac{(f_{T1}-f_{T2})}{2} \end{aligned} \tag{3.80}$$

One transverse polarization uses f_A, the other $-f_A$. For longitudinal modes there is only one component f_L. For brevity we continue to quote the results for transverse modes. Results for longitudinal modes can be obtained by interpreting f_S as f_L and putting $f_A = 0$. The relevant parameters for the acoustic and optical modes are obtained by noting that since we are assuming that all force constants are independent of material and therefore independent of position we have

$$\Phi(n, n') = \Phi(n - n') \tag{3.81}$$

which implies that the Fourier component is

$$\Phi(k, k') = \Phi(k)\delta(k - k') \tag{3.82}$$

where

$$\begin{aligned} \Phi(k) &= \frac{1}{a}\sum_n \Phi(n)e^{-ikna} = \Phi_0 + ik\Phi_1 - k^2\Phi_2 \ldots \\ \Phi_0 &= \frac{1}{a}\sum_n \Phi(n) \\ \Phi_1 &= -\sum_n n\Phi(n) \\ \Phi_2 &= \frac{a}{2}\sum_n n^2\Phi(n) \end{aligned} \tag{3.83}$$

We are now in a position to relate the parameters Γ_0, Γ_2, χ_0, χ_1, γ_0, γ_2, which enter into the acoustic and optical equations of motion to the nearest-neighbour force constants. Invariance with respect to translation eliminates Γ_0 and χ_0, but not γ_0. In evaluating Γ_2, χ_1, γ_0 and γ_2 in terms of f_S and f_A it is necessary to transform from the representation we have been using, which is based on relating quasi-continuum functions to the position of the unit cell, to a representation based on the actual positions of the atoms in the unit cell, so that nearest-neighbour forces can be correctly identified. The result for the bulk (Foreman, 1995) is

$$\begin{aligned}\Gamma_2 &= -\frac{1}{2}af_S\\ \chi_1 &= -f_A\\ \gamma_0 &= \frac{2}{a}f_S\\ \gamma_2 &= \frac{1}{2}\frac{\mu}{M}af_S\end{aligned} \tag{3.84}$$

From these relationships the relevant bulk parameters can be deduced. Thus

$$\begin{aligned}C_{2b} &= -\Gamma_2 - \frac{\chi_1^2}{\gamma_0 - \omega^2\mu_0} \xrightarrow{\omega\to 0} \frac{a}{2}\frac{f_S^2 - f_A^2}{f_S} \qquad c_{0b} = \gamma_0 = \frac{2}{a}f_S\\ c_{2b} &= \gamma_2 - \frac{\chi_1^2}{\omega^2 M_0} = \frac{a}{M}\left(\mu f_S - \frac{f_A^2}{\omega^2}\right)\end{aligned} \tag{3.85}$$

For longitudinal modes $f_A = 0, f_S = f_L$. The elastic constant describing spatial dispersion for transverse modes is thus always smaller than that for longitudinal modes (provided that $f_S \approx f_L$).

In order to obtain $c_0(x)$, we can retain the approximation that the force constants are the same throughout and that only nearest-neighbour interactions are important. Any interface term then arises solely as a consequence of the spatial variation of mass. It turns out that by far the bigger contribution comes from the mass dependence of γ_0. According to Eq. (3.57), γ depends on the mass ratios $r_1(n)$ and $r_2(n)$. Since $r_2(n) = 1 - r_1(n)$, γ_0 can be expressed entirely in terms of $r_1(n)$. The bulk term, constant throughout the interface, is $2f_S/a$, and the difference can be written

$$c_0(x) = \frac{1}{a}\left\{\begin{array}{ll} \dfrac{2f_S}{\mathrm{a}} + \begin{array}{l} f_S(1-2r_{1-})(r_{1i}-r_{1-}) - f_A(r_{1i}-r_{1-}) \\ f_S(1-2r_{1-})(r_{1i}-r_{1-}) - f_A(r_{1i}-r_{1-}) \\ f_S(1-2r_{1-})(r_{1i}-r_{1-}) - f_A(r_{1i}-r_{1-}) \end{array} & \begin{array}{l} -\frac{3a}{4} < x < -\frac{a}{4} \\ -\frac{a}{4} < x < \frac{a}{4} \\ \frac{a}{4} < x < \frac{3a}{4} \end{array}\end{array}\right\} \tag{3.86}$$

giving the area under this function to be the interface term

$$c_{0i} = f_S\left[1 + (r_{1i} - r_{1-})^2 + (r_{1i} - r_{1+})^2\right] - f_A(r_{1+} - r_{1-}) \tag{3.87}$$

Mass coefficients for some semiconductors are shown in Table 3.1.

Table 3.1 *Ionic masses*[a]

	GaAs	AlAs	InAs	InP	InSb	GaP	GaSb	Ge	Si
M	144.64	101.90	189.74	145.79	236.57	100.69	191.47	145.18	56.172
μ	36.114	19.837	45.338	24.393	59.092	21.446	44.333	36.295	14.043
μ/M	0.24968	0.19468	0.23895	0.16732	0.24979	0.21299	0.23154	0.2500	0.2500
r_1	0.48202	0.26478	0.60514	0.78757	0.48535	0.69242	0.36413	0.5	0.5
r_2	0.51799	0.73522	0.39486	0.21243	0.51465	0.30758	0.63587	0.5	0.5

Note: [a] M = mass of unit cell (molecular weight), μ = reduced mass, $r_1 = m_c/M$ (cation), $r_2 = m_a/M$ (anion).

The relevant quantity in the boundary conditions is $c_{0i} - \omega^2 \mu_{0i}$. As mentioned previously, the importance of interface effects enters only when the zone-centre frequencies of the two media are close together; otherwise the condition $u = 0$ is usually satisfactory. In order to assess quantitatively the importance of interface effects we may in this case take $\omega^2 = 2f_S/\bar{\mu}$, where $\bar{\mu}$ is the average reduced mass. Let us put $c_{0i=f_s} + \Delta c_{0i}$ and $\mu_{0i} = \bar{\mu}/2 + \Delta\mu^{0i}$, so that

$$c_{0i} - \omega^2\mu_{0i} = f_S + \Delta c_{0i} - \frac{2f_S}{\bar{\mu}}\left(\tfrac{\mu}{2} + \Delta\mu_{0i}\right) = \Delta c_{0i} - \omega^2\Delta\mu_{0i} \tag{3.88}$$

The effective interface term thus depends on the differences between the interface and bulk terms, viz.:

$$\Delta\mu_{0i} = \frac{1}{2}(\mu_{0i} - \bar{\mu})\Delta c_{0i} = f_S\left[(r_{1i} - r_{1-})^2 + (r_{1i} - r_{1+})^2\right] - f_A(r_{1+} - r_{1-}) \tag{3.89}$$

Associated with these effective interface terms are the difference functions

$$\Delta\mu_0(x) = \mu_0(x) - \mu_{0b}(x) \quad \text{and} \quad \Delta c_0(x) = c_0(x) - c_{0b}(x) \tag{3.90}$$

In general, these functions are not symmetric about $x = 0$ and, therefore, the effective interface position, x_i, is not at $x_i = 0$, but rather at the centre of gravity of the function $\Delta c_0(x) - \omega^2\Delta\mu_0(x)$. It is straightforward to show that the contributions $x_{i\mu}$ and x_{ic} from Δc_0 and $\Delta\mu_0$ are

$$\begin{aligned} x_{i\mu} &= \frac{a(\mu_- + \mu_+)}{32\Delta\mu_{0i}} \\ x_{ic} &= \frac{a}{4\Delta c_{0i}}[f_S\{(1 - 2r_{1+})(r_{1i} - r_{1+}) - (1 - 2r_{1-})(r_{1i} - r_{1-})\} \\ &\quad -f_A\{r_{1+} + r_{1-} - 2r_{1i}\}] \end{aligned} \tag{3.91}$$

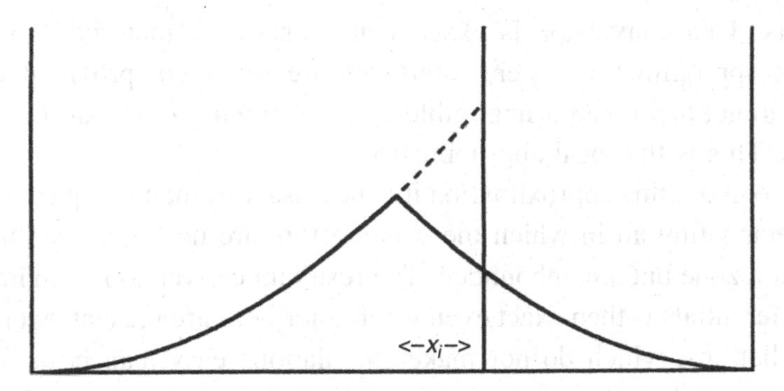

Fig. 3.5 Relationship between choice of position of the interface and discontinuities.

and, therefore

$$x_i = \frac{x_{ic}\Delta c_{0i} - x_{i\mu}\omega^2\Delta\mu_{0i}}{\Delta c_{0i} - \omega^2\Delta\mu_{0i}} \tag{3.92}$$

The shift from zero is very small and can be ignored in long-wavelength analysis. However its presence can go some way to explain, at least qualitatively, why other envelope-function treatments (e.g., Akero and Ando, 1989) have predicted discontinuities in both amplitude and stress. Fig. 3.5 shows how discontinuities can arise in the present case if the solutions are extrapolated to the nominal interface at $x = 0$.

These effective interface terms $\Delta\mu_{0i}$ and Δc_{0i} are useful to illustrate in a qualitative way when interface effects are important, but for quantitative estimates we must use μ_{0i} and c_{0i}, since these are the parameters that appear in the equation of motion. Foreman (1995) has applied this interface model to the case of InAs/GaSb and obtained a good agreement with the results of discrete theory.

3.8 Summary

In a multitude of practical cases it is essential to have reliable theories of macroscopic physics that encapsulate the microscopic properties of solids. In the case of lattice vibrations the coupled motion of individual atoms must be represented in terms of a wave whose spatially varying amplitude is the envelope of the discrete amplitudes of the individual atoms. This can be done rigorously and uniquely through the techniques of quasi-continuum theory, as this chapter has shown, at any rate, for the one-dimensional situation.

An important step is the conversion of the non-local integrals that appear in the discrete theory of the dynamics of the lattice into an infinite sum of local differentials. This conversion is exact in homogeneous material, but it can be used as an approximation where interfaces are involved, provided the quasi-continuum functions contain negligible spectral intensity outside the first Brillouin zone. This is the local approximation.

In order to avoid this approximation it is necessary to map the quasi-continuum on to a metacontinuum in which the wavevectors are no longer confined to the first Brillouin zone but are unbounded. The resultant conversion to an infinite sum of local differentials is then exact even when interfaces are present. Moreover, the Gibbs oscillations, which do not make calculations easy, can be avoided. The price to pay is that the metacontinuum functions are not quasi-continuum envelope functions, though they are mathematically related (Foreman, 1995a).

For long wavelengths in bulk homogeneous material, all derivatives above second order may safely be neglected. This proves to be the case also at an interface as far as deriving boundary conditions is concerned. In bulk material the 1D equations of motion for acoustic and optical vibrations are

$$\omega^2 MU + \frac{\partial}{\partial x}\left(C_2 \frac{\partial U}{\partial x}\right) = 0 \tag{3.93a}$$

$$(\omega^2\mu - c_0)u - \frac{\partial}{\partial x}\left(c_2 \frac{\partial u}{\partial x}\right) \tag{3.93b}$$

In the nearest-neighbour approximation $C_2 = (f_S^2 - f_A^2)a/2f_s$, $c_2 = (\mu f_s - f_A^2/\omega^2)a/M$ and $c_0 = 2f_s/a$. For longitudinally polarized modes $f_A = 0$.

At an interface M, C_2, μ, c_0 and c_2 all become functions of x. Because ω^2 is small for the acoustic modes there is no interface component and the acoustic boundary conditions are the classical ones

$$\Delta U = 0 \quad \text{and} \quad \Delta\left(C_2 \frac{\partial U}{\partial x}\right) = 0 \tag{3.94}$$

Interface components may arise in the case of optical modes if $c_0 - \omega^2\mu$ is not negligible at the interface. In general, the boundary conditions are

$$\Delta u = 0 \quad \text{and} \quad (c_{0i} - \omega^2\mu_{0i})u(x_i) + \Delta\left(C_{2b}\frac{\partial u}{\partial x}\right) = 0 \tag{3.95}$$

For acoustic modes the boundary conditions predicted by both linear-chain theory and quasi-continuum theory coincide with the classical ones. With respect to optical modes there are differences and similarities. The main difference is that u is continuous in quasi-continuum theory but not in linear-chain theory. Another difference is that the results of quasi-continuum theory are independent of interface position (within $a/2$), whereas those of linear-chain theory are not. Indeed, in quasi-continuum theory only an interface *region* is defined. Such an independence on precise interface position is consistent with the spirit of continuum theory. Quasi-continuum theory also extends the validity of the envelope-function approach from merely long wavelengths to variations in which the spectral intensity lying outside the first Brillouin zone is small. It also provides a basis for more accurate calculations that go beyond the abrupt-step approximation. The main similarity to linear-chain theory is the emergence of delta-function-like terms at the interface arising from mass discontinuities, though in quasi-continuum theory they affect only the slope of the envelope function. Foreman has shown that, nevertheless, good numerical agreement with discrete calculations is obtained for modes, including local ones in the system GaSb/InAs.

In systems where there is a shared anion (such as GaAs/AlAs) the interface terms are relatively weak and the classical boundary conditions

$$\Delta u = 0 \quad \text{and} \quad \Delta\left(C_{2b}\frac{\partial u}{\partial x}\right) = 0 \tag{3.96}$$

hold to a good approximation.

The application of quasi-continuum theory to the 3D case is slightly complicated by the lower symmetry of optical, compared with acoustic, vibrations. If rotational strains are added to the familiar compressional and shear strains the nine stress components are related to this full complement of strain by 81 elastic constants. In cubic crystals the number of independent constants for optical strains is reduced by symmetry to 5 or 4, and in the case of an isotropic cubic crystal the number reduces to 3. (For acoustic strains the corresponding number is 2.) We discuss this further in the next chapter. However, this causes no problems and the elastic boundary conditions remain of the same form. The principal additions in 3D are the electromagnetic boundary conditions and the inclusion of piezoelectricity.

Although the assumption of isotropy is useful for the purpose of estimating the strength of the electron–phonon interaction, it cannot be used to model the results of detailed spectroscopy. Crystals such as GaAs are elastically anisotropic. This manifests itself in the different dispersion of LO and TO vibrations in different directions and this can be revealed experimentally via the technique of inelastic neutron scattering (e.g., Strauch and Dorner, 1990). Indeed, in certain directions the dispersion can be of opposite sign to the prevailing situation. A rigorous description of the electron–phonon interaction has to take this anisotropy into account, even though what is normally required is some directional average. Fortunately, the anisotropy is often small and an isotropic approximation is then reasonably accurate. A theoretical investigation of the effects of acoustic anisotropy on the GaAs optical modes in the AlAs/GaAs/AlAs system indicates that the frequency corrections are small but spectroscopically observable (Chamberlain *et al.*, 1994).

It is fortunate that in many cases of practical interest an accurate knowledge of mechanical boundary conditions is not required; mismatch of frequencies is so large that the assumption that the optical displacement vanishes is usually valid.

Appendix: The Local Approximation

We take $m(x, x')$ to be a non-local operator. It is useful to introduce new variables, following Kunin (1982) (see Fig. 3.6).

$$y = \frac{1}{\sqrt{2}}(x - x'), \quad y' = \frac{1}{\sqrt{2}}(x + x') \tag{A3.1}$$

and their associates in **k**-space

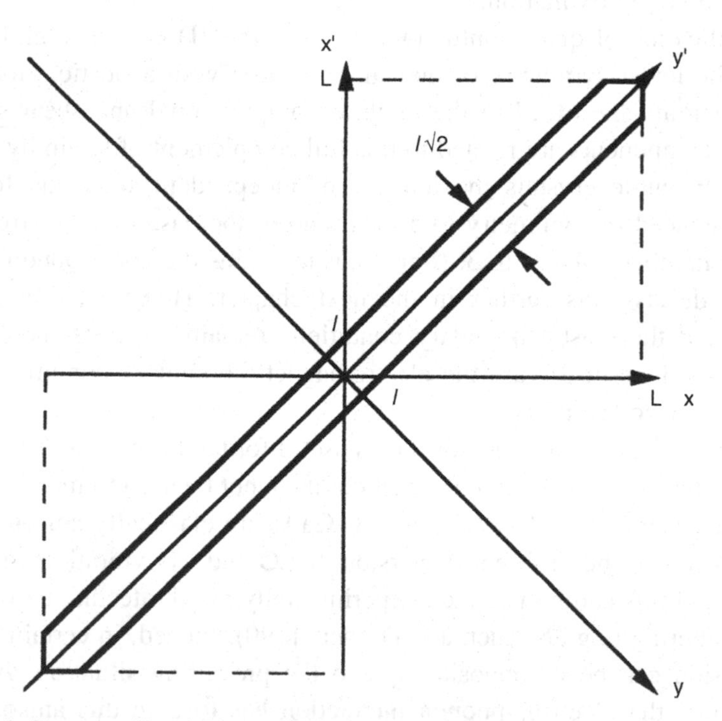

Fig. 3.6 Change of variables. Medium bounded by a characteristic dimension 2L, with l equal to characteristic range of interaction (Kunin, 1982).

$$K = \frac{1}{\sqrt{2}}(k + k'), \quad K' = \frac{1}{\sqrt{2}}(k - k') \tag{A3.2}$$

Thus $m(x, x')$ transforms to $m(y, y')$, and $m(k, k')$ to $m(K, k')$, where

$$m(K, K') = \frac{1}{2\pi}\iint e^{-iKy} m(y, y') e^{iK'y'} dy\, dy' \tag{A3.3}$$

This equation can be written

$$\begin{aligned} m(K, K') = &\frac{1}{2\pi}\iint \cos Ky\, m^S(y, y') e^{iK'y'} dy\, dy' \\ &- \tfrac{i}{2\pi}\iint \sin Ky\, m^A(y, y') e^{iK'y'} dy\, dy' \end{aligned} \tag{A3.4}$$

where $m(y, y')$ has been decomposed into its symmetric and antisymmetric parts:

$$\begin{aligned} m^S(y, y') &= \frac{1}{2}[m(y, y') + m(y', y)] \\ m^A(y, y') &= \frac{1}{2}[m(y, y') - m(y', y)] \end{aligned} \tag{A3.5}$$

Expanding the cosine gives a series in K^2 that can be converted into a series in $K'^2 + 2kk'$. A similar series is obtained from the sine if a factor K is taken outside and replaced by $\sqrt{2k' - K'}$. Collecting together terms of the same order in kk' eventually yields

$$m(k, k') = \frac{1}{\sqrt{2\pi}}\sum_{n=0}^{\infty}(kk')^n[m_{2n}(k' - k) + ik'm_{2n+1}(k' - k)] \tag{A3.6}$$

where $m_{2n}(k' - k)$ and $m_{2n+1}(k' - k)$ are integrals over y and y' of $m(y, y')e^{ik'y'}$ multiplied by a power series in K'.

Thus

$$\begin{aligned} m_{2n}(k' - k) = &\frac{(-2)^n}{\sqrt{2\pi}}\iint e^{iK'y'} m(-y, y')(k')^{-2n} \\ &\times \sum_{j=n}^{\infty}\tfrac{j!}{n!(j-n)!}\left[\frac{(-iK'y)^{2j}}{(2j)!} + \frac{(-iK'y)^{2j+1}}{(2j+1)!}\right] dy\, dy' \end{aligned} \tag{A3.7a}$$

$$\begin{aligned} m_{2n+1}(k' - k) = &i\sqrt{2}\frac{(-2)^n}{\sqrt{2\pi}}\iint e^{iK'y'} m^A(-y, y')(K')^{-2n-1} \\ &\times \sum_{j=n}^{\infty}\tfrac{j!}{n!(j-n)!}\frac{(-iK'y)^{2j+1}}{(2j+1)!} dy\, dy' \end{aligned} \tag{A3.7b}$$

with $m^S(y, y') = m^S(-y, y')$, $m^A(y, y') = -m^A(-y, y')$.

In the long-wavelength approximation the displacement is slowly varying and so $k' \to 0$. Thus, in Eq. (A3.6), only the lowest terms are of importance, i.e.

$$m(k,k') \approx \frac{1}{\sqrt{2\pi}}[m_0(k'-k) + ik'm_1(k'-k)$$
$$- kk'm_2(k'-k) - ikk'^2 \times m_3(k'-k) + k^2k'^2m_4(k'-k)] \quad \text{(A3.8)}$$

where, for example

$$m_0(k'-k) = \frac{1}{\sqrt{2\pi}} \iint e^{iK'(y'-y)} m(-y,y')dy\,dy' \quad \text{(A3.9a)}$$

$$m_1(k'-k) = \frac{1}{\sqrt{\pi}} \iint e^{iK'y'} m^A(-y,y') \frac{\sin K'y}{K'} dy\,dy' \quad \text{(A3.9b)}$$

$$m_2(k'-k) = -\sqrt{\frac{2}{\pi}} \iint e^{iK'y'} m(-y,y')(K')^{-2}[e^{ik'y'} - 1 + iK'y]dy\,dy' \quad \text{(A3.9c)}$$

Conversion to the x-representation gives

$$W(x) = \sum_{n=0}^{\infty} (-1)^n \left[m_{2n}(x) \frac{\partial^n u}{\partial x^n} \frac{\partial^n v}{\partial x^n} + m_{2n+1}(x) \frac{\partial^n u}{\partial x^n} \frac{\partial^{n+1} v}{\partial x^{n+1}} \right] \quad \text{(A3.10)}$$

When $m(x', x)$ is symmetric, only the even-numbered terms are non-zero, and therefore

$$\int m(x,x')u(x')dx' = \sum_{n=0}^{\infty} \frac{\partial^n}{\partial x^n} \left[m_{2n}(x) \frac{\partial^n}{\partial x^n} u(x) \right] \quad \text{(A3.11)}$$

In particular

$$m_0(x) = \int e^{iK'x} m(K')dK' = \int m(x',x)dx' \quad \text{(A3.12)}$$

4

Bulk Vibrational Modes in an Isotropic Continuum

He sobbed and he sighed, and a gurgle he gave,
Then he plunged himself into the billowy wave.
The Mikado, *W. S. Gilbert*

4.1 Elasticity Theory

We now consider the situation in three dimensions. Classical elasticity theory successfully describes static stress and strain in crystals considered as continua. It is also successful in describing the propagation of long-wavelength acoustic waves and the way in which these waves reflect and transmit at a boundary. The same cannot be said to be true of optical vibrations, the description of which has focussed almost entirely on dispersionless models (see Born and Huang, 1954). A continuum description of dispersion is, however, essential if reflection and transmission are to be treated, and where this is the case it has always been assumed that the acoustic form can be used (see, e.g., Babiker, 1986; Ridley, 1991; Trallero-Giner *et al.*, 1992; Pérez-Alvarez *et al.*, 1993). It is by no means obvious that this is correct, as the analysis of the previous chapter has shown. Indeed, the analysis to be presented suggests that it is wrong, quite apart from problems connected with interface potentials.

To approach this problem it is useful to be reminded of classical elasticity (Love, 1927; Landau and Lifshitz, 1986). Hooke's law relates a stress component σ_{ij} with a strain component e_{kl} via a set of elastic constants c_{ijkl} such that

$$\sigma_{ij} = \sum_{kl} c_{ijkl} e_{kl} \tag{4.1}$$

The labels i, j, k, l represent the x, y and z directions with $i = 1$, 2 or 3 and j, k and l similarly. The stress σ_{ij} refers to the stress exerted by a force parallel to the ith direction acting on the face of an elementary cube of the material whose normal is parallel to the jth direction. The strain e_{kl} is the change of displacement

U in the kth direction with change of position x in the lth direction, viz:

$$e_{kl} = \frac{1}{2}\left(\frac{\partial \mathbf{U}_k}{\partial x_l} + \frac{\partial \mathbf{U}_l}{\partial x_k}\right) \tag{4.2}$$

There are 9 stress components and 9 strain components and there are therefore 81 elastic constants.

In a continuum theory, stresses and strains are specified at a point. In the case of acoustic vibrations that point is the centre of mass of a primitive unit cell and it is assumed that there can be no net torque as a consequence. This means that

$$\sigma_{ij} = \sigma_{ji} \tag{4.3}$$

and since there are no rotations (we exclude rigid rotations)

$$e_{kl} = e_{lk} \tag{4.4}$$

Both the second-rank tensors describing stress and strain are therefore symmetric. Because $c_{ijkl} = c_{jilk}$ the number of elastic constants reduces to 36. The necessity of there being an appropriate quadratic form for the stress–strain energy density imposes a further condition on the elastic constants, namely

$$c_{ijkl} = c_{klij} \tag{4.5}$$

and this reduces the number to 21. Subsequent reductions occur as a consequence of symmetry restrictions. In particular, complete isotropy reduces the number to 2 (Table 4.1).

The stress–strain energy density, W_{ac}, is given by

$$W_{ac} = \frac{1}{2}\sum c_{ij}e_ie_j \tag{4.6}$$

Table 4.1 *Acoustic elastic constants for a cubic crystal* (O, O_h, T, T_h, T_d) *(isotropy* $c_{11} - c_{12} = 2c_{44}$*)*

	e_1 exx	e_2 eyy	e_3 ezz	e_4 $\frac{1}{2}(e_{zx} + e_{yz})$	e_5 $\frac{1}{2}(e_{xz} + e_{zx})$	e_6 $\frac{1}{2}(e_{xy} + e_{yz})$
σ_1, σ_{xx}	c_{11}	c_{12}	c_{12}			
σ_2, σ_{yy}	c_{12}	c_{11}	c_{12}			
σ_3, σ_{zz}	c_{12}	c_{12}	c_{11}			
σ_4, σ_{zy}				$2c_{44}$		
σ_5, σ_{xz}					$2c_{44}$	
σ_6, σ_{xy}						$2c_{44}$

where the elastic constants and strain components are now depicted in so-called reduced notation ($c_{11} = c_{1111}$, $c_{44} = c_{2323}$, etc). For simplicity, we will write W explicitly only for the isotropic case, and for reasons that will become clear, we choose to write it as follows:

$$W_{ac} = \frac{1}{2}c_{11}(\nabla.\mathbf{U})^2 + \frac{1}{2}c_{44}(\nabla \times \mathbf{U})^2 + c_{44}\sum_{kl}(e_{kl}e_{lk} - e_{ll}e_{kk}) \tag{4.7}$$

with, now

$$e_{kl} = \frac{\partial U_k}{\partial x_l} \tag{4.8}$$

Optical vibrations involve the relative motion of the two atoms in the primitive unit cell. The energy density will be different from the acoustic one in two respects. In the limit of long wavelength there will be an energy density proportional to the square of the displacement **u** and nothing else. Waves of finite length will have a reduced energy density and the amount of reduction will be some function of optical strains defined now by Eq. (4.8).

The standard assumption has been that the form is identical to that for acoustic waves, and so we can write

$$W_{op}^A = \frac{1}{2}fu^2 - \frac{1}{2}c_{11}(\nabla.\mathbf{u})^2 - \frac{1}{2}c_{44}(\nabla \times \mathbf{u})^2 - c_{44}\sum_{kl}(e_{kl}e_{lk} - e_{ll}e_{kk}) \tag{4.9}$$

where f is a force constant, the elastic constants are optical elastic constants and the strains are optical strains. We refer to this as the acoustic model and label energy density with a superscript.

To take over the concepts of "acoustic" elasticity and apply them uncritically to "optical" elasticity is clearly inappropriate. Long-wavelength acoustic displacements involve the unit cell moving as a unit and experiencing dilatational and shear stresses exerted by the surrounding medium, whereas optical displacements involve the relative motion of the two atoms in the unit cell. In this case restoring forces are generated only as a response to the atoms changing their distance apart or by rotation. Including rotation means that the medium supporting optical vibrations must be regarded as rotationally elastic (to use the phrase coined by Larmor, 1894, for the aether). This means that additional elastic constants related to rotation exist for optical modes and, consequently, the stress tensor acquires antisymmetric components. Maintaining the mirror symmetry about the diagonal imposed by the form for the energy reduces the number of independent constants from 81 to 45. This number comes down to 4 in the case of a cubic crystal with O, O_h or T_d symmetry as Table 4.2 shows.

Table 4.2 *Optical elastic constants for a cubic crystal (O, O_h, T_d) (isotropy $c_{11} - c_{12} = d_1 + d_2$)*

	e_{xx}	e_{yy}	e_{zz}	e_{zy}	e_{yz}	e_{zx}	e_{xz}	e_{xy}	e_{yx}
σ_{xx}	c_{11}	c_{12}	c_{12}						
σ_{yy}	c_{12}	c_{11}	c_{12}						
σ_{zz}	c_{12}	c_{12}	c_{11}						
σ_{zy}				d_1	d_2				
σ_{yz}				d_2	d_1				
σ_{zx}						d_1	d_2		
σ_{xz}						d_2	d_1		
σ_{xy}								d_1	d_2
σ_{yx}								d_2	d_1

Source: Foreman, 1994.

In order to resolve these problems associated with optical modes, resort can be made to elasticity theory derived from three-dimensional microscopic theory (Forman and Ridley, 1999). The starting point is the two-parameter Keating potential for zinc blende semiconductors (Keating, 1966). In this model there are two force constants a and β, where a is the central force constant quantifying the nearest-neighbour interaction, and β is the non-central force constant quantifying the second-neighbour interaction. Following Martin (1970), different non-central force parameters, β^A and β^B, can be used for anions and cations, and with the focus on purely elastic properties we neglect Coulomb effects. A labelling convention for the atoms in the tetrahedral lattice is as follows: for the four nearest neighbours:

any A atom $\bar{1} = [1\,\bar{1}\,\bar{1}]$, $\bar{2} = [\bar{1}\,1\,\bar{1}]$, $\bar{3} = [\bar{1}\,\bar{1}\,1]$, $\bar{4} = [1\,1\,1]$
any B atom $1 = [\bar{1}\,1\,1]$, $2 = [1\,\bar{1}\,1]$, $3 = [11\bar{1}]$, $4 = [\bar{1}\,\bar{1}\,\bar{1}]$

The 12 second-nearest neighbours are labelled:

for A $\bar{1}2, \bar{1}3, \bar{1}4, \bar{2}1, \bar{2}3, \bar{2}4, \bar{3}1, \bar{3}2, \bar{3}4, \bar{4}1, \bar{4}2, \bar{4}3$
for B $1\bar{2}, 1\bar{3}, 1\bar{4}, 2\bar{1}, 2\bar{3}, 2\bar{4}, 3\bar{1}, 3\bar{2}, 3\bar{4}, 4\bar{1}, 4\bar{2}, 4\bar{3}$

The total potential involving atom 0 can then be written as follows:

$$\Phi_0 = \frac{1}{8a^2}\left\{\sum_{i=0}^{4} a_{0\bar{\imath}}\left[\Delta\left(\mathbf{x}^{i0}.\mathbf{x}^{i0}\right)\right]^2 + \sum_{i,j1} \beta^A_{\bar{\imath}0j}\left[\Delta\left(\mathbf{x}^{i0}.\mathbf{x}^{0j}\right)\right]^2 + \sum_{i,k\neq i} \beta^B_{0\bar{\imath}\kappa}\left[\Delta\left(\mathbf{x}^{0i}.\mathbf{x}^{ik}\right)\right]^2\right\} \tag{4.10}$$

where $\mathbf{x}^{i0} = \mathbf{x}^{\mathbf{B}}(\bar{\mathbf{i}}) - \mathbf{x}^{\mathbf{A}}(0)$, the nearest-neighbour separation is $\sqrt{3}a$, and $\mathbf{x}$ is the instantaneous position of the atom related to its equilibrium position $\mathbf{X}$ by its displacement $\mathbf{u}$ such that $\mathbf{x}=\mathbf{u}+\mathbf{X}$. To first order $\Delta(\mathbf{x}^p.\mathbf{x}^q)=\mathbf{u}^p.\mathbf{X}^q+\mathbf{u}^q.\mathbf{X}^p$. The first sum is over the nearest neighbours of atom A. The second sum is over pairs of nearest neighbours with atom A as the central atom. The third sum is over nearest neighbour (i) and second-neighbours (k) with the central atom being B.

The equation of motion for atom 0 is:

$$\omega^2 m^A u_\alpha^0 = \frac{\partial \Phi_0}{\partial u_\alpha^0}$$
$$= \frac{1}{4a^2}\left\{4\alpha \sum_i u_\lambda^{0i} X_\lambda^{\bar{i}0} X_\alpha^{\bar{i}0} + \beta^A \sum_{i,j\neq i} u_\lambda^{0j} X_\lambda^{\bar{i}0}\left(X_\alpha^{\bar{j}0}+X_\alpha^{\bar{i}0}\right) + \beta^B \sum_{i,k\neq i} X_\alpha^{k\bar{i}}\left(u_\lambda^{0\bar{i}} X_\lambda^{k\bar{i}} - u_\lambda^{k\bar{i}} X_\lambda^{\bar{i}0}\right)\right\} \tag{4.11}$$

where m^A is the mass of atom A. It is useful to rearrange the terms so that the sums are unrestricted so that we can use the identities:

$$\sum_i X_\alpha^{\bar{i}0} = 0, \quad \sum_i X_\alpha^{\bar{i}0} X_\beta^{\bar{i}0} = 4a^2 \delta_{\alpha\beta} \tag{4.12}$$

to obtain:

$$\omega^2 m^A u_\alpha^0 = \frac{1}{4a^2}\left\{4(\alpha-\beta)\sum_i u_\lambda^{0i} X_\lambda^{\bar{i}0} X_\alpha^{\bar{i}0} + 8a^2\beta \sum_i u_\alpha^{0\bar{i}} - \beta^B \sum_{i,k} u_\lambda^{k\bar{i}} X_\alpha^{k\bar{i}} X_\lambda^{\bar{i}0}\right\} \tag{4.13}$$

A second-order Taylor series expansion about the equilibrium location of the atom converts Eq. (4.13) into a second-order differential equation:

$$u_\lambda^{\bar{i}0} = u_\lambda^B(\bar{i}) - u_\lambda^A(0) = u_\lambda^B + X_\sigma^{\bar{i}0} u_{\lambda,\sigma}^B + \frac{1}{2} X_\sigma^{\bar{i}0} X_\tau^{\bar{i}0} u_{\lambda\sigma\tau}^B - u_\lambda^A$$
$$u_\lambda^{ki} = u_\lambda + X_\sigma^{i0} u_{\lambda,\sigma} + \frac{1}{2} X_\sigma^{\bar{i}0} X_\tau^{\bar{i}0} u_{\lambda,\sigma} + X_\sigma^{k\bar{i}} u_{\lambda,\sigma}^A + \frac{1}{2}\left(X_\sigma^{ki} X_\tau^{ki} + X_\sigma^{k\bar{i}} X_\tau^{\bar{i}0} + X_\sigma^{\bar{i}0} X_\tau^{ki}\right) u_{\lambda,\sigma\tau}^A \tag{4.14}$$

Here we adopt compact notation in which the subscript comma denotes differentiation $\left(u_{\lambda,\sigma} = \partial u_\lambda / \partial x_\sigma\right)$. Substitution in Eq. (4.13) and evaluating sums:

$$\sum_i X_\alpha^{\bar{i}0} X_\beta^{\bar{i}0} X_\gamma^{\bar{i}0} = 4a^3 \left|\varepsilon_{\alpha\beta\gamma}\right|, \quad \sum_i X_\alpha^{\bar{i}0} X_\beta^{\bar{i}0} X_\sigma^{\bar{i}0} X_\tau^{\bar{i}0} = 4a^4 \Delta_{\alpha\beta\sigma\tau},$$
$$\left|\varepsilon_{\alpha\beta\gamma}\right| = 1, \quad \alpha \neq \beta \neq \gamma, \quad \Delta_{\alpha\beta\sigma\tau} = \delta_{\alpha\beta}\delta_{\alpha\tau} + \left(1-\delta_{\alpha\beta}\right)\left(\delta_{\alpha\sigma}\delta_{\beta\tau} + \delta_{\alpha\tau}\delta_{\beta\sigma}\right) \tag{4.15}$$

we get an explicit second-order envelope-function equation for the motion of atom A. Repeating the procedure for atom B gives a similar equation for atom B.

These form a pair of coupled second-order differential equations that can then be transformed into long-wavelength centre-of-mass (acoustic) and relative (optical) coordinates. The equation of motion for acoustic modes is:

$$\omega^2 \rho U_\alpha = -\left(C_{\alpha\lambda\beta\mu} U_{\beta,\mu}\right)_{,\lambda} \tag{4.16}$$

Here, ρ is the mass density $(m^A+m^B)/16a^3$ and U is the displacement of the unit cell. Cubic symmetry reduces the number of elastic constants to three. Explicitly, these are:

$$C_{xxxx} = \frac{\alpha+3\beta}{4a}, \quad C_{xxyy} = \frac{\alpha-\beta}{4a}, \quad C_{xyyx} = C_{xyxy} = \frac{\alpha\beta}{a(\alpha+\beta)} \tag{4.17}$$

where $\beta = (\beta^A+\beta^B)/2$. These are exactly those derived by Keating. We note that the last equality of Eq. (4.17) is a manifestation of rotational invariance. The boundary conditions for acoustic modes are well known. This is not the case for optical modes.

The equation of motion for the optical modes is:

$$\omega^2 \rho_r u_\alpha = c_{\alpha\beta} u_\beta + \left(c_{\alpha\lambda\beta\mu} u_{\beta,\mu}\right)_{,\lambda} \tag{4.18}$$

Here $\rho_r = \mu/16a^3$ is the reduced mass density with $\mu = m^A m^B/M$, $M = m^A+m^B$. In this case there is no rotational invariance and so the number of elastic constants is four plus the interatomic force constant:

$$\begin{aligned} c_{\alpha\beta} &= \frac{\alpha+\beta}{4a^3}\delta_{\alpha\beta}, \quad c_{xxxx} = \frac{\mu}{\mathrm{M}}\frac{\alpha+3\beta}{4a} - \frac{\beta'}{4a}, \quad c_{xxyy} = \frac{\mu}{\mathrm{M}}\frac{\alpha}{4a} - \frac{\beta'}{8a}, \\ c_{xyyx} &= \frac{\mu}{\mathrm{M}}\frac{\alpha}{4a} - \frac{\beta'}{8a} - \frac{(\alpha-\beta)^2}{a\omega^2 M}, \quad c_{xyxy} = \frac{\mu}{M}\frac{\alpha+\beta}{4a} - \frac{(\alpha-\beta)^2}{a\omega^2 M} \end{aligned} \tag{4.19}$$

where $\beta' = \beta + \frac{1}{2}r(\beta^A - \beta^B)$, $r = \frac{1}{2}\frac{(m^A-m^B)}{M}$. The stress tensor is not symmetric.

Thus, in the long-wavelength limit, the acoustic and optical vibrations are completely decoupled.

Referring to Table 4.2, we see that $c_{11} = c_{xxxx}$, $c_{12} = c_{xxyy}$, $d_1 = c_{xyxy}$, $d_2 = c_{xyyx}$. In the light of the tetrahedral symmetry, we can recast Eq. (4.18) as follows:

$$\omega^2 \rho_r u_x = c_0 u_x + \left[c_{11} u_{x,x} + (c_{12}+d_1+d_2)\left(u_{y,y}+u_{z,z}\right)\right]_{,x} - [\nabla\times d_1 \nabla\times\mathbf{u}]_x \tag{4.20}$$

(Note that the subscript x on the last term denotes the x component.) If the crystal is isotropic, $c_{12}+d_1+d_2 = c_{11}$ and the equation becomes:

$$\omega^2 \rho_r u_x = c_0 u_x + [c_{11}\nabla.\mathbf{u}]_{,x} - [\nabla\times d_1 \nabla\times\mathbf{u}]_x \tag{4.21}$$

(Note that the subscript on the last term denotes the x component!)

The microscopic theory outlined above can be generalized to the case where the crystal is inhomogeneous in, say, the z direction. In this case the force constant becomes dependent on the spatial variation of the mass factor r:

$$
\begin{aligned}
c_{xx} &= c_{yy} = c_0 - \frac{a+\beta}{8a}\left[(\mu/\mathrm{M})_{,zz} + 2(r_{,z})^2\right] \quad c_0 = \frac{a+\beta}{4a^3}, \\
c_{zz} &= c_0 + \frac{a+\beta}{8a}\left[(\mu/M)_{,zz} + 2(r_{,z})^2\right] - \frac{\Delta\beta}{8a}(r_{,zz}) \qquad \Delta\beta = \frac{\beta^{A} - \beta^{B}}{2} \\
c_{xy} &= c_{yz} = \frac{a-\beta}{4a^2}(r_{,z})
\end{aligned}
\tag{4.22}
$$

In terms of the optical-stress components Eq. (4.18) becomes:

$$
\omega^2 \rho_r u_x = c_{xx}u_x + c_{xy}u_y + c_{xz}u_z + \sigma_{xx,x} + \sigma_{xy,y} + \sigma_{xz,z} \tag{4.23}
$$

and similarly for y and z. The stress components acting across the interface are:

$$
\begin{aligned}
\sigma_{zz} &= c_{11}u_{z,z} + c_{12}\left(u_{x,x} + u_{y,y}\right) \\
\sigma_{xz} &= d_1 u_{x,z} + d_2 u_{z,x} \\
\sigma_{yz} &= d_1 u_{y,z} + d_2 u_{z,y}
\end{aligned}
\tag{4.24}
$$

As an example of the procedure for determining the interface conditions, we focus on σ_{xz}. The variation of this stress component across the interface from z_1 to z_2 can be obtained by integrating Eq. (4.23). Thus:

$$
\sigma_{xz}(z_2) = \sigma_{xz}(z_1) + \int_{z_1}^{z_2} \left(\omega^2 \rho_r u_x - c_{xx}u_x - c_{xy}u_y - c_{xz}u_z - \sigma_{xx,x} - \sigma_{xy,y}\right) dz \tag{4.25}
$$

with similar expressions for the other stress components. A further integration can give the variation of the components of displacement. Notice that the first and second derivatives of the mass parameters (μ/M and r) give rise to delta-like functions and their derivatives. Compared with the contribution to the integral of the other more slowly varying terms, the role of the mass parameters is dominant. The Keating parameters a and β themselves can vary over the interface, but that is a variation that can be neglected in many cases in comparison with the variation of mass factors – as, in fact, we have done here. The importance of the variation of the mass factor in boundary-condition problems was first pointed out by Akero and Ando (1989) in their linear-chain model. In many cases, where the disparity of the material properties is large, the boundary condition reduces simply to $\mathbf{u} = \mathbf{0}$. More general boundary conditions for optical modes have been discussed by Foreman (1998).

4.2 Polar Material

When the two atoms in each unit cell have opposite charges the optical vibrations are accompanied by electric fields and a continuum description must accommodate the elastic field and the electromagnetic field and their interaction. The Lagrangian density in an isotropic continuum, to lowest order, can be written in the optical model (Babiker, 1994)

$$L = L_{em} + L_{el} + L_{inter} \tag{4.26}$$

where

$$L_{em} = \frac{1}{2}\varepsilon_0\left[(\dot{\mathbf{A}}+\nabla\phi)^2 - c^2(\nabla\times\mathbf{A})^2\right] \tag{4.27}$$

$$L_{el} = \frac{1}{2}\left[\mu(\dot{\mathbf{u}}^2+\omega_{TO}^2\mathbf{u}^2) + c_L(\nabla.\mathbf{u})^2 + c_T(\nabla.\times\mathbf{u})^2\right] \tag{4.28}$$

$$L_{inter} = -\left[\mathbf{P}(\dot{\mathbf{A}}+\nabla\phi) + \frac{1}{2}(\varepsilon_\infty - \varepsilon_0)\left(\dot{\mathbf{A}}+\nabla\phi\right)^2\right] \tag{4.29}$$

The electromagnetic Lagrangian density has been written in terms of the vector and scalar potentials, $\mathbf{A}$ and ϕ, and is standard. The elastic Lagrangian is derived from the optical model of the previous section with a slight change of notation, $\mu\omega_{\mathrm{TO}}^2$ replacing c_0 and c_L, c_T replacing c_{11} and d_1. The interaction Lagrangian consists of terms associated with the electronic and ionic polarization of the lattice, with

$$\mathbf{P} = e^*\mathbf{u} - (\varepsilon_\infty - \varepsilon_0)\left(\dot{\mathbf{A}} - \nabla\phi\right) \tag{4.30}$$

$$e^* = \left[\mu\omega_{\mathrm{TO}}^2(\varepsilon_s - \varepsilon_\infty)\right]^{1/2} \tag{4.31}$$

where ε_s, ε_∞ are the static and high-frequency permittivities and ε_0 is the permittivity of the vacuum. (I am indebted to M. Babiker and N.A. Zakhleniuk for illuminating the problem of the interaction terms.)

The equations of motion associated with the variable $\mathbf{X}$ are obtained from

$$\frac{\partial}{\partial t}\frac{\partial L}{\partial \dot{X}_i} + \sum_j \nabla_j \frac{\partial L}{\partial(\nabla_j X_i)} - \frac{\partial L}{\partial X_i} = 0 \tag{4.32}$$

where $\mathbf{X}$ stands for $\mathbf{A}$, $\mathbf{u}$ or ϕ, and these are

$$\varepsilon_\infty\frac{\partial}{\partial t}(\dot{\mathbf{A}}+\nabla\phi) + \varepsilon_0 c^2(\nabla\times\nabla\times A) - e^*\dot{\mathbf{u}} = 0 \tag{4.33}$$

$$\mu(\ddot{\mathbf{u}} + \omega_{\mathrm{TO}}^2\mathbf{u}) + \nabla(c_L\nabla.\mathbf{u}) - \nabla \times (c_L\nabla \times \mathbf{u}) + e^*(\mathbf{A}+\nabla\phi) = \mathbf{0} \tag{4.34}$$

$$\nabla.[\varepsilon_\infty(\mathbf{A}+\nabla\phi) - \mathbf{e}^*\mathbf{u}] = 0 \tag{4.35}$$

The last of these equations is identifiable as Gauss's theorem. The latter suggests that a permittivity function $\varepsilon(\omega)$ can be defined by introducing the electric field $\mathbf{E}$

$$\mathbf{E} = -\dot{\mathbf{A}} - \nabla\phi \tag{4.36}$$

whence

$$(\varepsilon(\omega) - \varepsilon_\infty)\mathbf{E} = e^*\mathbf{u} \tag{4.37}$$

which establishes a proportionality between $\mathbf{E}$ and $\mathbf{u}$. Recasting the equations of motion accordingly gives

$$-\omega^2\varepsilon(\omega)\mathbf{E} + \varepsilon_0 c^2\nabla \times \nabla \times \mathbf{E} = 0 \tag{4.38}$$

$$\mu\left(-\omega^2\mathbf{u} + \omega_{\mathrm{TO}}^2\left(\frac{\varepsilon(\omega) - \varepsilon_s}{\varepsilon(\omega) - \varepsilon_\infty}\right)\mathbf{u}\right) + \nabla(c_L\nabla.\mathbf{u}) - \nabla \times (c_T\nabla \times \mathbf{u}) = 0 \tag{4.39}$$

$$\nabla.\varepsilon(\omega)\mathbf{E} = 0 \tag{4.40}$$

4.3 Polar Optical Waves

From Eq. (4.40) it can be deduced that if $\mathbf{E}$ is a longitudinal field ($\nabla \times \mathbf{E} = 0$), then $\varepsilon(\omega) = 0$. It then follows that the equations of motion reduce to

$$\mu(-\omega^2\mathbf{u} + \omega_{\mathrm{LO}}^2\mathbf{u}) + \nabla(c_L\nabla.\mathbf{u}) = 0 \tag{4.41}$$

where

$$\omega_{\mathrm{LO}}^2 = \omega_{\mathrm{TO}}^2\frac{\varepsilon_s}{\varepsilon_\infty} \tag{4.42}$$

which is the Lyddane–Sachs–Teller relation, and

$$\mathbf{E} = -\frac{e^*}{\varepsilon_\infty}\mathbf{u} \tag{4.43}$$

Eq. 4.41 has plane-wave solutions of the form

$$\mathbf{u} = \mathbf{u}_0 e^{i(\mathbf{k}.\mathbf{r}-\omega t)} \tag{4.44}$$

which obey a dispersion relation of the form

$$\omega^2 = \omega_{\mathrm{LO}}^2 - v_L^2 k^2 \tag{4.45}$$

where $v_{\mathrm{L}}^2 = c_L/\mu$.

Transversely polarized waves obey the following dispersion relations:

$$\omega^2 \varepsilon(\omega) - \varepsilon_0 c^2 k^2 = 0 \tag{4.46a}$$

$$\omega^2 = \omega_{\mathrm{TO}}^2 \left(\frac{\varepsilon(\omega) - \varepsilon_s}{\varepsilon(\omega) - \varepsilon_\infty} \right) - v_T^2 k^2 \tag{4.46b}$$

where $v_{\mathrm{T}}^2 = c_{\mathrm{T}}/\mu$. From Eq. (4.46b) we obtain an explicit frequency dependence for the permittivity

$$\varepsilon(\omega) = \varepsilon_\infty \left(\frac{\omega^2 - \omega_{\mathrm{LO}}^2 + v_T^2 k^2}{\omega^2 - \omega_{\mathrm{TO}}^2 + v_T^2 k^2} \right) = r\varepsilon_\infty \tag{4.47}$$

where r is known as the permittivity factor.

For $\omega^2 < (\omega_{\mathrm{TO}}^2 - v_T^2 k^2)$ or $\omega^2 > (\omega_{\mathrm{LO}}^2 - v_T^2 k^2)$ the permittivity function is a positive quantity and the waves are essentially electromagnetic waves propagating in a medium with a frequency-dependent permittivity. In the regions where ω is close to either ω_{TO} or ω_{LO}, the waves are known as bulk polaritons. When

$$\omega_{\mathrm{TO}}^2 - v_T^2 k^2 < \omega^2 < \omega_{\mathrm{LO}}^2 - v_T^2 k^2 \tag{4.48}$$

the permittivity is negative. From Eq. (4.46a) this means that k^2 is negative, corresponding to a surface or interface polariton. Thus, if the surface has its normal parallel to z and a surface polariton propagates along the surface with a wavevector k_x, say, then

$$k^2 = k_x^2 + k_z^2 = k_x^2 - a^2 \tag{4.49}$$

where $k_z = \pm ia$. Fig. 1.4(b) depicts the spectrum.

The electric fields associated with these transverse modes can be described in terms of a field factor s given by

$$\mathbf{E} = s\mathbf{E}_{\mathrm{LO}} \tag{4.50a}$$

$$s = -\frac{\varepsilon_\infty}{\varepsilon(\omega) - \varepsilon_\infty} = \frac{\omega^2 - \omega_{\mathrm{TO}}^2 + v_T^2 k^2}{\omega_{\mathrm{LO}}^2 - \omega_{\mathrm{TO}}^2} \tag{4.50b}$$

Thus waves for which

$$\omega^2 = \omega_{TO}^2 - v_T^2 k^2 \tag{4.51}$$

which are the TO modes, have $s = 0$ ($\varepsilon(\omega) \rightarrow \infty$), whereas surface polaritons and interface modes have field factors lying between zero and unity.

The isotropic continuum model that we have been describing is only a rough approximation to reality. Figures 4.1, 4.2 and 4.3 show that the dispersion of real crystals exhibits marked anisotropy. The use of the isotropic model for determining scattering rates etc. without resorting to intense numerical computations is too attractive to abandon on this count, but clearly the parameters that enter the model must be appropriate directional averages.

4.4 Energy Density

In order to quantize the coupled electromagnetic/mechanical system of polar optical modes we need to define an appropriate Hamiltonian. The Hamiltonian density which is derived from the Lagrangian in Eqs. (4.27–29) is

$$H = \frac{1}{2}\left[\mu(\dot{\mathbf{u}}^2 + \omega_{TO}^2 \mathbf{u}^2) - c_L(\nabla.\mathbf{u})^2 - c_T(\nabla \mathbf{x} \mathbf{u})^2\right] + \frac{1}{2}\left(\varepsilon_\infty \mathbf{E}^2 + \frac{1}{\mu_0}\mathbf{B}^2\right) + \nabla.(\mathbf{D}\phi) \tag{4.52}$$

where $\mathbf{D}$ is the electric displacement and $\mathbf{B} = \nabla \times \mathbf{A}$. This expression properly accounts for spatial dispersion and the effects of purely electronic polarization and is therefore a generalization of the Hamiltonian found in Born and Huang (1954). As written, the Hamiltonian appears to consist of a sum of purely mechanical components and purely electrical components, but in fact, in the "unretarded" approximation applicable here, it is essentially a mechanical–energy density. Using the relation between electric field and mechanical displacement in Eq. (4.37) and the generalized expression for the permittivity, viz.:

$$\varepsilon = \varepsilon_\infty \frac{\omega_{LO}^2 - \omega^2 + \Omega^2}{\omega_{TO}^2 - \omega^2 + \Omega^2} \tag{4.53}$$

where

$$\Omega^2 = (\mu \mathbf{u})^{-1}.[\nabla(c_L \nabla.\mathbf{u}) - \nabla \mathbf{x} c_T \nabla \mathbf{x} u] \tag{4.54}$$

we can write the (unretarded) Hamiltonian density in the form:

$$H = \frac{1}{2}\mu(\dot{\mathbf{u}}^2 + \omega^2 \mathbf{u}^2) + \frac{1}{2}\left(-\varepsilon(\omega)\mathbf{E}^2 + \frac{1}{\mu_0}\mathbf{B}^2\right) \tag{4.55}$$

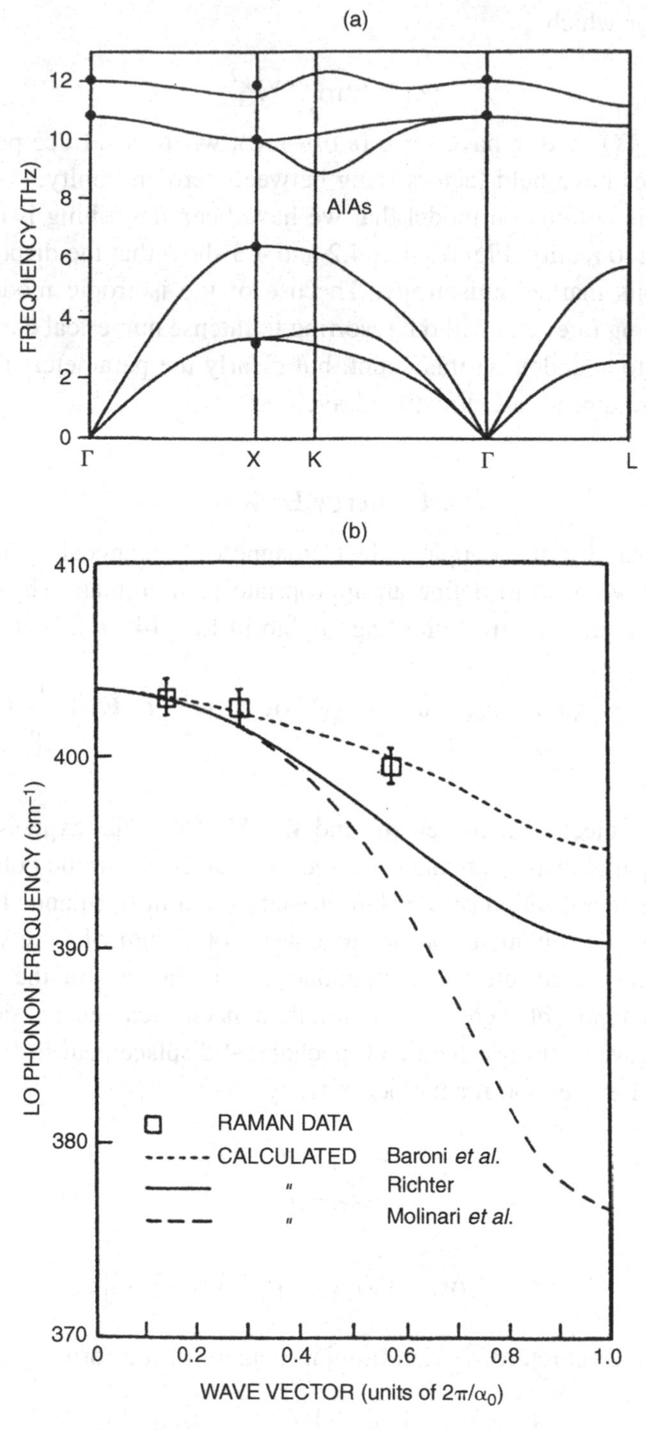

Fig. 4.1 Phonon dispersion for Al compounds: (a) AlAs: numerical (Ren *et al.*, 1989); (b) AlAs: experiment (confined modes) (Mowbray *et al.*, 1991), numerical (Baroni *et al.*, 1990, Richter, 1986; and Molinari *et al.*, 1987); (c) AlSb: numerical (Landolt–Börnstein, 1987).

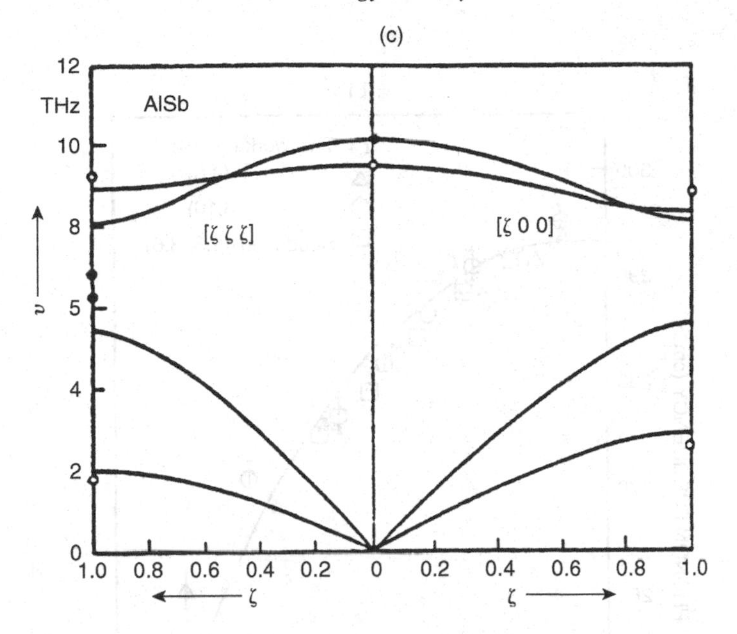

Fig. 4.1 (*cont.*)

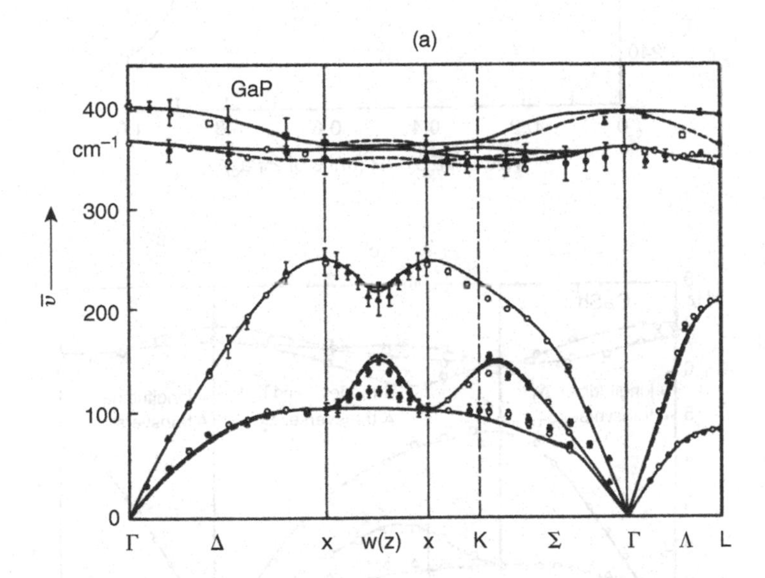

Fig. 4.2 Phonon dispersion for Ga compounds: (a) GaP (Landolt–Börnstein, 1987); (b) GaAs: experimental points (confined modes) (Mowbray *et al.*, 1991; Sood *et al.*, 1985); solid lines: neutron scattering data (Strauch and Dorner, 1990); (c) GaSb (Landolt–Börnstein, 1987).

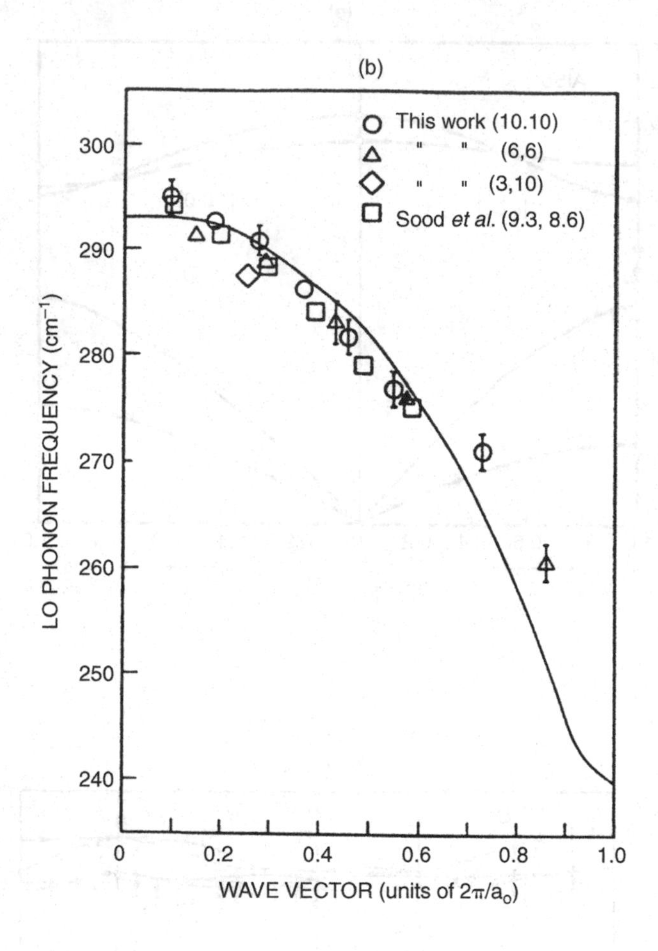
(b)
This work (10.10)
" " (6,6)
" " (3,10)
Sood et al. (9.3, 8.6)
300
290
280
270
260
250
240
LO PHONON FREQUENCY (cm⁻¹)
0
0.2
0.4
0.6
0.8
1.0
WAVE VECTOR (units of 2π/a₀)

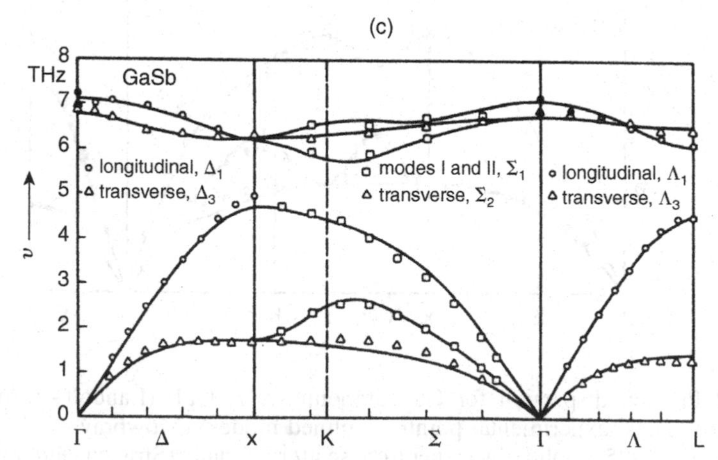
(c)
8
THz
GaSb
7
6
5
4
3
2
1
0
longitudinal, Δ₁
transverse, Δ₃
modes I and II, Σ₁
transverse, Σ₂
longitudinal, Λ₁
transverse, Λ₃
ν
Γ
Δ
X
K
Σ
Γ
Λ
L

Fig. 4.2 (*cont.*)

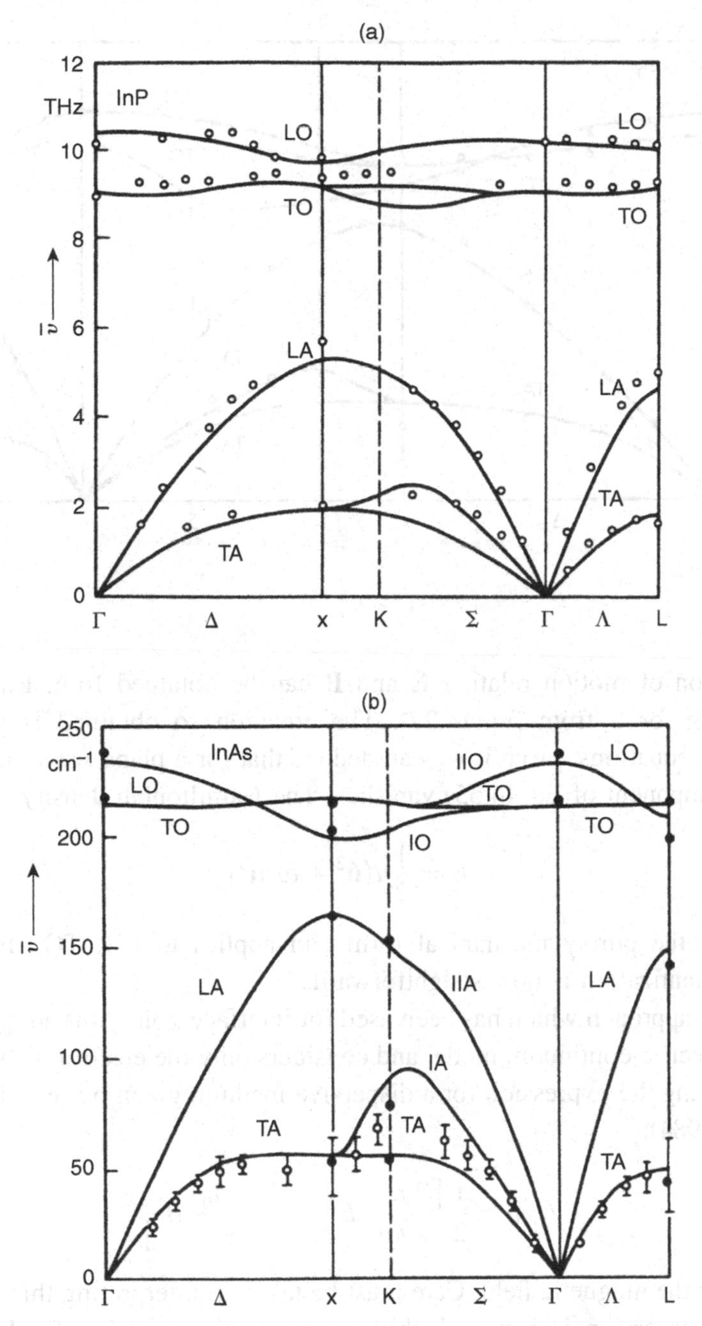

Fig. 4.3 Phonon dispersion for In compounds: (a) InP; (b) InAs: experiment (neutron scattering, open circles; Raman scattering, full circles); (c) InSb: solid lines; numerical (Landolt–Börnstein, 1987).

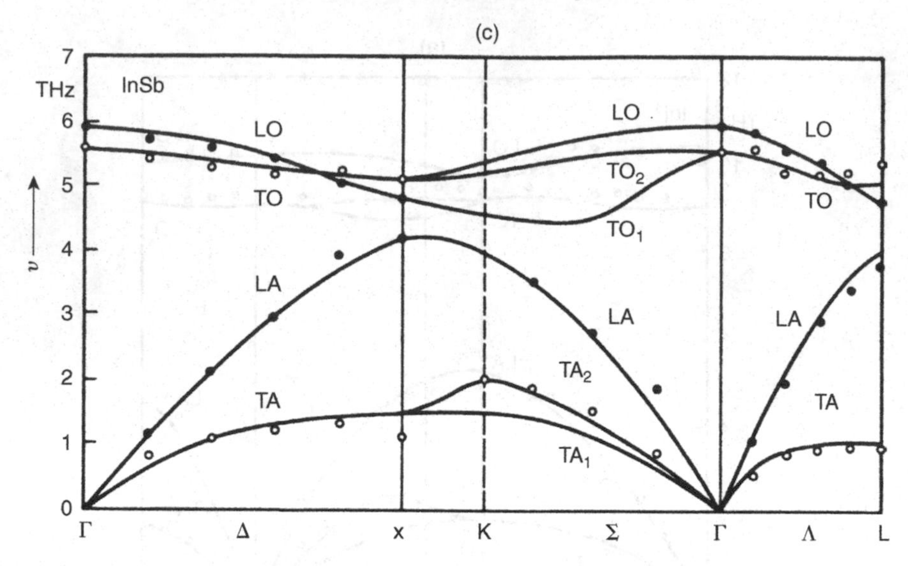

Fig. 4.3 (*cont.*)

An equation of motion relating **E** and **B** can be obtained from Eq. (4.33) by substituting for u from Eq. (4.37). The equation so obtained is just one of Maxwell's equations. From it we can deduce that for a plane wave the electrical energy component of Eq. (4.55) vanishes. The Hamiltonian density is therefore:

$$H = \frac{1}{2}\mu(\dot{\mathbf{u}}^2 + \omega^2\mathbf{u}^2) \tag{4.56}$$

which has the purely mechanical form and applies to LO, TO and interface modes. Quantization is now staightforward.

Another approach which has been used for interface polaritons adopts the spirit of the dielectric-continuum model and considers only the electromagnetic energy density, using the expression for a dispersive medium given by (e.g. Landau and Lifshitz, 1984):

$$H_{EM} = \frac{1}{2}\left[\frac{\partial(\omega\varepsilon)}{\partial\omega}E^2 + \frac{\partial(\omega/\mu_0)}{\partial\omega}B^2\right] \tag{4.57}$$

where **B** is the magnetic field. Care must be taken in interpreting this expression. If spatial dispersion is assumed, this energy-density vanishes for LO and TO modes. The LO modes do not possess electromagnetic energy because $\varepsilon = 0$ and $B = 0$. As a consequence of the dependence of ω on k, $\partial\varepsilon/\partial\omega = 0$, and $E = 0$ for TO modes, so once again the electrical energy disappears. In order to obtain something sensible, spatial dispersion must be ignored, and ω and k taken as independent parameters. Thus, as for unretarded interface modes, the permittivity function is:

$$\varepsilon = \varepsilon_{\infty} \frac{\omega_{\text{LO}}^2 - \omega^2}{\omega_{\text{TO}}^2 - \omega^2} \tag{4.58a}$$

and hence

$$\omega \frac{d\varepsilon}{d\varepsilon} E^2 = \frac{2\omega^2 e^{*2} u^2}{\varepsilon_{\infty}(\omega_{\text{LO}}^2 - \omega_{\text{TO}}^2)} = 2\mu\omega^2 u^2 \tag{4.58b}$$

which is obtained using Eq. (4.31) and the Lyddane–Sachs–Teller relation, Eq. (4.42). In the case of a frequency independent permeability Eq. (4.57) becomes

$$H_{EM} = \mu\omega^2 u^2 + \frac{1}{2}\left(\varepsilon E^2 + \frac{1}{\mu_0} B^2\right) = \mu\omega^2 u^2 + \varepsilon E^2 \tag{4.58c}$$

which is just the sum of mechanical and electromagnetic energy densities for a travelling wave. (Note that for a travelling wave, $B^2 = E^2(k/(\omega)^2$ and potential and kinetic contributions are equal.) Thus in this scheme, the electrical energy vanishes for LO and TO modes as before, but interface modes have a finite electric component of the electrical-energy density, in contradiction to our previous conclusion.

This approach based purely on electromagnetic considerations is rather unsatisfactory. This is hardly surprising since its Hamiltonian is not rigorously derived from the Lagrangian for the system. Nevertheless, the differences turn out to be small in practice. The ratio of electromagnetic to mechanical energy densities can be shown to be very small, so in effect the conclusion that the energy density is purely mechanical in the unretarded limit is, in practice, not violated by the approach. In any case, spatial integration can be shown to remove the interface electrical-energy density entirely.

Other treatments may be found in the literature, where it is often the case that Hamiltonian densities have been either merely quoted or have been derived somewhat informally. Our final expression for the Hamiltonian density in Eq. (4.56) has been derived rigorously from the Lagrangian of Eq. (4.27–29) and it can be confidently used in the quantization procedure.

Standard theory of field quantization expresses the relative displacement field $\mathbf{u}$ in terms of phonon annihilation and creation operators as follows:

$$\mathbf{u} = \frac{1}{\sqrt{V}} \sum_{\mathbf{k}} \left(\frac{\hbar}{2\mu\omega}\right)^{1/2} \left(\mathbf{e} e^{i\mathbf{k}.\mathbf{r}} a_{\mathbf{k}} + \mathbf{e} * e^{-i\mathbf{k}.\mathbf{r}} a_{\mathbf{k}}^{\dagger}\right) \tag{4.59}$$

where **e** is a unit polarization vector, and N is the number of unit cells. The conjugate momentum is

$$\pi = \frac{1}{\sqrt{V}} \sum_{\mathbf{k}} \left(\frac{\hbar \omega \mu}{2} \right)^{1/2} \left(-\mathbf{e} e^{i\mathbf{k}.\mathbf{r}} a_{\mathbf{k}} + \mathbf{e} * e^{-i\mathbf{k}.\mathbf{r}} a_{\mathbf{k}}^{\dagger} \right) \quad (4.60)$$

and the Hamiltonian density operator is

$$H = \frac{\pi^2}{2\mu} + \frac{f}{2} k^2 u^2 \quad (4.61)$$

where f is a force constant such that $k^2 f = \mu\omega^2$. Inserting Eq. (4.59) and Eq. (4.60) in Eq. (4.61) leads to the Hamiltonian

$$\mathcal{H} = \int H d\mathbf{r} = \sum_{\mathbf{k}} \hbar\omega \left(a_{\mathbf{k}}^{\dagger} a_{\mathbf{k}} + \frac{1}{2} \right) \quad (4.62)$$

where the relation ${}_k a_{k'}^{\dagger} - a_k^{\dagger} a_{k'} = \delta_{kk'}$ has been used.

4.5 Two-Mode Alloys

Polar-optical modes in mixed crystals have received the attention of a number of authors (e.g., Chang and Mitra, 1971; Kim and Spitzer, 1979; Nash *et al.*, 1987). In the alloy $A_xB_{1-x}C$ the two modes will be AC-like (label 1) and BC-like (label 2), but the AC-like mode will involve some motion of B ions and, likewise, the BC-like mode will involve some motion of A ions (Figs. 4.4 and 4.5). If the frequencies in the corresponding binary compounds are very different, this mixing will be small, and to a first approximation we can assume that the polarizabilities of the bonds are those of the respective binaries. In a binary compound the equation of motion, Eq. (4.41), neglecting the effect of macroscopic elastic stresses, can be written

$$\mu \frac{\partial^2 \mathbf{u}}{\partial t^2} = -\mu\omega_{\mathrm{TO}}^2 \mathbf{u} + e^* \mathbf{E} \quad (4.63)$$

where **u** is the relative ionic displacement, μ is the reduced mass density, e^* is the ionic charge density and **E** is the electric field. The electrostatic equations are

$$\mathbf{D} = \varepsilon\mathbf{E} = \varepsilon_0\mathbf{E} + \mathbf{P} \quad (4.64)$$

where **D** is the electric displacement, ε is the permittivity and **P** is the polarization given by

$$\mathbf{P} = (\varepsilon_\infty - \varepsilon_0)\mathbf{E} + e^*\mathbf{u} \quad (4.65)$$

where ε_∞ is the high-frequency permittivity. From these equations we can deduce a lattice polarizability χ, defined by $\mathbf{P}_L = \varepsilon_0 \ \chi\mathbf{E}$ where $\mathbf{P}_L = e^* \ \mathbf{u}$ as follows:

$$\chi = \frac{e^{*2}}{\mu\varepsilon_0(\omega_{\mathrm{TO}}^2 - \omega^2)} \tag{4.66}$$

From Eqs. (4.64) and (4.65)

$$\mathbf{E} = -\frac{e^*\mathbf{u}}{\varepsilon_\infty - \varepsilon} \tag{4.67}$$

At zero frequency, from Eq. (4.63), $\mathbf{u} = e^*\mathrm{E}/\mu\omega_{\mathrm{TO}}^2$, whence, from Eq. (4.67)

$$e^{*2} = \mu\omega_{\mathrm{TO}}^2(\varepsilon_s - \varepsilon_\infty) = \mu\varepsilon_\infty(\omega_{\mathrm{LO}}^2 - \omega_{\mathrm{TO}}^2) \tag{4.68}$$

where ε_s is the static permittivity, and we have used the Lyddane–Sachs–Teller relation. Thus, the polarizability can be written

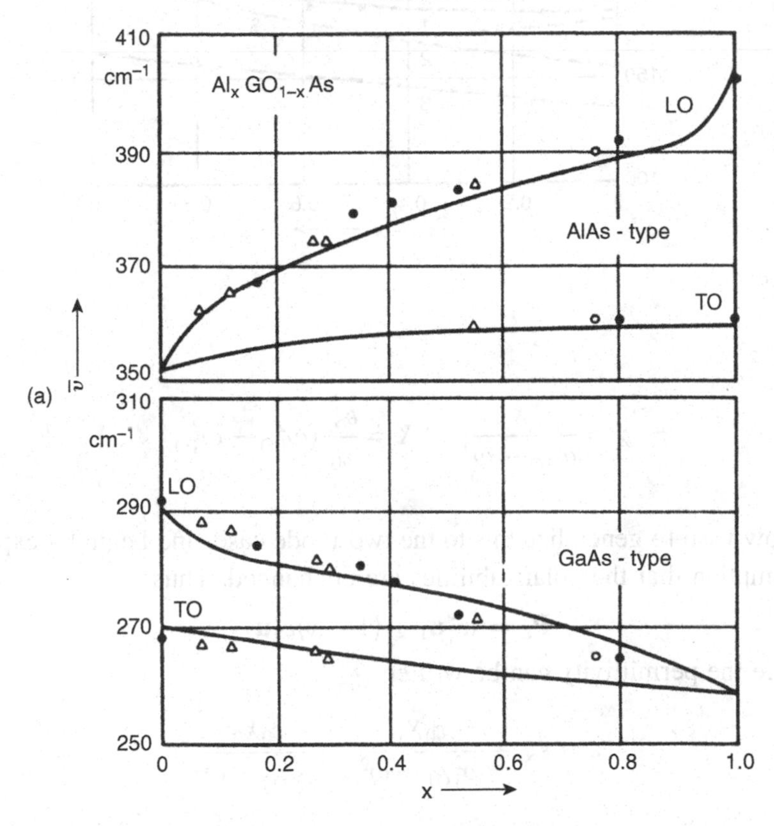

Fig. 4.4 Composition dependence in tertiary alloys of LO and TO zone-centre frequencies: (a) AlGaAs; (b) GaInAs; (c) GaInSb (Landolt–Börnstein, 1987).

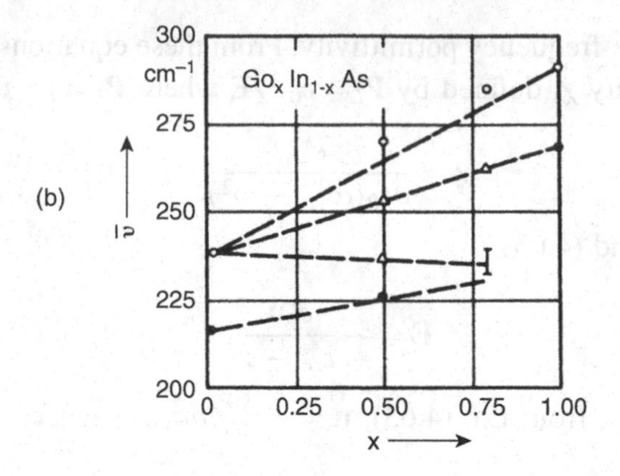

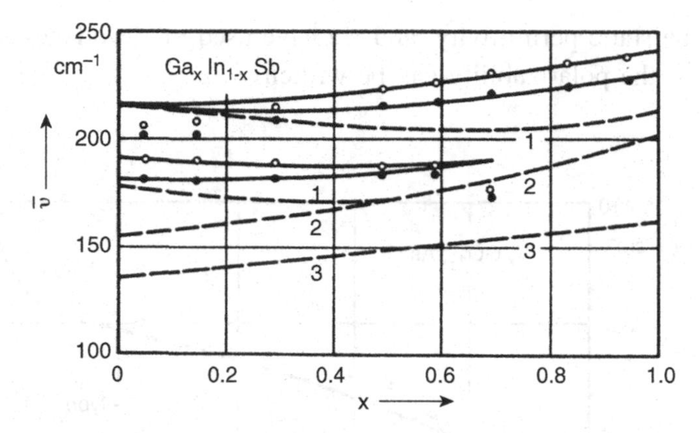

Fig. 4.4 (*cont.*)

$$\chi = \frac{X}{\omega_{\mathrm{TO}}^2 - \omega^2}, \quad X = \frac{\varepsilon_\infty}{\varepsilon_0}(\omega_{\mathrm{LO}}^2 - \omega_{\mathrm{TO}}^2) \tag{4.69}$$

We now wish to generalize this to the two-mode case, and begin by exploiting our assumption that the polarizabilities are unchanged. Thus

$$\mathbf{P}_L = xe_1^*\mathbf{u}_1 + (1-x)e_2^*\mathbf{u}_2 \tag{4.70}$$

and hence the permittivity can be written

$$\varepsilon = \varepsilon_\infty + \frac{\varepsilon_0 X_1}{\omega_{\mathrm{TO1}}^2 - \omega^2} + \frac{\varepsilon_0 X_2}{\omega_{\mathrm{TO2}}^2 - \omega^2} \tag{4.71}$$

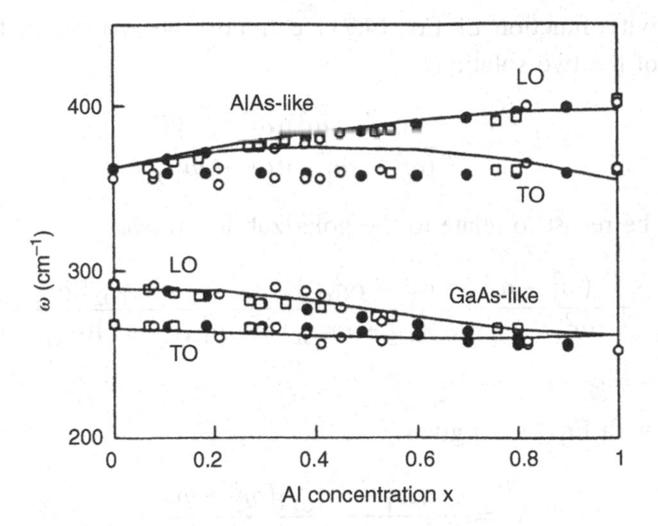

Fig. 4.5 LO and TO frequencies in AlGaAs calculated by microscopic model. Experimental data points indicated (Lee *et al.*, 1995).

where

$$X_1 = x\frac{\varepsilon_{\infty 1}}{\varepsilon_0}(\omega_{LO1}^2 - \omega_{TO1}^2)$$
$$X_2 = (1-x)\frac{\varepsilon_{\infty}}{\varepsilon_0}(\omega_{LO2}^2 - \omega_{TO2}^2) \tag{4.72}$$

The LO frequencies are now determined by the quadratic equation obtained from Eq. (4.71) by putting $\varepsilon = 0$. The solutions are

$$\omega^2 = \frac{1}{2}\left[\omega_{TO1}^2 + X_1' + \omega_{TO2}^2 + X_2' \pm \left\{(\omega_{TO1}^2 + X_1' - \omega_{TO2}^2 - X_2')^2 + 4X_1'X_2'\right\}^{1/2}\right] \tag{4.73}$$

where $X' = X\ \varepsilon_0/\varepsilon_\infty$. When the term 4 X_1', X_2' is negligible compared with the frequency-difference term the two solutions are approximately

$$\omega_1^2 = \omega_{TO1}^2 + X_1'$$
$$\omega_2^2 = \omega_{TO2}^2 + X_2' \tag{4.74}$$

Note that the high-frequency permittivity of the alloy is often connected simply with the corresponding permittivities of the binaries via the linear relation

$$\varepsilon_\infty = x\varepsilon_{\infty 1} + (1-x)\varepsilon_{\infty 2} \tag{4.75}$$

The permittivity function of Eq. (4.71) can now be written in terms of the frequencies of the two solutions:

$$\varepsilon = \varepsilon_\infty \frac{(\omega^2 - \omega_1^2)(\omega^2 - \omega_2^2)}{(\omega^2 - \omega_{\mathrm{TO1}}^2)(\omega^2 - \omega_{\mathrm{TO2}}^2)} \tag{4.76}$$

and this can be recast to relate to the polarizability factors

$$\varepsilon = \varepsilon_\infty \left(1 + \frac{(\omega_1^2 - \omega_{\mathrm{TO1}}^2)(\omega_2^2 - \omega_{\mathrm{TO1}}^2)}{(\omega_{\mathrm{TO1}}^2 - \omega^2)(\omega_{\mathrm{TO2}}^2 - \omega_{\mathrm{TO1}}^2)} - \frac{(\omega_1^2 - \omega_{\mathrm{TO2}}^2)(\omega_2^2 - \omega_{\mathrm{TO2}}^2)}{(\omega_{\mathrm{TO2}}^2 - \omega^2)(\omega_{\mathrm{TO2}}^2 - \omega_{\mathrm{TO1}}^2)}\right) \tag{4.77}$$

Comparison with Eq. (4.75) gives

$$X_1 = \frac{\varepsilon_\infty}{\varepsilon_0} \frac{(\omega_1^2 - \omega_{\mathrm{TO1}}^2)(\omega_2^2 - \omega_{\mathrm{TO1}}^2)}{(\omega_{\mathrm{TO2}}^2 - \omega_{\mathrm{TO1}}^2)} \tag{4.78a}$$

$$X_2 = \frac{\varepsilon_\infty}{\varepsilon_0} \left(\frac{(\omega_{\mathrm{TO2}}^2 - \omega_1^2)(\omega_2^2 - \omega_{\mathrm{TO2}}^2)}{(\omega_{\mathrm{TO2}}^2 - \omega_{\mathrm{TO1}}^2)}\right) \tag{4.78b}$$

Note that Eq. (4.76) implies the following generalization of the Lyddane–Sachs–Teller relation:

$$\frac{\varepsilon_s}{\varepsilon_\infty} = \prod_i \frac{\omega_{\mathrm{LOi}}^2}{\omega_{\mathrm{TOi}}^2} \tag{4.79}$$

We will need to refer to the discussion in this section when, in Chapter 8, we look at the interaction with an electron.

5

Optical Modes in a Quantum Well

"But they were *in* the well," Alice said to the Dormouse. . .
"Of course they were," said the Dormouse, "well in."
Alice in Wonderland, *L. Carroll*

5.1 Non-Polar Material

A very simple case is obtained by assuming that the layer is confined by infinitely rigid material. This corresponds roughly to quantum-well systems where the materials making up the well and the barriers have very different zone-centre optical-mode frequencies, such as Ge/Si (Fig. 5.1). In such a case the appropriate, and largely uncontroversial, mechanical boundary condition is

$$u = 0, \qquad z = \pm\frac{a}{2} \tag{5.1}$$

This condition can be satisfied without hybridization by the s-polarized TO mode (Fig. 5.2), but now the mode pattern is (with the usual coordinate system)

$$u_y(z) = A\sin k_z\left(\frac{z-a}{2}\right) \tag{5.2}$$

with $k_z a = n\pi$, and it is symmetric for n odd, antisymmetric for n even.

For p-polarized modes the boundary condition cannot be satisfied by either TO or LO mode on its own. We therefore take the linear combination

$$u_x(z) = \frac{1}{2}k_x(Ae^{ik_L z} + Be^{-ik_L z}) + \frac{1}{2}k_T(Ce^{ik_T z} + De^{-ik_T z}) \tag{5.3a}$$

$$u_z(z) = \frac{1}{2}k_L(Ae^{ik_L z} - Be^{-ik_L z}) - \frac{1}{2}k_x(Ce^{ik_T z} - De^{-ik_T z}) \tag{5.3b}$$

where k_L, k_T are the z-components of the wavevector component in the z-direction for the LO and TO modes, respectively. The amplitudes have been

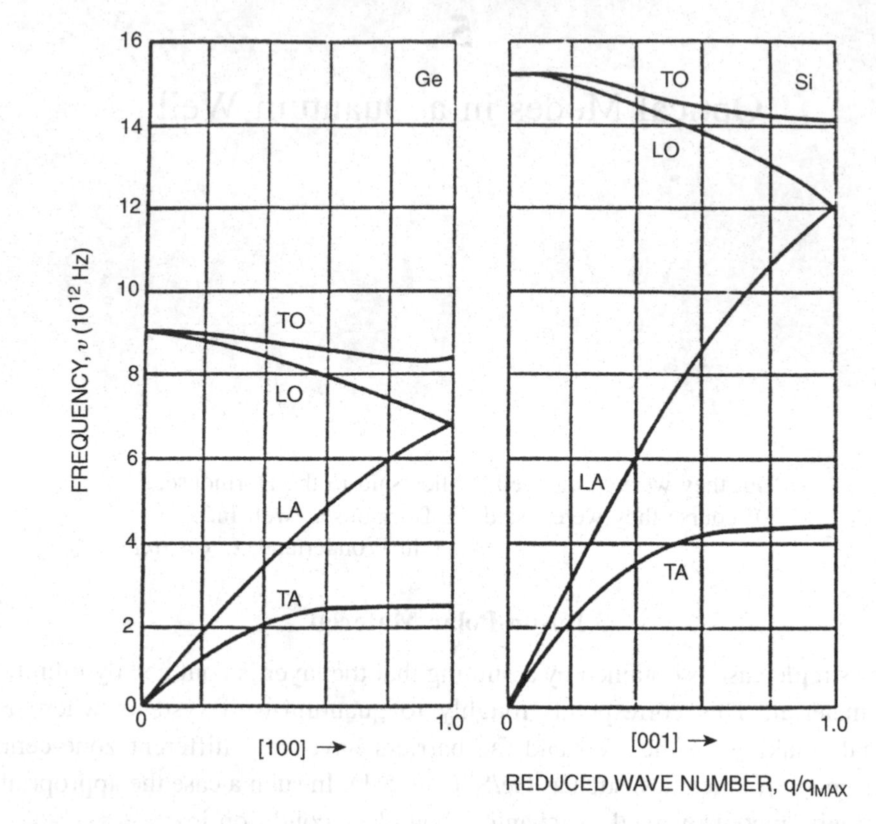

Fig. 5.1 Measured phonon spectra in Ge, Si (Brockhouse and Iyengar, 1958; Brock-house, 1959).

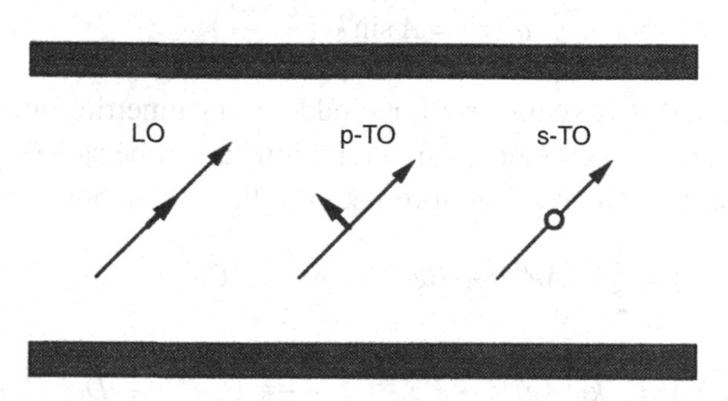

Fig. 5.2 *s*- and *p*-modes.

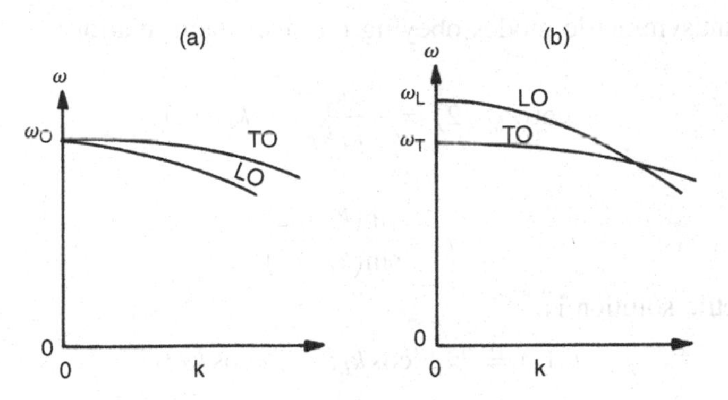

Fig. 5.3 General forms of LO and TO dispersion: (a) non-polar material; (b) polar material.

modified by the wavevector components in order to make explicit that $\nabla \times \mathbf{u} = \mathbf{0}$ for the LO mode and $\nabla . \mathbf{u} = 0$ for the TO mode. Note that the common frequency and in-plane wavevector component imply a relation between k_L and k_T of the form

$$\omega^2 = \omega_{TO}^2 - v_L^2(k_x^2 + k_T^2) = \omega_{TO}^2 - v_T^2(k_x^2 + k_T^2) \tag{5.4}$$

where ${v_L}^2 = c_L/\mu$ and ${v_L}^2 = c_T/\mu$. The general form of the dispersion is depicted in Fig. 5.3(a).

Applying the boundary condition leads to four equations for the four amplitudes. Solutions exist provided the following dispersion relation is satisfied:

$$(qr + pt)^2 \sin k_T a \sin k_L a - 2qrpt[\cos[(k_L - K_T)a] - 1] = 0 \tag{5.5}$$

or alternatively

$$\begin{aligned}&\left(qr \sin\left(\frac{k_L a}{2}\right) \cos\left(\frac{k_T a}{2}\right) + pt \cos\left(\frac{k_L a}{2}\right) \sin\left(\frac{k_T a}{2}\right)\right)\\ &\times\left(qr \cos\left(\frac{k_L a}{2}\right) \sin\left(\frac{k_T a}{2}\right) + pt \sin\left(\frac{k_L a}{2}\right) \cos\left(\frac{k_T a}{2}\right)\right) = 0\end{aligned} \tag{5.6}$$

the parameters p, q, r and t have the values

$$p = k_x, \quad q = -k_T, \quad r = k_L, \quad t = -k_x \tag{5.7}$$

The guided-wave patterns are as follows:

$$u_x(z) = ik_x A(\sin k_L z - s_T \sin k_T z) \tag{5.8a}$$

$$u_z(z) = k_L A\left(\cos k_L z + \frac{k_x^2}{k_L k_T} s_T \cos k_T z\right) \tag{5.8b}$$

These are antisymmetric modes obeying the dispersion relation

$$\cot(k_L a/2) = -\frac{k_x^2}{k_L k_T}\cot(k_T a/2) \tag{5.9}$$

with

$$s_T = \frac{\sin(k_L a/2)}{\sin(k_T a/2)} \tag{5.10}$$

The symmetric solution is

$$u_x(z) = ik_x A(\cos k_L z - c_T \cos k_T z) \tag{5.11a}$$

$$u_z(z) = -k_L A\left(\sin k_L z + \frac{k_x^2}{k_L k_T} c_T \sin k_T z\right) \tag{5.11b}$$

$$\tan(k_L a/2) = -\frac{k_x^2}{k_L k_T}\tan(k_T a/2) \tag{5.11c}$$

$$c_T = \frac{\cos(k_L a/2)}{\cos(k_T a/2)} \tag{5.11d}$$

In all cases the wavevectors k_L and k_T are related according to Eq. (5.4). Fig. 5.4 illustrates some mode patterns.

Examining the possibility of interface waves, we put $k_L = \mathrm{i}\alpha_L$ and $k_T = \mathrm{i}\alpha_T$. A solution that satisfies the dispersion relations and the frequency condition is $\alpha_L = \alpha_T = \pm k_x$, but that is a null mode, i.e. $\mathbf{u} = 0$ for all z. Thus, interface waves do not exist, a result that is not surprising in view of a boundary condition that forbids motion at the interface. Rigid boundaries thus eliminate two of the possible modes of long-wavelength vibration.

When $k_x \rightarrow 0$ the antisymmetric mode is characterized roughly by $k_L a/2 = (2n_L - 1)\,\pi/2$ or $k_T a/2 = n_T \pi$, and the symmetric mode by $k_L a/2 = n_T \pi$ or $k_T a/2 = (2n_T - 1)\,\pi/2$, corresponding to almost pure LO or TO modes.

5.2 Polar Material

In polar material it is necessary to satisfy both mechanical and electromagnetic continuity conditions at each interface (Ridley, 1992; Trallero–Giner *et al.*, 1992). The condition $\mathbf{u} = 0$ and the electromagnetic connection rules can be satisfied by Eq. (5.2) for the *s*-polarized TO mode and by a linear combination of LO, TO and IP for the *p*-polarized modes. For frequencies above ω_{TO} the *p*-polarized TO mode is evanescent, and the solution for *p*-polarized modes,

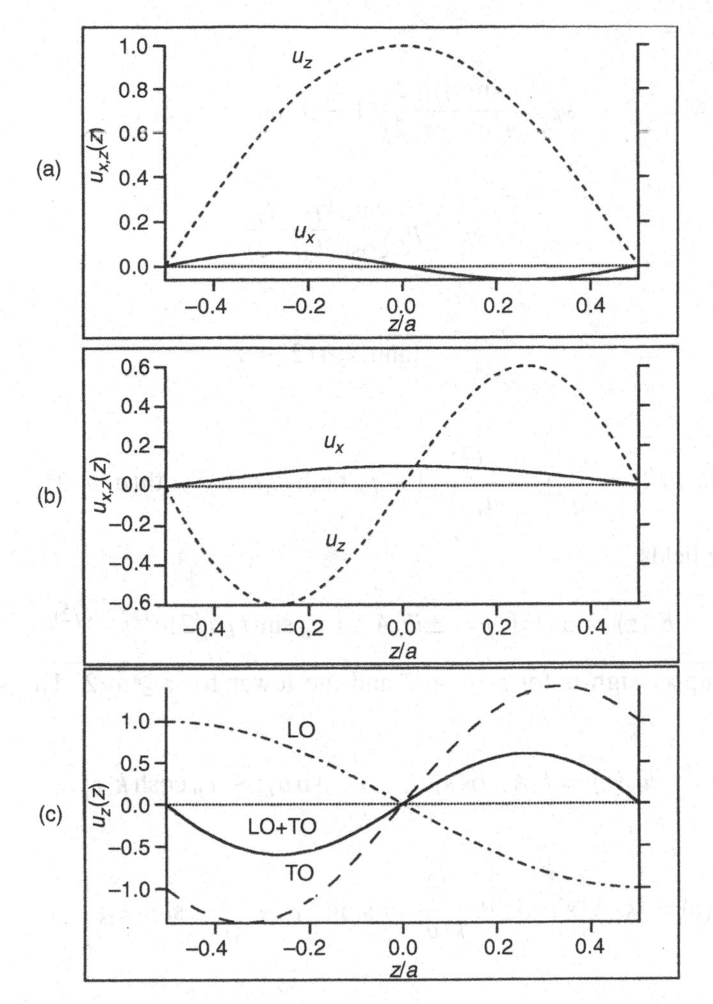

Fig. 5.4 Confined non-polar modes ($u(\pm\, a/2) = 0$, $k_x \rightarrow 0$): (a) antisymmetric; (b) symmetric; (c) LO and TO components of the symmetric $u_z(z)$. The modes are LO-like in the sense that $k_L a \approx \pi$ (asymmetric), 2π (symmetric); equivalent TO-like modes would have $k_T a \approx 2\pi$ and π.

given here, is for the case when the LO dispersion is strong enough to cover the frequency range of the interface polaritons. The antisymmetric solution is

$$u_x(z) = ik_x A(\sin k_L z - s_T \sinh a_T z - s_p \sinh k_x z) \tag{5.12a}$$

$$u_z(z) = k_L A\left(\cos k_L z - \frac{k_x^2}{k_L a_T} s_T \cosh a_T z - \frac{k_x}{k_L} s_p \cosh k_x z\right) \tag{5.12b}$$

where

$$s_T = \frac{\sin(k_L a/2)}{\sin(\alpha_T a/2)}(1 - p_a \tan hk_x a/2) \tag{5.12c}$$

$$s_p = p_a \frac{\sin(k_L a/2)}{\cosh(k_x a/2)} \tag{5.12d}$$

$$p_a = \frac{1}{s(\tanh(k_x a/2) + r)} \tag{5.12e}$$

with

$$\cot(k_L a/2) = \frac{k_x}{k_L} p_a + \frac{k_x^2}{\alpha_T k_L}(1 - p_a \tanh(k_x a/2)) \coth(\alpha_T a/2) \tag{5.13}$$

and barrier fields

$$E_x(z) = \pm i E_z(z) = \pm i k_x A r s \rho_0 p_a \sin(k_L a/2) e^{\pm k_x (z \pm a/2)} \tag{5.14}$$

where the upper sign is for $z \leq -a/2$ and the lower for $z \geq a/2$. The symmetric solution is

$$u_x(z) = k_x A\left[\cos k_L z - c_T \cosh \alpha_T z - c_p \cosh k_x z\right] \tag{5.15a}$$

$$u_z(z) = i k_L A\left[\sin k_L z + \frac{k_x^2}{k_L \alpha_T} c_T \sinh \alpha_T z + \frac{k_x}{k_L} c_p \sinh k_x z\right] \tag{5.15b}$$

where

$$c_T = \frac{\cos(k_L a/2)}{\cosh(\alpha_T a/2)}(1 - p_s \coth(k_x a/2)) \tag{5.16a}$$

$$c_p = \frac{p_s \cos(k_L a/2)}{\sinh(k_x a/2)} \tag{5.16b}$$

$$p_s = \frac{1}{s(\cos(k_L a/2) + r)} \tag{5.16c}$$

with

$$\tan k_L a/2 = -\frac{k_x}{k_L} p_s - \frac{k_x^2}{\alpha_T k_L}(1 - p_s \coth(k_x a/2)) \tanh(\alpha_T a/2) \tag{5.17}$$

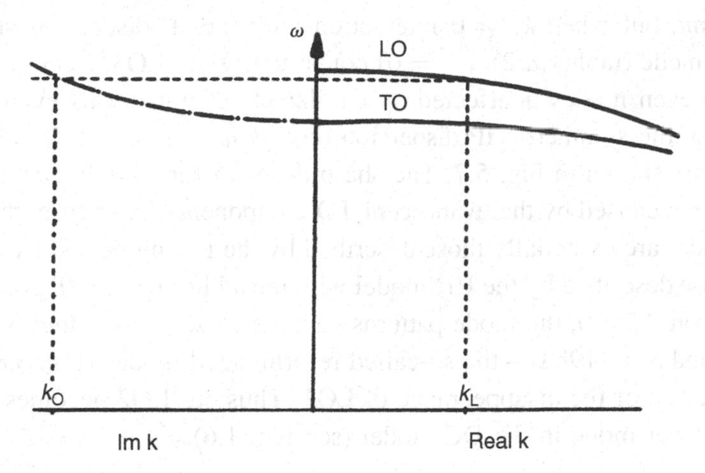

Fig. 5.5 LO and TO dispersion showing that the TO component of the hybrid is highly localized near the interface (a_T»k_L).

and barrier fields

$$E_x(z) = \pm iE_z(z) = -k_x Ars\rho_0 p_s \cos(k_L a/2)e^{\pm k_x(z\pm a/2)} \tag{5.18}$$

where, once more, the upper sign is for $z \leq -a/2$, the lower for $z \geq a/2$. In these equations the field, permittivity and charge parameters are, as before

$$s = \frac{\omega^2 - \omega_{TO}^2}{\omega_{LO}^2 - \omega_{TO}^2}, \quad r = \frac{\varepsilon_\infty}{\varepsilon_B}\frac{\omega^2 - \omega_{LO}^2}{\omega^2 - \omega_{TO}^2}, \quad \rho_0 = \frac{e^*}{\varepsilon_\infty} \tag{5.19}$$

where ε_B is the permittivity (assumed to be frequency-independent consistent with the assumption of rigidity) of the barrier. The amplitude A is determined by the quantization procedure of Section 5.5.

For frequencies below ω_{TO} the TO wavevector becomes real and we must substitute $a_T \rightarrow ik_T$ in the above equations. a_T is related to k_L via the frequency condition; see Fig. 5.3(b) and Fig. 5.5.

$$\omega^2 = \omega_{LO}^2 - v_L^2(k_x^2 + k_L^2) = \omega_{TO}^2 - v_T^2(k_x^2 + a_T^2) \tag{5.20}$$

and the unretarded limit is taken for the interface polariton.

As in the case of non-polar material, rigid boundaries suppress pure interface modes. Interface polaritons appear only as coherent components of a hybrid. The case of weak dispersion, where this allows interface modes to appear virtually unadorned, is discussed in Section 5.4.

Dispersion is described by Eqs. (5.13) and (5.17), and shown for GaAs quantum wells in Fig. 5.6. For $k_x = 0$ the confined LO wavevectors are described by $k_L a = n\pi$, but when $k_x \neq 0$ interaction with the IP dispersion at the antisymmetric mode ($\tanh(k_x a/2) + r = 0$) converts LO1 to LO3, LO3 to LO5, etc., leaving the even modes unaffected. A similar effect transforms even modes in the region of the symmetric IP dispersion ($\coth(k_x a/2) + r = 0$). Mode patterns for $k_x \approx 0$ are shown in Fig. 5.7. The sharp drops to zero exhibited by u_x of the odd modes are caused by the evanescent TO component. Apart from this feature, the odd modes are essentially those described by the HD model. The even modes are also those described by the HD model with the addition of an IP component in u_x. Away from $k_x = 0$, the mode patterns become close to the patterns described by Huang and Zhu (1988) – the so-called reformulated modes (Fig. 5.8) – if due account is taken of the disappearance of LO1. Thus our LO2 becomes similar to the lowest-order mode in the DC model (see Fig. 1.6).

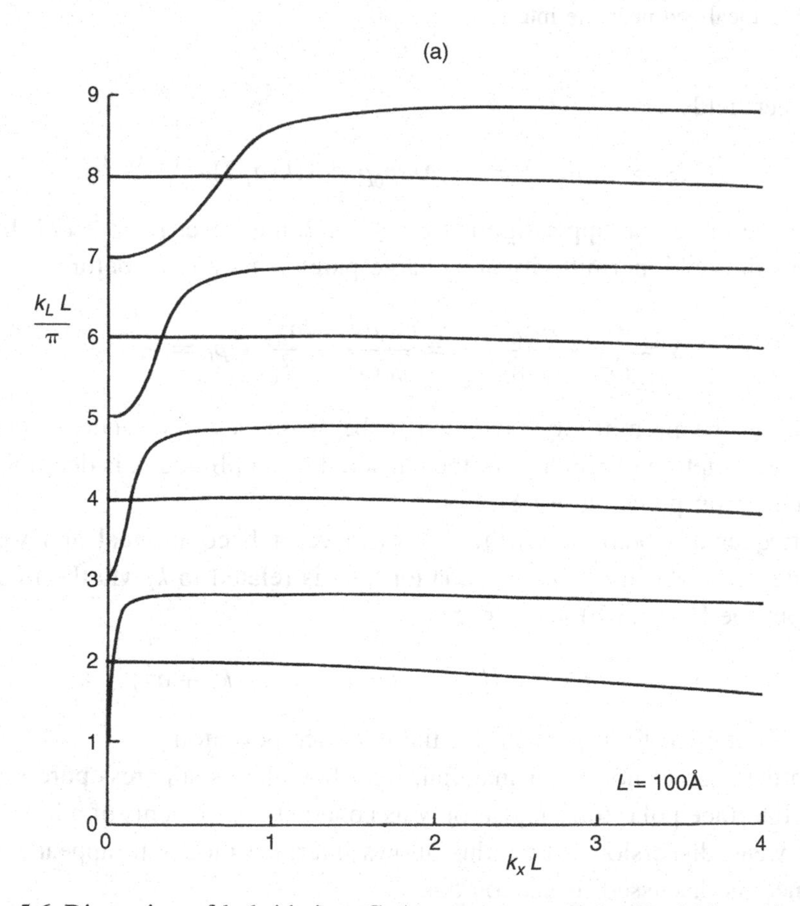

Fig. 5.6 Dispersion of hybrids in a GaAs quantum well (well-width = L): (a) $L = 100$ Å; (b) $L = 50$ Å.

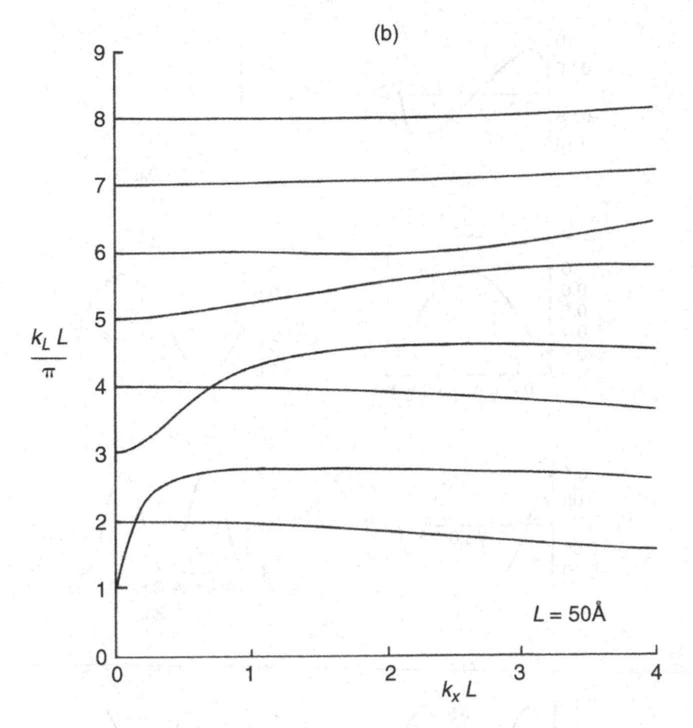

Fig. 5.6 (*cont.*)

The mode patterns depicted in Figs. 5.7 and 5.8 are LO-like in that their frequencies and wavevectors do not coincide with the IP dispersion. When such a coincidence occurs the modes become IP-like (Figs. 5.9 and 5.10). This division of hybrid modes into LO-like and IP-like is, of course, a rough one. Clearly LO, IP and TO features are evident in all modes.

We postpone a discussion of how the hybrid model compares with the results of numerical lattice dynamics, and of experiments, until the next chapter, where superlattice hybrids are described.

5.3 Barrier Modes: Optical-PhononTunnelling

The modes we have been describing are those indigenous to the material that forms the quantum well. The barrier material that surrounds the quantum well also supports optical modes in general, and under certain circumstances these may penetrate the quantum-well material. A full description of optical modes in a quantum well must therefore include those contributed by the barrier material (Ridley, 1994).

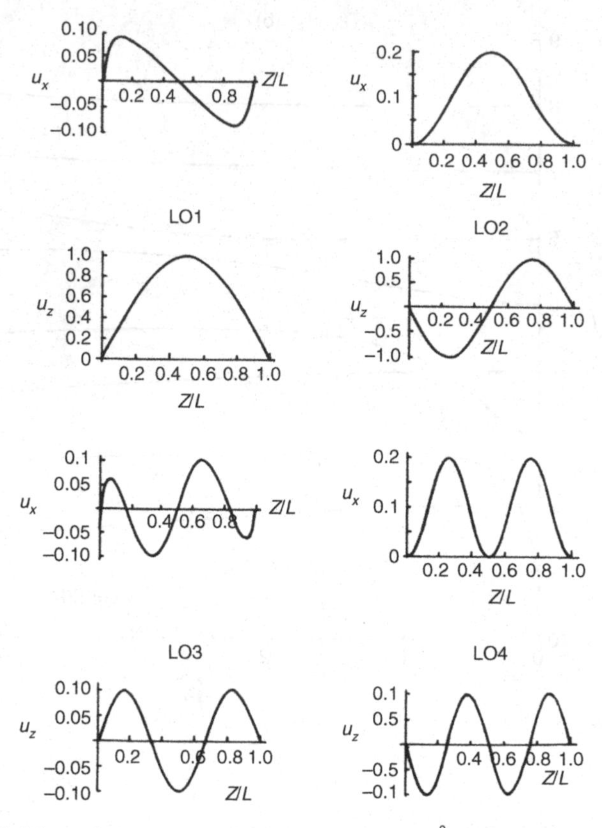

Fig. 5.7 Hybrid-mode patterns for $k_x \approx 0$ (a = 38 Å) ($L = a$ = well-width).

We assume that the material on either side of the well is the same and that it occupies the space from $|z| = a/2$ to ∞. Were the mechanical mismatch relatively small we would expect incident LO and TO bulk barrier modes to produce evanescent waves in the well, which would allow a certain amount of transmission probability. When the mismatch is large, and in the limit when the boundary is absolutely rigid at the barrier-mode frequencies, there can be no penetration if the barrier material is non-polar. Thus, in non-polar material the influence of barrier modes in the well can safely be neglected. The interesting case is when the barrier is polar. Then, although TO modes cannot penetrate significantly, LO modes are able to do so because of their long-range electric fields. Thus in spite of the rigidity of the interfaces an incident LO mode can be transmitted on the other side of the well via the excitation in the well of an interface polariton.

Fig. 5.11 depicts the modes excited by an LO mode incident on the interface at $z = -L/2$ whose relative ionic displacement can be written

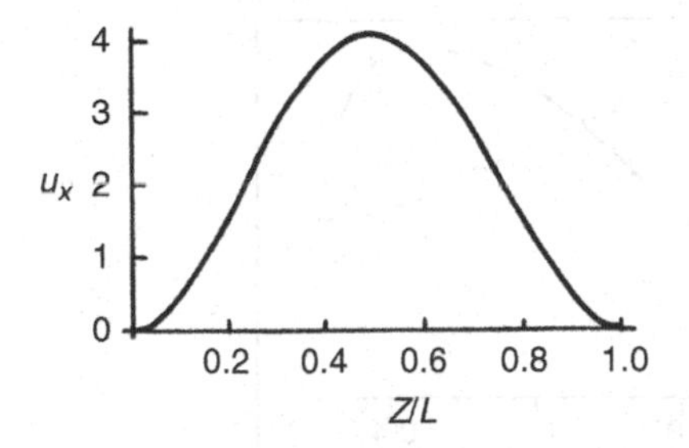

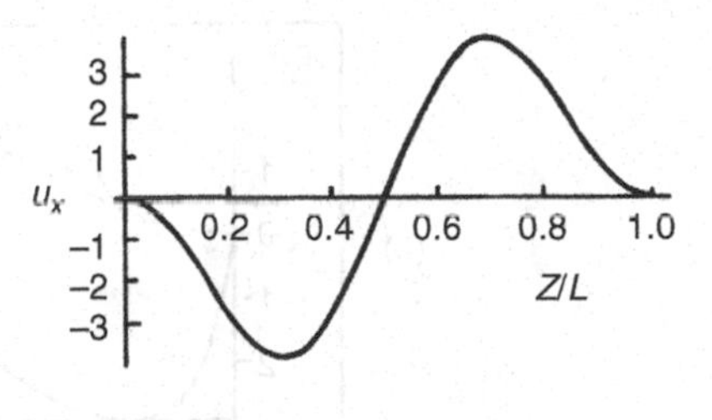

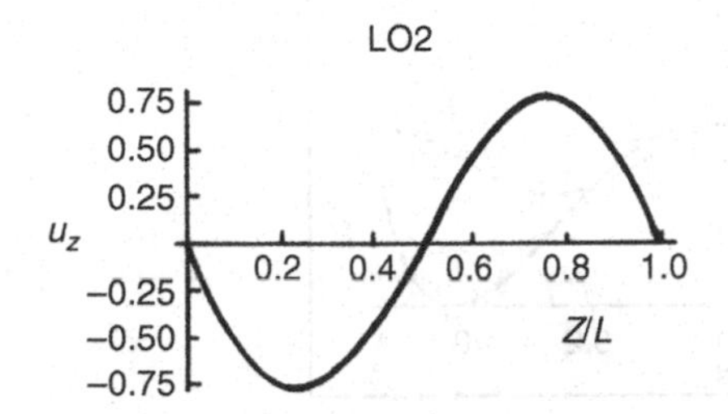

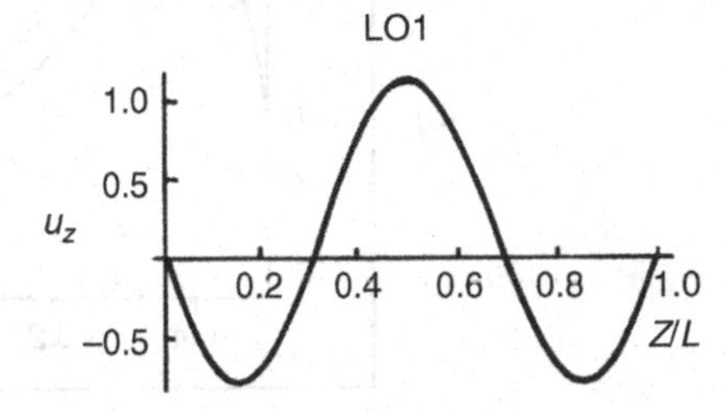

Fig. 5.8 Hybrid-mode patterns for $k_x a = 3$ ($a = 38$ Å). The pattern labelled LO1 is the mode that has converted from $k_L a = \pi$ to $k_L \mathrm{a} \approx 3\pi$ ($L = a =$ well-width).

$$u_L = (k_x, 0, k_L)Ae^{i(k_x x + k_z z - \omega t)} \tag{5.21}$$

This excites a reflected LO mode and evanescent TO and IP modes:

$$u_L = (k_x, 0, -k_L)Be^{i(k_x x - k_z z - \omega t)} \tag{5.22a}$$

$$u_T = (k_T, 0, -ik_x)Ce^{i(k_x x - \omega t) + k_T z} \tag{5.22b}$$

$$u_P = (k_x, 0, -ik_x)De^{i(k_x x - \omega t) + k_x z} \tag{5.22c}$$

and, in the well, IP fields of the form

$$\mathbf{E} = (k_x(Ee^{-k_x z} + Fe^{k_x z}), 0, ik_x(Ee^{-k_x z} - Fe^{k_x z}))\ e^{i(k_x x - \omega t)} \tag{5.23}$$

There are no ionic displacements in the well at the frequency of the barrier mode by hypothesis. For $z \geq \mathrm{L}/2$ the modes are

$$u_L = (k_x, 0, k_L)Ge^{i(k_x x + k_L z - \omega t)} \tag{5.24a}$$

$$u_T = (k_T, 0, ik_x)He^{i(k_x x - \omega t) - k_T z} \tag{5.24b}$$

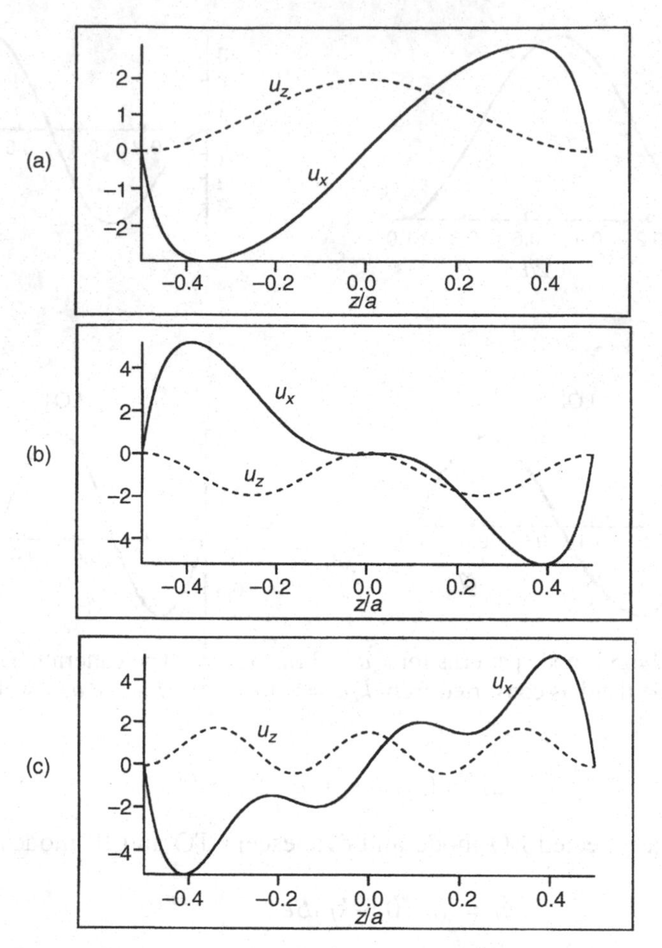

Fig. 5.9 IP-like hybrids for the three highest frequencies in a GaAs quantum well ($a = 50\text{Å}$): (a) $k_La \approx 2\pi$; (b) $k_La \approx 4\pi$; (c) $k_La \approx 6\pi$. All are antisymmetric.

$$u_p = (k_x, 0, ik_x)Ie^{i(k_xx-\omega t)-k_xz} \tag{5.24c}$$

at each boundary $\mathbf{u} = 0$ and the usual electric conditions pertain. This gives eight equations for the eight unknowns. The results are

$$Be^{ik_LL/2} = \left[\left(\frac{\gamma_+\eta_+e^{k_xL} - \gamma_-\eta_-e^{-k_xL}}{d}\right)\right]Ae^{-ik_LL/2} \tag{5.25a}$$

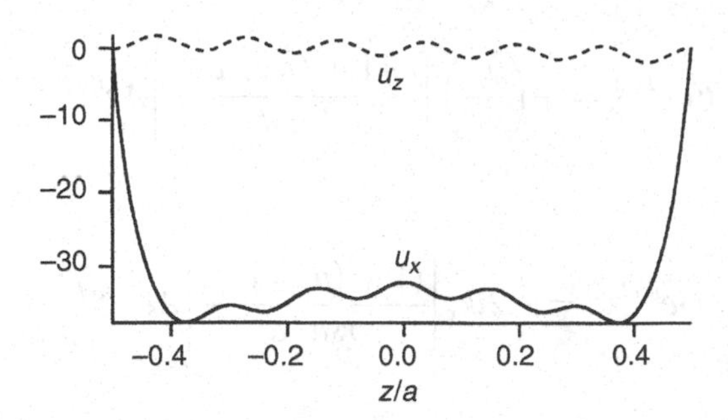

Fig. 5.10 A symmetric IP-like hybrid.

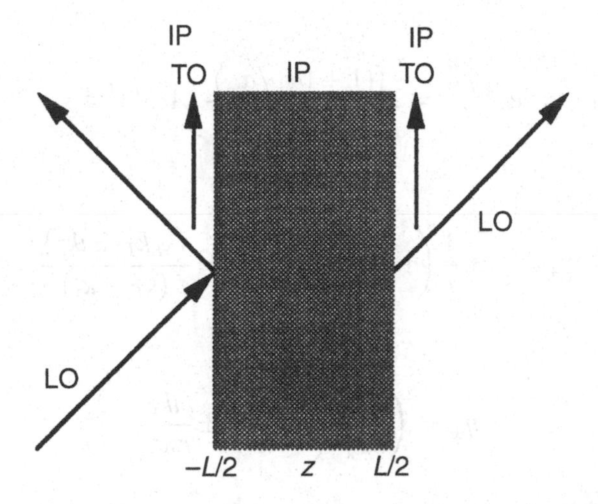

Fig. 5.11 Polar-optical modes generated by an incident barrier LO mode.

$$Ce^{-k_T L/2} = \frac{i\left[k_L + ik_x - (k_L - ik_x)\left(\gamma_+\eta_+ e^{k_x L} - \gamma_-\eta_- e^{-k_x L}\right)/d\right]\left[Ae^{-ik_L L/2}\right]}{(k_T - k_x)} \tag{5.25b}$$

$$De^{-k_x L/2} = \left[\frac{(1 + (a_+/a_-))}{rsd}\right]\left[\gamma_+ e^{k_x L} + \gamma_- e^{-k_x L}\right]Ae^{-ik_L L/2} \tag{5.25c}$$

$$Ee^{-k_x L/2} = -\left[\frac{e_i(1 + (a_+/a_-))}{\varepsilon_\infty d}\right]Ae^{-ik_L L/2} \tag{5.25d}$$

$$Fe^{k_xL/2} = +\left(\frac{\gamma_-}{\gamma_+}\right)\left[\frac{e_i(1+(a_+/a_-))}{\varepsilon_\infty d}\right]Ae^{ik_LL/2} \tag{5.25e}$$

$$Ge^{ik_LL/2} = -2ik_x\left[\frac{(1+(a_+/a_-))}{rsa_-d}\right]Ae^{-ik_LL/2} \tag{5.25f}$$

$$He^{-k_TL/2} = -2\left[\frac{(k_L-ik_x)}{(k_T-k_x)}\right]k_x\left[\frac{(1+(a_+/a_-))}{rsa_-d}\right]Ae^{-ik_LL/2} \tag{5.25g}$$

$$Ie^{-k_xL/2} = 2\left[\frac{(1+(a_+/a_-))}{rsd}\right]Ae^{-ik_LL/2} \tag{5.25h}$$

$$\gamma_\pm = 1 \pm \frac{1}{r}\left(1-\frac{ik_x}{sa_-}\right), \quad a_\pm = \frac{(k_Lk_T \pm ik_x^2)}{(k_T-k_x)} \tag{5.26a}$$

$$\eta_\pm = \left(\frac{a_+}{a_-}\right)\left(1\pm\frac{1}{r}\right)\pm\frac{ik_x}{rsa_-} \tag{5.26b}$$

$$d = \gamma_+^2e^{k_xL} - \gamma_-^2e^{-k_xL} \tag{5.26c}$$

In these expressions there are two frequency-dependent terms r and s. The first is the well-known permittivity factor for the IP mode, viz.:

$$r = r_0\frac{\omega^2-\omega_{\mathrm{LO}}^2}{\omega^2-\omega_{\mathrm{TO}}^2} \tag{5.27}$$

where $r_0 = \varepsilon_\infty/\varepsilon_w$ (ε_∞ is the high-frequency permittivity of the surrounding medium and ε_w is the permittivity of the well), and ω_{LO}, ω_{TO} are the zone-centre LO and TO frequencies. The second is the field factor for IP modes, viz.:

$$s = \frac{\omega^2-\omega_{\mathrm{TO}}^2}{\omega_{\mathrm{LO}}^2-\omega_{\mathrm{TO}}^2} \tag{5.28}$$

Note that an electron in the well will experience a potential given by

$$\begin{aligned}\varphi &= i(Ee^{-k_x z} + Fe^{k_x z})e^{i(k_x x-\omega t)} \\ &= -i\left[\frac{e^*(1+(a_+/a_-))}{\varepsilon_\infty d}\right]\left[\gamma_+ e^{k_x(L/2-z)} - \gamma_- e^{-k_x(L/2-z)}\right]Ae^{-ik_L L/2}e^{i(k_x x-\omega t)}\end{aligned} \tag{5.29}$$

In order to understand what is going on, it is useful to focus attention on the factor d, Eq. (5.26c), that inhabits the denominator of each of the amplitudes. This can be reexpressed as follows:

$$d = 2\sinh k_x L\left[\tanh(k_x L/2) + \frac{1}{r}\left(1 - \frac{ik_x}{sa_-}\right)\right]\left[\coth(k_x L/2) + \frac{1}{r}\left(1 - \frac{ik_x}{sa_-}\right)\right] \tag{5.30}$$

Now d is complex and can never be zero but it can be minimized if either of the conditions

$$\tanh(k_x L/2) + \frac{1}{r} = 0 \tag{5.31a}$$

$$\coth(k_x L/2) + \frac{1}{r} = 0 \tag{5.31b}$$

holds. But these are the well-known dispersion relations for classical IP modes (classical in this context meaning modes satisfying only electrical but not mechanical boundary conditions). This means that transmission will be maximized when the frequency of the incident LO mode is such that it coincides with that of a classical IP mode. The form of the classical IP dispersion and that of the LO mode are depicted in Fig. 5.12(a). A given k_x determines the two frequencies of the classical IP mode and this plus the given k_x determine two values of k_L, the z-component of the LO wave, which satisfies this resonance, Fig. 5.12(b).

More clarification can be had by making the simplification of letting k_T increase without limit, which is equivalent to neglecting the role of shear stress entirely. In real systems this will be a good approximation for all frequencies except those close to ω_{TO}. In these circumstances reflection and transmission coefficients are defined as follows:

$$R = \frac{Be^{ik_L L/2}}{Ae^{-ik_L L/2}} \equiv \frac{1}{d}\left(\left[\left(1+\frac{1}{r}\right)^2 + u^2\right]e^{k_x L} - \left[\left(1-\frac{1}{r}\right)^2 + u^2\right]e^{-k_x L}\right) \tag{5.32}$$

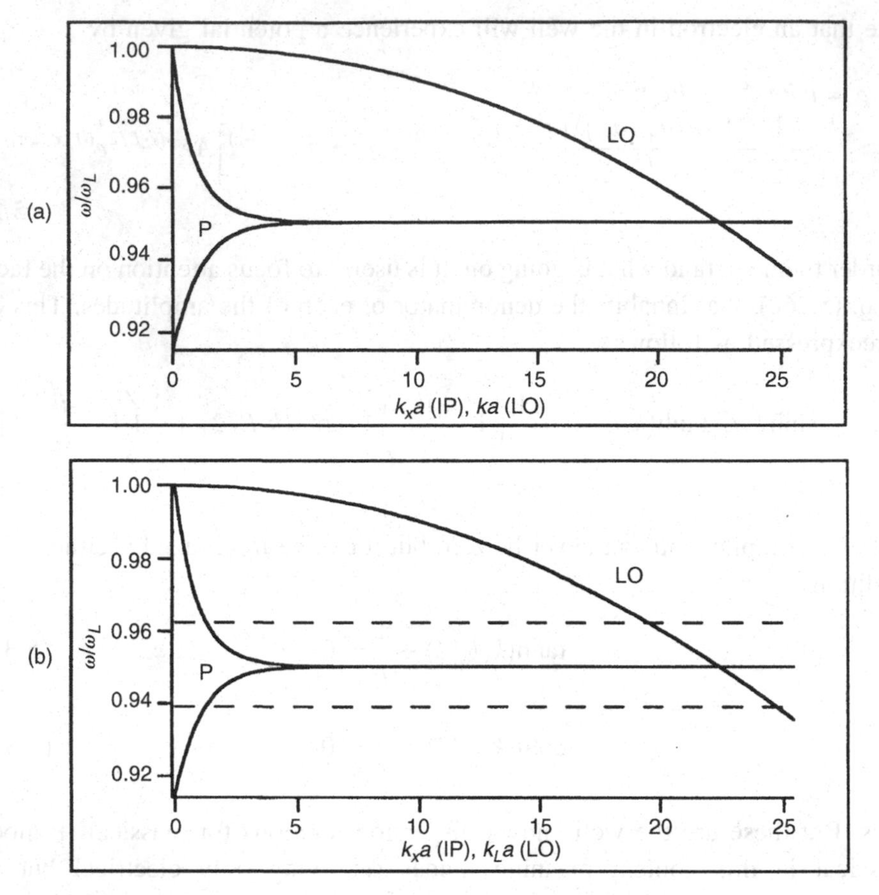

Fig. 5.12 (a) LO and IP dispersion; (b) resonant frequencies for $k_x a = 1.26$.

$$T = \frac{Ge^{ik_L L/2}}{Ae^{-ik_L L/2}} = -\frac{4u}{d} \tag{5.33}$$

where $u = k_x/rsk_L$. Surface wave generation coefficients can be defined as follows:

$$S_1 = \frac{k_x De^{-k_x L/2}}{k_L Ae^{-ik_L L/2}} = -\frac{2u}{d}\left[\sqrt{(1+1/r)^2 + u^2 e^{k_x L}} + \sqrt{(1-1/r)^2 + u^2 e^{ik_x L}}\right] \tag{5.34}$$

$$S_2 = \frac{k_x Ie^{-k_x L/2}}{k_L Ae^{-ik_L L/2}} = \frac{4u}{d} \tag{5.35}$$

For the remote electron–phonon interaction it is useful to have a measure of the coupling strength relative to that of a local electron–phonon interaction. In general, the strength of the Fröhlich coupling $e\varphi$ is inversely proportional to the wavevector, with the latter determined by the conservation of momentum and energy. For a confined electron in the well undergoing an intrasubband transition the relevant wavevector for momentum and energy conservation is k_x but the interaction strength depends upon the total wavevector, i.e. $(K_x^2 + K_1^2)^{1/2}$. In bulk local interactions there is only k_x. Thus we can define a coupling coefficient for the remote interaction as follows:

$$\Gamma = \left[\frac{2k_x}{d\left(k_x^2 + k_L^2\right)^{1/2}}\right]\left[\sqrt{(1+1/r)^2 + u^2 e^{k_xL/2}} - \sqrt{(1-1/r)^2 + u^2 e^{-k_xL/2}}\right] \tag{5.36}$$

using the potential, Eq. (5.29), at the centre of the well.

These results are illustrated in Fig. 5.13 for a purely hypothetical system ($\varepsilon_\infty = 10.8\varepsilon_0$, $v_L = 3.8 \times 10^5$ cm^{-1}, $k_x = 2.52 \times 10^6$ cm^{-1}, $\varepsilon_B = 16\varepsilon_0$, L = 50 Å). The in-plane wavevector, k_x, is fixed and the normal wavevector, k_L, varied. The fixed k_x determines the classical IP frequencies, which in turn determine the values of k_L necessary to achieve 100 percent transmission. In the example chosen, resonance occurs at rather large values of k_L. Increasing the well-width would cause the two resonances to merge at even larger values of k_L. Decreasing the width would increase the separation and increase the coupling strength for the resonance at smaller k_L.

Figure 5.14 shows the electric potential in the well at the two resonances. Relative to the midplane of the well the potential consists of a symmetric and an antisymmetric component. For the resonance at smaller k_L the symmetric component dominates and vice versa for the larger k_L. These symmetries are

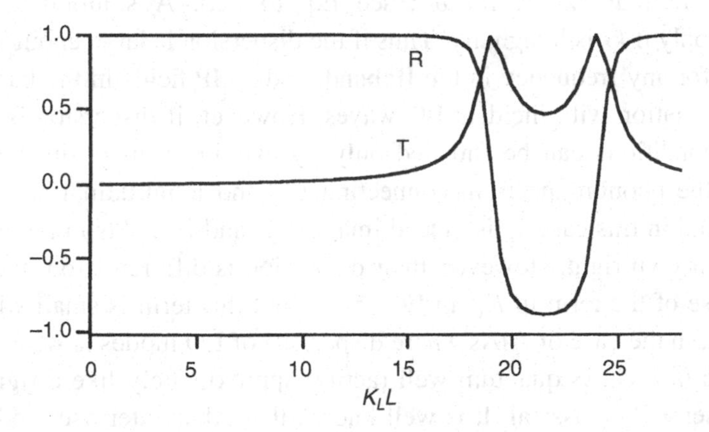

Fig. 5.13 Reflection and transmission coefficients.

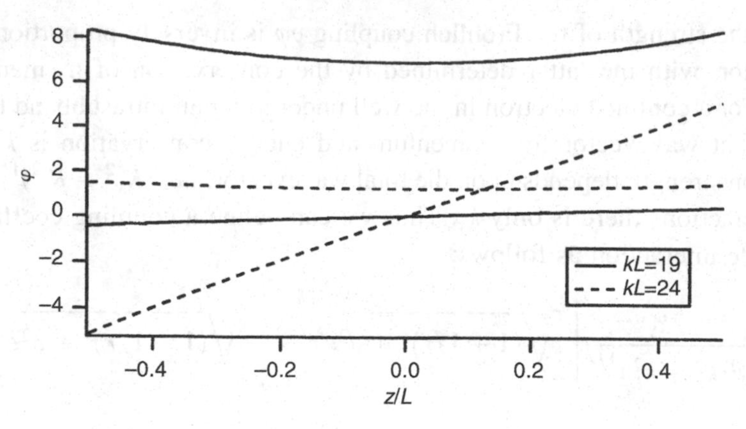

Fig. 5.14 Potential profiles in the well at the two resonances. Each is made up of a symmetric and an antisymmetric part, which are shown.

important for determining intra- and intersubband transitions in a quantum well and they are determined largely by the classical symmetries of the IP modes.

Finally, it is interesting to ask whether IP modes can be supported by the surrounding material without the incidence of LO waves. This situation can be analysed in the same way as before, but now there are only outgoing waves on either side (i.e., $A = 0$). The loss of one variable means that the eight equations resulting from the boundary conditions have solutions only if the following dispersion relationship is satisfied:

$$(\tanh(k_x L/2) + r)(\coth(k_x L/2) + r) + \frac{k_x^2}{s a_-^2} = 0 \qquad (5.37)$$

All terms are real except for a_-; see Eq. (5.26a). A solution is therefore obtainable only if k_L is imaginary. Thus if the dispersion is large enough, k_L is not imaginary for any frequency in the IP band, and so IP fields in the barrier occur only in association with incident LO waves. However, if dispersion is weak, the boundary conditions can be satisfied only by evoking a mode in the complex branch of the phonon spectrum connecting LO and longitudinal acoustic (LA) branches, and in this case k_L is indeed imaginary, and IP fields in the barrier can exist in their own right. However, their dispersion is different from the classical case because of the term in K_x^2 in Eq. (5.39), but this term is small when $|k_L|$ is large. Thus, in the case of AlAs where dispersion of LO modes is weak there will be IP fields in a GaAs quantum well (acting approximately like a rigid barrier) that are essentially classical. It is well-known that AlAs interface modes indeed contribute strongly to the scattering of electrons in a GaAs quantum well (Rucker *et al.*, 1991, 1992; Tsen *et al.*, 1991; and Ozturk *et al.*, 1994). Our discussion shows that these modes lie almost entirely in the frequency range *below* that of the LO band.

In conclusion, we have shown that the bulk polar LO modes incident on a rigid barrier can resonantly be transmitted with 100% efficiency, provided the frequencies of the LO and IP branches overlap in the region of interest defined by the in-plane wavevector.

5.4 The Effect of Dispersion

The foregoing discussion in Section 5.2 concerning well modes explicitly assumes that an overlap of LO, TO and IP can take place, but this depends upon the dispersion of LO and TO modes. In non-polar material, overlap of the real band structure of the LO and TO modes usually fails at short wavelengths, and in this case the LO mode must hybridize with a mode drawn from the TO/TA complex branch. Of more practical importance is the effect in polar material. In the case of the more polar semiconductors, the zone-centre splitting of LO and TO frequencies can be large enough for a gap to appear between the zone-edge frequency of the LO mode and the zone-centre frequency of the TO mode. This occurs, for instance, in AlAs, but not in GaAs. When this happens the IP modes can satisfy the required boundary conditions by hybridizing with an evanescent TO mode, as usual, and with a mode drawn from the LO/LA complex branch of the phonon spectrum. Over much of the frequency range in this gap, the TO and LO components will have amplitudes that fall off rapidly from the interface over one or two unit cells, and, consequently, the IP mode will appear in virtually pure dielectric continuum (DC) form over the quantum well. This may appear to explain the important role played by the AlAs interface modes in the AlAs/GaAs system as observed in experiment. However, it turns out that the degree of dispersion of LO barrier modes is not important. Whatever the dispersion, the scattering rate due to barrier modes is close to the DC rate, as we will show in Section 8.7.

Figures 4.1–4.3 show dispersion curves for a number of semiconductors. The topic of dispersion arises again in Section 6.6 in connection with superlattices, where a somewhat fuller discussion may be found.

5.5 Quantization of Hybrid Modes

The presence of interfaces between materials with different vibrational properties introduces boundary conditions that can profoundly affect the vibrational spectrum. In the context of an isotropic elastic continuum this effect manifests itself, as we have seen, in the mixing of LO, TO and, in polar material, IP (interface polariton) modes, whose polarization lies in the plane of incidence – the so-called p-polarized modes. To summarize, these hybrid optical modes consist of a coherent linear combination of components, each of which is a solution of the basic equations of motion, having the same frequency and the same in-plane wavevector. For example, in polar materials the relative ionic displacement in a single layer of width a takes the form, for symmetric modes

$$u_x = Ak_x e^{i(k_x x - \omega t)}(\cos k_L z + c_T \cos k_T z - c_p \cosh k_x z) \quad (5.38a)$$

$$u_z = iAk_L e^{i(k_x x - \omega t)}\left(\sin k_L z - c_T \frac{k_x^2}{k_L k_T} \sin k_T z + c_P \frac{k_x}{k_L} \sinh k_x z\right) \quad (5.38)$$

and for antisymmetric modes

$$u_x = Ak_x e^{i(k_x x - \omega t)}(\sin k_L z + s_T \cos k_T z - s_p \sinh k_x z) \quad (5.39a)$$

$$u_z = -iAk_L e^{i(k_x x - \omega t)}\left(\cos k_L z - s_T \frac{k_x^2}{k_L k_T} \cos k_T z + s_P \frac{k_x}{k_L} \cosh k_x z\right) \quad (5.39b)$$

which represent hybrids propagating in the x-direction in the plane parallel to the interface with in-plane wavevector k_x, and consisting of LO and TO components with real wavevector components k_L and k_T in the z-direction, perpendicular to the interface, and at IP components, assumed to be unretarded. These describe displacements within the region $-a/2 \leq z \leq a/2$. Outside this region there are solutions typically depicting evanescent modes with amplitudes falling away from the interfaces. For a given frequency k_L, k_T and k_x are determined, and either k_L or k_T or both may turn out to be imaginary, but what frequencies are allowed is determined by dispersion relations that derive from the boundary conditions. Note that the case of non-polar material is obtained by putting $c_p = s_p = 0$.

A preliminary to the quantization of hybrid modes is to write the classical energy, which, following the discussion of Section 4.5, is predominantly mechanical. Ignoring electrical energy entirely we obtain

$$H = \frac{1}{2M}\int \mathbf{p}^2(\mathbf{r},t)\frac{d\mathbf{r}}{V_0} + \frac{1}{2}M\omega^2 \int \mathbf{u}^*(\mathbf{r},t).\mathbf{u}(\mathbf{r},t)\frac{d\mathbf{r}}{V_0} \quad (5.40)$$

where **u** (r, t) is the total displacement field, M is the mass of the oscillator (reduced mass for optical modes) and V_0 is the volume of the unit cell. We can express the total displacement in terms of annihilation and creation operators a and $a\dagger$ as follows:

$$\mathbf{u}(r,t) = \sum_{\mathbf{k}} (u_k a + u_k^* a^\dagger) \tag{5.41}$$

$$\mathbf{p}(r,t) = M\dot{u}(\mathbf{r},t)$$

and hence obtain for the hybrid modes in Eqs. (5.38) and (5.39)

$$A_i = \left(\frac{2}{NK_i^2}\right)^{1/2} \cdot \left(\frac{\hbar}{2M\omega_k}\right)^{1/2} \tag{5.42}$$

$$\begin{aligned} K_s^2 =& k_L^2 + k_x^2 - (k_L^2 - k_x^2)\frac{\sin k_L a}{k_L a} + c_T^2 \frac{k_x^2}{k_T^2}\left(k_T^2 + k_x^2 + (k_T^2 - k_x^2)\frac{\sin k_T a}{k_T a}\right) \\ & + 2c_p^2 k_x^2 \frac{\sinh k_x a}{k_x a} + 4c_T k_x^2 \frac{\sin(k_T a/2)}{k_T a/2}\cos(k_L a/2) \\ & - 4c_T c_p k_x^2 \frac{\sin(k_T a/2)}{k_T a/2}\cosh(k_x a/2) - 4c_p k_x^2 \frac{\sin(k_x a/2)}{k_x a/2}\cos(k_L a/2) \end{aligned} \tag{5.43a}$$

$$\begin{aligned} K_a^2 =& k_L^2 + k_x^2 + (k_L^2 - k_x^2)\frac{\sin k_L a}{k_L a} + s_T^2 \frac{k_x^2}{k_T^2}\left(k_T^2 + k_x^2 - (k_T^2 - k_x^2)\frac{\sin k_T a}{k_T a}\right) \\ & + 2s_p^2 k_x^2 \frac{\sin k_x a}{k_x a} - 4s_T k_x^2 \frac{\cos k_T a/2}{k_T a/2}\sin(k_L a/2) \\ & + 4s_T s_p k_x^2 \frac{\cos k_T a/2}{k_T a/2}\sin(k_x a/2) - 4s_p k_x^2 \frac{\cosh k_x a/2}{k_x a/2}\sin(k_L a/2) \end{aligned} \tag{5.43b}$$

where N is the total number of unit cells and the subscripts s and a refer to symmetric and antisymmetric solutions, respectively.

Note that in the dielectric-continuum model the corresponding normalizing factors for the LO modes are

$$K_s^2 = K_a^2 = K_L^2 + K_x^2 \tag{5.44}$$

and for the IP modes are

$$K_s^2 = K_a^2 = 2k_x^2 \frac{\sinh k_x a}{k_x a} \tag{5.45}$$

In the DC model, LO and IP modes remain separate and distinct entities, but not so when hybridization is taken into account.

6

Superlattice Modes

One deep calleth another. . ..
Psalms xlii

6.1 Superlattice Hybrids

We now consider the situation in a superlattice of two different polar materials where the mismatch of optical-mode frequencies is large enough for the interfaces to be regarded as infinitely rigid. Applying the triple hybridization scheme to the superlattice is straightforward. We assume that the superlattice is perfectly periodic, consisting of a well of width a and a barrier of width b (Fig. 6.1). Patterns in adjacent periods are assumed to differ only through a phase factor exp $ik_z(a+b)$, and so k_z is the wavevector describing propagation in the z-direction.

The general case leads to rather unwieldy expressions (see Chamberlain *et al.*, 1993). Rather than quote these here we focus attention on modes with frequencies not too close to ω_{TO} and assume that $a_T a/2$ is large. In other words, we assume that the mode patterns are well represented by double, rather than triple, hybrids. The hybrids obtained then obey the following dispersion relation:

$$\sin k_L a\left(1+\frac{k_x^2}{4k_L^2 r^2 s^2 d^2}(\beta^2-\alpha^2+\gamma^2)\right)+\frac{k_x}{k_L rsd}(\beta-\alpha\cos k_L a)=0 \quad (6.1a)$$

or, alternatively

$$\tan(k_L a/2)=\frac{k_L rsd}{k_x(\alpha+\beta)}\left[-\left(1+\frac{k_x^2}{4k_L^2 r^2 s^2 d^2}(\beta^2-\alpha^2+\gamma^2)\right)\right.$$
$$\left.\pm\left\{\left[1+\frac{k_x^2}{4k_L^2 r^2 s^2 d^2}(\beta^2-\alpha^2+\gamma^2)\right]^2-\frac{k_x^2}{4k_L^2 r^2 s^2 d^2}(\beta^2-\alpha^2)\right\}^{1/2}\right] \quad (6.1b)$$

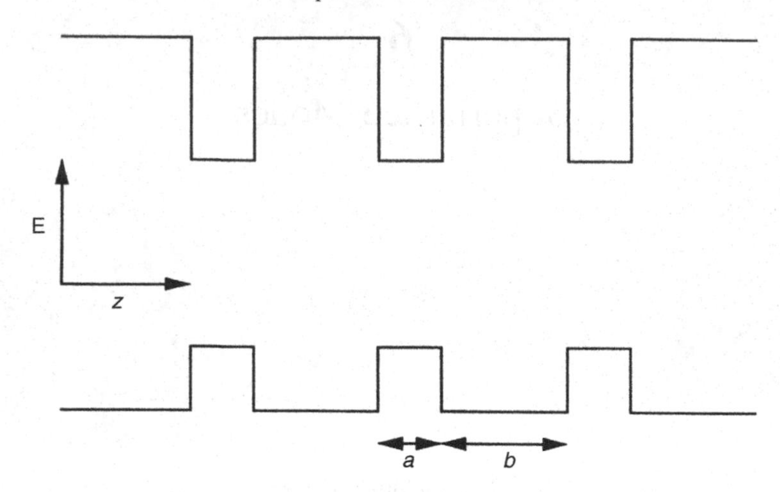

Fig. 6.1 Superlattice.

where

$$a = \sinh k_x b \cosh k_x a + r \cosh k_x b \sinh k_x a \tag{6.2a}$$

$$\beta = \sinh k_x b + r \sinh k_x a \cos k_z(a+b) \tag{6.2b}$$

$$d = \cos k_z(a+b) - Z \sinh k_x a \sinh k_x b - \cosh k_x a_x b \tag{6.2c}$$

$$Z = \frac{(1+r^2)}{2r} \tag{6.2d}$$

$$\gamma^2 = r^2 \sinh^2 k_x a \sin^2 k_z(a+b) \tag{6.2e}$$

$$\beta^2 - a^2 + \gamma^2 = 2rd \sinh k_x b \sinh k_x a \tag{6.2f}$$

The permittivity parameter r and the field parameter s are as before, Eq. (5.19). Note that $d=0$ describes the superlattice IP dispersion (discussed later).

The double-hybrid approximation ($\tanh(a_T a/2) \approx 1$) leads to displacement patterns in the well that are a linear combination of symmetric and antisymmetric forms, viz.:

$$\mathbf{u} = (u_x(z), 0, u_z(z))e^{i(k_x x + k_z z - \omega t)} \tag{6.3}$$

$$\begin{aligned} u_x(z) = k_x A \Bigg[X_1 \left(\sin k_L z - \frac{k_L}{K_x} \frac{\sinh k_x z}{\cosh(k_x a/2)} \cos(k_L a/2) \right) \\ + X_2 \left(\cos k_x z + \frac{k_L}{K_x} \frac{\cosh k_x z}{\sinh(k_x a/2)} \sin(k_L a/2) \right) \Bigg] \end{aligned} \tag{6.4a}$$

$$u_z(z) = -ik_L A\left[X_1\left(\cos k_L z - \frac{\cosh k_x z}{\cosh(k_x a/2)}\cos(k_L a/2)\right)\right.$$
$$\left.-X_2\left(\sin k_x z - \frac{\sinh k_x z}{\sinh(k_x a/2)}\sin(k_L a/2)\right)\right] \quad (6.4b)$$

where

$$X_1 = -\sin\left(\frac{k_L a}{2}\right) + \left(\frac{k_x}{2rsdk_L}\right)\cos\left(\frac{k_L a}{2}\right)(\alpha - \beta + i\gamma) \quad (6.5a)$$

$$X_2 = \cos\left(\frac{k_L a}{2}\right) + \left(\frac{k_x}{2rsdk_L}\right)\sin\left(\frac{k_L a}{2}\right)(\alpha + \beta - i\gamma) \quad (6.5b)$$

and $-a/2 \leq z \leq a/2$. The barrier fields are (suppressing the phase factor)

$$E_x = rs\rho_0 k_L A\left(Y_1\frac{\sin k_x z'}{\cosh(k_x b/2)} + Y_2\frac{\cosh k_x z'}{\sinh(k_x b/2)}\right)e^{ik_z(a+b)/2} \quad (6.6a)$$

$$E_z = -irs\rho_0 k_L A\left(Y_1\frac{\cosh k_x z'}{\cosh(k_x b/2)} + Y_2\frac{\sinh k_x z'}{\sinh(k_x b/2)}\right)e^{ik_z(a+b)/2} \quad (6.6b)$$

where

$$Y_1 = X_1\cos\left(\frac{k_L a}{2}\right)\cos\left(\frac{k_z(a+b)}{2}\right) - iX_2\sin\left(\frac{k_L a}{2}\right)\sin\left(\frac{k_z(a+b)}{2}\right) \quad (6.7a)$$

$$Y_2 = iX_1\cos\left(\frac{k_L a}{2}\right)\cos\left(\frac{k_z(a+b)}{2}\right) + X_2\sin\left(\frac{k_L a}{2}\right)\cos\left(\frac{k_z(a+b)}{2}\right) \quad (6.7b)$$

and $-b/2 \leq z' \leq b/2$.

To obtain the hybrid mode patterns for modes at the barrier frequency we merely regard a as now the barrier width and b the well-width, and define the parameters r and s in terms of the barrier properties.

Purely symmetric or antisymmetric modes are obtained only when $k_z(\mathrm{a}+\mathrm{b})=0$ or π. In this case the parameter $\gamma = 0$ and the dispersion relation simplify to

$$\cot(k_L a/2) = \frac{k_x}{k_L}p_a, \quad p_a = -\frac{(\alpha+\beta)}{2rsd} \quad (6.8a)$$

$$\tan(k_L a/2) = -\frac{k_x}{k_L} p_s, \quad p_s = \frac{(\beta - a)}{2rsd} \tag{6.8b}$$

where the subscripts a and s refer to antisymmetric and symmetric modes, respectively. Thus for antisymmetric modes $X_2 = 0$, and for symmetric modes $X_1 = 0$. A further simplification is that the parameter d, which describes polariton dispersion, factorizes as follows:

$$d = \frac{\sin k_x a \sin k_x b}{2r} \begin{cases} (\tanh(k_x a/2) + r\tanh(k_x b/2))(\coth(k_x a/2) \\ \quad + r\coth(k_x b/2)) \quad Q = 0 \\ (\tanh(k_x a/2) + r\coth(k_x b/2))(\coth(k_x a/2) \\ \quad + r\tanh(k_x b/2)) \quad Q = \pi \end{cases} \tag{6.9}$$

where $Q = k_z(a + b)$. Consequently

$$p_a = \begin{cases} [s(\tanh(k_x a/2) + r\tanh(k_x b/2))]^{-1} & Q = 0 \\ [s(\tanh(k_x a/2) + r\coth(k_x b/2))]^{-1} & Q = \pi \end{cases} \tag{6.10a}$$

$$p_s = \begin{cases} [s(\coth(k_x a/2) + r\coth(k_x b/2))]^{-1} & Q = 0 \\ [s(\coth(k_x a/2) + r\tanh(k_x b/2))]^{-1} & Q = \pi \end{cases} \tag{6.10b}$$

Letting the barrier width b increase without limit leads to the simple dispersion relations of Eq. (6.8) for all Q. These are identical to the dispersion relations of a single quantum well (Section 5.2).

Note that in the double-hybrid approximation u_x is not zero at the interfaces. It is straightforward but tedious to recover this condition by adding the TO contributions of appropriate symmetry as was done for the case of a single quantum well. Except for frequencies close to ω_{TO} it is usually admissable to regard the TO component as acting implicitly in reducing u_x to zero at the interfaces, but otherwise having negligible effect on the energy normalization.

6.2 Superlattice Dispersion

In the quantum-well case the frequencies of hybrids can vary rapidly with in-plane wavevector k_x, as a consequence of polariton dispersion. The same feature occurs in the case of a superlattice, but with the added degree of freedom

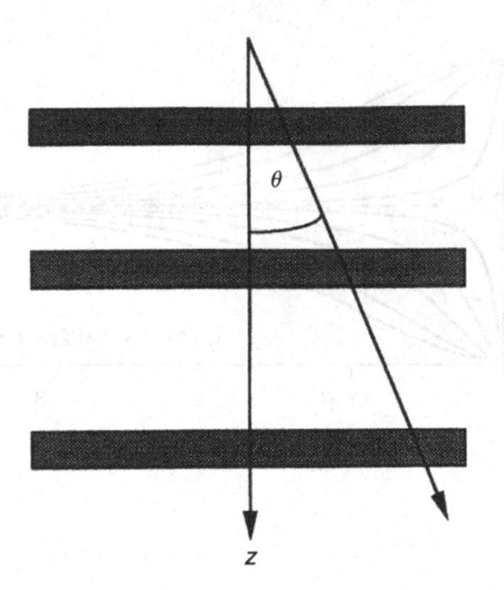

Fig. 6.2 Angle of propagation through a superlattice.

that propagation in the z-direction affords (Fig. 6.2). As a result, there is marked anisotropy in the dispersion. That this must be so is easily seen from the dispersion curves for pure superlattice interface polaritons (Fig. 6.3), where the dispersion is a strong function of both k_x and k_z (Section 6.4). We can therefore anticipate anticrossings of the pure LO and pure IP dispersions associated with the same symmetry, and simple crossings when the symmetry is opposite.

In order to discuss this matter further let us limit our attention to hybrids with wavevectors near the centre of the superlattice Brillouin zone, i.e., $k_x \to 0$, $k_z \to 0$. The dispersion of Eq. (6.1) becomes

$$\cot(k_L a/2) = -\frac{k_x(a+\beta)}{2k_L rsd} \approx \frac{2k_x^2(b+ra)}{sk_L\left(rk_z^2(a+b)^2 + [(1+r^2)ab + r(a^2+b^2)]k_x^2\right)} \tag{6.11}$$

which describes the antisymmetric modes, and

$$\tan(k_L a/2) = -\frac{k_x(\beta - a)}{2k_L rsd} \approx -\frac{k_x^2 a\left[rk_z^2(a+b)^2 + b(a+rb)k_x^2\right]}{2sk_L\left(rk_z^2(a+b)^2 + [(1+r^2)ab + r(a^2+b^2)]k_x^2\right)} \tag{6.12}$$

which describes the symmetric modes. We define an angle θ between the direction of propagation and the superlattice axis (z-direction) (Fig. 6.2) so that

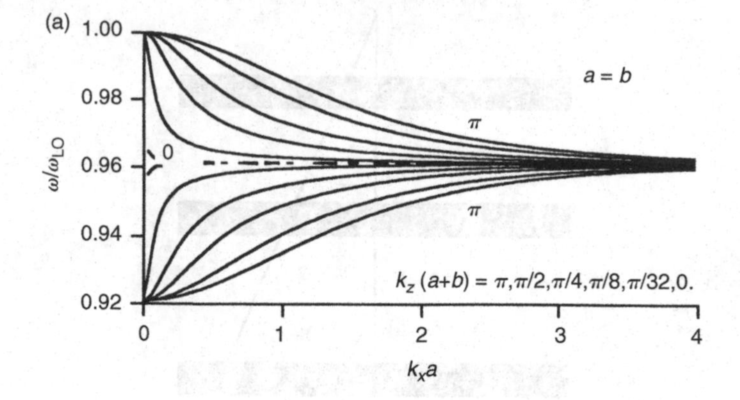

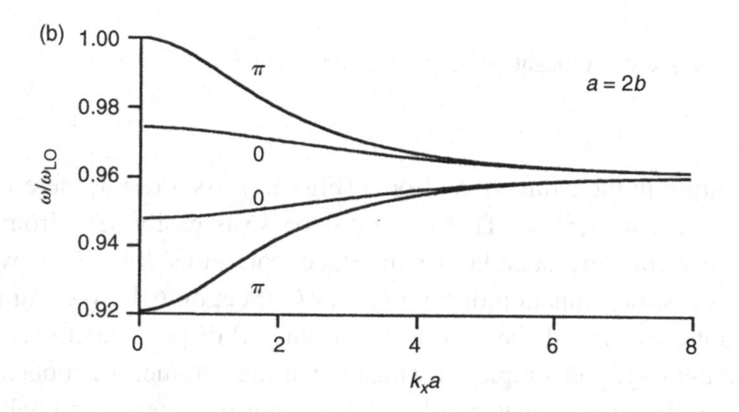

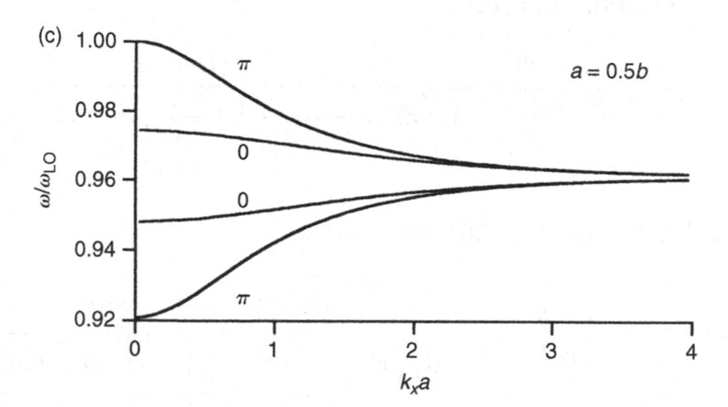

Fig. 6.3 Dispersion of pure IP modes in a superlattice: (a) $a=b$; (b) $a=2b$; (c) $a=0.5b$.

$$\tan\theta = \frac{k_x}{k_z} \tag{6.13}$$

The condition $\theta = 0$ represents motion along the superlattice axis. In this case cot $(k_La/2) = 0$ and thus $k_La/2 = (2n - 1)\,\pi/2$ for the antisymmetric modes, and tan $(k_La/2) = 0$ and thus $k_La/2 = n\pi$ for the symmetric modes. Odd numbers define antisymmetric modes and even numbers symmetric modes.

As θ increases, the condition $d = 0$ is approached. The effect on the modes depends upon the permittivity factor, r, which is small for frequencies near ω_{LO} but large for frequencies near ω_{TO}, and on the field factor s, which is unity for $\omega = \omega_{LO}$, and zero for $\omega = \omega_{TO}$. Thus according to Eqs. (6.11) and (6.12) the antisymmetric mode responds more rapidly than the symmetric mode at frequencies near ω_{LO}, and vice versa for frequencies near ω_{TO}. Both r and d are negative and so both cot $k_La/2$ and tan $k_La/2$ become more negative. This corresponds to a shift in k_L upward for the antisymmetric mode, but downward for the symmetric mode.

The polariton dispersion is reached at the critical angle θ_C, where

$$\tan^2\theta_c = -\frac{r(a+b)^2}{(1+r^2)ab + r(a^2+b^2)} \tag{6.14}$$

Passing through this, d changes sign and increases, and this continues the trend of k_L to increase for antisymmetric modes when $r \longrightarrow 0$ and to decrease for symmetric modes when r $\rightarrow \infty$. For θ sufficiently beyond θ_c (which we will denote $\theta >> \theta_C$), the dispersion relations become

$$\cot(k_La/2) = \frac{2(b+ra)}{k_Ls[(1+r^2)ab + r(a^2+b^2)]} \quad \text{antisymmetric} \tag{6.15a}$$

$$\tan(k_La/2) = -\frac{k_x^2 ab(a+rb)}{2sk_L[(1+r^2)ab + r(a^2+b^2)]} \quad \text{symmetric} \tag{6.15b}$$

For $r \rightarrow 0$, $s \approx 1$, the dispersion relation for the antisymmetric mode can be written

$$\tan(k_La/2) = \frac{k_La}{2} \tag{6.16}$$

which has solutions $k_La/2 = 2.86\pi/2, 4.92\pi/2, 6.94\pi/2$, etc. With $\theta = 0$ these solutions were $k_La/2 = \pi/2, 3\pi/2, 5\pi/2$, etc. Thus as θ increases through θ_C the $n = 1$ mode converts almost to the $n = 3$ mode, $n = 3$ to $n = 5$, and so on.

For $\theta > \theta_C$ there is no n = 1 mode. Meanwhile, the symmetric mode has the dispersion relation

$$\tan(k_L a/2) \approx -\frac{k_x^2 a}{2k_L} \tag{6.17}$$

which is the same as at $\theta = 0$. In this case $k_L a/2 \approx n\pi$ for all θ – the symmetric mode is virtually unaffected. For $r \to \infty$, $s \to 0$ (note that rs $\to -1$) the antisymmetric mode has the dispersion (since k_L is large)

$$\cot(k_L a/2) \approx 0 \tag{6.18}$$

for $\theta = 0$ and $\theta >> \theta_c$, so this mode is not strongly affected in this regime. The symmetric mode, on the other hand, is strongly affected. For $\theta >> \theta_c$

$$\tan(k_L a/2) \approx 0 \tag{6.19}$$

Although the solution is still $k_L a/2 = n\pi$, a mode with frequency near ω_{TO} has moved from n at $\theta = 0$ to $n - 1$ at $\theta > \theta_C$.

There is, therefore, no $n = 1$ mode propagating in the plane ($\theta = \pi/2$). In fact, compared with the situation at $\theta = 0$, the in-plane modes are compressed into a smaller frequency span. Note that modes are not lost, merely transformed.

A lattice-dynamics calculation of the full phonon spectrum of a GaAs/AlAs superlattice is shown in Fig. 6.4, which shows the angular dispersion of the IP mode. It also clearly shows the difference between the GaAs and AlAs LO dispersions: in GaAs the LO spectral range covers the whole of the IP spectral range, whereas in AlAs the LO spectral range is very narrow in the $\langle 001 \rangle$ direction. A comparison of the angular dispersion of the hybrid and lattice-dynamics models, which takes into account that the effective well-width is the distance between Al ions immediately adjacent to each interface, is shown in Figs. 6.5a and b. The agreement is very satisfactory for the GaAs modes, and also for the AlAs modes provided the electrostatic effect of the mass approximation used in the microscopic calculations is taken into account by equating the high-frequency permittivities of the two materials to $10.6\varepsilon_0$. Figs. 6.6a and b show a comparison of the anisotropy predicted by continuum theory with the results of micro-Raman-scattering experiments showing reasonable agreement.

6.3 General Features

Over most of the frequency range between ω_{LO} and ω_{TO} the TO component is a short-range evanescent mode that influences the mode patterns very close to the interfaces but otherwise has little effect. Indeed, the approximation that we have

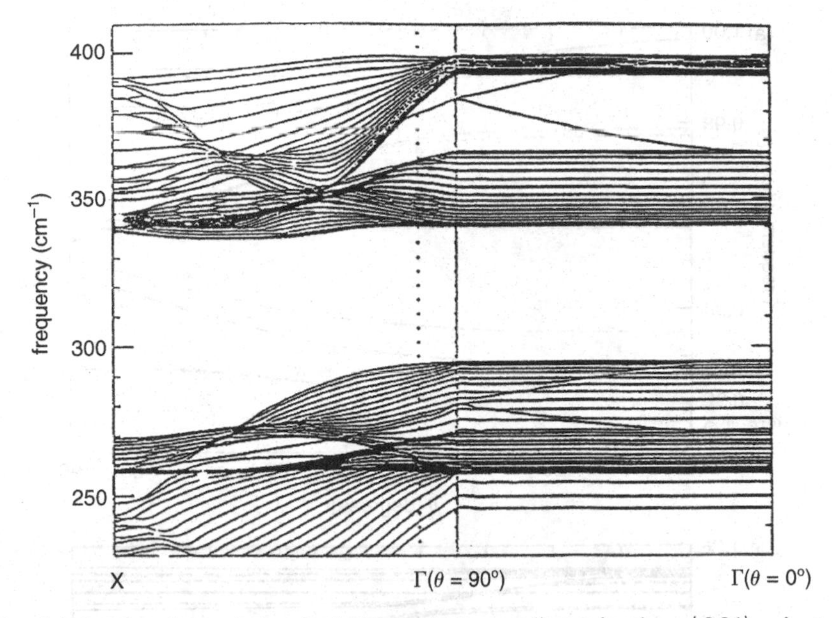

Fig. 6.4 Lattice dynamical calculations of phonon dispersion in a $\langle 001 \rangle$ oriented $(GaAs)_{20}(AlAs)_{20}$ superlattice along the in-plane direction (left panel), and as a function of θ (right panel) for a wavevector near the zone-centre. The vertical dotted lines mark the minimum and maximum wavevectors that can participate in the intrasubband scattering of a 0.25-eV electron (Rucker *et al.*, 1991).

already made is to ignore the TO component entirely in this frequency regime, which is equivalent to ignoring the condition $u_x = 0$. (This, in turn, is equivalent to neglecting shear stresses.) For most purposes we can regard these hybrids as consisting principally of an LO and an IP component, but when the frequency is very near ω_{TO} the neglect of the TO component is inadmissable.

Hybrids with a superlattice wavevector near the superlattice zone-centre or near the superlattice zone-edge ($Q_z = k_z(a+b) = 0$ or π) are easiest to describe since they have well-defined symmetry in each component of the superlattice. Thus, in the well, the modes are either antisymmetric, *A*, or symmetric, *S*, referring to the reflection properties about the centre of the well of u_x, the *x*-component of the displacement. For $k_z = 0$ the symmetry of the modes in the barrier must be the same as that of modes in the well, but the reverse is the case for $Q = \pi$. In between these extremes of superlattice wavevector the symmetry is mixed.

There are, naturally, two frequency bands, one associated with the material making up the well, the other with the barrier. By assuming the boundary condition $\mathbf{u} = 0$, we have envisaged these two bands to be far apart. We can thus speak of well modes and barrier modes. A well mode would have associated with

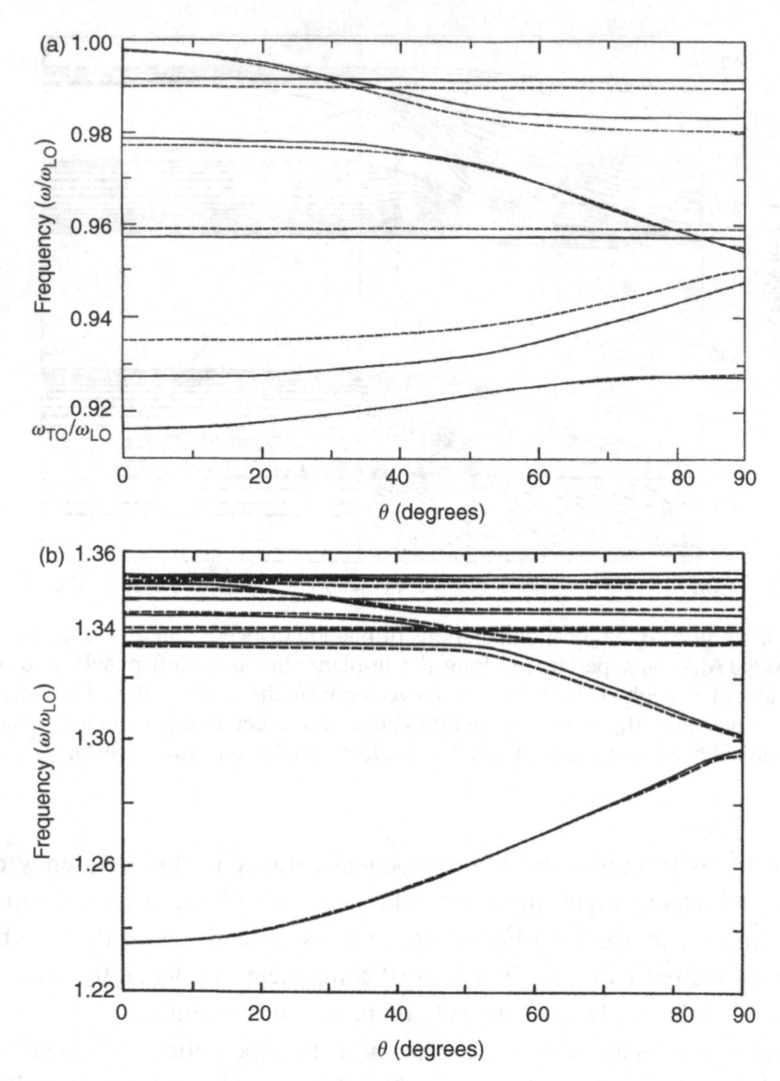

Fig. 6.5 Hybrid theory angular dispersion of a $(GaAs)_{20}((AlAs)_{20}$ superlattice: (a) GaAs modes (Chamberlain *et al.*, 1993); (b) AlAs modes (Chamberlain and Cardona, 1994) (dashed: hybird; continuous: lattice dynamic).

it a pattern of ionic displacement with accompanying electric field confined entirely to each well, connected to a mode pattern in the barrier consisting entirely of electric field and zero ionic displacement, and mutatis mutandis for the barrier modes. This, of course, is an idealized case. In real systems the ionic motion would never be quite zero in alternate segments.

As we have seen, the interplay of LO and IP properties introduces considerable anisotropy into the dispersion. Without this interplay the LO modes would be

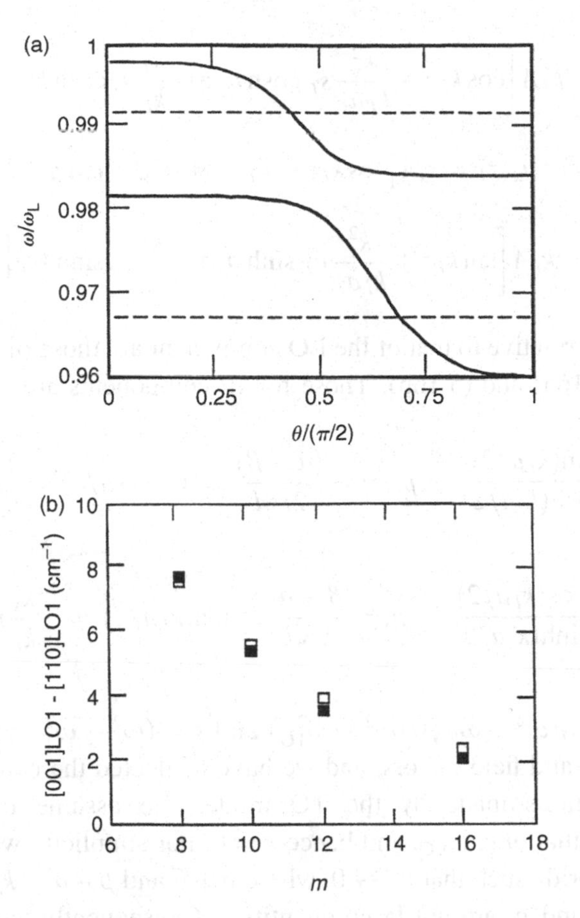

Fig. 6.6 Superlattice dispersion of hybrid modes: (a) angular dispersion of zone-centre modes (Constantinou *et al.*, 1993); (b) comparison with micro-Raman-scattering data of Haines and Scamarcio (1992): closed squares, experiment; open squares, theory.

comfortably defined by the condition $k_L a = n\pi$ (or $k_L b = n\pi$) with n an integer, odd numbers referring to A modes, even to S modes. This generally happy condition pertains only to long-wavelength modes ($k_z \approx 0$, $k_x \approx 0$) propagating nearly parallel to the superlattice axis. Where LO and IP dispersions nearly coincide, a hybrid can change its character from being basically LO-like to being basically IP-like.

It is useful to recall the expressions describing the mode patterns when $Q = 0$ or π and $\tanh \alpha_T a = 1$ ($\tanh \alpha_T b$ also), which are found in Eqs. (5.12) and (5.15) of Section 5.2, viz.:

$$\text{antisymmetric} \quad u_x(z) = ik_x A\left[\sin k_L z - s_T \sinh \alpha_T z - s_p \sinh k_x z\right] \tag{6.20a}$$

$$u_z(z) = k_L A \left[\cos k_L z - \frac{k_x^2}{k_L \alpha_T} s_T \cosh \alpha_T z - \frac{k_x}{k_L} s_p \cosh k_x z \right] \tag{6.20b}$$

$$\text{symmetric} \quad u_x(z) = k_x A \left[\cos k_L z - c_T \cosh \alpha_T z - c_p \cosh k_x z \right] \tag{6.21a}$$

$$u_z(z) = i k_L A \left[\sin k_L z + \frac{k_x^2}{k_L \alpha_T} c_T \sinh \alpha_T z + \frac{k_x}{k_L} c_p \sinh k_x z \right] \tag{6.21b}$$

The amplitudes relative to that of the LO component are those of Eqs. (5.12a) and (5.12b) and (5.16a) and (5.16b). Those for IP components are

$$s_p = p_a \frac{\sin(k_L a/2)}{\cosh(k_x a/2)}, \quad p_a = -\frac{(\alpha + \beta)}{2rsd}, \quad \cot(k_L a/2) = \frac{k_x}{k_L} p_a \tag{6.22a}$$

$$c_p = p_s \frac{\cos(k_L a/2)}{\sinh(k_x a/2)}, \quad p_s = \frac{\beta - \alpha}{2rsd}, \quad \tan(k_L a/2) = -\frac{k_x}{k_L} p_s \tag{6.22b}$$

with $r = (\varepsilon_\infty/\varepsilon_0)(\omega^2 - \omega_{\text{LO}}^2)/(\omega^2 - \omega_{\text{TO}}^2)$ and $s = (\omega^2 - \omega_{\text{TO}}^2)/(\omega_{\text{LO}}^2 - \omega_{\text{TO}}^2)$ as the permittivity and field factors, and we have neglected the contributions to the dispersion relations made by the TO mode. We assume that confinement always ensures that $\omega < \omega_{\text{LO}}$, and hence $r \neq 0$. For simplicity we consider long-wavelength hybrids such that $k_x \to 0$, where $\alpha + \beta$ and $\beta - \alpha \sim k_x \to 0$. Provided that $d \neq 0$, p_a and p_s are not large quantities. Consequently, $\cot k_L a/2 \approx 0$ for asymmetric modes and $\tan k_L a/2 \approx 0$ for symmetric modes, and it follows that both s_p and c_p are not large. The hybrids are thus more LO-like than IP-like. A look at the normalization coefficients, Eq. (5.43), confirms this. Neglecting the contribution from the TO mode, we obtain

$$K_a^2 = k_L^2 + k_x^2 + \left(k_L^2 - k_x^2\right) \frac{k_L a}{k_L a} - 4 p_a \frac{k_x}{a} \sin^2(k_L a/2)(2 - p_a \tanh(k_x a/2)) \tag{6.23a}$$

$$K_s^2 = k_L^2 + k_x^2 + \left(k_L^2 - k_x^2\right) \frac{\sin k_L a}{k_L a} - 4 p_s \frac{k_x}{a} \cos^2(k_L a/2)(2 - p_s \coth(k_x a/2)) \tag{6.23b}$$

When $d \neq 0$ and k_x is small, $K_a^2 \approx K_s^2 \approx k_L^2 + k_x^2$, which is the factor for pure LO modes. Most of the energy in the mode is associated with the LO component in this case.

When $d=0$ either p_a or p_s diverges, but not both. That it is one or the other but not both arises mathematically from the fact that when $d=0$, then $\beta^2 - a^2 = 0$, so that $\beta - a = 0$ or $\beta + a = 0$. Thus, we can deduce from Eq. (6.22) that

$$p_a \to \infty, \quad p_s \to \frac{\sinh k_x b \sinh k_x a}{s(a+\beta)} \tag{6.24a}$$

$$p_a \to \frac{\sinh k_x b \sinh k_x a}{s(\beta - a)}, \quad p_s \to \infty \tag{6.24b}$$

represent the two alternatives. In the first, we have $\tan(k_L a/2) = 0$ or nearly so, and in the second $\cot(k_L a/2) = 0$ or nearly so.

The amplitudes do not diverge. In the first case the product $p_a \sin(k_L a/2)$ remains finite, and so does the product $p_s \cos(k_L a/2)$ in the second case. In fact

$$s_p \to \frac{k_L}{k_x \cosh(k_x a/2)}, \quad c_p \to \frac{-k_L}{k_x \sinh(k_x a/2)} \tag{6.25}$$

and these imply that the amplitude of the IP component is much larger than that of the LO component. The normalization factors become

$$K_a^2 = K_L^2 + K_x^2 + 2K_L^2 \frac{\tanh(k_x a/2)}{k_x a/2} \tag{6.26a}$$

$$K_s^2 = k_L^2 + k_x^2 + 2k_L^2 \frac{\coth(k_x a/2)}{k_x a/2} \tag{6.26b}$$

If the amplitudes $k_x s_p$ and $k_x c_p$ are factorized out of the preceding expressions we get

$$K_a^2 = k_x^2 s_p^2 \left(\frac{(k_L^2 + k_x^2)}{k_x^2 s_p^2} + \frac{2 \sinh k_x a}{k_x a} \right) \tag{6.27a}$$

$$K_s^2 = k_x^2 c_p^2 \left(\frac{(k_L^2 + k_x^2)}{k_x^2 c_p^2} + \frac{2 \sinh k_x a}{k_x a} \right) \tag{6.27b}$$

The factor 2 sin k_xa/k_xa is the familiar normalizing factor for pure IP modes. The IP character is dominant in the case of symmetric modes, less so for antisymmetric modes. Thus when $d=0$ the hybrid is IP-like.

The condition $d=0$ can occur only for special combinations of k_L, k_x and k_z. We now turn to the study of these cases where pure LO and pure IP dispersions cross.

6.4 Interface Polaritons in a Superlattice

The cause of the strong anisotropic dispersion found in hybrid modes is to be traced to the properties of "pure" IP modes, i.e. those described by the DC model. Conventionally, interface polaritons have been described using only electromagnetic boundary conditions; in a superlattice this leads to the Camley–Mills (1984) dispersion relationship in the unretarded limit

$$\begin{aligned} d \equiv \cos k_z(a+b) - \left[\frac{(1+r^2)}{2r}\right] \sinh k_xa \sinh k_xb - \cosh k_xa \cosh k_xb \\ = 0 \end{aligned} \tag{6.28}$$

where $r=\varepsilon_a(\omega)/\varepsilon_b(\omega)$. When the LO/TO frequencies of well and barrier materials are well separated, IP modes are distinctly well-like or barrier-like. Fig. 6.3 shows the dispersion for the cases $b/a < 1$, $b/a=1$ and $b/a > 1$, for certain values of the superlattice phase factor, $Q=k_z(a+b)$. Also important for determining the interaction with electrons in the well are the symmetries of the interaction potential in the well, and these are depicted in Fig. 1.4(b).

Two features are worth noting. One is that the width and position of the allowed frequency bands are influenced by the ratio of barrier widths to well-widths. The other is that the density of modes is highest near $Q=\pi$. Both of these features are to be found in the properties of hybrid modes.

We can imagine superimposed on these diagrams a set of almost horizontal lines representing the dispersion of LO modes obeying the hydrodynamic boundary condition $u_z=0$. Where they intersect with IP bands, particularly those parts of the bands near $Q=\pi$, we can expect to find hybrids with strong IP-like character. Outside these areas the hybrids will be LO-like.

Broadly speaking, there are two areas probed by experiment defined by the magnitude of the wavevector involved in the interaction. The latter is typically of order 10^4–$10^5\,\text{cm}^{-1}$ in Raman-scattering experiments, whereas in electron scattering the in-plane wavevector is of order $10^6\,\text{cm}^{-1}$. Thus Raman-scattering experiments probe essentially zone-centre modes, whereas transport or energy-relaxation experiments reach out to larger wavevectors.

6.5 The Role of LO and TO Dispersion

The original use of the term *dispersion* described the variation with wavelength of the velocity of propagation of a sound wave or a light wave. Applied to optical modes it refers to the variation of frequency with wavelength, and it is this latter meaning we have been using here and in our earlier discussion (Section 5.4), following common usage.

The schemes of double hybridization in non-polar material and triple hybridization in polar material make the implicit assumption that vibrational modes of all types exist at the frequency of interest, even if some may be evanescent and belong to the complex branches of the overall spectrum. One of these modes is the evanescent TO component at a frequency near ω_{LO} that is required to form the triple hybrid in polar material. There are two other cases of importance.

In non-polar material real TO modes span the frequency band from ω_{TO} down to ω_{TZ} at the zone boundary, and LO modes span from ω_{LO} down to ω_{LZ}. In real crystals these zone-edge frequencies vary with crystallographic direction, but we tend to ignore this complication in the spirit of our isotropic-medium approximation. Linear-chain theory predicts that near the zone-centre the dispersion will be quadratic in wavevector with a curvature determined by v_{LA} where v_{LA} is the corresponding velocity of acoustic waves. Since $v_{LA} > v_{TA}$ we expect $\omega_{LZ} < \omega_{TZ}$ on the basis of the quadratic relation, and this is confirmed by more accurate formulations. There will, therefore, be a range of frequencies between ω_{TZ} and ω_{LZ} where hybridization entails the mixing of non-polar LO modes with modes from the TO/TA complex branch of the spectrum. Except at frequencies close to ω_{TZ}, such modes will be heavily evanescent and affect mode patterns only very close to the interfaces. In such a case the resultant hybrids are essentially pure LO modes obeying $u_z = 0$ but not $u_x = 0$. In the slab case they would obey the hydrodynamic stress condition but not the conditions involving shear. The involvement of the TO/TA complex branch may also occur in polar materials if the LO and TO branches cross. Note that insofar as the interaction with electrons involves only long waves, hybridization involving the TO/TA branch, which will occur typically at large wavevectors, need not be considered for the electron–hybrid interaction.

Another situation occurs in polar material when $\omega_{LZ} > \omega_{TO}$, and this will be encountered increasingly frequently with increasing polarity. This situation was encountered in our discussion of quantum-well modes. In order to satisfy all boundary conditions, TO modes and IP modes must hybridize with modes in the LO/LA complex branch. Remarks similar to those made previously apply. Except near ω_{LZ}, the LO/LA modes are heavily evanescent, and the resultant hybrids are very nearly double hybrids involving TO and IP components obeying $u_x = 0$ but

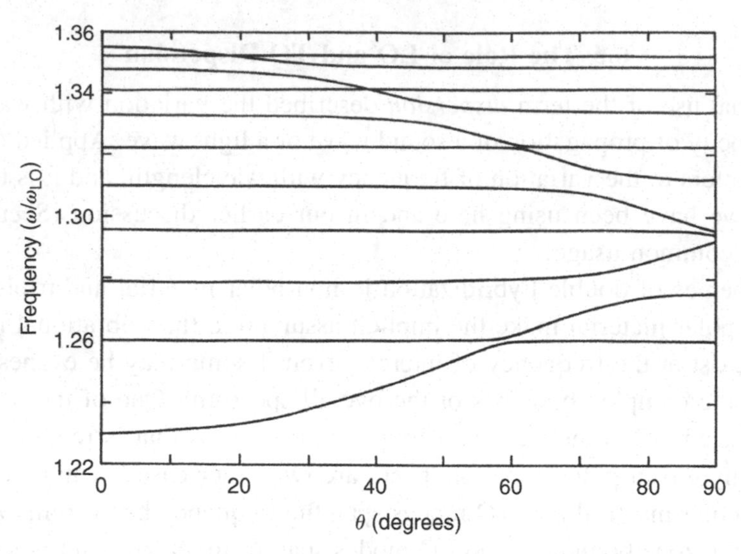

Fig. 6.7 Angular dispersion of AlAs modes in a ⟨111⟩ oriented GaAs (29.4-Å)/AlAs (32.6-Å) superlattice (Chamberlain and Cardona, 1994).

not $u_z = 0$. For frequencies above ω_{TO} the TO component is also evanescent, and if heavily so, the resultant hybrid is virtually a pure IP mode obeying electromagnetic boundary conditions only, the elastic boundary conditions having been taken care of by the TO and LO/LA disturbances localized at the interface.

In real crystals, whether LO and TO branches overlap or not often depends upon crystallographic direction, and this vitiates an isotropic model unless the model incorporates some directional averaging. Thus, in AlAs there is no overlap in the ⟨001⟩ direction, Fig. 6.5(b), but in the ⟨111⟩ direction the LO dispersion is much bigger, resulting in GaAs-like behaviour, as the angular dispersion of Fig. 6.7 shows. Moreover, the involvement of short-range components deriving from the complex branches of the vibrational spectrum stretches any continuum model to its limits and often beyond, particularly with its assumption of quadratic dispersion. Fortunately, it is not always vital to know how fast an evanescent mode decays in space. Provided that it is faster than some characteristic rate in the problem, its precise magnitude is not needed. Thus, for example, in the case of a hybrid of frequency near ω_{LO} in polar material the precise value of the decay constant for the TO component, a_T, is not important provided $a_T a/2$ is large. The use of a quadratic dispersion law to determine a_T is unlikely to be accurate in this situation, but in many cases this will not matter. An obvious simple alternative to the use of quadratic dispersion in these cases is to assume that a_T is just of the order of the reciprocal of the unit-cell dimension.

This simple assumption is also applicable to the description of modes from TO/TA and LO/LA complex branches.

Given the complexity of the problem of describing hybrid optical modes, it is certainly advisable to regard the analytic forms derived previously as envelope functions with wavevectors that are related by the real spectrum of the material rather than by a pure continuum model, especially where lattice-dynamics theory and experimental results coincide.

Finally, we point out that the assumption that the superlattice is regular and of infinite extent is not, of course, sacrosanct. Huang *et al.* (1995) have applied the transfer-matrix formalism to finite periodic and to Fibonacci superlattices and have obtained localization lengths for the barrier phonons.

6.6 Acoustic Phonons

A one-dimensional continuum theory of acoustic vibrations in layered material was given long ago by Rytov (1956) using the classic acoustic boundary conditions, namely the continuity of acoustic displacement and stress. The derivation of the dispersion relation for a superlattice is straightforward; the result (a familiar form) is

$$\cos k_z(a+b) = \cos k_a a \cos k_b b - \left[\frac{(1+\eta^2)}{2\eta}\right] \sin k_a a \sin k_b b \tag{6.29}$$

where k_z is the superlattice wavevector and k_a, k_b are the wavevectors in the adjacent layers corresponding to the same frequency, ω. The factor η is given by

$$\eta = \frac{\rho_b v_b}{\rho_a v_a} \tag{6.30}$$

where v_a, v_b are the phase velocities and ρ_a, ρ_b are the mass densities. Each velocity is determined by the elastic constant c and density ρ according to the usual equation $v=(c/\rho)^{1/2}$.

Superlattice zone boundaries occur when $k_z = \mathrm{n}\pi/d$ $(d=a+b)$ and at each of these points a gap may appear in the spectrum. If there is no mismatch, $\eta = 1$ and Eq. (6.26) reduces to

$$\cos k_z d = \cos(k_a a + k_b b) = \cos\left(\omega\left(\frac{a}{v_a}+\frac{b}{v_b}\right)\right) \tag{6.31}$$

In this case an average velocity is defined thus:

$$\bar{v} = \frac{v_a v_b d}{v_a a + v_a b} \tag{6.32}$$

and the zone boundaries occur at $\omega = \bar{v}n\pi/d$, i.e. at a single frequency, so no gaps occur. On the other hand, if the mismatch is large ($\eta << 1$ or $\eta >> 1$) Eq. (6.26) reduces to

$$\cos k_z(a+b) = -\left[\frac{(1+\eta^2)}{2\eta}\right] \sin k_a a \sin k_b b \qquad (6.33)$$

This has the following approximate solution at a zone boundary:

$$\omega = \frac{n\pi v_a}{a} \quad \text{or} \quad \omega = \frac{n\pi v_b}{b} \qquad (6.34)$$

There is now a frequency gap equal to $n\pi[(v_b/b) - (v_a/a)]$

Restricting k_z to the first zone results in a folded acoustic spectrum (Fig. 6.8). Raman scattering probing the spectrum near $k_z = 0$ shows this folding very clearly in the (x, x) configuration. For further reading on this topic the reader is referred to the reviews of Klein (1986), Menendez (1989) and Cardona (1989).

As we have seen in Chapter 3, boundary conditions for acoustic modes are not problematic and the continuity of displacement and stress can be applied to

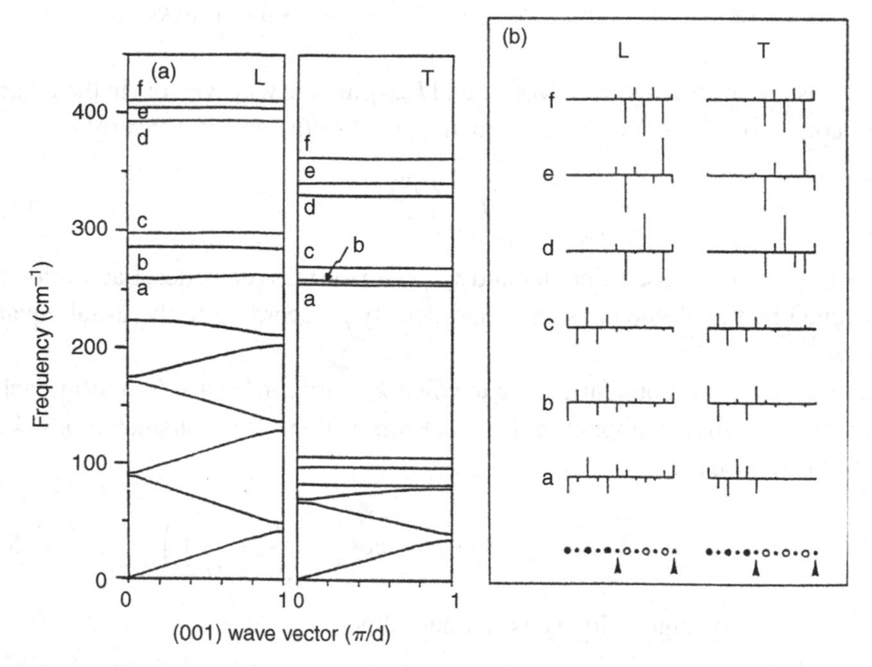

Fig. 6.8 Phonon models in a $(GaAs)_3(AlAs)_3$ superlattice: (a) dispersion; (b) ionic displacement predicted by lattice dynamics. Key • As, • Ga, Al (Molinari *et al.*, 1986).

transversely polarized modes as well as to longitudinally polarized modes. Modifications due to piezoelectricity are small in III-V compounds but can be readily incorporated, and Rayleigh and other types of surface wave can be described (e.g., Maradudin and Stegemann, 1991).

Because the gaps at the zone boundaries are usually small it is usually a good approximation, as far as the interaction with electrons is concerned, to ignore folding and to describe the acoustic phonons in terms of a bulk spectrum with properties obtained by averaging over the materials involved and ignoring surface or interface waves entirely.

7
Optical Modes in Various Structures

It cannot be thus long, the sides of nature will not sustain it.
Antony and Cleopatra, *W. Shakespeare*

7.1 Introduction

This chapter deals with several topics. There is considerable interest in fabricating quasi-2D structures in which the electron–phonon interaction is reduced. Optical-phonon engineering is in its infancy, but already there have been investigations of the effect of incorporating monolayers and conducting layers. One of the first quasi-2D systems to be studied was the thin ionic slab, yet there are still problems connected with the description of optical modes in such structures. The increasing sophistication of microfabrication techniques has led to the creation of quasi-one-dimensional (quantum wires) and quasi-zero-dimensional (quantum dots) structures that are expected to have interesting physical properties. It is important to establish the mode structure, both electron and vibrational, in these systems. In this chapter we consider some of these topics briefly.

7.2 Monolayers

The study of short-period superlattices in electronic and optical devices has received considerable attention and there are several reasons why this has been so. Ease of growth and reduction of interface roughness and residual impurities make for more perfect structures. Replacing random alloys, such as $Al_xGa_{1-x}As$, with their ordered superlattice counterparts $(GaAs)_m/(ALAs)_n$ eliminates alloy scattering. In the Al_xGa_{1-x} system there is the added advantage of avoiding the troublesome DX centre. The replacement of random alloys by equivalent superlattices in bandgap engineering is unproblematic. Thus the wavelength emitted by an $In_{0.2}Ga_{0.8}As/GaAs$ laser is not very different from that emitted when a $(InAs)_1/(GaAs)_4$ equivalent superlattice replaces the alloy. In view of the practical advantages of short-period superlattices it is important to assess how the electronic and vibrational properties of a quantum well are modified when it contains monolayers of a different semiconductor.

Significant changes of the optical-phonon spectrum in the well can be engineered by including one or more "foreign" monolayers, especially when the latter contains atoms either much lighter or much heavier than those of the host lattice. The condition is well-satisfied by an AlAs or an InAs monolayer in GaAs. Neither AlAs nor InAs can vibrate easily at the optical mode frequencies of GaAs. The frequency mismatch at an AlAs/GaAs interface is known to cause the optical displacement of a GaAs mode to vanish at the first Al atom, a result predicted by theory and confinement by experiment. The same result is expected to apply at a hypothetical InAs/GaAs interface. We can predict from this that a monolayer of AlAs in GaAs will inhibit optical displacement in its vicinity, and the same should be true for InAs. In the language of a continuum theory of optical modes this means that monolayers of this kind impose the approximate condition

$$\mathbf{u} = 0$$

where $\mathbf{u}$ is the optical displacement at the host frequency. (There may also be a local mode, and we will return briefly to this point later.)

The mechanical effect of reducing the displacement to zero is the major contribution of the monolayer. As regards electrical effects, it will be noted that within a bulk dielectric layer the variation of tangential field will be the linear combination $A \cosh k_x z + B \sinh k_x z$ (with the origin taken to be at the central plane of the layer). If the thickness of the layer is d, then provided $k_x d/2 << 1$, there can be only small changes of tangential field across the monolayer. Thus, the tangential field can safely be taken as continuous across the layer. The normal component of the field will induce an electronic polarization of the atoms of the monolayer, but this only affects the field within the atom and has no effect on the field on either side. Effectively, the monolayer has no electrical effect.

A thick layer, however, would contribute interface polaritons at frequencies determined by the properties of the layer. In the case of InAs in GaAs the interface frequency lies well below that of GaAs, so this would ensure that any modes were delocalized. The opposite is the case for AlAs in GaAs. Adding a thick layer of AlAs would contribute interface modes that would enhance the scattering rate. The question arises; How thick must a layer be to support recognizable interface modes? The latter arise as a consequence of negative permittivity, which can happen only through the antiscreening action of the ions.

A single layer of ions can indeed respond to in-plane fields in this way, but it obviously cannot antiscreen transverse fields effectively. As an interface mode has both types of field components of equal amplitude it is impossible for a single monolayer to support one. At any rate, we will assume that a monolayer is dielectrically neutral and encapsulate its total effect by the boundary condition $\mathbf{u} = 0$.

7.2.1 Single Monolayer

Let us look at the case of an AlAs or InAs monolayer centrally placed in a GaAs quantum well (Fig. 7.1). As we have found, assuming a double-hybrid model for the GaAs modes allows a good approximation. The most striking result for this case is that the antisymmetric modes, which are responsible for intersubband scattering, are significantly affected (Ridley, 1995).

The general form of the GaAs modes is

$$u = F(z)Ae^{i(k_x x - \omega t)}(u_x, u_z) \tag{7.1a}$$

$$u_x = k_x\left[d_1 \cos k_L z \mp d_3 \sin k_L z + \frac{1}{s} d_2 \cosh k_x z \pm \frac{k_L}{k_x} d_3 \sinh k_x z\right] \tag{7.1b}$$

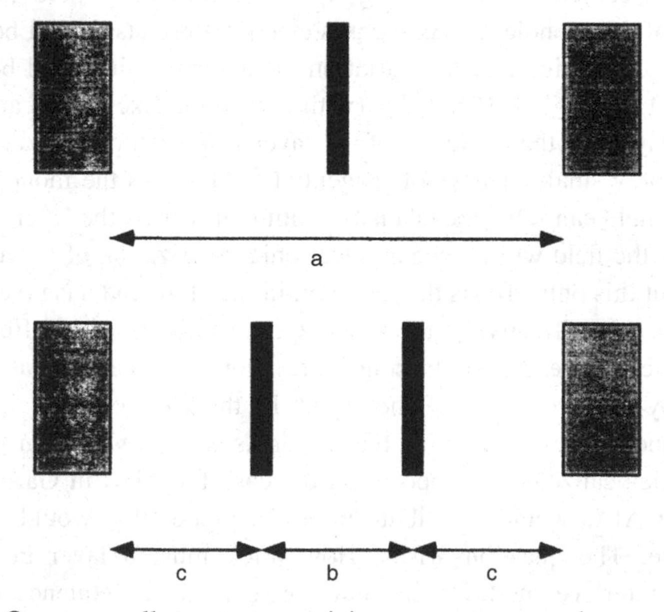

Fig. 7.1 Quantum-well structure containing one or two monolayers.

$$u_z = ik_L\left[d_1 \sin k_L \pm zd_3 \cos k_L z - \frac{k_x}{sk_L} d_2 \sinh k_x z \mp d_3 \cosh k_x z\right] \tag{7.1c}$$

where the upper sign is for $z > 0$, the lower for $z < 0$, and

$$d_1 = 1 - \gamma_1 \cos\left(\frac{k_L a}{2}\right) + \frac{k_x}{sk_L} \sin\left(\frac{k_L a}{2}\right) \sinh\left(\frac{k_x a}{2}\right) \tag{7.2a}$$

$$d_2 = 1 - \cos\left(\frac{k_L a}{2}\right) \cosh\left(\frac{k_x a}{2}\right) - \frac{sk_L}{k_x} \gamma_2 \sin\left(\frac{k_L a}{2}\right) \tag{7.2b}$$

$$d_3 = \gamma_1 \sin\left(\frac{k_L a}{2}\right) + \frac{k_x}{sk_L} \cos\left(\frac{k_L a}{2}\right) \sinh\left(\frac{k_x a}{2}\right) \tag{7.2c}$$

$$\gamma_1 = \cosh\left(\frac{k_x a}{2}\right) + r \sinh\left(\frac{k_x a}{2}\right) \tag{7.2d}$$

$$\gamma_2 = \sinh\left(\frac{k_x a}{2}\right) + r \cosh\left(\frac{k_x a}{2}\right) \tag{7.2e}$$

In general, there are exponential field variations in the barriers at $|z| > a/2$. In these equations, $s = (\omega^2 - \omega_{TO}^2)/(\omega_{LO}^2 - \omega_{TO}^2)$ is the field factor for the IP component, and $r = (\varepsilon_\infty/\varepsilon_B)(\omega^2 - \omega_{LO}^2)/(\omega^2 - \omega_{TO}^2)$ is the permittivity factor. Here ω_{LO} is the zone-centre LO frequency, ε_∞ is the high-frequency permittivity of GaAs and ε_B is the permittivity of the AlAs barrier (assumed to be essentially frequency-independent at the frequency of GaAs), and a is the well-width. Note that the designation "symmetric" refers to u_x, which is proportional to the potential. The potential seen by an electron is

$$\phi = -i\left(\frac{e^*}{\varepsilon_0 V_0}\right) F(z) A e^{i(k_x x - \omega t)} [d_1 \cos k_L z \mp d_3 \sin k_L z + d_2 \cosh k_x z \pm \frac{sk_L}{k_x} d_3 \sinh k_x z] \tag{7.3}$$

where $e^{*2} = \mu V_0 \omega_{LO}^2 \varepsilon_0^2 (\varepsilon_\infty^{-1} - \varepsilon_s^{-1})$, μ is the reduced mass, V_0 is the unit-cell volume, ε_0 is the permittivity of the vacuum and ε_s is the static permittivity of GaAs.

The dispersion relation is

$$(d_1 + d_2) d_3 = 0 \tag{7.4}$$

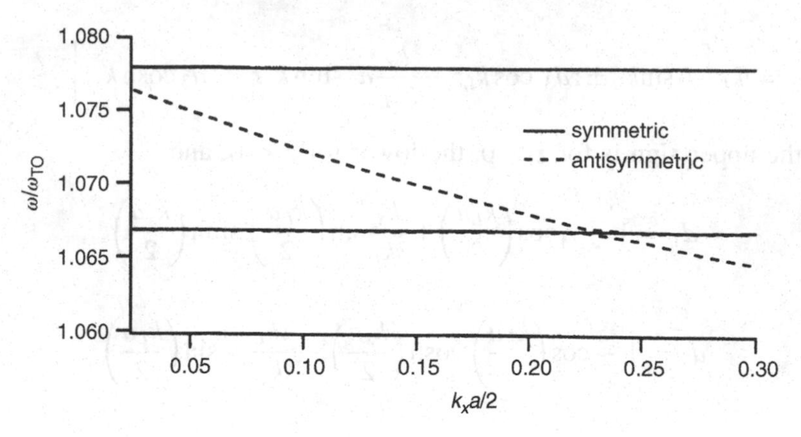

Fig. 7.2 Dispersion of low-order hybrid optical modes in a GaAs quantum well with a centrally placed monolayer (well-width 42Å).

with $d_3 = 0$ giving symmetric modes and $d_1 + d_2 = 0$, antisymmetric modes (Fig. 7.2). Symmetric modes have the form ($F(z) = 1$)

$$u = Ae^{i(k_x x - \omega t)}(u_x, u_z) \tag{7.5a}$$

$$u_x = d_1 k_x \left[\cos k_L z - \frac{\cos(k_L a/2)}{s\gamma 1}\cos k_x z\right], \quad -\frac{a}{2} \leq z \leq \frac{a}{2} \tag{7.5b}$$

$$u_z = id_1 k_L \left[\sin k_L z + \frac{k_x \cos(k_L a/2)}{sk_L \gamma_1}\sinh k_x z\right] \tag{7.5c}$$

For these modes the potential at the monolayer is non-zero and the monolayer has no electrical effect. The condition $u_z = 0$ at $z = 0$ is already satisfied by symmetric well-modes, so, in fact, the monolayer has no effect whatsoever. The symmetric modes are described exactly as if the monolayer were not there (see Section 5.2).

The antisymmetric modes, however, are significantly modified by the monolayer, since for these modes the potential vanishes at $z = 0$. Normal antisymmetric well-modes do not have this property, *in tandem with* $u_z = 0$. Thus antisymmetric modes are forced to have a symmetric component in order to satisfy $u_z = 0$ at $z = 0$. They have the more complicated form

$$\underline{u} = (\theta(z) - \theta(-z))Ae^{i(kx - \omega t)}(u_x, u_z) \tag{7.6a}$$

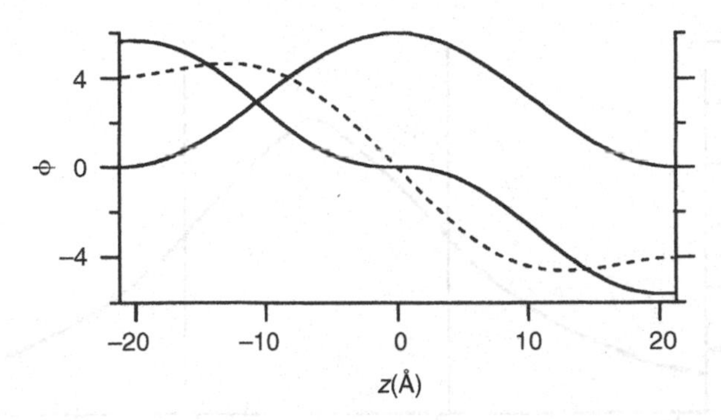

Fig. 7.3 Waveforms of scalar potential in a GaAs quantum well containing a centrally placed monolayer for the lowest-order symmetric and antisymmetric modes. The dashed curve is the lowest-order antisymmetric mode in the absence of monolayer. ($k_x a/2 = 0.1$ for all curves. The potential is in arbitrary units.)

$$u_x = k_x\left[d_1(\cos k_L z - \frac{1}{s}\cosh k_x z) \mp d_3(\sin k_L z - \frac{k_L}{k_x}\sinh k_x z)\right] \tag{7.6b}$$

$$u_z = ik_L\left[d_1(\sin k_L z + \frac{k_x}{sk_L}\sinh k_x z) \mp d_3(cos\, k_L z - \cosh k_x z)\right] \tag{7.6c}$$

where, again, the upper sign is for $z > 0$, the lower for $z < 0$ (Fig. 7.3).

Since symmetric modes are unaffected, any change of intrasubband scattering will be confined to that generated by the change of the electron potential introduced by the monolayer since this modifies the electron wavefunction. An AlAs monolayer would contribute a negative potential and this would repel electrons from the centre, leading to a weaker interaction with the symmetric modes. However, an InAs monolayer would tend to attract electrons (Fig. 7.4) and produce an enhancement of the intrasubband rate. Local monolayer modes cannot be expected to play a significant role in the electron–phonon interaction. Those of AlAs have frequencies far above those of GaAs and these modes will be heavily localized. The reverse will be true of InAs. In the system we have chosen to study the AlAs interface mode associated with the *barriers* will always be effective in scattering electrons, and will not be affected by the presence of the monolayer.

As regards adjacent intersubband scattering, by far the biggest contribution in a quantum well comes from the lowest-order antisymmetric mode, which involves an LO component for which, typically, $k_L a/2 \approx 4.49$. When a central monolayer is present, the corresponding quantity is about 6.2, i.e. shorter wavelengths are required to satisfy the boundary conditions. Roughly speaking, we can take the Frohlich potential to be inversely proportional to K_L^2, and hence we may expect the

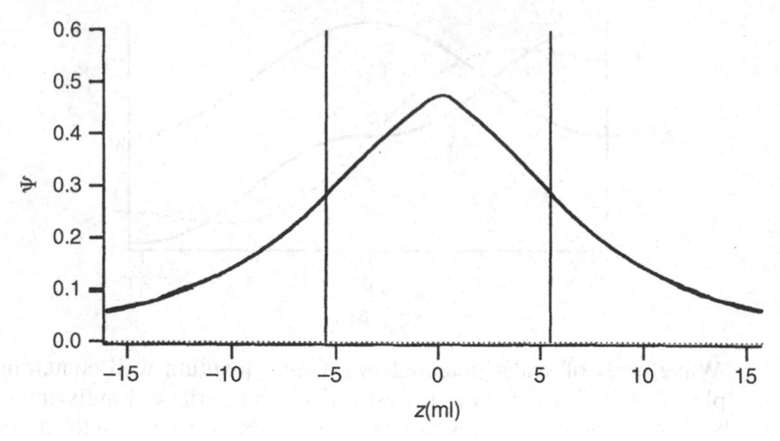

Fig. 7.4 Ground-state electron wavefunctions for $Al_{0.3}Ga_{0.7}As/(GaAs)_5$ $(InAs)_1(GaAs)_5/Al_{0.3}Ga_{0.1}As$.

intersubband rate to be reduced by a factor of about 2. However, the IP contribution to the hybrid is quite different, the shape is different and the electron wavefunctions will be affected by the monolayer potential, so there are many factors that enter to determine the scattering rate in addition to the size of the LO potential. However, a detailed analysis has shown that, in spite of such modifications that occur, there is no significant change in the scattering rate (Bennett *et al.*, 1998).

7.2.2 Double Monolayer

The analytic mode patterns and dispersion in a quantum well containing two symmetrically placed monolayers (Fig. 7.1) are more complicated. Regarding once again each monolayer as dielectrically neutral but imposing the boundary condition $u_z = 0$ in the double-hybrid approximation, we obtain

$$\mathbf{u} = Ae^{i(k_x x-\omega t)}(u_x, u_z)$$

$$\frac{b}{2} \leq |z| \leq \frac{a}{2} \left\{ \begin{array}{l} u_x = \pm(\chi_1 \cos(k_L b/2)) \pm (\chi_2 \sin(k_L b/2)) \frac{k_x}{d_3} [d_1 \cos k_L(z \pm (b/2)) \\ \qquad \pm d_3 \sin k_L(z \pm (b/2)) \\ \qquad + \frac{1}{s} \{d_2 \cosh k_x(z \pm (b/2)) \mp (\frac{sk_L}{k_x} d_3 \sinh k_x(z \pm (b/2))\}] \\ u_z = \pm(\chi_1 \cos(k_L b/2)) \pm (\chi_2 \sin(k_L b/2)) \frac{ik_L}{d_3} [d_1 \sin k_L(z \pm (b/2)) \\ \qquad \mp d_3 \cos k_L(z \pm (b/2)) \\ \qquad - \frac{k_x}{sk_L} d_2 \sinh k_x(z \pm (b/2)) \pm d_3 \cosh k_x(z \pm (b/2))] \\ \qquad \text{upper sign for } z < 0, \text{ lower sign for } z < 0 \end{array} \right. \tag{7.7}$$

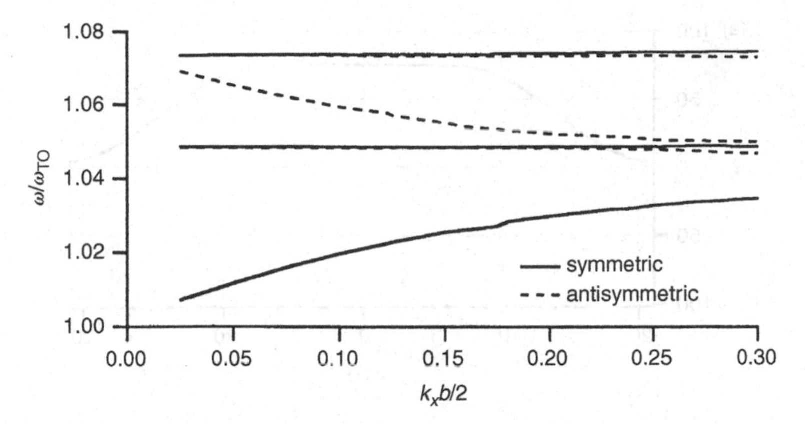

Fig. 7.5 Dispersion of low-order hybrid optical modes in a GaAs well with two uniformly placed monolayers (well-width 42Å).

$$|z| \leq \frac{b}{2} \left\{ \begin{aligned} u_x &= k_x\chi_1\left(\sin k_L z - \tfrac{k_L \cos(k_x b/2)}{k_x \cosh(k_x b/2)} \sinh k_x z\right) \\ &\quad + k_x\chi_2\left(\cos k_L z + \tfrac{k_L \sin(k_L b/2)}{k_x \sinh(k_x b/2)} \cosh k_x z\right) \\ u_z &= ik_L\chi_1\left(\cos k_L z - \tfrac{\cos(k_L b/2)}{\cosh(k_x b/2)} \cosh k_x z\right) \\ &\quad + ik_L\chi_2\left(\sin k_L z - \tfrac{\sin(k_L b/2)}{\sinh(k_x b/2)} \sinh k_x z\right) \end{aligned} \right.$$

In these equations

$$\chi_1 = \cos(k_L b/2) - i\sin(k_L b/2)\left(\frac{d_1 + d_2}{d_3} - \frac{sk_L}{k_x}\coth(k_x b/2)\right)$$

$$\chi_2 = \sin(k_L b/2) + \cos(k_L b/2)\left(\frac{d_1 + d_2}{d_3} - \frac{sk_L}{k_x}\tanh(k_x b/2)\right)$$

and d_1, d_2, d_3 (and their constituent parameters γ_1 and γ_2) are given by Eq. (7.2) with $a/2$ replaced by c (see Fig. 7.1). These waves satisfy the dispersion relation

$$\chi_1\chi_2 = 0 \tag{7.8}$$

with $\chi_1 = 0$ specifying symmetric, and $\chi_2 = 0$ antisymmetric solutions (Fig. 7.5).

Figs. 7.6a and b show lowest-order symmetric and antisymmetric modes. In contrast to the case of a single monolayer, the antisymmetric mode is now the quantum-well mode and it is the symmetric mode that is most affected. The stretching of the symmetric mode across the region between the two monolayers

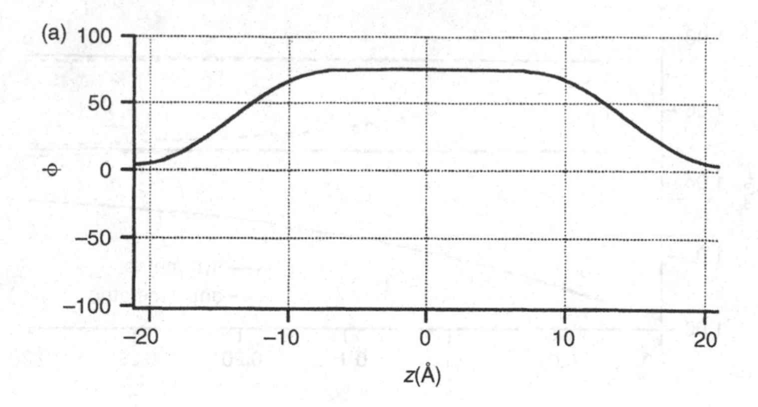

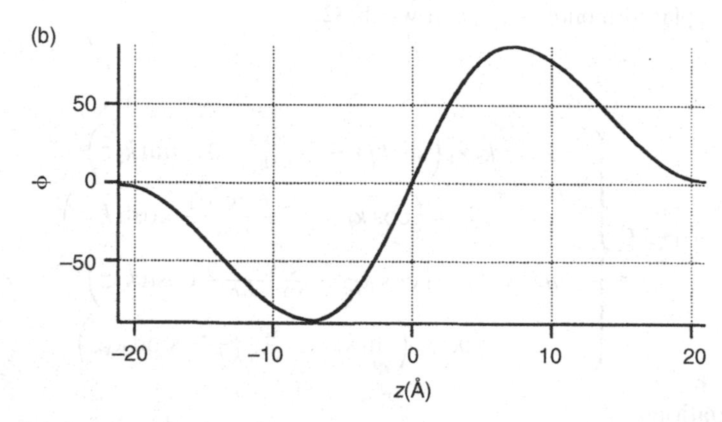

Fig. 7.6 Waveforms of scalar potential for lowest-order modes in a GaAs quantum well containing two symmetrically placed monolayers at $z = \pm 7$Å ($k_x b/2 = 0.25$): (a) symmetric; (b) antisymmetric.

is performed with the help of the symmetric interface mode of GaAs. Fig. 7.7 shows higher-order mode patterns.

A quantum well containing a superlattice of monolayers is also of interest. Specifying that the ionic displacement vanishes at each monolayer has the general effect of inhibiting long wavelengths of LO modes. The tendency is to reduce the coupling to electrons. Even if the limit is set by the interface modes of the well, a significant reduction in rates can be envisaged, unlikely in the light of the result for the single monolayer, evidence for the action of a sum rule.

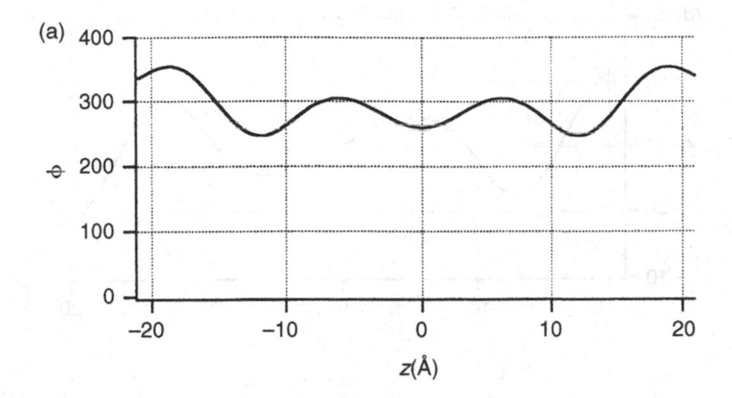

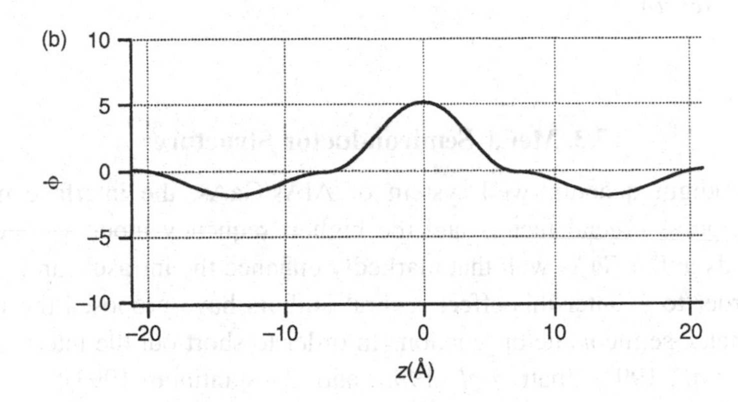

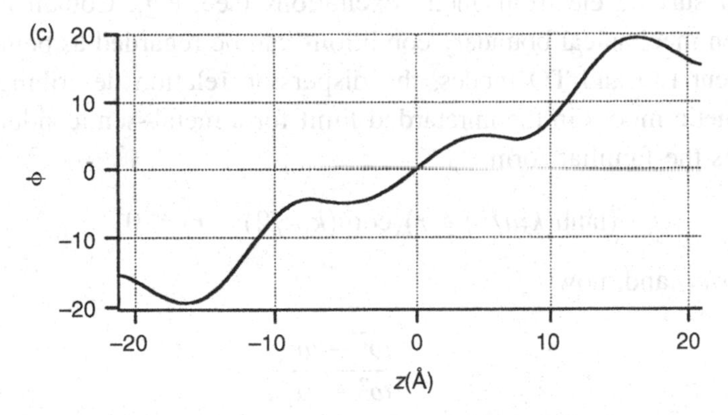

Fig. 7.7 High-order mode patterns in the double-monolayer system ($k_x b/2 = 0.25$).

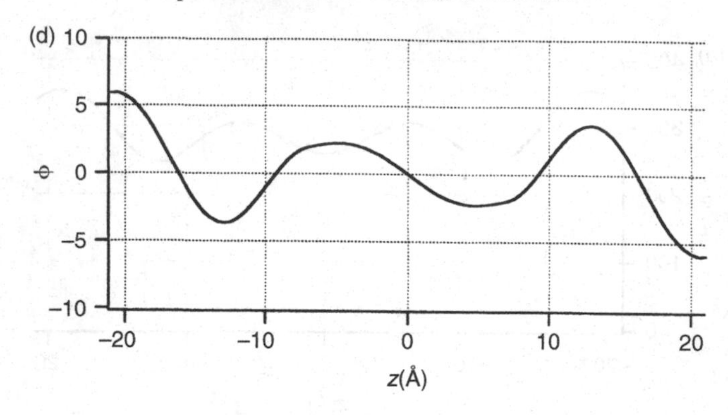

Fig. 7.7 (*cont.*)

7.3 Metal–Semiconductor Structures

In the paradigm quantum-well system of AlAs/GaAs, the interface modes in AlAs emerge as crucial factors and the higher-frequency mode produces symmetric fields in the GaAs well that markedly enhance the intrasubband scattering rate. In order to counter this effect several authors have proposed the incorporation of metal–semiconductor junctions in order to short out the interface modes (Stroscio *et al.*, 1992; Bhatt *et al.*, 1993; and Constantinou, 1993).

The interface between a metal and a polar semiconductor supports a rich spectrum of surface electromagnetic excitations (see, e.g., Cottam and Tilley, 1989). When mechanical boundary conditions can be regarded as being satisfied by evanescent LO and TO modes, the dispersion relation describing interface electromagnetic modes in the unretarded limit for a metal–semiconductor–metal structure has the familiar form

$$(\tanh(k_x a/2) + r)(\coth(k_x a/2) + r) = 0 \tag{7.9}$$

where $r = \varepsilon/\varepsilon_M$ and, now

$$\varepsilon = \varepsilon_\infty \frac{\omega^2 - \omega_{\mathrm{LO}}^2}{\omega^2 - \omega_{\mathrm{TO}}^2} \tag{7.10a}$$

$$\varepsilon_M = \varepsilon_0 \left(1 - \frac{\omega_p^2}{\omega^2}\right) \tag{7.10b}$$

Here we assume the metal to be a perfect conductor with a plasma frequency ω_p. The condition $\tanh k_x a/2 + r = 0$ refers to a symmetric

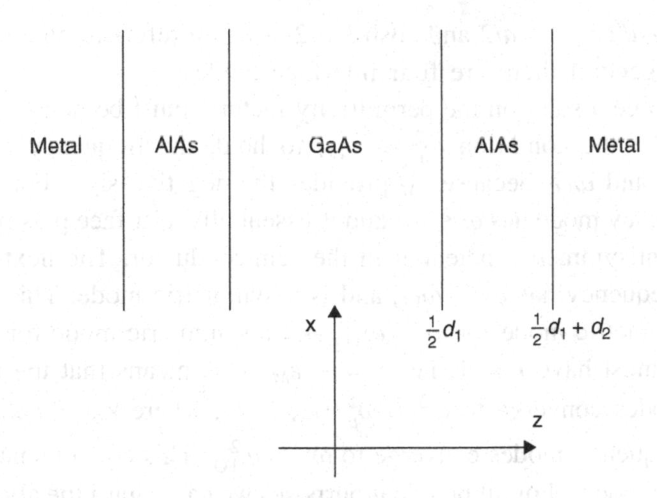

Fig. 7.8 Metal/AlAs/GaAs/AlAs/metal structure.

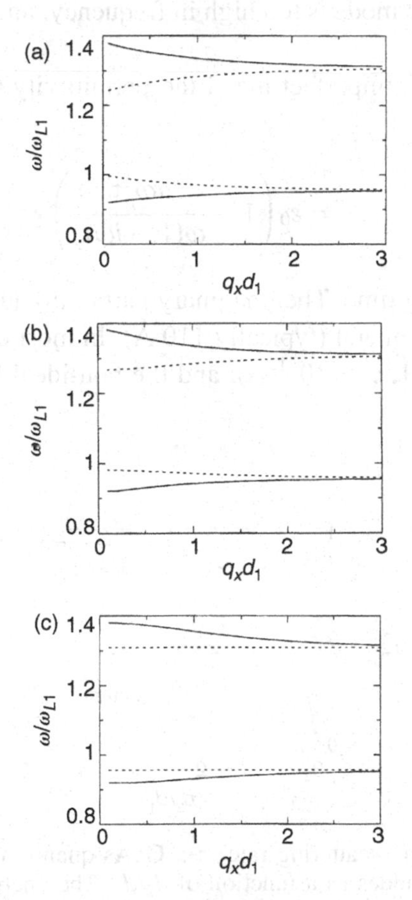

Fig. 7.9 Interface-polariton dispersion: (a) no metal; (b) $d_2/d_1 = 2$; (c) $d_2/d_1 = 0.5$ (ω_{L1} is the frequency of the GaAs LO phonon).

potential for $-a/2 \leqslant z \leqslant a/2$ and $\cosh k_x a/2 + r = 0$ refers to an antisymmetric potential. In general there are four interface modes.

For there to be a solution the permittivity factor r must be negative. Since we can safely take the condition $\omega_p^2 \gg \omega_{LO}$ to hold, the frequency must not lie between ω_{TO} and ω_{LO} because ε_M provides the negative sign. For $k_x \rightarrow 0$ the highest-frequency mode has $\omega \leqslant \omega_p$ and is essentially a surface plasma excitation yielding an antisymmetric potential in the semiconductor. The next in order of decreasing frequency has $\omega \leqslant \omega_{LO}$ and is a symmetric mode. This is followed by an antisymmetric mode for $\omega \leqslant \omega_{TO}$ and a symmetric mode for $\omega \approx 0$. For $k_x \rightarrow \infty$ we must have $r = 1$, i.e. $\varepsilon = -\varepsilon_M$. This means that the two higher-frequency modes converge to $\omega^2 \approx \omega_p^2/(\kappa_\infty + 1)$, where $\kappa_\infty = \varepsilon_\infty/\varepsilon_0$, and the two lower-frequency modes converge to $\omega^2 \approx \omega_{TO}^2$. This convergence is fast for the symmetric modes. For all practical purposes we can regard the allowed modes as plasma-like or TO-like. As far as the interaction with electrons is concerned, the presence of perfect metal interfaces eliminates the effect of interface modes, because the plasma-like mode is too high in frequency, and the TO-like mode has zero scalar potential.

In the real case of an imperfect metal the permittivity of Eq. (7.10b) must be replaced by

$$\varepsilon_M = \varepsilon_0 \left(1 + \frac{i\omega_p^2 \tau}{\omega(1 - i\omega\tau)} \right) \tag{7.11}$$

where τ is the scattering time. The imaginary part of ε_M quantifies the penetration depth of the field in the metal (typically 110 Å). In most cases of interest $\omega\tau$ » 1 (e.g., $\omega \approx 5 \times 10^{13}\ s^{-1}1$, $\tau \approx 10^{-12}\ s$), and the non-ideal behaviour of the metal can be ignored.

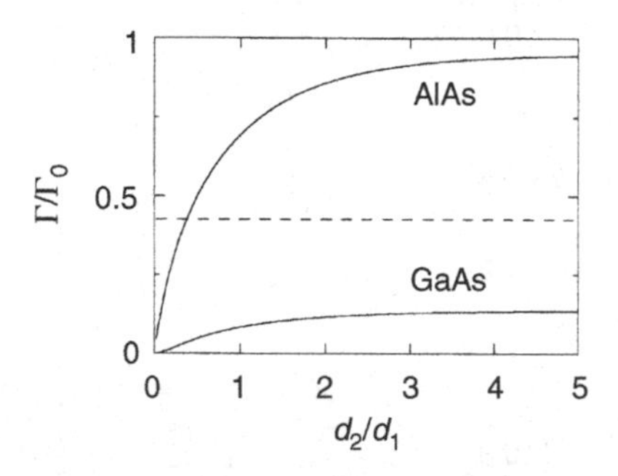

Fig. 7.10 Intrasubband scattering rate in a GaAs quantum well (d_1 =40Å) by AlAs and GaAs IP modes as a function of d_2/d_1. The energy of the electron is $2\hbar\omega_{L1}$. The dashed line is the DC rate for GaAs LO modes (Constantinou, 1993).

The reduction of the coupling strength of AlAs interface modes in the AlAs/GaAs/AlAs system by incorporating metal boundaries on the AlAs layers (Fig. 7.8) has been evaluated by Constantinou (1993). Fig. 7.9 shows the interface-mode dispersion and Fig. 7.10 shows the result for a 40 Å GaAs well – the intrasubband rate can be reduced significantly. A large reduction has also been found by Stroscio *et al.* (1992) in the case of a quantum wire.

7.4 Slab Modes

The study of combined optical modes began with attempts to describe optical modes in an NaCl slab. The most famous of these was the theory of Fuchs and Kliewer (1965), whose dielectric-continuum (DC) model, based on the assumption of purely electromagnetic boundary conditions, predicted the existence of long-wavelength modes of three distinct types: LO modes at the

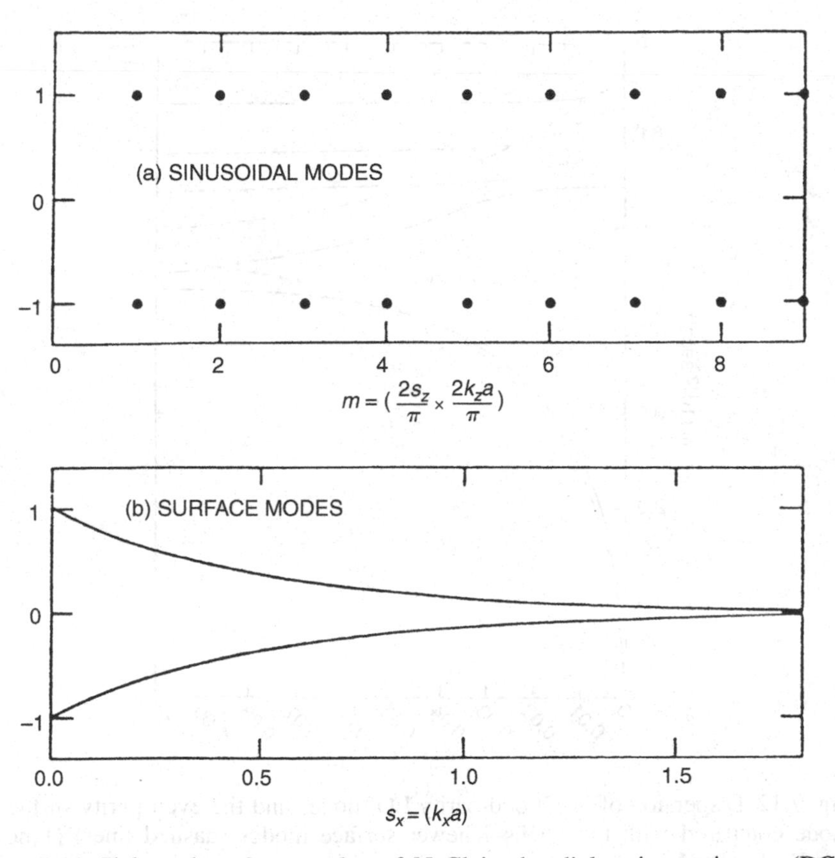

Fig. 7.11 Slab and surface modes of NaCl in the dielectric-continuum (DC) model (Fuchs and Kliewer, 1965).

frequency ω_{LO}, TO modes at the frequency ω_{TO} and two surface polaritons – one symmetric, the other antisymmetric – at frequencies lying between ω_{TO} and ω_{LO} (Fig. 7.11). It was never obvious, however, that a continuum model that ignored spatial dispersion, and therefore elastic stress, could be valid, quite apart from doubts stemming from the validity of *any* continuum description of optical modes. Indeed, as regards the latter point, Jones and Fuchs (1971) developed a microscopic theory of slab modes that indicated (with hindsight) that significant hybridization of optical modes occurred (Fig. 7.12), unlike what was found by Fuchs and Kliewer, and that the lack of hybridization in the latter's theory could be traced, at least in part, to neglect of the change of force constants at the surface. In the language of quasi-continuum theory this points to the role of delta-function components at the surface associated with change of force constant and change of mass. Indeed,

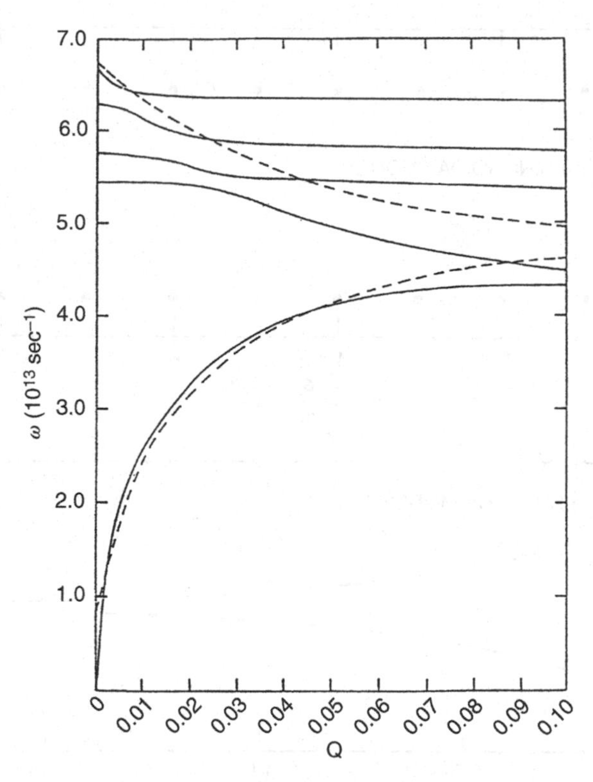

Fig. 7.12 Dispersion of NaCl odd-parity LO modes and the even-parity surface mode compared with the Fuchs–Kliewer surface modes (dashed lines) (Jones and Fuchs, 1971).

our discussion of linear-chain theory in Section 3.2.4 suggested that the mechanical boundary condition might be simply $\mathbf{u} = 0$.

But quite separate from this general question of the influence of interface factors, there remains the problem of the neglect of mechanical boundary conditions associated with macroscopic elastic stress. It turns out that taking optical stress to have antisymmetric components (following the discussion in Chapter 4) and demanding that it be zero at the surface justify the DC model. Thus taking the normal to the surface to be in the z direction the condition of optical stress means that $\nabla.\mathbf{u}$ and $(\nabla \mathbf{x} \mathbf{u})_{x,y}$ are all zero. LO modes have $\nabla \mathbf{x} \mathbf{u} = \mathbf{0}$ everywhere and $\nabla.\mathbf{u}$ is zero if $u_x = 0$ at the surface; this condition can be satisfied by a suitable choice of wavevector component in the z-direction, k_z. Since E_x is proportional to u_x this condition also satisfies the condition that the tangential component of the electric field is continuous. Thus LO modes as described by the DC model automatically satisfy the stress conditions and no hybridization is necessary. TO modes satisfy $\nabla.\mathbf{u} = \mathbf{0}$ everywhere and can satisfy $(\nabla \mathbf{x} \mathbf{u})_{x,y} = 0$ at the surface by suitable choice of k_z that makes $u_z = 0$. No electric boundary condition is involved, so TO modes, like LO modes, need no hybridization. In the unretarded limit, surface modes have the unusual property that both $\nabla.\mathbf{u}$ and $\nabla \mathbf{x} \mathbf{u}$ are zero everywhere, so they automatically satisfy the stress condition. We conclude that the DC model of Fuchs and Kliewer *implicitly assumes* that the optical stress tensor has antisymmetric components and that optical stress vanishes at the surface.

In general, however, we cannot rely on the optical stress tensor's having antisymmetric components and, moreover, it is not obvious that a free surface corresponds to the condition that optical stress should vanish. The latter point is part of the problem of interface components (which has yet to be handled by quasi-continuum theory). Our conclusion drawn from linear-chain theory is that the correct boundary condition is $\mathbf{u} = 0$, so that we would expect the optical modes in a GaAs slab to be hybridized in a similar way to those in a GaAs well enclosed by AlAs. This condition is qualitatively consistent with the lattice dynamic calculations of Jones and Fuchs (1971). On the other hand, if acoustic boundary conditions are insisted upon, hybridization turns out to be very weak, and the modes are essentially those described by Fuchs and Kliewer (Ridley *et al.*, 1994) and, moreover, optical-mode analogues of Rayleigh waves are predicted in non-polar material (Ridley, 1991). These matters are clearly resolvable in experiment. Hitherto, Raman scattering from ionic slabs has tended to lend qualitative support to the Fuchs–Kliewer DC model (see, for example, Ushioda and Loudon, 1982).

7.5 Quantum Wires

One of the first treatments of electron–phonon scattering in quasi-1D was that of Riddoch and Ridley (1984), who used a bulk phonon spectrum and assumed a rectangular cross-section for the wire. Their results have been given in Section 1.3.5, among which is the curious feature over a range of well-widths of a negative momentum-relaxation rate below the emission threshold of the polar LO mode, a manifestation of the preference of the polar interaction for small wavevector changes. The principal feature is, of course, the large rate at threshold in both polar and non-polar materials associated with large density of states near the bottom of each subband.

Recent work has focussed on the effects of phonon confinement using the DC model to describe the potentials. As we will show in Chapter 8, the DC model applied to the GaAs/AlAs system provides an excellent approximation for calculating scattering rates. As regards actual displacement patterns to date there are lattice-dynamic calculations for rectangular wires by Rossi *et al.* (1994) but no hybrid theory. Scattering in wires with circular cross-sections has been examined by Constantinou and Ridley (1994) and Gold and Gazali (1990), and extensions to wires with elliptical cross-sections have been made by Wang and Lei (1994) and Bennett *et al.* (1995). The more practical case of arbitrary cross-section has been examined by Knipp and Reinecke (1992, 1994). The advantages of circular and elliptical cross-sections are that there are no awkward corners, as there are in a rectangular cross-section, and the solutions are separable in the coordinates and describable in terms of standard functions – those of Bessel and Mathieu for the circular and elliptical cases, respectively. The main features of phonon confinement that emerge are that the potentials associated with the bulk confined modes tend to be concentrated, as are the electron wavefunctions, in regions of low curvature, whereas those associated with interface modes tend to be concentrated in regions of high curvature, a curious division first pointed out by Knipp and Reinecke (1994).

It is useful to focus attention on the case of elliptical cross-section, since this encompasses the case of circular cross-section, and when the major axis is relatively large an approach is made to the 2D situation. We consider a wire of elliptical cross-section, semimajor axis a and semiminor axis b, and let u be the radial coordinate, v the angular coordinate and z the axial coordinate. The solution of Schrödinger's equation for the electron wavefunction can be obtained in the form

$$\Psi(\mathbf{R}) = Ae^{ikz}U(u)V(v) \tag{7.12}$$

where $U(u)$ and $V(v)$ satisfy the Mathieu equations

$$\frac{d^2U(u)}{du^2} - (\beta - 2\lambda\cosh(2u))U(u) = 0 \tag{7.13a}$$

$$\frac{d^2V(v)}{dv^2} - (\beta - 2\lambda\cos(2v))V(v) = 0 \tag{7.13b}$$

The angular fluctuations are periodic ($V(v + 2\pi) = V(v)$); that means that the constant β is quantized with azimuthal quantum number m. In standard notation (see, for example, McLachlan, 1947) the solutions that are regular at the origin are

$$\Psi_{mnk}(\mathbf{R}) = A_{mn}e^{ikx}\begin{cases} Ce_m(u,\lambda_{mn})ce_m(v,\lambda_{mn}) & \text{even} \\ Se_m(u,\lambda_{mn})se_m(v,\lambda_{mn}) & \text{odd} \end{cases}$$
$$\lambda_{mn} = \frac{1}{4}f^2k_{mn}^2 \tag{7.14}$$

where $f = ae, e =$ eccentricity $= (1 - (b/a)^2)^{1/2}$ is the semifocal distance and k_{mn} is the confinement wavevector in the radial direction. Symmetry assignments are with respect to the parity of the angular functions with respect to v; if m is odd, the periodicity is π, and if m is even, it is 2π. The quantum number n is determined by the boundary condition at $u = u_0 = \cosh^{-1}(1/e)$, which, for simplicity, we take to be the vanishing of the wavefunction. The normalization factor is given by

$$A_{mn}^{-2} = L_z \int_0^{2\pi}\int_0^{u_0} U_m^2(u,\lambda_{mn})V_m^2(v,\lambda_{mn})f^2(\sinh^2 u + \sin^2 v)\,du\,dv \tag{7.15}$$

where L_z is the length of the wire.

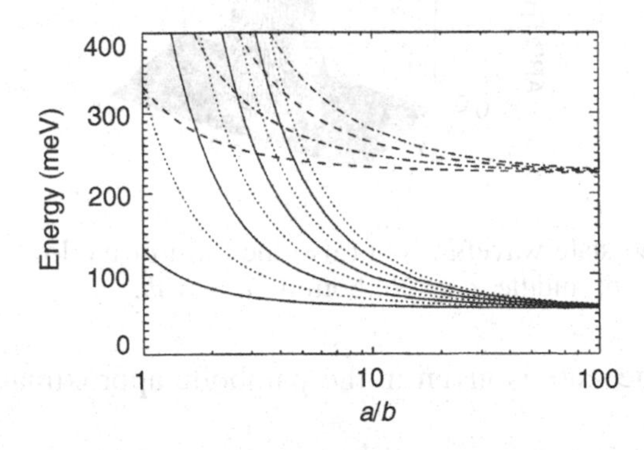

Fig. 7.13 Subband energies of electron states in an elliptical wire with semi-major axis a and semiminor axis b. Even states: solid line, period π, dotted, period 2π; odd states: dashed line, period π, dash–dot, period 2π. (GaAs, b = 50 Å)

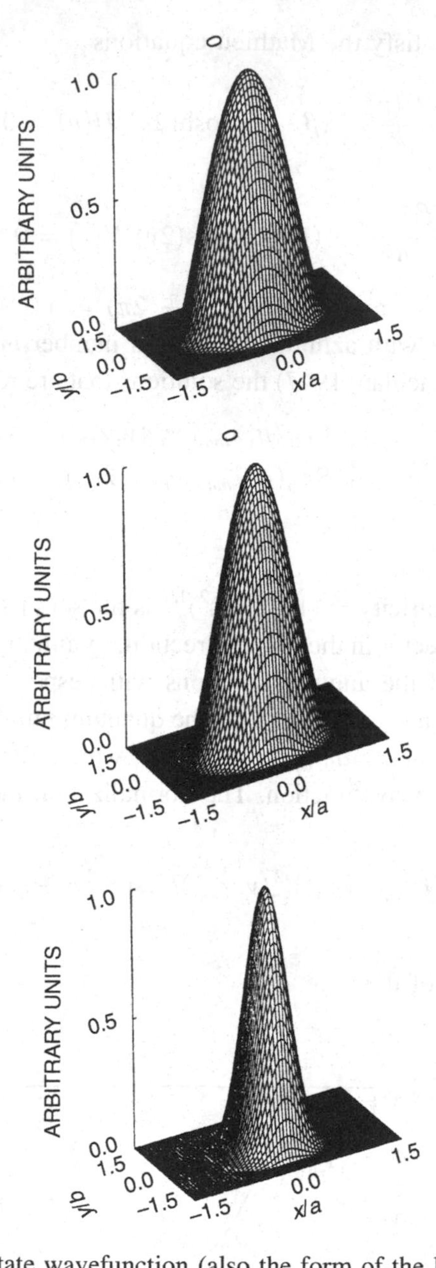

Fig. 7.14 Ground state wavefunction (also the form of the lowest confined LO mode). Top, $a = b$; middle, $a = 2b$; bottom, $a = 10b$.

The energy of the state is given in the parabolic approximation by

$$E_{mnk} = \frac{\hbar^2}{2m^*}(k_{mn}^2 + k^2) \tag{7.16}$$

Its variation in GaAs with a/b for fixed $b = 50$ Å is exhibited in Fig. 7.13. For $a/b = 1$ we have the case of a circular cross-section and for $a/b \to \infty$, that of 2D. Fig. 7.14 illustrates the ground-state wavefunction, which shows a tendency to concentrate in the region of smallest curvature.

Imposing the same boundary condition on the potential of confined optical modes (DC model) leads to identical forms. Fig. 7.14 serves also to illustrate the lowest-order confined mode.

The potentials associated with interface modes have the form

$$\Phi_{mq}(\mathbf{R}) = C_m e^{iqz} U_m(u) V_m(v) \tag{7.17a}$$

$$U_m = \begin{cases} Ce_m(u, \lambda_{mn}) Fek_m(u_0, \lambda_m) & u \le u_0 \\ Fek_m(u, \lambda_{mn}) Ce_m(u_0, \lambda_m) & u > u_0 \end{cases} \quad V_m(v) = ce_m(v, \lambda_m) \text{ even} \tag{7.17b}$$

$$U_m = \begin{cases} Se_m(u, \lambda_{mn}) Gek_m(u_0, \lambda_m) & u \le u_0 \\ Gek_m(u, \lambda_{mn}) Se_m(u_0, \lambda_m) & u > u_0 \end{cases} \quad V_m(v) = se_m(v, \lambda_m) \text{ odd} \tag{7.17c}$$

$$\lambda = -\frac{1}{4} f^2 q^2 \tag{7.17d}$$

These functions are finite at the origin and vanish at infinity. Imposing the usual electrical boundary conditions at $u = u_0$ leads to a dispersion relation that is most conveniently depicted in terms of the quantity P, independent of material parameters, where

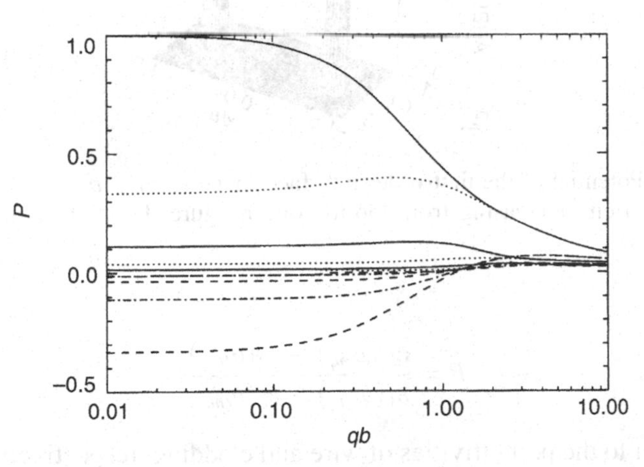

Fig. 7.15 P versus qb ($a = 2b$). Key as for Fig. 7.13.

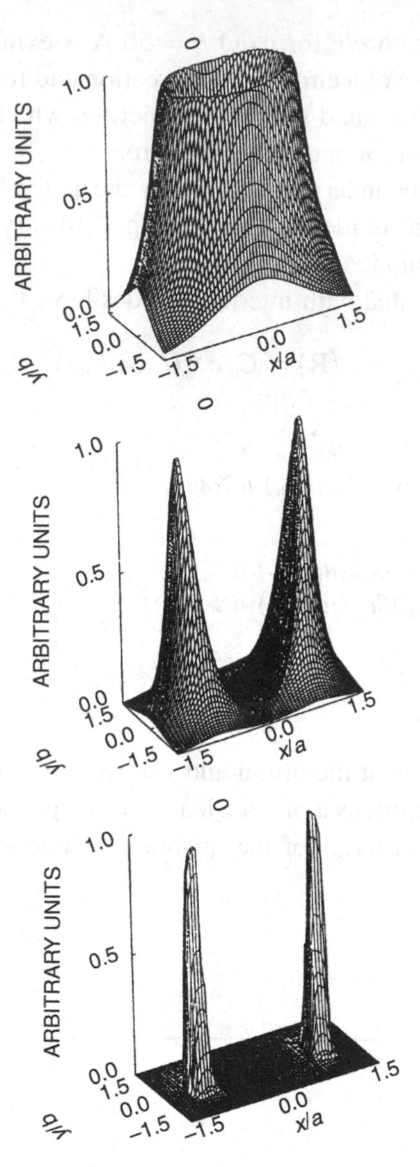

Fig. 7.16 Potential of the first-order interface mode for $a = b$, $a = 2b$, and $a = 10b$, eccentricity increasing from top to bottom figure (Bennett *et al.*, 1995).

$$P = \frac{\varepsilon_1(\omega_{mq}) + \varepsilon_2(\omega_{mq})}{\varepsilon_1(\omega_{mq}) - \varepsilon_2(\omega_{mq})} \tag{7.18}$$

and ε_1, ε_2 refer to the permittivities of wire and cladding, respectively. Thus $P = 1$ when ω_{mq} equals the zone-centre TO frequency in the wire or the zone-centre LO frequency of the cladding, and $P = -1$ when ω_{mq} equals the zone-centre LO frequency of the wire or the zone-centre TO frequency of the cladding. P versus qb is depicted in Fig. 7.15, where it is clear that the density of interface modes is highest where $P=0$.

The term proportional to λ in the defining equations for $U(u)$ and $V(v)$ acts as a potential energy and so it is not surprising that the change in sign of λ has an appreciable effect. Fig. 7.16 shows the potential profile for the first-order interface mode. Instead of preferring the trough in the confined-mode λ-potential, the interface modes prefer the peaks. In this respect the confined modes act as electrons and the interface modes as holes. The concentration of the interface mode in regions of high curvature is also strikingly evident in rectangular wires.

7.6 Quantum Dots

Interest in quantum dots stems mainly from their optical properties. Whereas in 2D and 1D it is usual to ignore the Coulomb interaction between carriers in determining subband energies, this is far less justifiable in quantum dots. In many practical cases electrons and holes are present as a result of photo excitation and therefore in equal concentrations, and one is usually more interested in excitonic behaviour than in single-particle processes. In the simplest picture, this means that there is negligible interaction with polar-optical modes as a consequence of the neutral nature of the exciton. More sophisticated models revive the polar interaction, which, in fact, must be allowed for in determining energy levels in the form of a Huang–Rhys factor and Franck–Condon effects. Multiphonon processes therefore become possible. There is much in the way of interesting work to be done here with hybrid modes, but these topics lead us too far away from the general thrust of this book to warrant coverage, even if the author were intimately familiar with the field; he is not. We will be content to refer the interested reader to the book by Banyai and Koch (1993) and references therein.

8
Electron–Optical Phonon Interaction in a Quantum Well

There's some that swear by whisky,
There's some that swear by rye,
There's some that swear by **A.p**,
And others by $e\phi$.
On Seeing the Light, *B. K. Ridley*

8.1 Introduction

An electron in a quantum-well subband can be scattered to another state in the same subband or into a state in another subband. Intrasubband and intersubband scattering rates have to be calculated separately since different wavefunction symmetries are involved in the two cases, and this implies correspondingly different symmetries of the optical mode. For simplicity we will assume that the electrons are completely confined within the well and that the interaction is with polar-optical modes. In the case of LO modes in a polar material this interaction is via a scalar potential. However, as we will see, it is possible in the unretarded limit (velocity of light is infinite) to replace the vector potential of the electromagnetic interface wave with a scalar potential via a unitary transformation (not a gauge transformation) and treat the IP mode on the same footing as an LO mode, but with a frequency-dependent scalar potential. We assume the TO mode has no interaction.

No fewer than four different scattering sources exist, in general. Two of these are associated with well modes, two with barrier modes. In general, the LO band of frequencies in either material does not span the range between the LO and TO zone-centre frequencies, ω_{LO} and ω_{TO}. Triple hybrids occur throughout this range, but, as mentioned several times, the hybrids below the LO band consist of an IP component and relatively rapidly varying TO and LO/LA components, and as such behave approximately as classical IP modes. In this they are very different from LO-like hybrids within the LO band but similar to IP-like hybrids.

Thus we distinguish two main categories of hybrid:

$$\omega_{\mathrm{LO}} > \omega > \omega_{\mathrm{LZ}} \quad \text{LO hybrids}$$
$$\omega_{\mathrm{LZ}} > \omega > \omega_{\mathrm{TO}} \quad \text{IP hybrids}$$

where ω_{LZ} is the frequency of the LO mode at the zone boundary. A similar division occurs in the case of barrier modes. We remind ourselves that LO hybrids are further subdivided into LO-like and IP-like.

For further simplicity we assume that the frequencies of well and barrier modes are so different that no penetration of ionic motion across the interface occurs. At first sight this may seem to rule out the interaction with barrier LO modes, but this is not the case because an incident LO wave can initiate an IP field in the well at the barrier frequency as described in Section 5.3.

In general, the evaluation of scattering rates has to be done numerically. Analytic expressions can be derived under certain rather restrictive conditions, namely when the well is an infinitely deep potential for carriers and when the initial state of the carrier corresponds to a threshold energy for the scattering process. These expressions are useful for providing magnitudes for the threshold rates in the limiting case of complete confinement of carriers and phonons, which can be used to compare different models and to obtain an insight into the dependence on LO dispersion. In real situations confinement is never total and a direct interaction with barrier modes can take place whose strength will depend upon the proportion of the electron wavefunction outside the well. There will also be an indirect interaction with well modes, so in general no fewer than eight sources of scattering have to be considered, four in the well and four in the barrier. Our analysis of the case of complete confinement will be sufficient to illustrate the principles involved.

8.2 Scattering Rate

The scattering between initial (i) and final (f) electron states is given to first order by

$$W = \frac{2\pi}{\hbar} \int |M(f,i)|^2 \delta(E_f - E_i) dN_f \tag{8.1}$$

where E_i, E_f are the initial and final state total energies and N_f is the number of final states. If the electron wavefunctions are ψ_i and ψ_f and the interaction energy is H_{int} the matrix element is

$$M(f,i) = \int \psi_f^* H_{\mathrm{int}} \psi_i d\mathbf{r} \tag{8.2}$$

where the integral is over all the space of the normalizing cavity. The triple-hybrid nature of the optical modes reflects itself in a triple interaction:

$$H_{\text{int}} = H_{\text{LO}} + H_{\text{TO}} + H_{\text{IP}} \tag{8.3}$$

In general, each component of the interaction would include a deformation potential, but in what follows we will focus solely on the polar interaction. The polar components are

$$H_{\text{LO}} = e\phi_{\text{LO}}, \quad H_{\text{TO}} = 0, \quad H_{\text{IP}} = -\frac{e}{m^*}\mathbf{A.p} \tag{8.4}$$

where ϕ_{LO} is a scalar potential associated with the LO component and $\mathbf{A}$ is the vector potential associated with the interface polariton. In the unretarded limit where propagation speeds of transverse modes are very small compared with that of light, the TO electric fields are vanishingly small and hence the polar interaction is negligible. This is not the case for IP modes whose frequencies lie between the zone-centre and LO and TO frequencies. IP modes are essentially transverse electromagnetic waves whose electric and magnetic fields can be described most conveniently in the Coulomb gauge by a single transverse vector potential. It is, however, possible to exploit the unretarded limit and associate the IP electric fields with a scalar potential, in which case the polar interaction takes the simpler form

$$H_{\text{int}}^{\text{pol}} = e(\phi_{\text{LO}} + \phi_{\text{IP}}) \tag{8.5}$$

The question of scalar and vector potentials for the IP mode is discussed in the Appendix.

8.3 Scattering Potentials for Hybrids

In what follows we will assume that the IP component possesses a scalar potential. To be specific we consider the interaction with hybrids in a quantum well with rigid boundaries. It is convenient to be reminded that the relative ionic displacements for the antisymmetric solution in the well are for $-a/2 \leq z \leq a/2$

$$\mathbf{u}_x = i\mathbf{q}_x A_a \left[\sin q_L z - s_T \sinh \alpha_T z - s_p \sinh q_x z\right] e^{i(\mathbf{q}_x.\mathbf{x}-\omega t)} \tag{8.6a}$$

$$u_z = q_L A_a \left[\cos q_L z - \frac{q_x^2}{q_L \alpha_T} s_T \cosh \alpha_T z - \frac{q_x}{q_L} s_p \cosh q_x z\right] e^{i(\mathbf{q}_x.\mathbf{x}-\omega t)} \tag{8.6b}$$

The associated electric fields are

$$\mathbf{E}_x = -i\left(\frac{e^*}{\varepsilon_\infty}\right)\mathbf{q}_x A_a\left[\sin q_L z - s s_p \sinh q_x z\right] e^{i(\mathbf{q}_x.\mathbf{x}-\omega t)} \tag{8.7a}$$

$$E_z = -\left(\frac{e^*}{\varepsilon_\infty}\right) q_L A_a \left[\cos q_L z - s\frac{q_x}{q_L} s_p \cosh q_x z\right] e^{i(\mathbf{q}_x.\mathbf{x}-\omega t)} \tag{8.7b}$$

where e^* is the ionic charge ($e^{*2} = \mu(\omega_{\mathrm{LO}}^2 - \omega_{\mathrm{TO}}^2)\varepsilon_\infty$, $\mu =$ reduced density) and s is the field factor $(\omega^2 - \omega_{\mathrm{TO}}^2)/(\omega_{\mathrm{LO}}^2 - \omega_{\mathrm{TO}}^2)$, and these can be related to a scalar potential

$$\phi_a = \left(\frac{e^*}{\varepsilon_\infty}\right) A_a\left[\sin q_L z - s s_p \sinh q_x z\right] e^{i(\mathbf{q}_x.\mathbf{x}-\omega t)} \tag{8.8}$$

In a similar way we can obtain the scalar potential for the symmetric solution

$$\phi_s = -i\left(\frac{e^*}{\varepsilon_\infty}\right) A_s\left[\cos q_L z - s c_p \cosh q_x z\right] e^{i(\mathbf{q}_x.\mathbf{x}-\omega t)} \tag{8.9}$$

The amplitudes A_a and A_s are those obtained through the energy normalization, and the fractional amplitudes of the IP components, s_p and c_p, along with the wavevector of the LO component, q_L, have been given previously, and are reproduced here in slightly different form using the relevant dispersion relation in the limit of $a_T \to \infty$

$$s_p = \frac{q_L \cos(q_L a/2)}{q_x \cosh(q_x a/2)}, \qquad c_p = \frac{q_L \sin(q_L a/2)}{q_x \sinh(q_x a/2)} \tag{8.10}$$

8.4 Matrix Elements for an Infinitely Deep Well

Simple analytic expressions can be obtained for transitions involving electrons in an infinitely deep quantum well, whose envelope wavefunctions are of the form

$$\psi(r) = \frac{1}{\sigma^{1/2}}\chi(z) e^{i\mathbf{k}_x.\mathbf{x}} \tag{8.11a}$$

$$\chi(z) = \left(\frac{2}{a}\right)^{1/2} \begin{cases} \cos k_s z \\ \sin k_a z \end{cases} \tag{8.11b}$$

where $k_s = (2n - 1)\pi/a$ and $k_a = 2n\pi/a$, $n = 1, 2$, etc., and σ is the area of the plane.

The matrix element is then

$$M(f,i) = \frac{1}{\sigma^{1/2}} \iint \chi_f(z)\, e^{-ik_{xf}\cdot\mathbf{x}} e\phi(z) e^{+i\mathbf{q}_x\cdot\mathbf{x}} \chi_i(z) e^{i\mathbf{k}_{xi}\cdot\mathbf{x}} dz d\mathbf{x}. I(q_f, q_i)$$
$$= \delta_{k_{xi}+q_x,\mathbf{k}_{xf}} e\phi_0 G(f,i) I(\mathbf{k}_f, \mathbf{k}_i) \tag{8.12}$$

where $I(\mathbf{k}_f,\mathbf{k}_i)$ is the overlap integral of the cell-periodic functions in a unit cell. Crystal momentum in the plane is conserved. The overlap integral, $G(f, i)$, is given by

$$G(f,i) = \int_{-a/2}^{a/2} \chi_f(z)\phi(z)\chi_i(z)dz \tag{8.13}$$

Symmetry considerations indicate that intrasubband transitions and also intersubband transitions involving subbands of the same parity can be effected only by symmetric hybrids, whereas transitions involving subbands of opposite parity require antisymmetric hybrids.

We will consider only two processes, arguably the most important, namely intrasubband transitions within the lowest subband and intersubband transitions involving the first and second subbands. For intrasubband processes in the lowest subband, $\phi \rightarrow \phi_S$, and

$$G(f,i) = -8\pi^2 i \frac{\sin(q_L a/2)}{a^3} \left[\frac{1}{q_L\left[(2\pi/a)^2 - q_L^2\right]} + \frac{sq_L}{q_x^2\left[(2\pi/a)^2 + q_x^2\right]} \right] \tag{8.14}$$

and for intersubband processes between the lowest subbands, $\phi \rightarrow \phi_a$, and

$$G(f,i) = 16\pi^2 \frac{q_L \cos(q_L a/2)}{a^3}$$
$$\left[\frac{1}{\left[(3\pi/a)^2 - q_L^2\right]\left[(\pi/a)^2 - q_L^2\right]} - \frac{s}{\left[(3\pi/a)^2 + q_x^2\right]\left[(\pi/a)^2 + q_x^2\right]} \right] \tag{8.15}$$

The LO wavevector is determined by the relevant dispersion relation, which in the limit $a_T \rightarrow \infty$ is for the antisymmetric mode

$$\cot(q_L a/2) = \frac{q_x}{sq_L(\tanh(q_x a/2) + r)} \tag{8.16}$$

and for the symmetric mode

$$\tan(q_L a/2) = -\frac{q_x}{s q_L(\coth(q_x a/2) + r)} \tag{8.17}$$

where r is the permittivity function $(\varepsilon_\infty/\varepsilon_B)(\omega^2 - \omega_{LO}^2)/(\omega^2 - \omega_{TO}^2)$. Essentially, q_L is determined by the dispersion relation as a function of the in-plane wavevector q_x, and this, in turn, is determined by the conservation of momentum in the plane.

8.5 Scattering Rates for Hybrids

The scattering rate can be written

$$W_{ij} = W_0 \hbar \left(\frac{2\hbar\omega}{m^* a^2}\right)^{1/2} \sum_{q_L,k} \int \left(n(\omega, q) + \frac{1}{2} \mp \frac{1}{2}\right) \delta_{\mathbf{k}_{xi} \pm \mathbf{q}_x, \mathbf{k}_{xf}} \frac{G_{ij}^2(q_x, q_L)}{Q_a^2(q_x, q_L)} \times I^2(\mathbf{k}_f, \mathbf{k}_i) \times \delta(E_f - E_i) q_x dq_x d\theta \tag{8.18}$$

where the upper sign is for absorption, the lower for emission, and

$$W_0 = \frac{e^2}{4\pi\hbar}\left(\frac{1}{\varepsilon_\infty} - \frac{1}{\varepsilon_s}\right)\left(\frac{2m^*\omega}{\hbar}\right)^{1/2} \tag{8.19}$$

$$\text{(intra)} \quad G_{11}(q_x, q_L) = -8\pi^2 i \frac{\sin(q_L a/2)}{a^3}\left[\frac{1}{q_L\left[(2\pi/a)^2 - q_L^2\right]} + \frac{s q_L}{q_x^2\left[(2\pi/a)^2 + q_x^2\right]}\right] \tag{8.20}$$

$$\text{(inter)} \quad G_{21}(q_x, q_L) = 16\pi^2 \frac{q_L \cos(q_L a/2)}{a^3} \left[\frac{1}{\left[(3\pi/a)^2 - q_L^2\right]\left[(\pi/a)^2 - q_L^2\right]} - \frac{s}{\left[(3\pi/a)^2 + q_x^2\right]\left[(\pi/a)^2 + q_x^2\right]}\right] \tag{8.21}$$

$$\text{(intra)} \quad Q_s^2 = q_L^2 + q_x^2 + (q_L^2 - q_x^2)\frac{\sin q_L a}{q_L a} - \frac{4 p_s q_x}{a}\cos^2\left(\frac{q_L a}{2}\right)(2 - p_s \coth(q_x a/2)) \tag{8.22}$$

$$\text{(inter)} \quad Q_a^2 = q_L^2 + q_x^2 - (q_L^2 - q_x^2)\frac{\sin q_L a}{q_L a} - \frac{4 p_a q_x}{a}\sin^2\left(\frac{q_L a}{2}\right)(2 - p_a \tanh(q_x a/2)) \tag{8.23}$$

$$p_s = [s(\coth(q_x a/2) + r)]^{-1} \tag{8.24}$$

$$p_a = [s(\tanh(q_x a/2) + r)]^{-1} \tag{8.25}$$

The last four expressions, Eqs. (8.22–8.25), have been derived assuming that the contribution of the TO mode to the total energy of the hybrid is negligible. The conservation of in-plane momentum implies that

$$k_{xf}^2 = k_{xi}^2 + q_x^2 \pm 2k_{xi}q_x \cos\theta \tag{8.26}$$

and the conservation of energy gives

$$\frac{\hbar^2 k_{xf}^2}{2m^*} + E_{n_f} = \frac{\hbar^2 k_{xi}^2}{2m^*} + E_{n_i} \pm \hbar\omega \tag{8.27}$$

where $E_{n_{i,f}}$ is the subband energy. Consequently the delta-function becomes

$$\delta(E_f - E_i) = \delta\left(\frac{\hbar^2 q_x^2}{2m^*} \pm \frac{\hbar^2 q_x}{m^*} k_{xi} \cos\theta + E_{n_f} - E_{n_i} \mp \hbar\omega\right) \tag{8.28}$$

and it becomes convenient to transform the integral over θ to an integral over $\cos\theta$ by the transformation

$$\int_0^{2\pi} d\theta \rightarrow \int_{-1}^{+1} \frac{2d(\cos\theta)}{\sin\theta} \tag{8.29}$$

Integration then gives

$$W_{ij} = W_0\hbar\left(\frac{2\hbar\omega}{m^* a^2}\right)^{1/2} \frac{2m^*}{\hbar^2 k_{xi}} \sum_{q_L} \int_{q_{x\min}}^{q_{x\max}} \left(n(\omega, q) + \frac{1}{2} \mp \frac{1}{2}\right) \times \frac{G_{ij}^2(q_x, q_L)}{Q_a^2(q_x, q_L)}$$
$$\times I^2(\mathbf{k}_f, \mathbf{k}_i) \frac{dq_x}{\sin\theta(q_x, k_{xi})} \tag{8.30}$$

where

$$\sin\theta(q_x, k_{xi}) = \left[1 - \left(\frac{m^*\omega^*}{\hbar k_{xi} q_x} \mp \frac{q_x}{2k_{xi}}\right)^2\right]^{1/2} \tag{8.31a}$$

$$\mp\hbar\omega^* = E_{n_f} - E_{n_i} \mp \hbar\omega \tag{8.31b}$$

$$q_{x\,\min} = k_{xi}\begin{bmatrix}\left(\sqrt{1+\frac{\hbar\omega^*}{E(k_{xi})}}-1\right) & \text{absorption}\\ \left(1-\sqrt{1-\frac{\hbar\omega^*}{E(k_{xi})}}\right) & \text{emission}\end{bmatrix} \quad (8.31c)$$

$$q_{x\,\max} = k_{xi}\begin{bmatrix}\left(\sqrt{1+\frac{\hbar\omega^*}{E(k_{xi})}}+1\right) & \text{absorption}\\ \left(1+\sqrt{1-\frac{\hbar\omega^*}{E(k_{xi})}}\right) & \text{emission}\end{bmatrix} \quad (8.31d)$$

When the scattered particle is an electron in a zone-centre band the cell overlap integral is unity to a good approximation, whereas for a hole the overlap integral is a function of scattering angle, i.e. the angle between $\mathbf{k}_{xi}$ and $\mathbf{k}_{xf}$. Thus (see Section 2.4) to lowest order

$$\begin{aligned} \mathrm{I}^2(\mathbf{k}_f,\mathbf{k}_i) &= 1 && \text{intraparabolic conduction valley}\\ &= \frac{1}{4}(1+3\cos^2\theta_q) && \text{intra-heavy-hole or intra-light-hole} \quad (8.32)\\ &= \frac{3}{4}\sin^2\theta_q && \text{inter-heavy-hole and light-hole} \end{aligned}$$

The scattering angle is readily related to the polar angle appearing in the scattering rate. As pointed out in Section 2.4, θ_q goes to the in-plane scattering angle when $k_x >> n\pi/a$, in which case the squared overlap integral is like the bulk. For strong confinement $k_x << n\pi/a$, the angle ϕ_q goes to zero. In this case, for intrasubband scattering the integral is near unity and near zero for the HH–LH transitions.

In general the integration of Eq. (8.30) must be done numerically, but in the case of electron scattering an analytical expression can be found for the particular case of threshold rates.

8.6 Threshold Rates

Here we focus attention on *emission processes* only, such that the initial energy of the electron is just large enough in the case of the intrasubband processes for a hybridon to be emitted, and in the case of intersubband transitions, where the electron is at the bottom of the upper subband. For intrasubband processes q_x is, therefore, fixed at the value

$$q_x = \left(\frac{2m^*\omega}{\hbar}\right)^{1/2} = q_0 \quad (8.33)$$

which in GaAs is 2.52×10^6 cm^{-1}, for $\hbar\omega_{LO} = 36$ meV. For the intersubband process where the subbands are ΔE apart q_x is determined by

$$q_x = \left(2m^* \frac{(\Delta E - \hbar\omega)}{\hbar^2}\right)^{1/2} = q_1 \tag{8.34}$$

This fixes $G(q_x,q_L)$ independent of θ and the sum over the final states just yields the number of single spin states in the quasi-2D subband.

The intrasubband and intersubband rates can, therefore, be written

$$W_{ij} = \sum_{q_L} W_0 \frac{2\pi^2}{a^2}\left(\frac{\hbar\omega}{E_0}\right)^{1/2} \frac{G_{ij}^2(q_x, q_L)}{Q_a^2(q_x, q_L)} \tag{8.35}$$

where $E_0 = \hbar^2\pi^2/2m^*a^2$ (the energy of the lowest subband in an infinitely deep well). For the intrasubband transition ($ij = 11$) $q_x = q_0$ and $Q_a = Q_s$. For the intersubband transition ($ij = 21$) $q_x = q_1$, and $Q_a = Q_a$.

As mentioned before, hybrids can be LO-like or IP-like. LO-like hybrids are characterized by having $q_L a \approx n\pi$, where n is an integer, such that sin $q_L a/2 \approx 0$ for symmetric modes and cos $q_L a/2 \approx 0$ for antisymmetric modes. In these cases

$$Q_s^2 \approx Q_a^2 \approx q_L^2 + q_x^2 \tag{8.36}$$

IP-like hybrids, on the other hand, are characterized by, again, $q_L a \approx n\pi$, but now with cos $q_L a/2 \approx 0$ for symmetric modes and sin $q_L a/2 \approx 0$ for antisymmetric modes. In these cases

$$Q_s^2 \approx q_L^2\left(1 + 4\frac{\coth(q_x a/2)}{q_x a}\right) + q_x^2 \tag{8.37}$$

and

$$Q_a^2 \approx q_L^2\left(1 + 4\frac{\tanh(q_x a/2)}{q_x a}\right) + q_x^2 \tag{8.38}$$

The form of the dispersion relation shows that for a fixed well-width a hybrid transforms from LO-like to IP-like and back to LO-like as a function of the in-plane wavevector. The same thing happens for a fixed in-plane wavevector as a function of well-width (Fig. 8.1), but this behaviour is scarcely noticeable in the total threshold emission rates (Figs. 8.2 and 8.3).

The principal feature is the reduction in the scattering rates with reducing well-width, which follows from the dependence of the Frohlich interaction on the inverse square of the wavevector. Increasing confinement entails increasing wavevector, even though the in-plane components remain fixed (at any rate, for

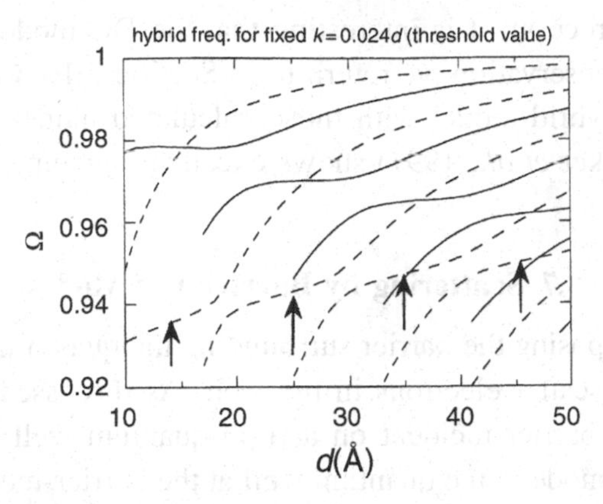

Fig. 8.1 Hybrid-mode frequency (ω/ω_{L1}) for a GaAs well as function of well-width d for a fixed threshold wavevector $q = 2.4 \times 10^6$ cm^{-1}. Full curves: odd modes; dashed, even. The arrows indicate the regions where the mode is IP-like.

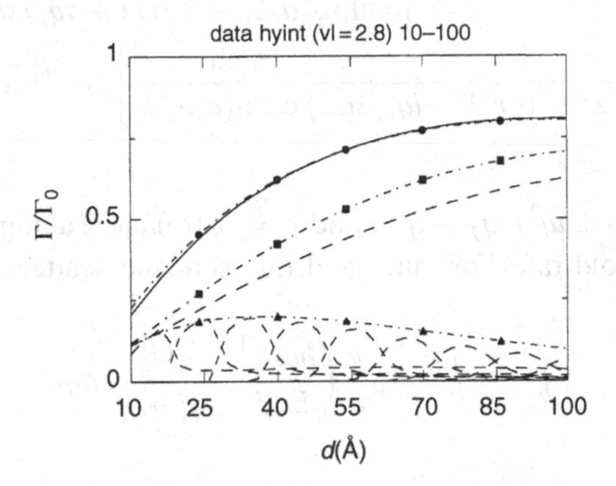

Fig. 8.2 Threshold intrasubband-scattering rate as a function of well-width d: solid curve, total rate with $v_L = 2.8 \times 10^3$ ms^{-1}; dashed curves, contributions from individual hybrids; dot-dash curves are DC model results: triangles GaAs interface; squares, confined GaAs modes; circles, total. The total hybrid and DC rates virtually coincide (Γ_0=W_0).

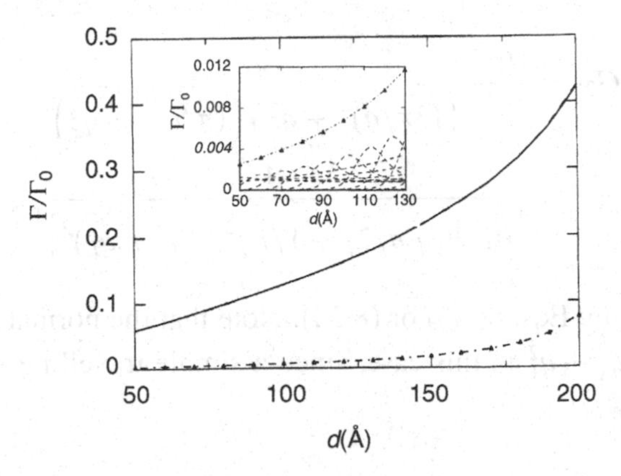

Fig. 8.3 Intersubband-scattering rate as a function of well-width. The key is the same as in Fig. 8.2. Once more the total hybrid and DC rates coincide.

the intrasubband process). It is interesting that the DC model predicts virtually identical rates, an observation we return to in Section 8.10. Comparison of rates calculated in the hybrid model with those calculated numerically using microscopic theory (Rucker *et al.*, 1991) shows excellent agreement.

8.7 Scattering by Barrier LO Modes

If the material composing the barrier surrounding the quantum well is polar, then barrier modes can scatter electrons in the well. As discussed in Chapter 5, LO bulk modes in the barrier incident on a rigid quantum well can tunnel via the excitation of an IP mode in the quantum well at the barrier-mode frequency. This IP mode has associated with it a scalar potential of the form, Eq. (5.29)

$$\phi = -i\frac{1}{2}\left(\frac{e^*}{\varepsilon_\infty}\right)(1+(a_+/a_-))\left[\frac{1}{\coth(q_x a/2)+1/r(1-iq_x/sa_-)}\frac{\cosh q_x z}{\sinh(q_x a/2)}\right.$$
$$\left.-\frac{1}{\tanh(q_x a/2)+1/r(1-iq_x/sa_-)}\frac{\sinh q_x z}{\cosh(q_x a/2)}\right]Ae^{-iq_L^{L/2}}e^{i(q_x x-\omega t)} \tag{8.39}$$

where $a\pm = (q_L q_T \pm iq_x^2/(q_T - q_x))$ and $r = \varepsilon_B(\omega)/\varepsilon_W$. Taking $a\pm \approx q_L$, we can express the threshold rates for intra- and intersubband scattering thus:

$$W_{i_j} = \int_0^{q_{L\max}} W_0 \frac{\pi}{a}\left(\frac{\hbar\omega}{E_0}\right)^{1/2} \frac{G_{ij}^2}{q_x^2+q_L^2}\, dq_L \tag{8.40}$$

where

$$G_{11} = \frac{8\pi^2}{a^3}\frac{1}{q_x\left((2\pi/a)^2+q_x^2\right)}\frac{1}{\left[(\coth(q_x a/2)+1/r)^2+(q_x/rsq_L)^2\right]^{1/2}} \tag{8.41}$$

$$G_{21} = \frac{16\pi^2}{a^3}\frac{q_x}{\left((3\pi/a)^2+q_x^2\right)\left((\pi/a)^2+q_x^2\right)}$$
$$\frac{1}{\left[(\tanh(q_x a/2)+1/r)^2+(q_x/rsq_L)^2\right]^{1/2}} \tag{8.42}$$

where q_x is given by Eqs. (8.33) or (8.34). Note that the normalizing factor Q^2 is given simply by $q_x^2 + q_L^2$ in this case, since a simple travelling bulk LO mode is

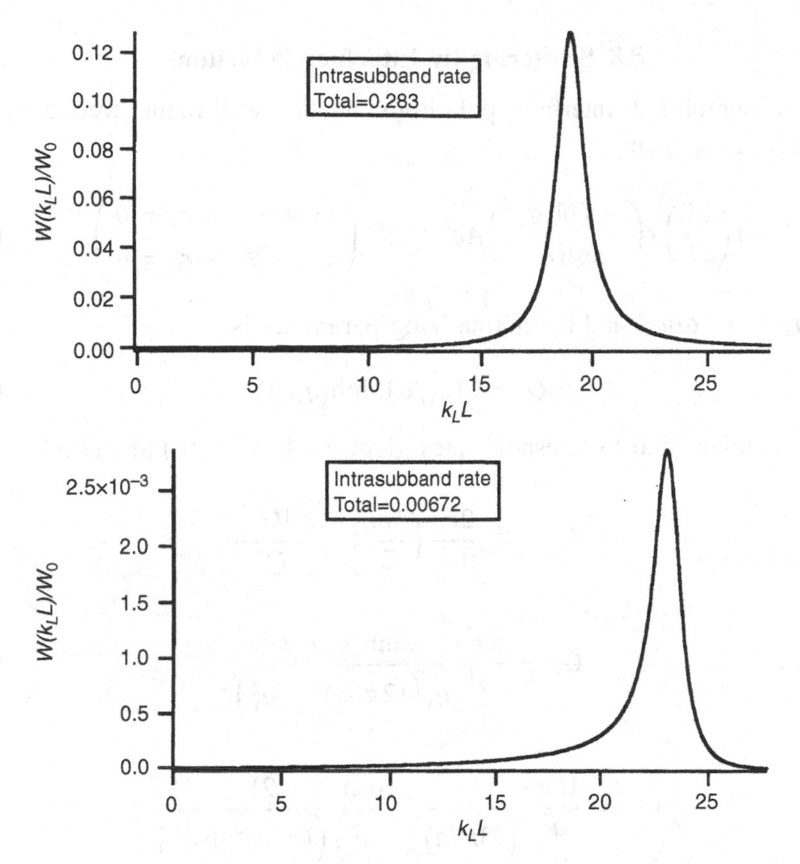

Fig. 8.4 Scattering rates in the well (width $L = 50$ Å) of fully hybridized LO modes in the barrier as a function of wavevector.

involved, and a factor of 2 has been included to account for modes on both sides of the well. The results for a situation in which the LO dispersion is large enough to preclude independent barrier IP modes are shown in Fig. 8.4. The resonances are clearly seen. For vanishing well-widths it is readily shown that the threshold intrasubband rate approaches that for bulk barrier phonons, i.e. $W = W_0\ \pi/2$. We conclude that scattering of electrons in a quantum well by fully hybridized bulk LO modes in the (polar) barrier can indeed take place.

When the LO dispersion is not large enough to cover the frequency range between ω_{LO} and ω_{TO}, IP hybrids can exist by hybridizing with an evanescent mode from the LO/LA branch. To a good approximation they can be treated as classical IP modes, except that their frequency range is restricted to lie below the LO band. This situation can be met in the case of well modes and in the case of barrier modes.

8.8 Scattering by Interface Polaritons

Classical unretarded interface polaritons at the well-mode frequency have potentials of the form

$$\phi = i\left(\frac{e^*}{\varepsilon_\infty}\right)s\begin{pmatrix} -\sinh q_x z \\ \cosh q_x z \end{pmatrix} Ae^{i(q_x x-\omega t)} \begin{pmatrix} \tanh\left(\frac{q_x a}{2}\right) + r_w = 0 \\ \coth\left(\frac{q_x a}{2}\right) + r_w = 0 \end{pmatrix} \tag{8.43}$$

where $r_w = \varepsilon_w(\omega)/\varepsilon_B$ and the normalizing parameter is

$$Q^2 = 2(q_x/a)\sinh(q_x a) \tag{8.44}$$

These potentials lead to threshold rates given by Eqs. (1.56) and (1.57), viz.:

$$W_{ij} = W_0 \frac{2\pi^2}{a^2}\left(\frac{\hbar\omega}{E_0}\right)^{1/2} \frac{s^2|G_{ij}|^2}{Q^2} \tag{8.45}$$

where

$$G_{11} = \frac{8\pi^2 i}{a^3} \frac{\sinh(q_x a/2)}{q_x\left((2\pi/a)^2 + q_x^2\right)} \tag{8.46}$$

$$G_{21} = \frac{16\pi^2}{a^3} \frac{q_x \cosh(q_x a/2)}{\left((3\pi/a)^2 + q_x^2\right)\left((\pi/a)^2 + q_x^2\right)} \tag{8.47}$$

The corresponding patterns of potential in the well contributed by barrier polaritons are

$$\phi = i\left(\frac{e^*}{\varepsilon_\infty}\right)_B s_B\begin{pmatrix} \cosh q_x z \\ -\sinh q_x z \end{pmatrix} Ae^{i(q_x x-\omega t)} \begin{pmatrix} \tanh\left(\frac{q_x a}{2}\right) + r_B = 0 \\ \coth\left(\frac{q_x a}{2}\right) + r_B = 0 \end{pmatrix} \tag{8.48}$$

The overlap factors G_{11} and G_{12} are, of course, the same as before, Eqs. (8.46) and (8.47), but the normalization factor is different. This factor is derived from the mechanical energy in the barrier (the electrical energy in the well and barrier is zero), where the mode patterns are of decaying exponentials. The normalization parameter is readily shown to be

$$Q^2 = \frac{4q_x}{a}\begin{cases} \cosh^2(q_x a/2) & \text{symmetric} \\ \sinh^2(q_x a/2) & \text{antisymmetric} \end{cases} \tag{8.49}$$

The rates are also given in Eqs. (1.58) and (1.59). The rate associated with barrier interface modes is compared with that of barrier hybridized modes in Fig. 8.5.

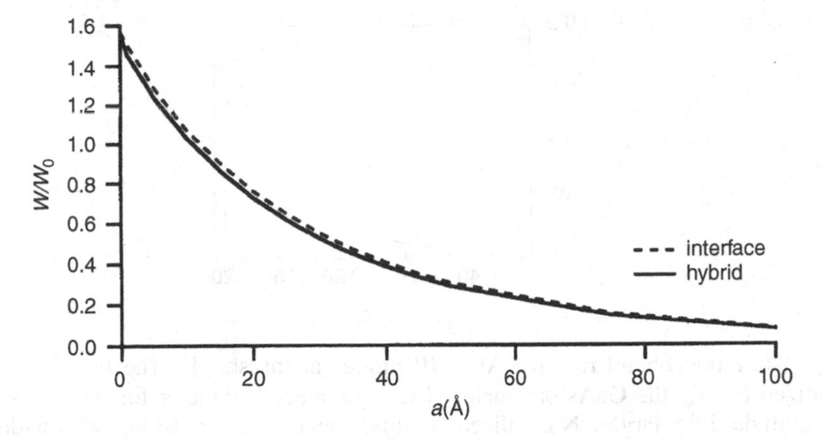

Fig. 8.5 Scattering rates in the well by barrier modes (hybridized and non-hybridized interface).

Once again, we find that the DC model gives the same result as the hybrid model, which strongly indicates that the rate is independent of the degree of dispersion of the LO modes in the barrier.

It should be noticed that well-interface modes and barrier-interface modes act in a complementary fashion. The higher-frequency well mode obeys $\tanh(q_x a/2) + r_w = 0$ and is antisymmetric. Its role is, therefore, in intersubband transitions, and having a frequency near to ω_{LO}, its field factor, s, is near to unity, corresponding to maximum coupling strength. The antisymmetric barrier mode, however, obeys $\cot h q_x a/2 + r_B = 0$, and is, therefore, the lower-frequency mode, with correspondingly weaker coupling strength. The reverse situation occurs in the case of intrasubband transitions initiated by the symmetric modes – the barrier mode has the larger field factor in this case, being the higher-frequency mode. Fig. 8.6 depicts the intrasubband rate and Fig. 8.7 the intersubband rate as a function of well-width. Note the resonance in the intersubband rate when the subbands are exactly $\hbar\omega$ apart, whence $q_x = 0$. In the limit of small well-widths $|G_{11}| \rightarrow 1$ and $|G_{21}| \rightarrow 0$. The intrasubband rate becomes

$$W_{11} \rightarrow W_0 \frac{2\pi q_x s^2}{Q^2 a} \tag{8.50}$$

For the well-mode it is easy to show that s becomes

$$s = \frac{\omega^2 - \omega_{\mathrm{TO}}^2}{\omega_{\mathrm{LO}}^2 - \omega_{\mathrm{TO}}^2} = \frac{r_0}{r_0 + \coth(q_x a/2)} \rightarrow r_0 \frac{q_x a}{2} \tag{8.51}$$

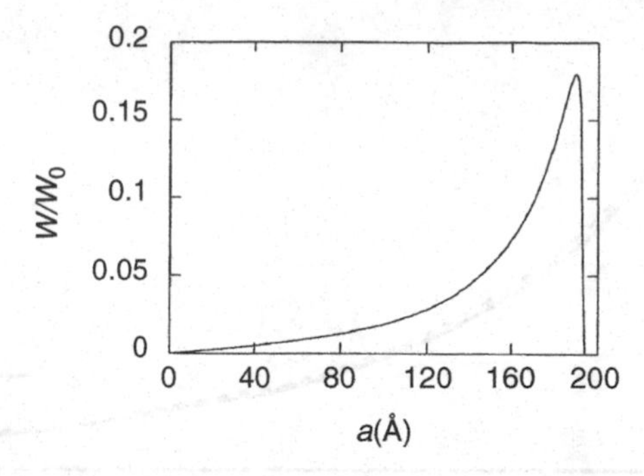

Fig. 8.6 Intrasubband rate for AlAs IP modes at threshold. (The rate is normalized by W_0, the GaAs parameter. The equivalent parameter for AlAs has a magnitude 1.65 larger. Normalized to this, the rate at vanishing well-width would be $\pi/2$.)

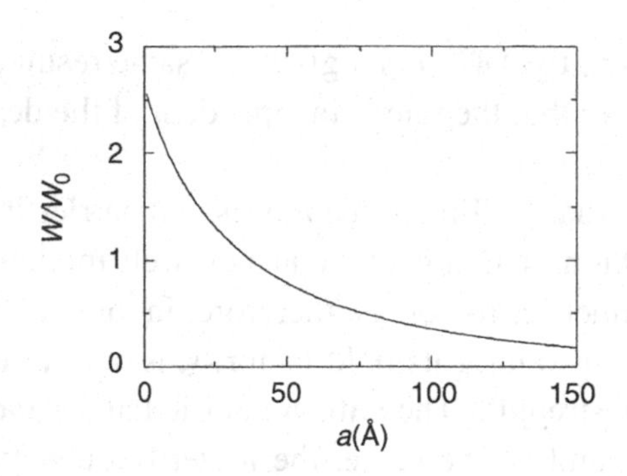

Fig. 8.7 Intersubband rate for AlAs IP modes.

whence

$$W_{11} \to W_0 \frac{\pi^2 r_0^2}{4} \left(\frac{\hbar\omega}{E_0}\right)^{1/2} \tag{8.52}$$

where $r_0 = \varepsilon_{\infty W}/\varepsilon_B$, and thus the rate vanishes, since $E_0 = \hbar^2\pi^2/2_{m*}\, a^2$. But note that if the barrier is vacuum, r can be large, rendering the reduction somewhat academic. The intersubband rate, likewise, vanishes. For the barrier mode $s \to 1$ and $Q^2 \to 4q_x/a$, thus

$$W_{11} \to W_0 \frac{\pi}{2} \tag{8.53}$$

which turns out to be exactly the rate obtained with bulk LO modes in the same limit. As in the case of the well mode, the intersubband rate vanishes as $a \to 0$.

8.9 Summary of Threshold Rates in an Infinitely Deep Well

An electron in the subband of a quantum well can be scattered by optical modes from both the well and the barrier. As we have seen, the structure of the optical modes in either well or barrier depends upon the degree of dispersion of the LO modes. Within the LO frequency band, all modes are hybrids of LO, TO and IP. If there is a gap between the LO band and ω_{TO}, the modes in this gap are hybrids of LO/LA, TO and IP, and they approximate to classical IP modes. In general, therefore, there are four ways in which electrons can scatter – via fully hybridized well-modes, fully hybridized barrier modes, interface well-modes and interface barrier modes. Expressions for these rates for intra- and intersubband transitions, ignoring contributions from the TO and, if present, from the LO/LA evanescent mode, are summarized next.

8.9.1 Intrasubband Rates

(a) Fully hybridized well-modes

$$W_{11} = \sum_{q_L} W_0 2(2\pi)^6 \left(\frac{\hbar\omega_L}{E_0}\right)^{1/2} \frac{\sin^2(q_L a/2)\left[\left(1/(4\pi^2-(q_L a)^2)+(sq_L^2/q_x^2)/(4\pi^2+(q_x a)^2\right)\right]^2}{(q_L a)^4\left[\left(1+3\frac{\sin q_L a}{q_L a}+4\frac{\sin^2(q_L a/2)\coth(q_x a/2)}{q_x a}\right)+\left(\frac{q_x}{q_L}\right)^2\left(1+\frac{\sin(q_L a)}{q_L a}\right)\right]} \tag{8.54}$$

The values of q_L are the solutions of

$$\tan(q_L a/2) = -\frac{q_x}{sq_L(\coth(q_x a/2)+r)} \tag{8.55}$$

also

$$q_x = \frac{(2m^*\hbar\omega_L)^{1/2}}{\hbar} \tag{8.56}$$

All parameters refer to well-modes, and $r = \varepsilon_w(\omega)/\varepsilon_B$, $E_0 = \hbar^2\pi^2/2m^*a^2$.

(b) Fully hybridized barrier modes

$$W_{11} = \int_{q_{L\min}}^{q_{L\max}} W_0 \frac{(8\pi^3)^2}{\pi a}\left(\frac{\hbar\omega_L}{E_0}\right)^{1/2} \frac{dq_L}{(q_x a)^2(q_x^2+q_L^2)(4\pi^2+(q_x a)^2)^2\left[(\coth(q_x a/2)+1/r)^2+(q_x/rsq_L)^2\right]} \tag{8.57}$$

Apart from E_0, all parameters refer to barrier modes, with $r = \varepsilon_B(\omega)/\varepsilon_w$.

(c) Interface well-mode

$$W_{11} = W_0(8\pi^3)^2\left(\frac{\hbar\omega_L}{E_0}\right)^{1/2} \frac{r_0^2 \tanh(q_x a/2)}{2(q_x a)^3(4\pi^2 + (q_x a)^2)^2(r_0 + \coth(q_x a/2))^2} \tag{8.58}$$

All parameters refer to the well-modes, and $r_0 = \varepsilon_{\infty W}/\varepsilon_B$.

(d) Interface barrier modes

$$W_{11} = W_0(8\pi^3)^2\left(\frac{\hbar\omega_L}{E_0}\right)^{1/2} \frac{r_0^2 \tanh^2(q_x a/2)}{2(q_x a)^3(4\pi^2 + (q_x a)^2)^2(r_0 + \tanh(q_x a/2))^2} \tag{8.59}$$

Apart from E_0, all parameters refer to the barrier mode, and $r_0 = \varepsilon_{\infty B}/\varepsilon_w$.

8.9.2 Intersubband Rates

(a) Fully hybridized well-modes

$$W_{21} = \sum_{q_L} W_0 2(16\pi^3)^2\left(\frac{\hbar\omega_L}{E_0}\right)^{1/2}$$

$$\frac{\cos^2 q_L a/2\left[\left[1/(9\pi^2 - (q_L a)^2)(\pi^2 - (q_L a)^2)\right] - \left[s/(9\pi^2 + (q_x a)^2)(\pi^2 + (q_x a)^2)\right]\right]^2}{(1 - 3(\sin q_L a/q_L a) + 4[\cos^2(q_L a/2)\tanh(q_x a/2)/q_x a]) + (q_x/q_L)^2(1 - (\sin(q_L a)/q_L a))} \tag{8.60}$$

The values of q_L are solutions of

$$\cot(q_L a/2) = \frac{q_x}{s q_L(\tanh(q_x a/2) + r)} \tag{8.61}$$

also

$$q_x = \frac{(2m^*(E_2 - E_1 - \hbar\omega_L))^{1/2}}{\hbar} \tag{8.62}$$

All parameters refer to well-modes.

(b) Fully hybridized barrier modes

$$W_{21} = \int_{q_{L\min}}^{q_{L\max}} \frac{W_0(16\pi^3)^2}{\pi a}\left(\frac{\hbar\omega_L}{E_0}\right)^{1/2}$$

$$\frac{(q_x a)^2}{(q_x^2 + q_L^2)(9\pi^2 + (q_x a)^2)^2\left[\pi^2 + (q_x a)^2\right]^2\left[(\tanh(q_x a/2) + (1/r))^2 + (q_x/r s q_L)^2\right]} \tag{8.63}$$

Apart from E_0, all parameters refer to barrier modes.

(c) Interface well-modes

$$W_{21} = W_0(16\pi^3)^2\left(\frac{\hbar\omega_L}{E_0}\right)^{1/2}\frac{q_x a r_0^2 \coth q_x a/2}{2(9\pi^2 + (q_x a)^2)^2\left[\pi^2 + (q_x a)^2\right]^2 (r_0 + \tanh(q_x a/2))^2} \tag{8.64}$$

All parameters refer to the well-mode.

(d) Interface barrier modes

$$W_{21} = W_0(16\pi^3)^2\left(\frac{\hbar\omega_L}{E_0}\right)^{1/2}\frac{q_x a r_0^2 \coth^2 q_x a/2}{2(9\pi^2 + (q_x a)^2)^2\left[\pi^2 + (q_x a)^2\right]^2 (r_0 + \coth(q_x a/2))^2} \tag{8.65}$$

All parameters, apart from E_0, refer to the barrier mode.

8.10 Comparison with Simple Models

The dielectric-continuum model for confined LO and interface modes uses purely electrical boundary conditions and gives mode patterns in GaAs quantum wells at variance with observation via Raman scattering and with microscopic theory. Nevertheless, the scattering rates by well-modes predicted by the DC model for both intra- and intersubband transitions in an infinitely deep well have been shown to give results very close to those of the hybrid model (Figs. 8.2, 8.3 and Fig. 8.5; Constantinou and Ridley, 1994). The threshold LO rates are given by the following; see Eqs. (1.51) and (1.53).

$$W_{\text{intra}} = W_0 8\left(\frac{\hbar\omega}{E_0}\right)^{1/2}\sum_n\left[\left(\frac{1}{n} + \frac{n}{4 - n^2}\right)^2\frac{1}{(q_x a)^2 + (n\pi)^2}\right] \quad n = 1, 3, 5\ldots \tag{8.66}$$

$$W_{\text{inter}} = W_0 8\left(\frac{\hbar\omega}{E_0}\right)^{1/2}\sum_n\left[\left(\frac{1}{n^2 - 9} - \frac{1}{n^2 - 1}\right)^2\frac{n^2}{(q_x a)^2 + (n\pi)^2}\right] \quad n = 2, 4, 6\ldots \tag{8.67}$$

The corresponding rates for the interface modes are those of Eqs. (8.58), (8.59), (8.64) and (8.65). In the well, adding LO and interface rates gives the hybrid rate to a good approxmation, provided it is permissible to neglect the role of the TO mode. Since the TO component of the triple hybrid can affect energy normalization of frequencies only near ω_{TO}, this neglect is justified in most cases. Thus, the simpler DC expressions may often be used to calculate the scattering rates associated with well-modes.

In fact, the DC expressions turn out to be excellent analytical approximations of those of the hybrid model. This can be seen most easily by focussing on the intrasubband threshold rates. A study of the effect of dispersion on scattering rate by well-modes (Constantinou and Ridley, 1994) showed that the rate was virtually invariant. Dispersion is the source of the complicated resonances in rate that occur as the well-width is varied (Fig. 8.1), and it entails solving transcendental equations to establish k_L, the wavevector of the LO component along the confinement direction. The invariance of the rate on dispersion suggests that a simple model is one in which k_L is given. The hybrid rate for well-modes (Eq. (8.54)) can be written as follows:

$$W_{11} = W_0 2(2\pi)^6 \left(\frac{\hbar\omega_L}{E_0}\right)^{1/2} \sum_{q_L} \frac{B}{(q_L a)^4 A} \tag{8.68}$$

The factors taken outside the summation are only very weakly dependent on q_L because of the weak LO dispersion. For LO-like hybrids ($(\coth(q_x a/2) + r) \neq 0$) we can take $q_L a = 2n\pi$ and $A = 1 + (q_x/q_L)^2$, whence

$$W_{11} = W_0 \left(\frac{\hbar\omega_L}{E_0}\right)^{1/2} \frac{1}{8} \sum_n \frac{1}{n^4\left\{1 + (q_x a/2n\pi)^2\right\}} \left[\delta_{n,1} + \frac{4n}{q_x a\{1 + (q_x a/2\pi)\}\{\coth(q_x a/2) + r\}}\right]^2 \tag{8.69}$$

In the limit $(q_x a/2\pi)^2 < 1$ the summation can be carried out using $\sum_n n^{-2} = \pi^2/6$ and we obtain

$$W_{11} = W_0 \left(\frac{\hbar\omega_L}{E_0}\right)^{1/2} \frac{1}{8}\left[9 + 4\left(\frac{\pi^2}{6} - 1\right)\right] = W_0 \left(\frac{\hbar\omega_L}{E_0}\right)^{1/2} 1.448 \tag{8.70}$$

The result of the DC model is identical except that the numerical factor is 1.447. For interface-like hybrids ($(\coth(q_x a/2) + r) = 0$) we may take the normalizing factor $A = 4\coth(q_x a/2)/q_x a$, $q_L a = (2n - 1)\pi$ and $B = s^2(q_L/q_x)^4/\{4\pi^2 + (q_x a)^2\}^2$. Noting that $s = (1 - r/r_0)^{-1}$ we obtain the DC expression (Eq. (8.58)) exactly.

Eq. (8.57) gives the intrasubband rate for fully hybridized barrier modes. The integrand is sharply peaked near $q_L = q_{Lc}$ where $\coth(q_x a/2) + r^{-1} \approx 0$.

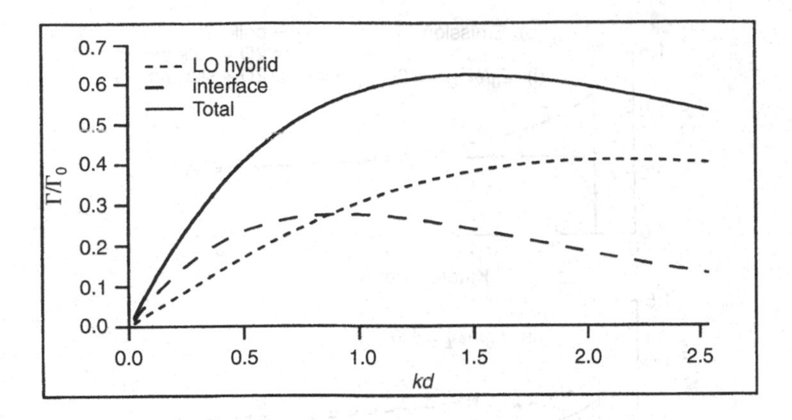

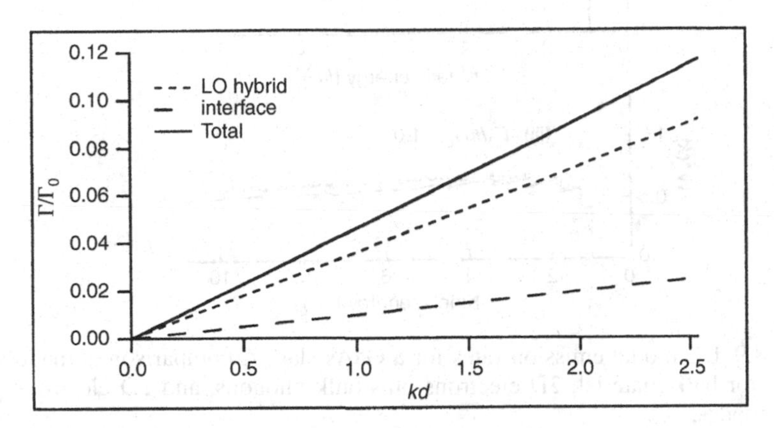

Fig. 8.8 Simplified model threshold rates: (top) intrasubband; (bottom) intersubband.

Exploiting this by putting $q_L = q_{Lc}$ in all terms other than the rapidly varying one, we obtain

$$W_{11} = W_0 \frac{(8\pi^3)^2}{2\pi a}\left(\frac{\hbar\omega_L}{E_0}\right)^{1/2} \frac{q_{Lc}^2 + q_x^2}{q_{Lc}(q_x a)^2\left\{4\pi^2 + (q_x a)^2\right\}^2} I \tag{8.71}$$

where I is the standard integral of the form

$$I = \int_{-\infty}^{+\infty} \frac{dx}{ax^2 + bx + c} \tag{8.72}$$

We thereby obtain the DC result (Eq. (8.59)) exactly.

Similar manipulations of the hybrid intersubband rates can be made to establish close numerical agreement with the DC model. Fig. 8.8 shows the

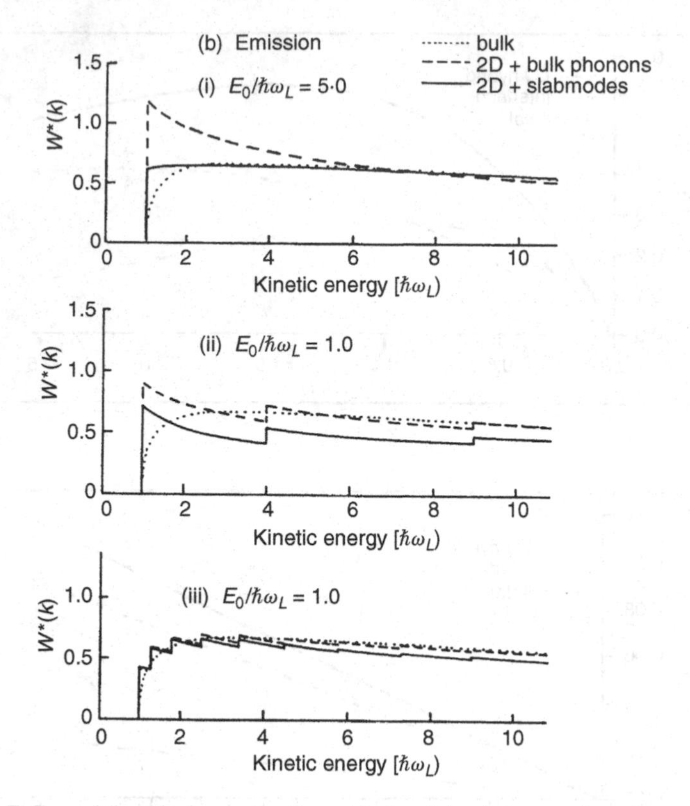

Fig. 8.9 DC model emission rates for a GaAs slab. A comparison is made with rates for bulk material, 2D electrons plus bulk phonons, and 2D electrons plus slab modes.

contribution of the lowest order LO-like hybrid and the interface-like hybrid to the intrasubband and intersubband rates.

In the case of a polar free-standing slab, the DC model has standardly been used. Scattering rates for an electron in a GaAs slab were calculated by Riddoch and Ridley (1985) and are shown in Fig. 8.9. This case is interesting in that there is no polar barrier to provide barrier-interface-mode scattering. As these authors pointed out, this implied that scattering should vanish in the limiting case of thin slabs, but, in practice, the large permittivity ratio (r_0) makes this virtually unachievable. The same would be true for systems involving non-polar barriers.

8.11 The Interaction in a Superlattice

The interaction with the hybrid modes of a superlattice is complicated by the mixed symmertry of both the electron and phonon envelope functions (see Sections 2.2 and 6.1), and by the miniband structure of electron energies. The initial state of the

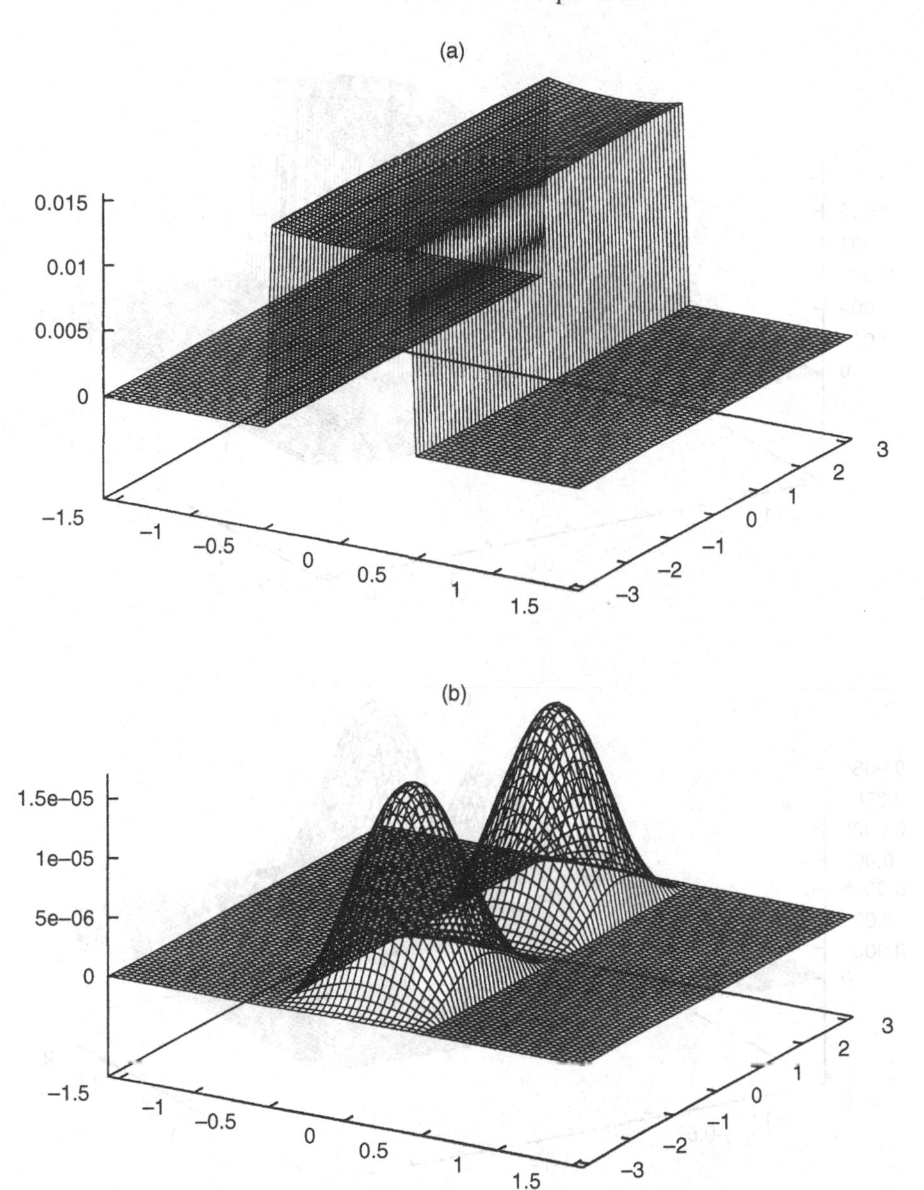

Fig. 8.10 Squared matrix element for intrasubband scattering by first- and second-order hybrid modes in a GaAs well. Initial electron state has $E_i = 1.25\hbar\omega_{\mathrm{LO}}, k_{zi} = 0$. Vertical axis: $|M|^2$; horizontal axis $-\pi/2 \leq \theta \leq \pi/2$, $-\pi \leq k_z(a+b) \leq \pi$: (a) mode 1; (b) mode 2 for $a = b = 56.6$ Å; (c) mode 1; (d) mode 2 for $a = b = 28.3$ Å.

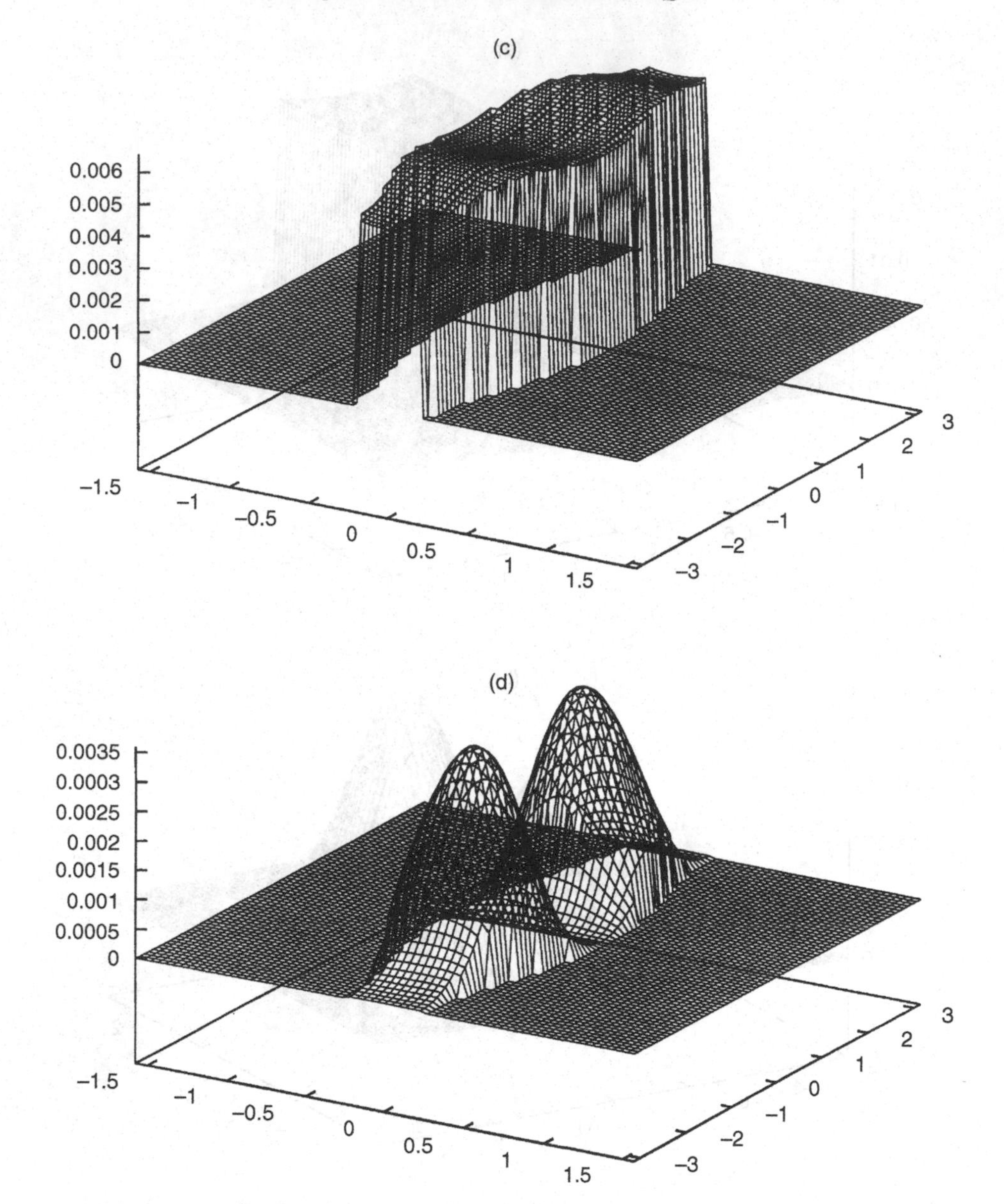

Fig. 8.10 (*cont.*)

electron is specified by the wavevector component parallel to the axis of the superlattice as well as by the in-plane component and subband number. Crystal momentum is conserved in all three directions as it is in the bulk. Except in especially simple situations the scattering rate must be calculated numerically. Fig. 8.10 shows examples of the variation of the squared matrix element.

The simplest situation regarding the electron wavefunctions occurs when the potential barriers are large, for then the wavefunction is just that for an isolated quantum well multiplied by the phase factor, $\exp(ik_z d)$. In the equivalent situation for hybrid optical modes the permittivity factor is very small ($r \approx 0$), as occurs for frequencies near ω_{LO}. Modes satisfying this condition are just the ones that interact most strongly with electrons. With the factor γ – see Section 6.1, Eq. (6.2e) ≈ 0 – the modes in the well are symmetric or antisymmetric, independent of k_z. Under these combined conditions the problem reduces to that for a single quantum well in the limit $r \approx 0$.

8.12 The Interaction in an Alloy

The coupling constant for the polar interaction in a two-mode alloy is a function of the polarization factors introduced in Section 4.5 and must be of the form

$$\alpha_i = \frac{e^2}{4\pi\hbar}\left(\frac{m^*}{2\hbar\omega_{\text{LO}i}}\right)^{1/2}\frac{\varepsilon_0 X_i^*}{\varepsilon_\infty^2\omega_{\text{LO}i}^2} \tag{8.73}$$

where X_i^* is an effective polarization factor for mode i that takes the polarization of both types of displacement fully into account.

We can discover what X_i^* is by noting that the total lattice polarization can be written in terms of the individual X factors as follows:

$$\mathbf{P}_L = \varepsilon_0^{1/2}(X_1^{1/2}\rho_1^{1/2}\mathbf{u}_1 + X_2^{1/2}\rho_2^{1/2}\mathbf{u}_2) \tag{8.74}$$

An effective X factor can now be defined using energy normalization

$$\begin{aligned} \varepsilon_0^{1/2}X^{*1/2}(\rho_1 u_1^2 + \rho_2 u_2^2)^{1/2} &= P_L \\ \therefore X^{*1/2} &= \frac{P_1 + P_2}{(P_1^2/X_1 + P_2^2/X_2)^{1/2}} \end{aligned} \tag{8.75}$$

Now, polarization depends upon frequency. From the discussion in Section 4.5 it is clear that P is proportional to $X/(\omega_{\text{TO}}^2 - \omega^2)$, so that we can write with $\omega = \omega_{\text{LO}i}$ ($= w_i$ in the notation of Section 4.5)

$$\frac{P_1}{P_2} = \frac{X_1(\omega_{\text{TO}2}^2 - \omega_{\text{LO}i}^2)}{X_2(\omega_{\text{TO}1}^2 - \omega_{\text{LO}i}^2)} \tag{8.76}$$

Thus the effective polarization factor is

$$X_i^* = \frac{\left[X_1(\omega_{\text{TO}2}^2 - \omega_{\text{LO}i}^2) + X_2(\omega_{\text{TO}1}^2 - \omega_{\text{LO}i}^2)\right]^2}{X_1(\omega_{\text{TO}2}^2 - \omega_{\text{LO}i}^2) + X_2(\omega_{\text{TO}1}^2 - \omega_{\text{LO}i}^2)^2} \tag{8.77}$$

so that

$$X_1^* = \frac{\varepsilon_\infty}{\varepsilon_0}\frac{(\omega_{\rm LO1}^2 - \omega_{\rm TO1}^2)(\omega_{\rm TO2}^2 - \omega_{\rm LO1}^2)}{(\omega_{\rm LO2}^2 - \omega_{\rm LO1}^2)}$$
$$X_2^* = \frac{\varepsilon_\infty}{\varepsilon_0}\frac{(\omega_{\rm LO2}^2 - \omega_{\rm TO2}^2)(\omega_{\rm LO2}^2 - \omega_{\rm TO1}^2)}{(\omega_{\rm LO2}^2 - \omega_{\rm LO1}^2)} \tag{8.78}$$

Note that with $\omega_{\rm LO2}, \omega_{\rm TO2} > \omega_{\rm LO1}, \omega_{\rm TO1}$ the higher-frequency mode has its strength enhanced whereas the lower-frequency mode has its strength diminished relative to the simple prescription $X_i^* = (\varepsilon_\infty/\varepsilon_0)(\omega_{\rm LOi}^2 - \omega_{\rm TOi}^2)$.

8.13 Phonon Resonances

When a two-mode alloy ABC forms the barrier to a quantum well BC there exists the possibility of so-called phonon resonances in the rates of intersubband scattering (Babiker *et al.*, 1987) and particularly in the rate of well-capture (Babiker *et al.*, 1989). The condition to be satisfied is that the frequency of the BC LO mode in the alloy is lower than that in the binary material, a condition that is fulfilled in the AlGaAs/GaAs system. The resonances appear as enhancements of the polar-optical-phonon rate at rather well-defined well-widths. The result of an early calculation of the effect for intersubband scattering is shown in Fig. 8.11 and for well-capture in Fig. 8.12.

The effect is, at base, caused by the wavelength dependence of the Frohlich interaction – the smaller the wavevector, the bigger the interaction. Phonon modes in the well whose frequencies lie above the frequency band of corresponding modes in the alloy are largely confined to the well, whereas modes whose frequencies overlap the alloy band are not. In the latter case the phonon modes that scatter electrons have a wavevector q_w in the well and q_B in the barrier, with $q_w > q_B$. For transitions within the lowest subband, the electron, being mostly confined to the well, is not strongly affected by the barrier component, but electrons in the second subband, where the wavefunction stretches further into the barrier, are more influenced. The influence is even more pronounced in the case of capture, where the lowest unbound state can actually be concentrated in the barrier region in a superlattice (Fig. 8.13). Resonances in the rates occur whenever the well-width is such that $q_B \approx 0$ for a mode, i.e. when the frequency of the confined mode coincides with the zone-centre frequency of the mode in the alloy.

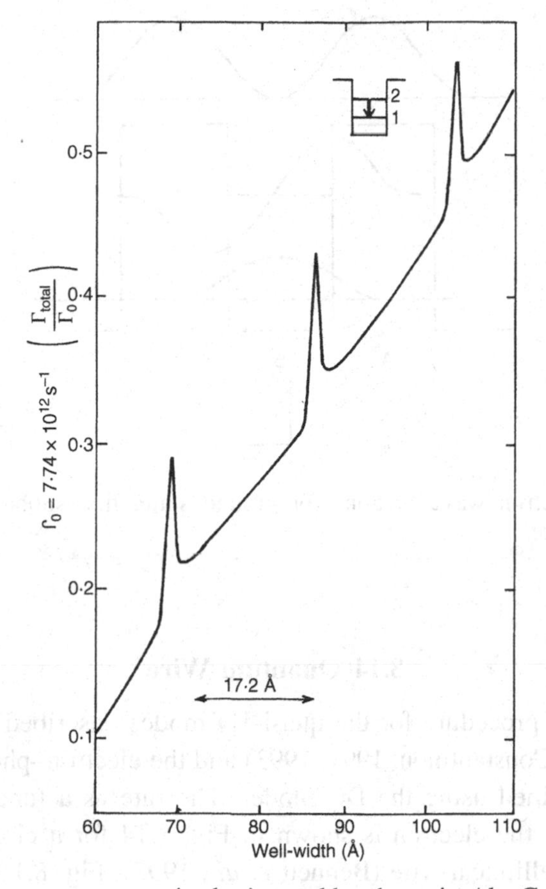

Fig. 8.11 Phonon resonances in the intersubband rate in $Al_{0.3}Ga_{0.7}As/GaAs$.

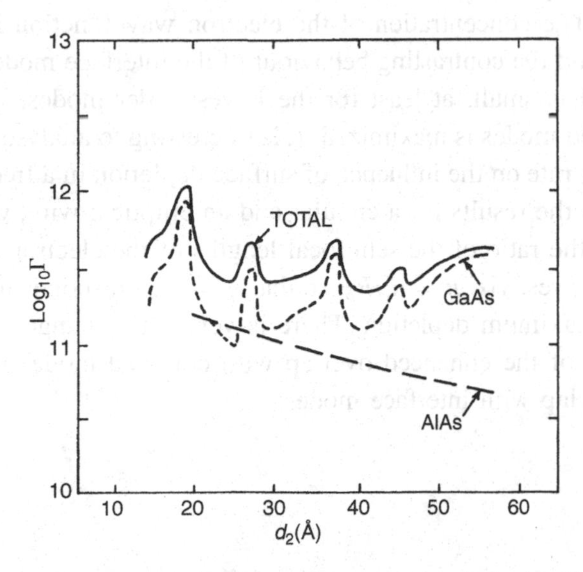

Fig. 8.12 Phonon resonances in the well-capture rate in a $Al_{0.3}Ga_{0.7}As/GaAs$ superlattice.

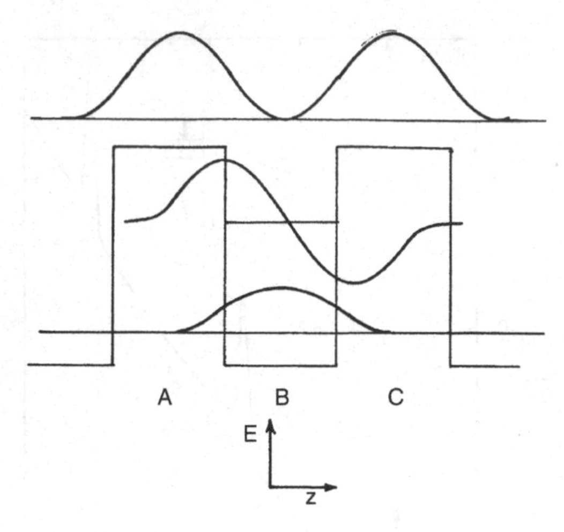

Fig. 8.13 Electron wavefunctions for ground state, first subband and lowest continuum state.

8.14 Quantum Wire

The quantization procedure for the quasi-1D modes described in Section 7.4 is straightforward (Constantinou, 1991, 1993) and the electron–phonon rates can be numerically obtained using the DC model. The rate as a function of the axial kinetic energy of the electron is shown in Fig. 8.14 for a circular wire and in Fig. 8.15 for an elliptical wire (Bennett *et al.*, 1995). Fig. 8.16 depicts the contributions made by individual modes and Fig. 8.17 shows the rate for a free-standing wire. The concentration of the electron wavefunction in the region of low curvature and the contrasting behaviour of the interface modes mean that the overlap integral is small, at least for the lowest-order modes, whereas overlap with the confined modes is maximized. It is interesting to analyse the dependence of the scattering rate on the influence of surface depletion in a free-standing wire. Fig. 8.18 shows the results for a circular and an elliptical wire, where the rate is plotted against the ratio of the semifocal lengths of the electron and the phonon confinement ellipses, viz. $\eta = R_e/R_p$, so that $\eta = 1$ corresponds to zero depletion and $\eta = 0$ to maximum depletion. There is very little change; as η approaches zero, the effect of the enhanced overlap with confined modes is off-set by the diminished overlap with interface modes.

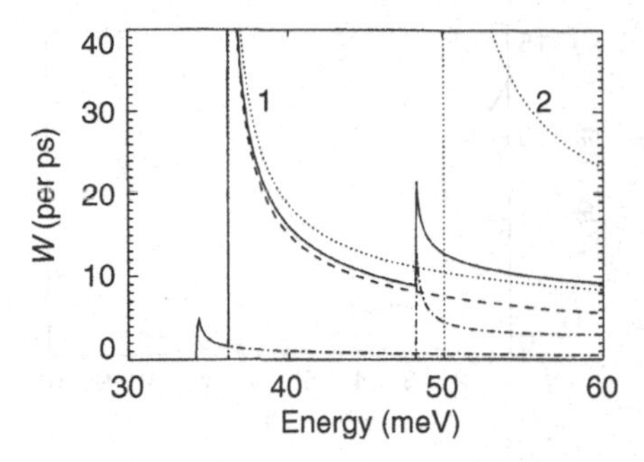

Fig. 8.14 The scattering rate as a function of axial energy for R = 70 Å GaAs circular wire embedded in AlAs: solid curve, total scattering rate from all allowed modes; dashed curve, contribution from the GaAs ($m = 0$) confined modes; dot–dashed curve, (which makes the smallest contribution), rate from $m = 0$ GaAs interface mode; other dot–dash curve (which makes a more significant contribution), due to the $m = 0$ AlAs interface mode. Dotted curves: 1, scattering rate assuming bulk GaAs phonons; 2, rate assuming AlAs bulk phonons.

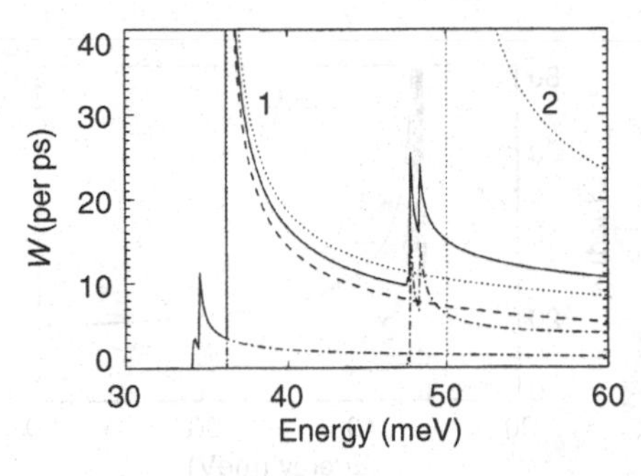

Fig. 8.15 The scattering rate for GaAs elliptical quantum wire with $b = 50$ Å and $a = 2b$. Solid curve, total rate; dashed curve, rate due to confined modes of the ellipse allowed via the selection rules; dot–dashed curve (which begins around 34 meV), contribution from the GaAs interface modes that are allowed to contribute by the selection rules; other dot–dashed curve, contribution from allowed AlAs interface modes; dotted curves: 1, scattering via bulk GaAs phonons; 2, scattering by bulk AlAs phonons.

8.15 The Sum-Rule

Register (1992) has described a sum-rule for polar LO-mode scattering of electrons in heterostructures that provides a microscopic basis to the similar sum-rule advanced by Mori and Ando (1989). Mori and Ando pointed out that because of the orthonormality of the optical eigenmodes the form factors that appear in the scattering rates sum to the form factor for bulk modes, and so, if the differences

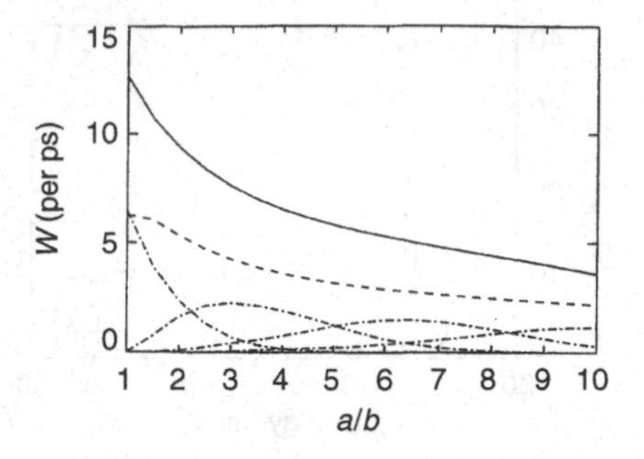

Fig. 8.16 The scattering rate for a GaAs elliptical quantum wire with $b = 50$ Å as a function of the a/b for fixed $E_k = 60$ meV: solid curve, total rate; dashed curve, contribution from allowed confined GaAs modes; dot–dashed curves, contribution from allowed AlAs interface modes. (It is noted that as a/b increases, the higher-order AlAs interface modes contribute more significantly. The contribution from the allowed GaAs interface modes is small and not depicted in the diagram for clarity, although their contribution is included in the total rate.)

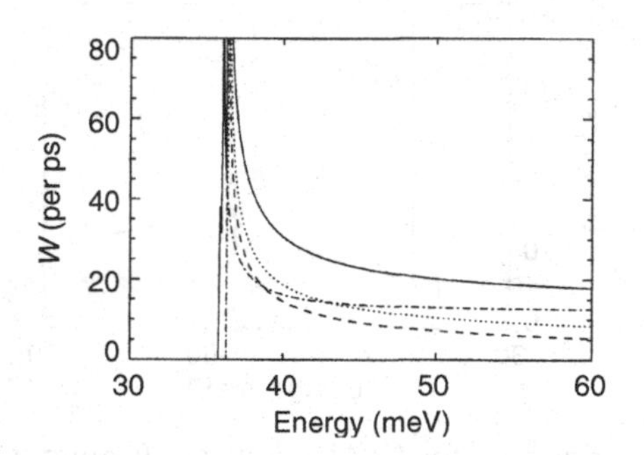

Fig. 8.17 The scattering rate for a free-standing elliptical GaAs wire with $b = 50$ Å and $a = 2b$ ($\eta = 1$): solid curve, total rate; dashed curve, contribution from allowed confined modes; dot–dashed curve, contribution from allowed GaAs interface modes; dotted curve, rate obtained by assumption of bulk GaAs modes.

of eigenfrequencies were ignored, the scattering rate would just be that for bulk phonons. This conclusion was reached long ago by Herbert (1973) in a different context.

Such a sum-rule is likely to be useful for quantum wells only in situations where the differences in frequency and ionicity of LO-modes in the well and barrier are negligible and interface modes do not play a significant role. The interaction strength of LO-modes can be regarded as being independent of frequency to a good approximation in most cases, but the interaction strength of

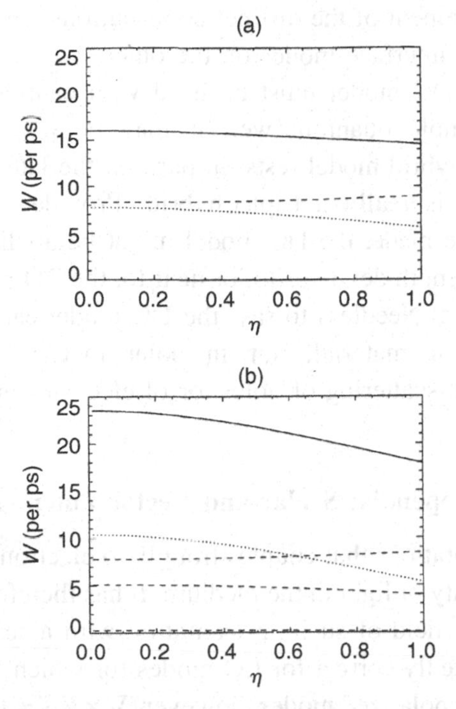

Fig. 8.18 (a) The scattering rate for a free-standing circular wire with $R = 70.7$ Å as a function of η for fixed $E_k = 60$ meV: solid curve, total rate with contribution from confined modes (dotted curve) and interface mode (dot–dashed curve); (b) the scattering rate for a free-standing elliptical wire with $b = 50$ Å and $a = 2b$ as a function of η for fixed $E_k = 60$ meV: solid curve, total rate with contributions from confined modes (dotted curve) and interface modes (dashed curves).

interface modes is a very sensitive function of frequency, ranging from LO-like to TO-like (i.e. zero). Moreover, their coupling strength reflects the difference in the polar nature of well and barrier. In general, therefore, it is necessary to take full account of the specific contributions to the scattering rate made by LO-modes and interface modes belonging to the well and LO-modes and interface modes belonging to the barrier.

A simple way of estimating scattering rates more reliably than using the sum-rule is to use the dielectric-continuum (DC) model. A glance at Figs. 8.2, 8.3 and 8.5 shows remarkable agreement with the results of the double-hybrid model. The explanation is straightforward. First, the envelope functions of the DC confined modes are not very different from from those of the LO-like hybrids (see Fig. 1.6). Second, the sum-rule is a good approximation for LO-like modes. Third, IP-like hybrids give a frequency-dependent contribution similar to that of pure IP-modes. The success of the DC model over the bulk-phonon model arises from its acknowledgement of the distinct contributions made by confined modes on the one hand and interface modes on the other.

Nevertheless, the DC model must be used with caution. In structures more complex than a simple quantum well it may be unreliable. The excellent agreement with the hybrid model rests, in part, on the latter's neglect of the TO component of what is really a triple hybrid. Provided that a double-hybrid approximation can be made, the DC model might be applicable, but where significant oscillator strength devolves to, or near to, the TO region of the spectrum it would be unreliable. Needless to say, the DC model cannot be applied to the situation in non-polar material, nor in polar material in connection with deformation-potential scattering of holes, or of electrons in [111] valleys.

Appendix: Scalar and Vector Potentials

The polar-optical vibrations that couple strongly to electrons propagate at speeds well below the velocity of light in the medium. It has therefore been customary to associate the electric field of such an excitation with a scalar potential ϕ. This procedure is undoubtedly correct for LO modes for which $\nabla \times \mathbf{E}_L = 0$ is strictly true. For transversely polarized modes, however, $\nabla \times \mathbf{E}_T \neq 0$ and this implies that $\mathbf{E}_T$ is associated in the radiation gauge with a vector potential $\mathbf{A}$. It then follows that the interaction energy is of the form $e\mathbf{A}.\mathbf{p}/m^*$, where $\mathbf{p}$ is the momentum operator and m^* is the effective mass of the electron, rather than $e\phi$. These interactions have very different forms and appear to be far from equivalent. The first calculation of scattering rates due to hybrids indeed assumed an $\mathbf{A}.\mathbf{p}$ interaction (Ridley, 1992, 1993). Nevertheless, the argument for using a scalar rather than a vector potential for waves travelling much more slowly than light seems plausible, even though $\nabla \times \mathbf{E}_T$ remains finite, though small, especially in view of the agreement of $\mathbf{A}.\mathbf{p}$ and $e\phi$ rates. The question is academic for TO modes for which $E_T \neq 0$, but is very pertinent for IP modes.

The Hamiltonian that describes an electron and its interaction with an electromagnetic field and with free charges, and the field itself, in an isotropic continuum is

$$H = \frac{1}{2m^*}(\mathbf{p} - e\mathbf{A}(\mathbf{r}_e))^2 + eV(\mathbf{r}_e) + \int H_f dr \qquad \text{(A8.1)}$$

in the effective-mass approximation. Here $\mathbf{r}_e$ and $\mathbf{p}$ are the position and momentum operators for the electron, $V(\mathbf{r}_e)$ is the electrostatic potential in which the electrons move, and the Hamiltonian density of the transverse electromagnetic field coupled to the polarization of the (non-magnetic) medium is

$$H_f = \tfrac{1}{2}\sum_{\omega}\left[\frac{\partial(\omega\varepsilon(\omega))}{\partial\omega}E_T^2(\omega,\mathbf{r}) + \tfrac{1}{\mu_0}\mathbf{B}^2(\omega,\mathbf{r})\right] \tag{A8.2}$$

See Eq. (4.48). For IP modes $\varepsilon(\omega) \neq 0$. All the electrostatic aspects are contained in $V(\mathbf{r}_e)$, so that the zero-order Hamiltonian for electrons is

$$H_e^0 = \frac{p^2}{2m^*} + V(\mathbf{r}_e) \tag{A8.3}$$

and the appropriate gauge for the electromagnetic field is

$$\nabla.\mathbf{A} = 0, \quad \phi = 0 \tag{A8.4}$$

which is referred to as the radiation gauge, in which

$$\mathbf{E} = -\dot{\mathbf{A}}, \quad \mathbf{B} = \nabla \times \mathbf{A} \tag{A8.5}$$

The choice of gauge in which the field has no scalar potential is a natural choice, but not the only one that can be made. Gauge transformations that preserve **E** and **B** can be made, but these would involve the introduction of a longitudinal vector potential A_L coupled with a scalar potential ϕ and would be plainly more cumbersome. Thus a gauge transformation cannot replace a vector potential with a scalar potential. Retaining the radiation gauge as the simplest, we can quantize the field in the usual way.

The field equation for IP modes is

$$\nabla^2\mathbf{A} + \frac{\omega^2}{\varepsilon(\omega)\mu_0}\mathbf{A} = 0 \tag{A8.6}$$

which translates into the wavevector equation

$$k_z^2 + \mathbf{k}_x^2 = \frac{\omega^2}{\varepsilon(\omega)\mu_0} \tag{A8.7}$$

where $\mathbf{k}_x$, k_z are the wavevector components in a direction in the plane and at right angles to the plane of the interface. For an IP mode k_z is pure imaginary, and the quantized vector potential is of the form

$$\mathbf{A}(\mathbf{r},t) = \sum_a \int d^2\mathbf{k}_x \left[\tilde{A}_a(\mathbf{k}_x, z)e^{i(\mathbf{k}_x\cdot\mathbf{x}-\omega t)}a_a(\mathbf{k}_x) + HC\right] \tag{A8.8}$$

where $\mathbf{r} = (\mathbf{x}, z)$, $\tilde{A}_a(\mathbf{k}x, z)$ are the mode vector functions associated with branch a, HC is the Hermitian conjugate and $a_a(\mathbf{k}_x)$ is a boson operator satisfying the commutation relation

$$\left[a_a(\mathbf{k}_x), a_\beta^\dagger(\mathbf{k}_x)\right] = \delta_{a\beta}\delta(\mathbf{k}_x - \mathbf{k}_x') \tag{A8.9}$$

and the Hermitian density is

$$H_f = \tfrac{1}{2}\sum_a \int d^2\mathbf{k}_x \hbar\omega(\mathbf{k}_x, a)(a_a(\mathbf{k}_x)a_a^\dagger(\mathbf{k}_x) + a_a^\dagger(\mathbf{k}_x)a_a(\mathbf{k}_x)) \tag{A8.10}$$

This formulation of the IP field and the interaction with an electron is entirely self-consistent and not reliant on any approximation.

Let us now consider how a scalar potential can be introduced to replace the vector potential. This replacement is only feasible in the limit of large $\mathbf{k}_x$ – the so-called unretarded limit. In this limit it is a good approximation to express the vector potential in terms of a field operator Λ thus:

$$\mathbf{A} = -\frac{\hbar}{e}\nabla\Lambda \tag{A8.11}$$

where

$$\Lambda = \frac{ie}{\hbar}\sum_a \int d\mathbf{k}_x \frac{1}{\omega}(a_a\tilde{\phi}_a + a_a^\dagger\tilde{\phi}_a^*) \tag{A8.12}$$

and the mode functions $\tilde{\phi}_a$ can be related to $\tilde{\mathrm{A}}_a$. The Hamiltonian is now

$$H = \frac{1}{2m^*}(\mathbf{p} + \hbar\nabla\Lambda)^2 + eV(\mathbf{r}_e) + \frac{1}{2}\sum_a \int d^2\mathbf{k}_x \hbar\omega(\mathbf{k}_x, a)(a_a(\mathbf{k}_x)a_a^\circ(\mathbf{k}_x) + a_a^\dagger(\mathbf{k}_x)a_a(\mathbf{k}_x)) \tag{A8.13}$$

We now make the unitary transformation U where

$$U = e^{-i\Lambda} \tag{A8.14}$$

The new operator can be expressed as the series

$$O' = e^{i\Lambda}Oe^{-i\Lambda} = O + i[\Lambda, O] + \frac{i^2}{2!}[\Lambda, [\Lambda, O]] + \cdots \tag{A8.15}$$

The new Hamiltonian will have the same form as before except that all operators are primed quantities, viz.:

$$H = \frac{1}{2m^*}(\mathbf{p}' + \hbar\nabla\Lambda\ )^2 + eV(\mathbf{r}_e)$$
$$+\frac{1}{2}\sum_a \int d^2\mathbf{k}_x \hbar\omega(\mathbf{k}_x, a)(a'_a(\mathbf{k}_x)a'^{\dagger}_a(\mathbf{k}_x) + a'^{\dagger}_a(\mathbf{k}_x)a'_a(\mathbf{k}_x)) \quad \text{(A8.16)}$$

where, using $\mathbf{p} = -i\hbar\nabla$, to first order

$$\mathbf{p}' \approx \mathbf{p} + i[\Lambda, \mathbf{p}] = \mathbf{p} - \hbar\nabla\Lambda \quad \text{(A8.17)}$$

and

$$a'_a \approx a_a + i[\Lambda, a_a] \quad \text{(A8.18)}$$

similarly for $a'^{\dagger}_a$. (Note that $\Delta\Lambda$ commutes with Λ.) The commutator gives

$$i[\Lambda a_a] = i\frac{(ie)}{\hbar}\sum_{a'}\int d\mathbf{k}'_x \frac{1}{\omega'}\left[a^{\dagger}_{a'}, a_a\right]\tilde{\phi}_a = \frac{e}{\hbar\omega}\tilde{\phi}_a \quad \text{(A8.19)}$$

and so the boson operators transform as follows:

$$a'_a = a_a + \frac{e}{\hbar\omega}\tilde{\phi}_a \quad \text{(A8.20)}$$

$$a'^{\dagger}_a = a^{\dagger}_a - \frac{e}{\hbar\omega}\tilde{\phi}_a \quad \text{(A8.21)}$$

Inserting these into the transformed Hamiltonian (Eq. (A.8.16)) leads to the field term

$$H'_f = H_f - e\phi + \Delta_{\text{self}} \quad \text{(A8.22)}$$

where H_f is the old field density. There are two new terms. One is a new operator given by

$$\phi = \sum_a \int d\mathbf{k}_x(a_a\tilde{\phi}_a - a^{\dagger}_a\phi_a) \quad \text{(A8.23)}$$

and this is exactly the scalar potential we have been looking for. The last term is a self-energy contributing to the mass of the electron given by

$$\Delta_{\text{self}} = -e^2\sum_a \int d\mathbf{k}_x \frac{\tilde{\phi}_a\tilde{\phi}_a}{\hbar\omega} \quad \text{(A8.24)}$$

Finally, the transformed Hamiltonian to first order is

$$
\begin{aligned}
H' &= e^{i\Lambda} H e^{-i\Lambda} \\
&= \frac{\mathbf{p}^2}{2m^*} + V(\mathbf{r}_e) - e\phi + \frac{1}{2}\sum_a \int d\mathbf{k}_x \hbar\omega (a_a a_a^\dagger + a_a^\dagger a_a) + \Delta_{\text{self}}
\end{aligned}
\qquad (A8.25)
$$

The Hamiltonian has come about via two approximations. One is the assumption of the unretarded limit, which has allowed the vector potential to be associated with the gradient of a scalar field. The other is the assumption that first-order perturbation theory is applicable to the problem. As far as the description of simple scattering processes involving first-order matrix elements on the energy shell is concerned the use of a scalar potential is valid and the result should be indifferent to whether a scalar or a vector potential is used. Considering the very different forms of the matrix elements that arise it is at first sight surprising that identical answers are obtained, but this is indeed the case for all inter- and intrasubband processes, as Babiker *et al.* (1993) have shown. But for second-order processes involving virtual transitions, and indeed for the determination of scattering profiles and spectroscopic line shapes, it is necessary to use the correct interaction defined by the vector potential.

9

Other Scattering Mechanisms

I conceive some scattered notions...
An Argument against Abolishing Christianity, *J. Swift*

9.1 Charged-Impurity Scattering

9.1.1 Introduction

Scattering of electrons by charged-impurity atoms dominates the mobility at low temperatures in bulk material and is usually very significant at room temperature (Fig. 9.1). The technique of modulation doping in high-electron-mobility field-effect transistors (HEMTs) alleviates the effect of charged-impurity scattering but by no means eliminates it. It remains an important source of momentum relaxation (but not of energy relaxation because the collisions are essentially elastic). Though its importance has been recognized for a very long time, obtaining a reliable theoretical description has proved to be extremely difficult.

There are many problems. First of all there is the problem of the infinite range of the Coulomb potential surrounding a charge, which implies that an electron is scattered by a charged impurity however remote, leading to an infinite scattering cross-section for vanishingly small scattering angles. Intuitively, we would expect distant interactions with a population of charged impurities to time-average to zero, leaving only the less frequent, close collisions to determine the effective scattering rate. This intuition motivated the treatment by Conwell and Weisskopf (1950) in which the range of the Coulomb potential was limited to a radius equal to half the mean distance apart of the impurities. Setting an arbitrary limit of this sort was avoided by introducing the effect of screening by the population of mobile electrons as was done by Brooks and Herring (1951) for semiconductors, following the earlier approach by Mott (1936). Obviously this scheme can work only when there are mobile electrons to do the screening, as is not the case in semiinsulating or insulating material. Building a bridge between

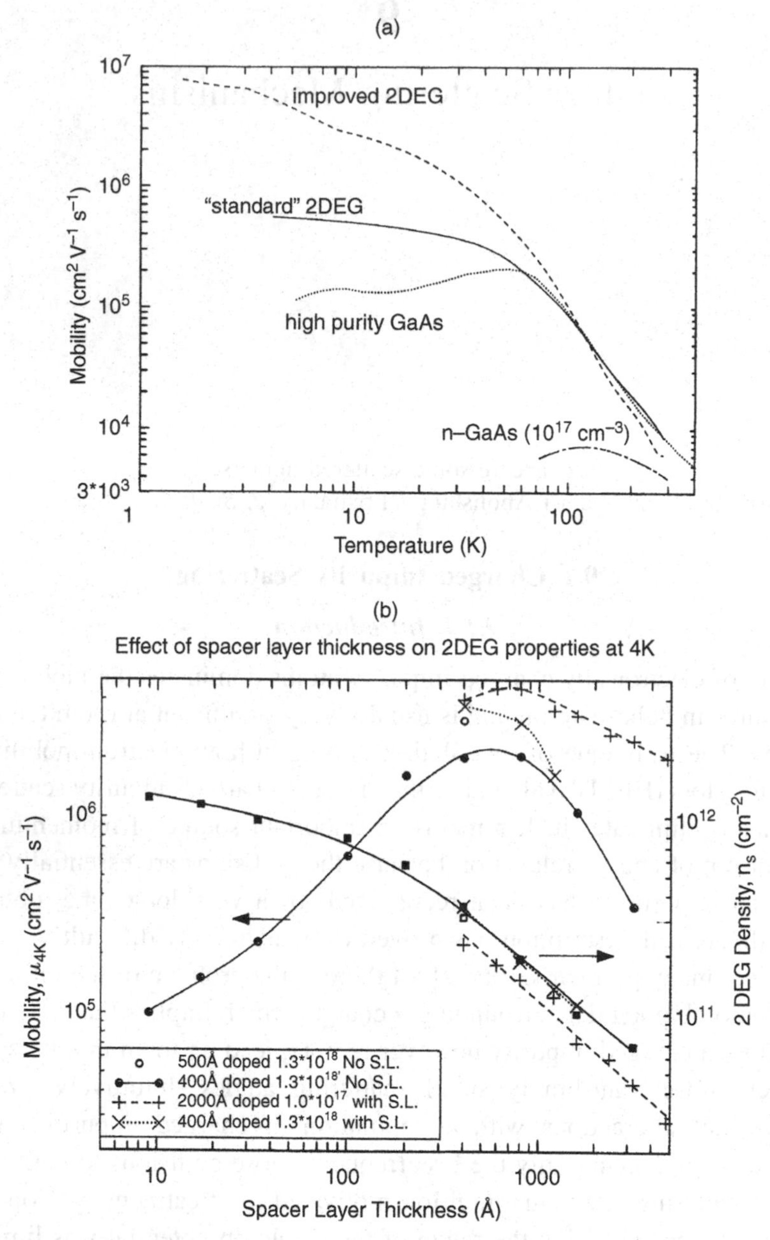

Fig. 9.1 (a) Temperature dependence of mobility in GaAs for doped, bulk material, high-purity bulk material, GaAs/AlGaAs heterojunction with 20-nm spacer (standard) and 40-nm spacer; (b) mobility and density at 4.2 K in GaAs/AlGaAs as a function of space-layer thickness (Harris *et al.*, 1989); also shown is the effect of replacing the AlGaAs with an AlAs/GaAs superlattice (SL).

the two approaches was done by arbitrarily limiting scattering to nearest-neighbour encounters (Ridley, 1977), and this scheme allowed a finite two-body scattering rate to be estimated whatever the carrier density.

Even so, estimates made along these lines relied on the validity of the Born approximation, which makes the assumption that both ingoing and outgoing wavefunctions in a two-body scattering event are plane waves. It is well known, however, from phase-shift analysis that the Born approximation represents only the first term of an infinite perturbation expansion and it can significantly over-estimate the scattering cross-section when the electron energy is small compared to the magnitude of the scattering potential.

An exact solution of Schrodinger's equation can be obtained for an electron being scattered by an unscreened charge in the 2D plane (Stern and Howard, 1967). The result is a cross-section

$$\sigma(\theta) = \frac{G \tanh \pi G}{2k \sin^2(\theta/2)}, \quad G = \frac{Ze^2}{4\pi\varepsilon\hbar v} \tag{9.1}$$

where v is the velocity of the electron, and G2 is the ratio of the effective Rydberg energy to electron kinetic energy. The Born approximation gives

$$\sigma_B(\theta) = \frac{\pi G^2}{2k \sin^2(\theta/2)} \tag{9.2}$$

i.e. when $\pi G << 1$, and the classical result is

$$\sigma_B(\theta) = \frac{G}{2k \sin^2(\theta/2)} \tag{9.3}$$

i.e. when $\pi G >> 1$. The Born approximation overestimates the scattering rate by the factor $\pi G/\tanh \pi G$, which becomes large at low temperatures for non-degenerate gas.

Corrections to the Born approximation can be made, but the problem goes beyond that. The interaction between an electron and a charged-impurity population is not a simple two-body encounter. The scattering process will include coherent scattering from two or more impurities, plus the effects of band-structure modifications induced by the impurity population. In other words, there are multiple-scattering and impurity-dressing effects to be taken into account. Following Moore (1967) we can express the overall scattering rate as follows:

$$W = W_0(1 + \delta_B + \delta_M + \delta_D) \tag{9.4}$$

where W_0 is the Born-approximation rate, δ_B is a correction incorporating high-order terms in the perturbation expansion, δ_M is the contribution made by multiple scattering and δ_D is the effect of dressing. Born corrections have been advanced by Moore (1967).

But there is still another problem. All of the foregoing work has assumed that the impurity distribution is random. In compensated material it is vital to know the degree to which oppositely charged impurities are paired by their electrostatic attraction, for, if they are paired, the scattering is dipolelike (Stratton, 1962). A pair-correlation function in this case has been considered by Falicov and Cuevas (1967). In uncompensated material the impurity distribution may be far from random. In the extreme case repulsion between donors would produce a regular array, and such a distribution would introduce a superlattice potential that would modify the motion of electrons through its minibands but would not scatter. Screening of the repulsion by the high-temperature plasma of ions and electrons during growth would be expected to limit the tendency to form a regular array. One might confidently expect that the resultant potential experienced by an electron would be somewhere in-between random and periodic. Scattering by a potential obeying Gaussian statistics has been treated by Yussouff and Zittartz (1973), Yanchev *et al.* (1979), and Lassnig (1988), and Poisson statistics have been used by Schubert *et al.* (1989). Small deviations from a regular array have been treated by Van Hall *et al.* (1988), and we will look at this in more detail later, representing, as it does, an extreme case to set against the case of a random distribution.

In addition to all this there are central-cell corrections to be made. These are related to the deviation from a purely Coulombic potential at the core of the impurity, which accounts for its bound state being shifted downward from the predicted Bohr energy. Because of the effect of the central cell, the scattering rate differs from one chemical species of donor to another, a phenomenon that has been studied by Ralph *et al.* (1975). It can also give rise to resonant scattering of the Breit–Wigner type, familiar in nuclear physics (El-Ghanem and Ridley, 1980; Ridley, 1993 (book)).

It is beyond the scope of this book to consider the details of all these processes and their successes and failures. Some aspects of the role of screening are discussed in Chapter 10. Other topics discussed in more depth can be found in the literature already cited and in, for example, Lancefield *et al.* (1987) and Abrams *et al.* (1978). In what follows we concentrate on simple estimates of scattering rate based on the Born approximation.

9.1.2 The Coulomb Scattering Rate

Let $\rho(\mathbf{R})$ be the distribution of space charge that scatters an electron, and let the associated potential be $\phi(\mathbf{R})$, related by Poisson's equation thus:

$$\nabla^2\phi(\mathbf{R}) = -\frac{\rho(\mathbf{R})}{\varepsilon} \tag{9.5}$$

When $\phi(\mathbf{R})$ and $\rho(\mathbf{R})$ are expanded in a Fourier series

$$\phi(\mathbf{R}) = \int \phi(\mathbf{Q})e^{i\mathbf{Q}.\mathbf{R}}d\mathbf{Q}, \qquad \rho(\mathbf{R}) = \int \rho(\mathbf{Q})e^{i\mathbf{Q}.\mathbf{R}}d\mathbf{Q}, \tag{9.6}$$

we have

$$\phi(\mathbf{Q}) = \frac{\rho(\mathbf{Q})}{\varepsilon Q^2} \tag{9.7}$$

Screening is accommodated by the **Q**-dependence of the permittivity, i.e. $\varepsilon \rightarrow \varepsilon(\mathbf{Q})$. The scattering rate W is usually calculated in the Born approximation with

$$W = \frac{2\pi}{\hbar}\int |M|^2\delta(E_f - E_i)dN_f \tag{9.8}$$

where f, i denote final and initial states. Usually what is required is the momentum-relaxation rate, which for elastic collisions is

$$W_m = \frac{2\pi}{\hbar}\int |M|^2(1-\cos\theta)\delta(E_f - E_i)dN_f \tag{9.9}$$

where θ is the scattering angle. If $\psi_f(\mathbf{R})$ and $\psi_i(\mathbf{R})$ are the final and initial state wavefunctions, the matrix element is given by

$$\begin{aligned} M &= \int \psi_f^*(\mathbf{R})e\phi(\mathbf{R})\psi_i(\mathbf{R})d\mathbf{R} \\ &= \iint \psi_f^*(\mathbf{R})e\frac{\rho(\mathbf{Q})}{\varepsilon(Q)Q^2}e^{i\mathbf{Q}\cdot\mathbf{R}}\psi_i(\mathbf{R})d\mathbf{R}d\mathbf{Q} \end{aligned} \tag{9.10}$$

9.1.3 Scattering by Single Charges

The 3D case is well known for the problems connected with the infinite range of the Coulomb potential surrounding a single scattering centre, and the same problem occurs in two dimensions. Additional problems arise in 1D in connection with localization, which is beyond the scope of this book to discuss, and therefore we will focus here on the 3D and 2D cases. A single scattering charge Ze at the origin is described by a charge density

$$\rho(\mathbf{R}) = Ze\delta(\mathbf{R}) \tag{9.11}$$

Since $\delta(x) = (1/2\pi)\int_{-\infty}^{\infty} e^{iqx}dq$, we have

$$\rho(\mathbf{Q}) = \frac{Ze}{(2\pi)^3} \tag{9.12}$$

and

$$M = \frac{Ze^2}{(2\pi)^3}\iint \psi_f^*(\mathbf{R})\frac{e^{i\mathbf{Q}.\mathbf{R}}}{\varepsilon(Q)Q^2}\psi_i(\mathbf{R})d\mathbf{R}d\mathbf{Q} \tag{9.13}$$

We can take the wavefunctions to be of the form

$$\psi(\mathbf{R}) = \begin{cases} \Omega^{-1/2}e^{i\mathbf{K}.\mathbf{R}} & \text{3D} \\ \sigma^{-1/2}e^{i\mathbf{k}.\mathbf{r}}\psi(z) & \text{2D} \end{cases} \tag{9.14}$$

The scheme is that capitals denote 3D vectors, lowercase letters 2D vectors. The matrix element is then

$$M = Ze^2\begin{cases} \dfrac{\delta_{\mathbf{Q},\mathbf{K}_f-\mathbf{K}_i}}{\varepsilon(Q)Q^2\Omega} & \text{3D} \\ \dfrac{\delta_{\mathbf{q},\mathbf{k}_f-\mathbf{k}_i}}{2\varepsilon(q)q\sigma}\displaystyle\int_{z_1}^{z_2}\psi_f(z)\psi_i(z)e^{-q|z|}dz & \text{2D} \end{cases} \tag{9.15}$$

where z_1, z_2 denote the boundaries of the quantum well. Umklapp processes are unimportant, and there is wavevector conservation in the unconfined directions. The matrix element heavily emphasizes small magnitudes of Q, q, thus tending to make intersubband processes unimportant. Restricting attention to intrasubband processes only and exploiting the elastic nature of the scattering we find that

$$\begin{aligned} Q &= 2K\sin(\theta/2) & \text{3D} \\ q &= 2K\sin(\theta/2) & \text{2D} \end{aligned} \tag{9.16}$$

where θ is the scattering angle in the scattering plane.

The scattering rate is given by

$$W = \frac{(Ze^2)^2m^*}{2\pi\hbar^3\varepsilon_s^2}\begin{cases} \dfrac{K}{\Omega}\displaystyle\int_0^{\pi}\frac{\sin\theta}{(Q^2+Q_s^2)^2}d\theta & \text{3D} \\ \dfrac{1}{2\sigma}\displaystyle\int_0^{\pi}\frac{F(q)}{(q+q_s)^2}d\theta & \text{2D} \end{cases} \tag{9.17}$$

and the momentum-relaxation rate by

$$W_m = \frac{(Ze^2)^2m^*}{\pi\hbar^3\varepsilon_s^2}\begin{cases} \dfrac{K}{\Omega}\displaystyle\int_0^{\pi}\frac{\sin^2(\theta/2)\sin\theta}{(Q^2+Q_s^2)^2}d\theta & \text{3D} \\ \dfrac{1}{2\sigma}\displaystyle\int_0^{\pi}\frac{F(q)\sin^2(\theta/2)}{(q+q_s)^2}d\theta & \text{2D} \end{cases} \tag{9.18}$$

In the preceding, $F(q)$ is the usual form function, viz.:

$$F(q) = \int_{z_1}^{z_2} \psi_f(z')\psi_i(z')\psi_f(z)\psi_i(z)e^{-q|z'-z|}dz'dz \tag{9.19}$$

and we have assumed a static dielectric screening function of the form

$$\begin{aligned} \varepsilon(Q) &= \varepsilon_s\left(1+\frac{Q_s^2}{Q^2}\right) && \text{3D} \\ \varepsilon(q) &= \varepsilon_s\left(1+\frac{q_s}{q}\right) && \text{2D} \end{aligned} \tag{9.20}$$

where Q_s and q_s are reciprocal screening lengths.

The incorporation of screening eliminates the divergence at small scattering angles. Note that the expressions for the 2D case incorporate remote impurity scattering since the origin is at the scattering centre (Price, 1984). When there is more than one centre a sum over the positions must be carried out to give the total rate, assuming each scatters independently.

9.1.4 Scattering by Fluctuations in a Donor Array

If the scattering charges were arranged in a regular array such that the electron experienced a periodic potential, it is clear that the foregoing treatment would be inapplicable. It would be necessary, instead, to solve a band-structure problem and to describe the effect on the motion of the electron in terms of effective mass, minibands and Bragg reflection. With normal scattering densities the extent of each superimposed Brillouin zone would be very small in comparison with those of the host lattice, and the minibands would be correspondingly narrow. Except at very small electron energies we cannot expect the effect on the electron's motion to be large. At least to a first approximation, it will be reasonable to ignore entirely the perturbative effects of a regular array of charges. In this view, scattering would result only from fluctuations in the positions of the charges from the lattice points of the array.

This leads to a radically different picture of charged-impurity scattering, which has the advantage of eliminating that troublesome divergence in 2D without relying on screening, as shown by van Hall *et al.* (1988). If the fluctuations are small, the scattering potential can be obtained from the single-charge case by a Taylor expansion. Thus, in place of the matrix element of Eq. (9.13) we have

$$M_{\text{fluc}} = \frac{Ze^2}{(2\pi)^3}\iint \psi_f^*(\mathbf{R})\frac{(1-e^{-i\mathbf{Q}.\mathbf{D}})}{\varepsilon(Q)Q^2}e^{i\mathbf{Q}.\mathbf{R}}\psi_i(\mathbf{R})d\mathbf{R}d\mathbf{Q} \tag{9.21}$$

In this way the interaction is converted to a dipole form in lowest order, where **D** is a vector describing the displacement of the charge from the lattice point of the array. Assuming the fluctuations to be random with a Gaussian distribution, we get

$$\left|1 - e^{-i\mathbf{Q}.\mathbf{D}}\right|^2 = 2(1 - e^{-Q^2\Delta^2/2}) \equiv G(Q) \tag{9.22}$$

where Δ is the standard deviation, expected to be of the order of the root mean square spacing between the donors.

In 2D with $\mathbf{D} = (\delta, \delta_z)$ we have

$$M_{\text{fluc}} = \frac{Ze^2}{(2\pi)^3}\iint \psi_f^*(\mathbf{R})\frac{(1 - e^{-i(\mathbf{q}.\delta + q_z\delta_z)})e^{i\mathbf{Q}.\mathbf{R}}}{\varepsilon(q)(q^2 + q_z^2)}\psi_i(\mathbf{R})d\mathbf{R}d\mathbf{Q} \tag{9.23}$$

With donors concentrated in a plane at $z = z_0$, so that $\delta_z = 0$,

$$\left|1 - e^{-i\mathbf{q}.\delta}\right|^2 = 2(1 - e^{-q^2\Delta^2/2}) \equiv G(q) \tag{9.24}$$

Therefore

$$W_{\text{fluc}} = \frac{(Ze^2)^2 m^*}{\pi\hbar^3\varepsilon_s^2}\begin{cases} \dfrac{K^2}{\Omega}\displaystyle\int_0^\pi \frac{G(Q)\sin\theta}{(Q^2+Q_s^2)^2}d\theta & \text{3D} \\ \dfrac{1}{2\sigma}\displaystyle\int_0^\pi \frac{G(q)e^{-2qz_0}F(q)}{(q+q_s)^2}d\theta & \text{2D} \end{cases} \tag{9.25}$$

and

$$W_{m\,\text{fluc}} = \frac{(Ze^2)^2 m^*}{\pi\hbar^3\varepsilon_s^2}\begin{cases} \dfrac{2K^2}{\Omega}\displaystyle\int_0^\pi \frac{G(Q)\sin^2(\theta/2)\sin\theta}{(Q^2+Q_s^2)^2}d\theta & \text{3D} \\ \dfrac{1}{\sigma}\displaystyle\int_0^\pi \frac{G(q)e^{-2qz_0}F(q)\sin^2(\theta/2)}{(q+q_s)^2}d\theta & \text{2D} \end{cases} \tag{9.26}$$

where Q,q are functions of θ given by Eq. (9.16) and the dependence on spacer thickness z_0 has been made explicit. When the donors are spread randomly over a region $z_0 \leq z \leq z_0 + w$, van Hall (1989) replaces $\Delta^2/2$ in Eq. (9.24) by $(w^2/24 + \Delta^2/2)$ for spacer thicknesses greater than 100 Å.

The theory describes fluctuations in the donor density that are essentially defined during growth and frozen during cooling. Other authors have used Gaussian statistics or Poisson statistics to describe the fluctuations, but the basic idea is always the same – the average potential of the impurities does not scatter; only charge fluctuations scatter. Another source of charge fluctuation is associated with the occupation of the donor levels, which can be important at low temperatures, especially in GaAs-based materials, where there exists the deep donor known as the DX centre. This situation has been analysed by Buks *et al.* (1994), who provide good evidence that correlations among the DX and ordinary donors do enhance the mobility. At room temperature in systems such as AlInAs/GaInAs that do not suffer from DX centres, only fluctuations in density need be considered.

9.1.5 An Example

In heavily doped material with substantial Fermi energies we can expect the Born approximation to be valid, and either of the models just described to hold. Their predictions of remote-impurity-scattering mobility at 300 K as a function of spacer thickness in a highly doped AlInAs/GaInAs MODFET for the special case $q\,\Delta << 1$ are shown in Fig. 9.2(a), and compared with experiment. The mobilities predicted by the normal model assuming a random planar array of Si donors are all too low, whereas the mobilities predicted by the fluctuation model in the dipole approximation are all too high. A crude hybrid model based on a Poisson probability distribution of fluctuations predicts a mobility given by

$$\mu^{-1} = 0.632\mu_{\text{norm}}^{-1} + 0.368\mu_{\text{fluc}}^{-1} \tag{9.27}$$

where μ_{norm}, μ_{fluc} are the mobilities predicted by the two extreme models. The argument here is simply that the probability that a region, in which the average occupation is one atom, actually contains one atom is just $e^{1} = 0.368$. This is taken to be the probability that scattering is described by the dipole fluctuation model. The corresponding probability of the region containing no atom or more than one is then $1 - e^{-1}$, and this is taken to be the probability that the normal model is applicable. The agreement with experiment is better but by no means perfect; see Fig. 9.2(b).

Another approach has been to restrict collisions to those involving nearest neighbours only – so-called statistical screening – and this has been shown to explain the high mobilities observed at low temperatures. The effect is to multiply the normal rate by $\exp(-2/\pi)$ (Ridley, 1996a). The model of van Hall (1989) also can give good agreement with experiment with a suitable choice of Δ. This quantity is going to be a function of the conditions of growth and of doping and can be expected to vary from structure to structure, and, indeed, from laboratory to laboratory, making predictions of the strength of remote-impurity scattering somewhat hazardous. An example of the sensitivity to the magnitude of Δ is illustrated in Fig. 9.3 for the case of an AlInAs/GaInAs 100-Å quantum well with a 100-Å spacer thickness.

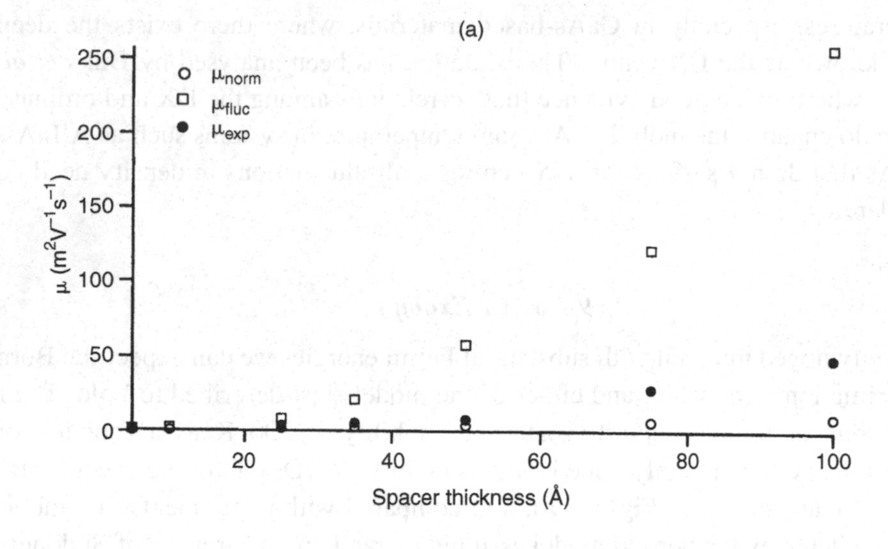

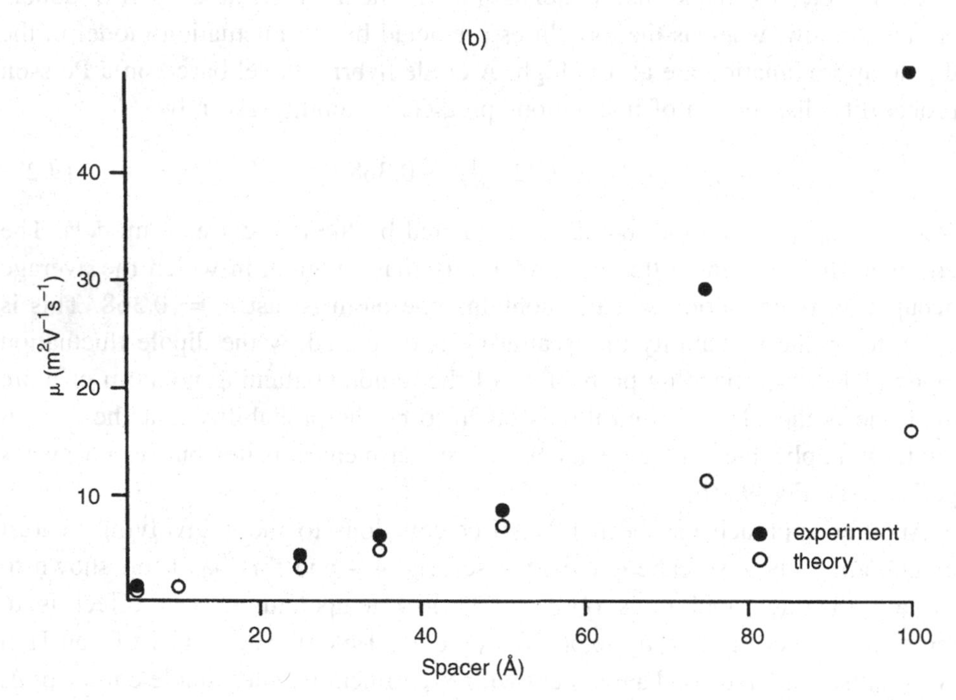

Fig. 9.2 (a) Component of mobility of electrons at 300 K associated with remote impurity scattering in a δ-doped AlInAs/GaInAs modulation-doped field effect transistor (MODFET) as a function of spacer thickness (experimental points from Seaford *et al.*, 1995). (b) Comparison with the hybrid model described in the text.

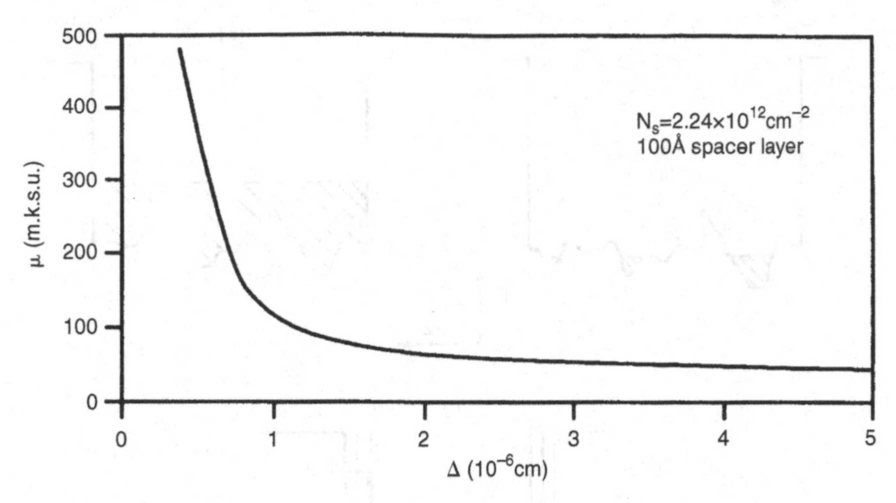

Fig. 9.3 Illustration of the sensitivity of mobility to choice of fluctuation parameter Δ.

9.2 Interface-Roughness Scattering

In spite of spectacular progress in crystal growing over recent decades it is not always possible to grow multilayered structures with perfect interfaces or free of potential fluctuations (Fig. 9.4). Being confined close to interfaces the electrons are highly sensitive to irregularities, which can be impurities trapped at the interface or monolayer steps. Scattering by charged impurities sitting on the interface can be treated by the methods discussed in the previous section. Here we look at the problem of scattering by geometrical irregularities in the interface, which can become the dominant mechanism at low temperatures and is seldom ignorable.

Let $\Delta(\mathbf{r})$ be the deviation in the z-direction (normal to the interface) of the interface from its average position. Any deviation will affect the subband energies and the wavefunctions and thereby affect the motion of the electron in the plane. If $\Delta(\mathbf{r})$ is small and the subband separation is large, the wavefunction will not change appreciably, so we can use the unperturbed solutions in the matrix element quantifying the scattering rate. If the deviation produces a change of energy $\Delta H(\mathbf{r}, z)$, the scattering rate will be

$$W = \frac{2\pi}{\hbar} \int |\langle \mathbf{k}'|\Delta H(\mathbf{r}, z)|\mathbf{k}\rangle|^2 \delta(E' - E) dN_f \tag{9.28}$$

with

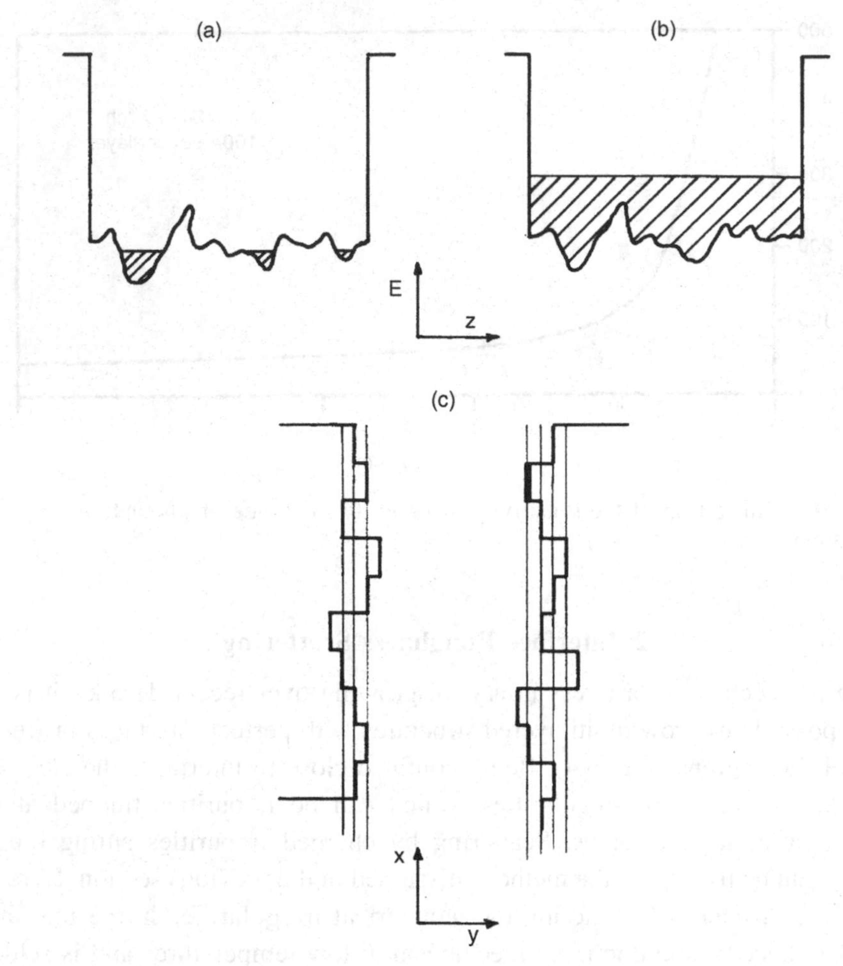

Fig. 9.4 Structural sources of scattering: (a) localization by potential fluctuations; (b) effect of potential fluctuations, less for a degenerate gas (or energetic electrons); (c) well-width fluctuations in units of molecular monolayers–interface roughness.

$$\Delta H(\mathbf{r}, z) = \Delta T(\mathbf{r}, z) + \Delta \mathrm{V}(\mathbf{r}, z) \tag{9.29}$$

where T is kinetic and V is potential energy.

The case of electrons confined electrostatically at a single heterojunction is complicated by space-charge effects, in general involving image charges. The situation has been analysed by Ando. The expression for the rate is somewhat lengthy and we will not reproduce it here. Instead, we look at the simpler case of a quantum well free of space-charge effects. Following Sakaki *et al.* (1987) we can express the energy change for an intrasubband process in terms of the dependence of subband energy E on well-width L, viz.

$$\Delta H(\mathbf{r}, z) = \Delta(\mathbf{r}) \frac{dE}{dL} \tag{9.30}$$

After expanding $\Delta(\mathbf{r})$ in a Fourier series, the squared matrix element reduces to

$$|<\mathbf{k}'|\Delta H(\mathbf{r}, z)|\mathbf{k}>|^2 = |\Delta(q)|^2 \left(\frac{dE}{dL}\right)^2, \mathbf{q} = \mathbf{k}' - \mathbf{k} \tag{9.31}$$

Only the power spectrum of $\Delta(\mathbf{r})$ is required. The usual approach is to assume, for mathematical simplicity, that the autocovariance function is isotropic and Gaussian. Thus:

$$<\Delta(\mathbf{r}')\Delta(\mathbf{r}' - \mathbf{r})> = \Delta^2 e^{-r^2/\Lambda^2} \tag{9.32}$$

and so

$$|\Delta(q)|^2 = \pi\Delta^2\Lambda^2 e^{-q^2\Lambda^2/4} \tag{9.33}$$

Including static 2D screening leads, with $x = \sin(\theta/2)$, θ is the scattering angle, to the momentum-relaxation rate

$$W_k = 4\left(\frac{dE}{dL}\right)^2 \frac{\Delta^2\Lambda^2 m^*}{\hbar^3} \int_0^1 \frac{x^4 e^{-k^2\Lambda^2 x^2}}{[x + (q_s/2k)]^2 (1 - x^2)^{1/2}} dx \tag{9.34}$$

The factors Δ and Λ specify the roughness, where Δ is the average deviation and Λ is a length of order of the range of the deviation in the plane of the interface (Fig. 9.5). Mobility measurements suggest that Δ is 1 to 3 ml (monolayers) (1 monolayer in GaAs = 2.83 Å) and Λ is 30 to 70 Å.

The assumption of a Gaussian autocovariance function is quite arbitrary and adopted for mathematical convenience. Experimental evidence from the study of Si/SiO_2 interfaces points rather to a simple exponential form (Goodnick *et al.*, 1985), in which case

$$|\Delta(\mathbf{q})|^2 = \pi\Delta^2\Lambda^2 \left[1 - \left(\frac{q^2\Lambda^2}{2}\right)\right]^{-3/2} \tag{9.35}$$

In either situation, the rate is very sensitive to well-width. For an infinitely deep well, $W \propto L^{-6}$. W is proportional to L^{-6}, where L is the well-width. A variation with well-width of this rapidity was observed by Sakaki *et al.* (1987) in the GaAs/AlAs system.

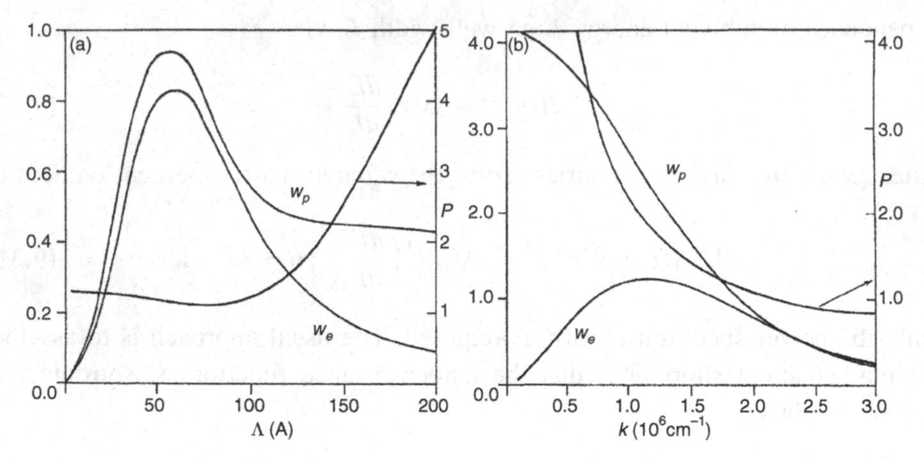

Fig. 9.5 Interface-roughness momentum-relaxation rates (10^{12} s^{-1}) in a 50–Å GaAs deep well. W_e and W_p are electron and interface phonon rates, respectively, and $P = Wp$: W_e: (a) as a function of correlation length Λ with $\Delta = 1$ ml and $k = 2 \times 10^6$ cm^{-1}; (b) as a function of wavevector with $\Delta = 1$ ml and $\Lambda = 60$ Å.

Similar effects can arise from fluctuations in the band-edge discontinuity.

9.3 Alloy Scattering

The simplest picture of a ternary alloy $A_xB_{1-x}C$ is of a virtual crystal having properties that are some average of the properties of the two binaries AC and BC (a scheme that is obviously extendable to quaternaries). Thus if V_A, V_B are the potentials experienced by the electron associated with the cations A and B in the corresponding binaries, the virtual-crystal average, in the simplest model, can be taken to be

$$V = V_A(x) + V_B(1 - x) \tag{9.36}$$

There are bound to be fluctuations of concentration during growth. If in a certain region the concentration of A is x', rather than the average x, then the deviation in potential will be

$$\Delta V(\mathbf{r}, z) = (V_A - V_B)(x' - x) \tag{9.37}$$

and this produces scattering. The root mean square (rms) deviation is just that for a binomial distribution

$$<\Delta V> = |V_A - V_B|\left(\frac{x(1-x)}{N_c}\right)^{1/2} \tag{9.38}$$

where N_c is the number of cation sites.

Expanding $\Delta V(\mathbf{r}, z)$ in a Fourier series we can write the quasi-2D rate as

$$W_{n',n} = \frac{2\pi}{\hbar}(V_A - V_B)^2 x(1-x)N(E)\Omega_0 F_{n',n} \tag{9.39}$$

where $N(E) = m^*/2\pi\hbar^2 L$, Ω_0 is the volume of the unit cell and

$$F_{n',n} = \int_{-\infty}^{\infty}\left(\int_0^L \psi n'(z)e^{iq_z z}\psi_n(z)dz\right)^2 \frac{L}{2\pi}dq_z \tag{9.40}$$

Here n, n' denote subbands. Alloy scattering is elastic and isotropic and, therefore, scattering and momentum-relaxation rates are identical.

The potentials V_A and V_B are not firmly known. The difference $|V_A - V_B|$ is obtained from the respective pseudopotentials of the band structure, or from the band-edge discontinuities, or from the electron affinities. Values of 0.5 eV for GaInAs, and 1.0 eV for AlInAs are often used.

9.4 Electron–Electron Scattering

9.4.1 Basic Formulae for the 2D Case

Scattering of one electron by another involves the exchange of both energy and momentum and it, therefore, acts to randomize those dynamic quantities over the electron gas, but it cannot relax either. If strong enough, it can maintain or establish quickly a Maxwellian or Fermi–Dirac distribution – or a drifted form of either distribution in the presence of a directional force. Establishing such a distribution is often referred to as thermalization, though the end point may be characterized by an electron temperature very different from that of the lattice.

The usual approach to the description of electron–electron scattering is to consider it as a two-body, screened (usually statically screened) interaction that can be treated within the Born approximation. Practically all the caveats that pertain to charged-impurity scattering pertain here, some less so, some more so. Compared with phase-shift analysis the Born approximation often overestimates the scattering rate. Screening is really dynamic, and not static – in the electron–electron (or electron–hole) case, with an effective frequency $w = \mathbf{q}.\mathbf{v}_{cm}$, where $\mathbf{v}_{cm}$ is the velocity of the centre of mass (see Section 10.9). Regarding each electron as scattering independently of the others overestimates the effect of the distant interactions.

With these reservations in mind we can write down the two-dimensional screened version of the Mott formula (see Messiah, 1966; Ridley, 1993) for the electron–electron differential cross-section in the Born approximation as follows:

$$\sigma(\theta) = \frac{e^4 \mu^{*2}}{8\pi\varepsilon_\infty^2 \hbar^4 g_{12}} \left[\frac{|F_{ijmn}(q_{12})|^2}{q_{12}^2 S(q_{12})^2} + \frac{|F_{ijmn}(q_{21})|^2}{q_{21}^2 S(q_{21})^2} - \frac{|F_{ijmn}(q_{12}) F_{ijmn}(q_{21})|}{q_{12} S(q_{12}) q_{21} S(q_{21})} \right] \tag{9.41}$$

where μ^* is the reduced mass ($= m^+/2$), ε_∞ is the high-frequency permittivity, the $F_{ijmn}(q)$ are form factors and the $S(q)$ are screening factors. (Note that the dimension of cross-section is now length.)

This gives the rate for the collision of an electron in subband i having wavevector $\mathbf{k}_1$, with an electron in subband j having wavevector $\mathbf{k}_2$, in the centre-of-mass frame of reference. In this reference frame

$$\mathbf{g} = \mathbf{k} - \mathbf{k}_{cm}, \quad \mathbf{k}_{cm} = \frac{1}{2}(\mathbf{k}_1 + \mathbf{k}_2) \tag{9.42}$$

and so

$$\mathbf{g}_{12} = \frac{1}{2}(\mathbf{k}_1 - \mathbf{k}_2) \tag{9.43}$$

After scattering, one electron is in subband m wavevector $\mathbf{k'}_1$, and the other is in subband n, wavevector $\mathbf{k'}_2$. Conservation of energy and momentum for parabolic bands and normal processes leads to

$$\begin{gathered} \mathbf{k}_1 + \mathbf{k}_2 = \mathbf{k'}_1 + \mathbf{k'}_2 \\ \frac{\hbar^2}{2m^*}({k_1}^2 + {k_2}^2) + E_i + E_j = \frac{\hbar^2}{2m^*}({k'_1}^2 + {k'_2}^2) + E_m + E_n \end{gathered} \tag{9.44}$$

Putting

$$\mathbf{q}_{12} = \mathbf{g}_{12} - \mathbf{g'}_{12} \tag{9.45}$$

and noting that

$${k_1}^2 + {k_2}^2 - {k'_1}^2 - {k'_1}^2 = 2(g_{12}^2 - g_{12}^{\prime 2}) \tag{9.46}$$

we translate the energy/momentum conservation into centre-of-mass wavevectors

$$g_{12}^2 = g_{12}^2 - {g_0}^2 \tag{9.47}$$

where

$$g_0^2 = \frac{m^*}{\hbar^2}(E_m + E_n - E_i - E_j) \tag{9.48}$$

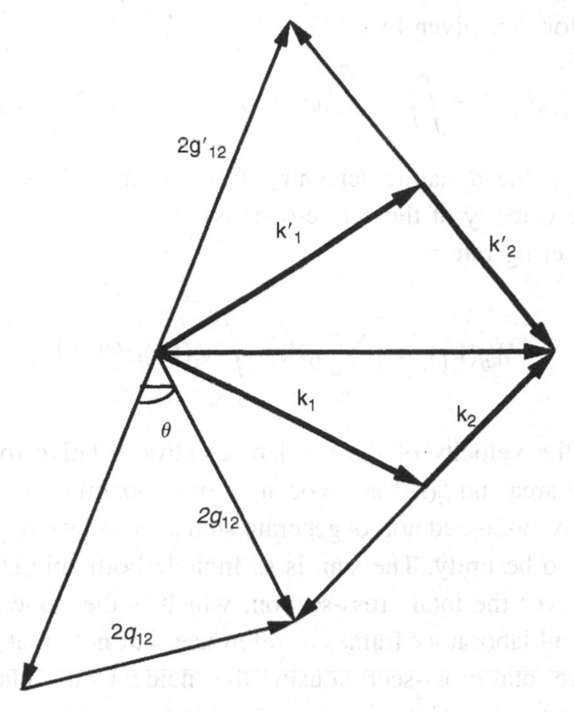

Fig. 9.6 Vector diagram for electron–electron scattering.

and therefore

$$q_{12}^2 = 2g_{12}^2 - g_0^2 - 2g_{12}(g_{12}^2 - g_0^2)^{1/2}\cos\theta \tag{9.49}$$

where θ is the angle between $\mathbf{g}_{12}$ and $\mathbf{g}'_{12}$. The vector $\mathbf{q}_{12}$ acts in exactly the same way as the vector $\mathbf{q}$ in the formula for charged-impurity scattering (Fig. 9.6).

The electrons emerging from the collision are indistinguishable and there would be no observable change if the electron with wavevector $\mathbf{k}'_1$ were to be exchanged with the electron whose wavevector was $\mathbf{k}'_2$. This gives rise to the second term in the expression for the cross-section, Eq. (9.41). In this case

$$g'_{21} = \frac{1}{2}(\mathbf{k}'_2 - \mathbf{k}'_1), \quad \mathbf{q}_{21} = \mathbf{g}_{12} - \mathbf{g}'_{12} \tag{9.50}$$

whence

$$q_{12}^2 = 2g_{12}^2 - g_0^2 - 2g_{12}(g_{12}^2 - g_0^2)^{1/2}\cos\theta \tag{9.51}$$

The third term in the cross-section arises as a result of quantum interference when the two electrons that collide have identical spins.

The form factors are given by

$$|F_{ijmn}(q)|^2 = \iint \psi_m^*(z)\psi_n^*(z)\psi_i(z')\psi_j(z')e^{-q|z-z'|}dzdz' \tag{9.52}$$

The factor $S(q)$ is the dynamic screening function at a frequency $\omega = \mathbf{q}.\mathbf{v}_{cm}$, where $\mathbf{v}_{cm}$ is the velocity of the centre-of-mass.

The total scattering rate is

$$W_{ij}(\mathbf{k}_1) = \frac{1}{A}\sum_{j,\mathbf{k}_2} f_j(\mathbf{k}_2)\int_0^{2\pi} v_i(\mathbf{k}_1)\sigma(\theta)d\theta \tag{9.53}$$

where $v_i(\mathbf{k}_1)$ is the velocity of the incident electron relative to the bombarded electron, A is the area and $f_j(\mathbf{k}_2)$ is the occupation probability of the state $|\mathbf{k}_2>$ in subband j. We have assumed non-degenerate statistics and we have taken the cell-overlap integral to be unity. The sum is to include both spin states. Integrating over the angle gives the total cross-section, which is the same quantity in both centre-of-mass and laboratory frames of reference, but note that the rate must be deduced from the total cross-section using the incident flux relative to the electron doing the scattering. Thus

$$v_i(\mathbf{k}_1) = \frac{2\hbar g_{12}}{m^*} = \frac{\hbar g_{12}}{\mu^*} \tag{9.54}$$

9.4.2 Discussion

Treatments of electron–electron scattering in 2D have almost all been numerical in Monte Carlo or molecular dynamical simulations. Even so it has been usual to make several approximations, some of them severe. A reasonable first step is to neglect intersubband processes on the grounds that these vanish anyway because of the orthogonality of the wavefunctions when $q = 0$ and the large wavevector changes that are generally required render the rates small. Limiting to intrasubband scattering (both electrons remaining in their subbands) allows us to put $g_0 = 0$ and hence

$$\begin{aligned} q_{12} &= 2g_{12}\sin(\theta/2) \\ q_{21} &= 2g_{12}\cos(\theta/2) \end{aligned} \tag{9.55}$$

Note that without intersubband transfers electrons in different subbands can still scatter off one another and contribute to the thermalization process.

A second step is to neglect the interaction involving parallel spins. The squared matrix element is of the form

$$M^2 = \tfrac{1}{2}\left(M_{12}^2 + M_{21}^2 + \left|M_{12} - M_{21}\right|^2\right) = M_{12}^2 + M_{21}^2 - \left|M_{12} - M_{21}\right| \tag{9.56}$$

where the factor 1/2 arises because in half of the collisions on average the spins are aligned and in the other half they are opposed. The third term in the brackets is most significant for scattering angles near $\pi/2$, i.e. when $q_{12} \approx q_{21}$. If its effect is ignored, we obtain

$$\sigma(\theta) = \frac{e^4\mu^{*2}}{16\pi\varepsilon_\infty^2\hbar^4 g_{12}}\left[\frac{\left|F_{ijij}(q_{12})\right|^2}{q_{12}^2 S(q_{12})^2} + \frac{\left|F_{ijij}(q_{21})\right|^2}{q_{21}^2 S(q_{21})^2}\right] \tag{9.57}$$

which implies that we are assuming that quantum interference eliminates half of the collisions. It is not clear how good the approximation is in general, though it is used widely. Adopting a classical approximation instead, i.e. ignoring quantum interference altogether, would *double* the magnitude of the cross-section. The rate corresponding to Eq. (9.57) can be written

$$W_{ij}(\mathbf{k}_1) = \frac{e^4\mu^*}{8\pi\varepsilon_\infty^2\hbar^4 A}\sum_{j,\mathbf{k}_2} f_j(\mathbf{k}_2)\int_0^{2\pi}\frac{\left|F_{ijij}(q_{12})\right|^2}{q_{12}^2 S(q_{12})^2}\,d\theta \tag{9.58}$$

where the sum is over spin as well as wavevector and subband.

This is a factor 4 less than the hitherto standard expression, used, for example, by Goodnick and Lugli (1992), Blom *et al.* (1993) and Mosko and Moskova (1994). The difference was first pointed out by Mosko *et al.* (1995). If indeed this is an error, it opens up the whole question of plasma thermalization that is observed to occur in a GaAs quantum well within about 100 fs (Knox *et al.*, 1986), and hitherto satisfactorily explained by Monte Carlo simulations using the standard expression with static screening – i.e. $S(q_{12})$ is taken to be

$$S(q_{12}) = 1 + \sum_i f_i(0)\frac{q_i}{q_{12}}, \quad q_i = \frac{e^2 m_i^*}{2\pi\hbar^2\varepsilon_\infty} \tag{9.59}$$

where the sum is taken over all the subbands (including holes) and $f_i(0)$ is the occupation probability at each subband edge and assumed to be spherically symmetric (Goodnick and Lugli, 1992). Screening is difficult to treat in general even in the static limit, and it may be the case that dynamic effects significantly reduce the effect of screening. Parallel-spin collisions will also add to the rate to some degree. Taken together they may explain the observed rapidity of thermalization. It is evident that this topic will bear a good deal more investigation.

Dynamically screened electra–electra collisions in 2D have been analysed by Ridley (2001) and Tripathi and Ridley (2002, 2003).

As regards electron–electron scattering in 1D, it turns out that the restriction of motion to one dimension completely eliminates any effect of the interaction, at least to the lowest order, as the reader can prove to himself in a few lines.

9.4.3 Electron–Hole Scattering

Electrons and holes are distinguishable particles so there is no question of quantum interference, nor of exchange. The operative matrix element is thus what we have called M_{12}, and this applies to all collisions irrespective of spin alignment. The scattering rate is, therefore, given by Eq. (9.58) with suitable interpretation of symbols. Note that it is also equal to the 2D rate for charged-impurity scattering, Eq. (9.17), when the hole mass is much larger than the electron mass (so that $\mu = m^*_e$ the electron mass).

9.5 Phonon Scattering

9.5.1 Phonon–Phonon Processes

The strength of hot-phonon effects is determined by phonon lifetime and scattering rate. These quantities for acoustic modes have received a good deal of attention in the past, principally because the associated processes are much more accessible to experiment than those for optical phonons. The area needs revisiting in the light of the zone-folding induced by superlattice structures. Here, we will limit our attention to optical phonons since much less attention has been given to scattering processes other than those involving electrons in which optical phonons are involved. What follows is based on the work of Ridley and Gupta (1991).

The lifetime of an optical phonon is determined by the anharmonicity of the lattice. The Hamiltonian for the lattice can be written to lowest order as follows:

$$\mathbf{H} = \sum_{r,i,j} M_i^{1/2}\omega_i M_j^{1/2}\omega_j u_i \cdot u_j \tag{9.60}$$

where $\mathbf{u}_i$ denotes the optical (or short-wavelength acoustic) displacement for the mode of vibration i and M_i is the corresponding mass of the oscillator. The sum is over all lattice sites and $\mathbf{u}_i$ can be written in second-quantization notation thus:

$$\mathbf{u}_i = \sum_q \left(\frac{\hbar}{2NM_i\omega_i}\right)^{1/2} \left(e_i a_q i e^{iq\cdot r} + e_i{}^* a_q i^\dagger e^{-iq\cdot r}\right) \tag{9.61}$$

where N is the number of unit cells in the cavity, e is a unit polarization vector, and a, a_+ are the usual annihilation and creative operators. A small perturbation can be regarded as changing the frequencies by an amount $d\omega_i$ so that

$$H = H_0 + H' \tag{9.62}$$

$$H_0 \sum_{qi} \hbar\omega_{qi}\left(a_{qi}^{\dagger}a_{qi} + \frac{1}{2}\right) \tag{9.63}$$

$$H' = \sum_{r,i,j} M_i^{1/2}\omega_i M_j^{1/2}\omega_j\left(\frac{\delta\omega_i}{\omega_i} + \frac{\delta\omega_j}{\omega_j}\right)\mathbf{u}_i \cdot \mathbf{u}_j \tag{9.64}$$

This form of the perturbation will be applied to a number of interactions in what follows.

The lifetime of the optical phonon can be calculated by taking the phonon to cause a frequency change in all other modes by virtue of the intrinsic anharmonicity of the lattice, viz.

$$\frac{\delta\omega}{\omega} = \Gamma.\mathbf{u} \tag{9.65}$$

where we have introduced the quantity Γ, which is the optical-phonon analogue of the Gruneisen constant. For simplicity we assume an isotropic medium that allows us to replace $\Gamma.\mathbf{u}$ by Γu. Eq. (9.64) becomes[†]

$$H' = 2\Gamma \sum_{r,i,j} M\omega_i\omega_j \mathbf{u}_i \cdot \mathbf{u}_j u_k \tag{9.66}$$

where we have $M_i = M_j = M$ in anticipation that modes i, j are acoustic modes and we use the subscript k to denote the optical mode. Energy and momentum conservation is satisfied when a long-wavelength LO phonon decays to two almost oppositely propagating LA modes such that $\omega_i = \omega_j = \omega_L/2$. The rate from first-order perturbation theory is

$$W = \frac{\Gamma^2\hbar\omega_L^3[2n(\omega_L/2)+1]}{32\pi\rho v_L^3} \tag{9.67}$$

which is the expression derived by Klemens (1966) if we take $\Gamma = \gamma\,\omega_L/v_L$, where γ is the Gruneisen constant and v_L is the velocity of LA modes, and ignore a factor 4/3. The quantity $n(\omega_L/2)$ is the Bose–Einstein factor. For the case of GaAs, experiment gives us a lifetime of about 8 ps at 10 K (Tsen *et al.*, 1988, 1989), which means that F (otherwise to be calculated in an a priori theory) for

GaAs is 2.4×10^8 cm^{-1}. This value leads to agreement with results of experiments at other temperatures (4 ps at 300 K, von der Linde *et al.*, 1980; 7 ps at 77 K, Kash *et al.*, 1985). In binary compounds in which the disparity in mass of the two ionic components is unusually large, such as occurs in GaN, for example, this channel of decay is not available, and other processes must be invoked (Ridley, 1996b).

Higher-order phonon–phonon processes turn out to be much weaker than the three-phonon interaction that ends the life of an LO phonon. Thus, the lifetime is also the momentum-relaxation time for phonon–phonon processes. A calculation taking into account phonon confinement and mode mixing (which would also affect the acoustic modes involved) has not been carried out, but it is unlikely that this general conclusion will prove incorrect for layered material. Hence it is unlikely, if phonon–phonon processes are the only ones active, that hot phonons relax electron momentum appreciably. But, of course, this is not the case.

An excellent review of phonon properties in semiconductors as observed in experiments on light scattering has been written recently by Tsang and Kash (1990).

As mentioned above, the Klemens process cannot work for GaN and other binaries. Recent work has identified a number of alternative decay routes. The main decay route is still a three-phonon process in which the LO mode decays into two lower frequency phonons with conservation of crystal momentum. Barman and Srivastrava (2004) have identified the following channels:

Klemens (1966): LO $\rightarrow$ 2LA
Ridley (1996b): LO $\rightarrow$ TO + LA or TA
Vallee–Bogani (1991): LO $\rightarrow$ LO + LA or TA
Barman–Srivastrava (2004): LO $\rightarrow S'O + S''$ $S', S'' = LorT$

The Barman–Srivastrava channel is possible only in wurtzite structures. In the wurtzite structure there are two zone-centre LO modes, A_1(LO) and E_1(LO) with slightly different frequencies and similarly the TO modes A_1(TO) and E_1(TO) and others (the E_2 and B modes) (see Chapter 13). It turns out that the Ridley channel is the dominant one for GaAs, GaN and indeed for many of the other binaries. The decay rate with TA emission is given by:

$$\frac{1}{\tau_p} = \frac{\hbar\gamma^2}{4\pi\rho\bar{c}^2 c_{\mathrm{TA}}^3}\omega_{\mathrm{TA}}^3\omega_{\mathrm{LO}}\omega_{\mathrm{TO}}\frac{\bar{n}(\omega_{\mathrm{TO}})\bar{n}(\omega_{\mathrm{TA}})}{\bar{n}(\omega_{\mathrm{LO}})} \tag{9.67a}$$

γ is the mode-average Grüneisen's constant, ρ is the mass density, $\bar{c}$ is the average acoustic mode velocity, C_{TA} is the velocity of the TA mode, the bar over the n denotes thermal equilibrium values at the lattice temperature. This expression is valid provided that the daughter phonons remain at thermodynamic equilibrium at the lattice temperature. Conservation of energy means that $\omega_{TA} = \omega_{LO} - \omega_{TO}$, and hence the rate increases with LO frequency as a greater density of states becomes accessible.

9.5.2 Charged-Impurity Scattering

The Fröhlich interaction allows a novel interaction to take place between charged-impurity atoms and LO modes. A charged impurity polarizes its surrounding medium so that its effective charge is

$$e_{\text{eff}} = \frac{Ze}{\kappa_s} \tag{9.68}$$

where Z is the charge number and κ_s is the static dielectric constant. The LO mode interacts coulombically with this charge and anharmonically with the surrounding polarization. Assuming that the internal dynamics of the impurity remains unaffected and that the impurity absorbs crystal momentum, we can calculate the scattering rate for the Fröhlich interaction in second order, i.e.

$$W = \frac{2\pi}{\hbar} \int \sum_n \left| \frac{<f|e_{eff}\phi|n><n|e_{eff}\phi|i>}{E_i - E_n} \right|^2 d(E_f - E_i) dN_F \tag{9.69}$$

in usual notation. The result is

$$W = \frac{Z^4 e^4 (n(\omega_L) + 1)\omega_L N_I}{4\pi \mathbf{k}_s^4 \hbar^2 v_L^2 q^3} \left(\frac{1}{\varepsilon_\infty} - \frac{1}{\varepsilon_s} \right)^2 \tag{9.70}$$

We have assumed a dispersion relation for the LO mode to be of quadratic form, which is why v_L, the velocity closely associated with the velocity of the LA mode, enters the preceding formula. The smaller v_L is, the weaker the dispersion and the greater the number of available final states that fall within the energy uncertainty. For GaAs with $Z = 1$ and $q = 2 \times 10^6$ cm^{-1}, which is a typical wavevector involved in electron–phonon interaction, the rate becomes 1.1×10^7 $N_I s^{-1}$ with N_I per cubic centimeters (cm^{-3}). It matches the lifetime when $N_I = 10^{18}$cm^{-3}. As is usual for the Fröhlich interaction the rate is stronger for smaller wavevectors. Typical wavevectors involved in Raman scattering are 10^5 cm^{-1} or less, increasing the rate by a factor of 10^3 or more! Well-defined long-wavelength LO modes cannot exist in these circumstances.

The anharmonic interaction with the surrounding polarization actually leads to an infinite cross-section as a consequence of the purely coulombic nature of the polarization field. However, weighting for momentum relaxation reduces the rate far below that for the direct Fröhlich one.

Both interactions will be screened if free carriers are available. Only bulk interactions have been considered to date.

9.5.3 Alloy Fluctuations and Neutral Impurities

A foreign atom alters the phonon frequency by virtue of its different mass and different bond strength (different force constant). The effect is well-known in the form of isotope scattering of acoustic waves. In the Born approximation, Eq. (9.64) with Eq. (9.61) leads to

$$H'_{ij} = \frac{1}{N}\sum_{r} \hbar\delta\omega \mathbf{e}_i.\mathbf{e}_j e^{i(\mathbf{q}_i-\mathbf{q}_j)\cdot\mathbf{r}} a_i a_j^{\dagger} \tag{9.71}$$

which describes phonon i being annihilated and phonon j being created, with $\mathbf{r}$ such that the zero of spatial coordinates coincides with the atom. When both modes have the same polarization $\mathbf{e}_i.\mathbf{e}_j = \cos\theta$, where θ is the angle between $\mathbf{q}_i$ and $\mathbf{q}_j$.

The material will be regarded as an alloy of form $A_xB_{1-x}C$, and we can use the alloy-scattering arguments to write

$$<\delta\omega>^2 = (\omega_A - \omega_B)^2 x(1-x) \tag{9.72}$$

where ω_A is the LO frequency for a particular mode when x approaches zero, ω_B for the same mode when x approaches unity in a real alloy, or, for neutral impurities including isotopes

$$\omega_A - \omega_B \rightarrow \frac{\Delta M}{2M}\omega_L \tag{9.73}$$

where ΔM is the mass difference. (Strictly, another term $(\Delta F/2F)\ \omega_L$ should be added to denote the effect of differing force constants, but the contribution is usually small.) The rate in 3D is

$$W = \frac{(\omega_A - \omega_B)^2 \omega_L q V_0 x(1-x)(n(\omega_L)+1)}{3\pi v_L^2} \tag{9.74}$$

and in 2D

$$W = \frac{(\omega_A - \omega_B)^2 \omega_L q V_0 x(1-x)(n(\omega_L)+1)}{2\pi v_L^2 L} \tag{9.75}$$

Here V_0 is the volume of the unit cell. Rates for realistic impurity concentrations are negligible. Isotopes exist in relatively high concentrations but ω is rather small and in the present context the isotopic scattering rate is usually negligible.

Table 9.1. *Phonon momentum-relaxation time associated with alloy scattering*[a]

Material	Phonon mode	$\omega_L(meV)$	$\Delta\omega_{AB}$ (meV)	t(ps)	
				3Dq*	2D (50 Å)
$Al_{0.3}Ga_{0.7}As$[b]	GaAs	34.8	5.27	30	8.0
	AlAs	47.4	5.54	20	5.3
$Al_{0.53}In_{0.47}As$[c]	InAs	29.4	0	∞	∞
	AlAs	45.5	8.7	6.6	1.8
$Ga_{0.52}In_{0.48}As$[d]	InAs	28.9	1.24	484	141
	GaAs	33.8	5.47	21	7.0

Note: [a]$q = 2.5\times10^6$ cm^{-1}.
Source: [b]Kim and Spitzer (1979). [c]Emyra *et al.* (1987). [d]Lucovsky and Chen (1970).

In alloys the rate depends on the mode, which can be AC-like or BC-like. Table 9.1 lists some magnitudes. The InAs mode in AlInAs has the same frequency for all x and so does not suffer scattering by alloy fluctuations. Note that the 2D rate is q-independent and larger than the 3D rate by a factor $3\pi/(2qL)$ and many of the time-constants are of order of the GaAs phonon life time.

9.5.4 Interface-Roughness Scattering

Applying the spirit of the approach in Section 9.2 to guided phonon modes we obtain a frequency shift given by

$$\delta\omega = \frac{d\omega}{dq_z}\cdot\frac{dq_z}{dL}\Delta = \frac{n^2 v_L^2 \pi^2}{\omega_L L^3}\Delta \tag{9.76}$$

with $q_z = n\pi/L$. Choosing a Gaussian autocovariance (for mathematical convenience) we obtain the rate

$$W = \frac{n^4\pi^4 v_L^2 \Delta^2 \Lambda^2}{2\omega_L L^6} I \tag{9.77a}$$

$$I = \int_0^{2\pi} (1-\cos\theta)\cos^2\theta \exp\left[q^2\Lambda^2 \sin^2\left(\frac{\theta}{2}\right)\right] d\theta \tag{9.77b}$$

where q is the wavevector in the plane. For realistic values of the parameters, this rate turns out to be relatively small.

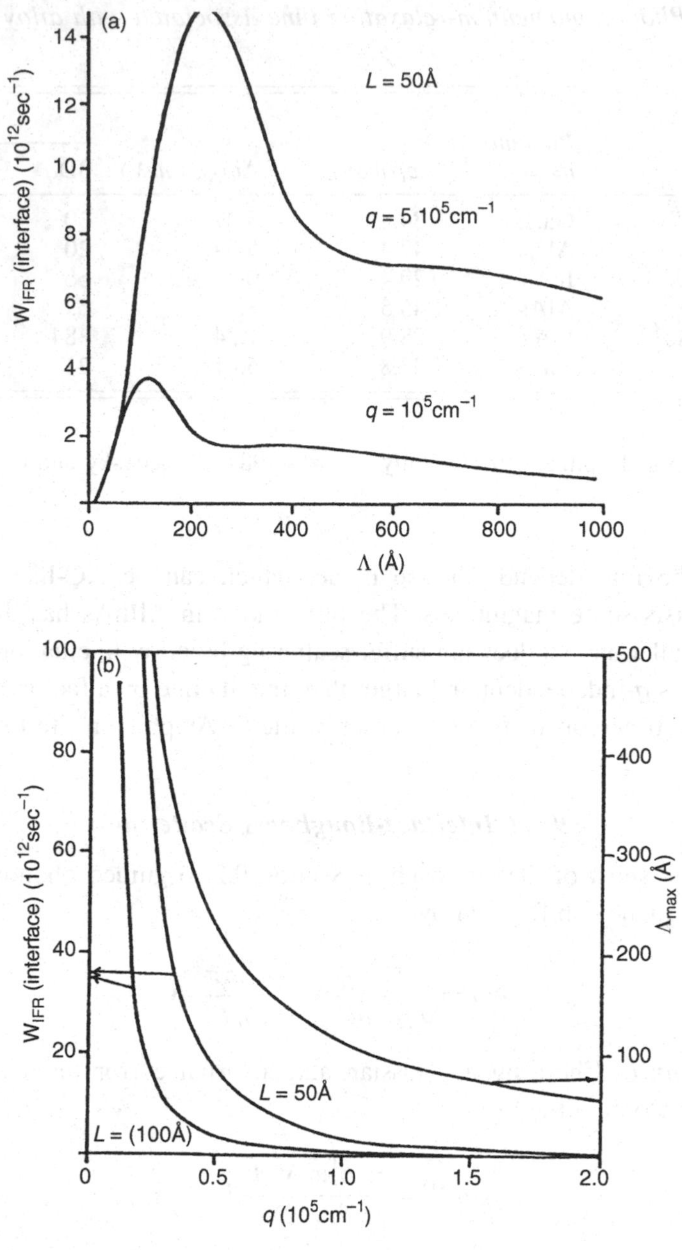

Fig. 9.7 Interface-roughness momentum-relaxation rates for interface modes in a GaAs well: (a) as a function of correlation length Λ for $q = 5 \times 10^5$ cm^{-1} and 1×10^6 cm^{-1}; (b) as a function of wavevector for $L = 50$ Å and 100 Å with Λ chosen to maximize the rate; for all cases $\Delta = 1$ ml. (Gupta and Ridley, 1990)

It might be expected, however, that interface modes will be more sensitive to interface quality. Instead of Eq. (9.76), which is only weakly dependent on interface roughness (IFR) because of the weak dispersion, we take $\delta\omega$ to be proportional to the difference in LO frequencies in the adjacent material. Thus we take

$$\delta\omega = \delta\omega_{AB}\frac{\Delta}{L} \tag{9.78}$$

whence

$$W = \frac{(\Delta\omega_{\mathrm{AB}})^2\omega_L\Delta^2 I^2}{2v_L^2L^2} \tag{9.79}$$

where I is given by Eq. (9.77b). Fig. 9.7 shows some features of this rate for the GaAs/$Al_{0.3}Ga_{0.7}As$ interface taking $\hbar\delta\omega_{AB}$ =1.84 meV and $\Delta = 1$ monolayer. Clearly, significant rates emerge, especially for small wavevectors, just those, in fact, that the Fröhlich interaction favours and that are copiously present in the hot-phonon spectrum. Unlike the case for electrons, IFR scattering of phonons is not screened (see Fig. 9.5).

Eq. (9.79) has been used to explain the experimental results on hot electrons and hot phonons in quantum wells by Gupta *et al.* (1991). In these experiments there is good agreement found between the temperature-field curves measured at a fixed lattice temperature using photoluminescence techniques and those deduced from the variation of mobility with field and with lattice temperature. This agreement suggests that hot phonons must be randomized, and the preceding theory of IFR scattering helps to account for that.

NOTES

† Strictly speaking, Γ is a third-order coupling coefficient to which symmetry restrictions apply, but this is usually ignored.

10
Quantum Screening

O polished perturbation!
Henry IV, Pt II, W. Shakespeare

10.1 Introduction

The presence of a potential $V(\mathbf{R}, t)$ that induces transitions between the single electronic states of a system also perturbs the whole electron gas, leading to the appearance of a screening potential $V_s(\mathbf{R}, t)$. The interaction potential that actually induces a transition is, therefore, the sum of the unscreened and screened potentials, viz.

$$V(\mathbf{R}, t) = V_0(R, t) + V_s(R, t) \tag{10.1}$$

If the perturbation of the electron gas is weak, the screening potential will be linearly related to the self-consistent potential $V(\mathbf{R}, t)$, and so we have

$$V(\mathbf{R}, t)(1 - \chi) = V_0(\mathbf{R}, t) \tag{10.2}$$

where $V_s = \chi V$, and the quantity $1 - \chi$ is the dielectric function $\varepsilon/\varepsilon_0$.

In order to calculate the dielectric function it is necessary to describe the effect of a perturbation on the whole electron gas, and this is most conveniently done using the concept of the density matrix. Essentially, we solve the Schrodinger equation to obtain the disturbance to the screening potential via Poisson's equation, following the approach of Ehrenreich and Cohen (1959). The reader is referred to the texts by Kittel (1963), Harrison (1970) and Landau and Lifshitz (1977) for fuller accounts of the topic as regards bulk screening.

A point regarding notation – we use capitals for 3D vectors and lowercase letters for 2D and 1D vectors.

10.2 The Density Matrix

We conceive the electron gas to be an assembly of N electrons each of which is described by a one-electron wavefunction $\psi(\mathbf{R}, t)$. The total wavefunction is $\psi_N(\mathbf{R}_1, \mathbf{R}_2 \ldots \mathbf{R}_{N,t})$ and is taken to be the Slater determinant

$$\psi_N(\mathbf{R}_1, \ldots, t) = \frac{1}{\sqrt{N!}} \begin{vmatrix} \psi_1(\mathbf{R}_1, t) & \psi_1(\mathbf{R}_2, t) \cdots \psi_1(\mathbf{R}_N, t) \\ \psi_2(\mathbf{R}_1, t) & \psi_2(\mathbf{R}_2, t) \cdots \psi_2(\mathbf{R}_N, t) \\ \vdots & \\ \psi_N(\mathbf{R}_1, t) \cdots \psi_N(\mathbf{R}_N, t) & \end{vmatrix} \tag{10.3}$$

$$= \frac{1}{\sqrt{N}} \left[\psi_1(\mathbf{R}_i, t)\psi_{N-1}^{(1)} - \psi_2(\mathbf{R}_i, t)\psi_{N-1}^{(2)} + \cdots \right]$$

An operator $\hat{O}$ that can be expressed as a sum of one-electron operator $\hat{O}\,(R_i)$ has an expectation value

$$<\hat{O}> = \sum_i \int \psi_N^* \hat{O}(R_i)\psi_N d\mathbf{R}_1, d\mathbf{R}_2, \ldots, d\mathbf{R}_N \tag{10.4}$$

Integrating over the rest of the coordinates for each $\mathbf{R}_i$ and noting that $\psi_k^*(\mathbf{R}_j)\psi_{k'}\,(R_{j'})_{=\delta_{kk'}\delta_{jj'}}$ give

$$<\hat{O}> = \frac{1}{N}\sum_i \int \psi_1^*(\mathbf{R}_i)\hat{O}(\mathbf{R}_i)\psi_1(\mathbf{R}_i) + \psi_2^*(\mathbf{R}_i)\hat{O}(\mathbf{R}_i)\psi_2(\mathbf{R}_i) \ldots d\mathbf{R}_i \tag{10.5}$$

The sum gives N identical terms, whence

$$<\hat{O}> = \sum_n \int \psi_n^*(\mathbf{R})\hat{O}(\mathbf{R})\psi_n(\mathbf{R})d\mathbf{R} \tag{10.6}$$

Now operators operate meaningfully on electrons and not states, so an integral must be suitably weighted by the probability $f(n)$ of the state's being occupied. Moreover a product of $\psi_n^*(\mathbf{R})$ and $\psi_n(\mathbf{R})$ is an electron density, so it is natural to seek to express expectation values of many-electron operators in terms of such a density. A step toward this can be made by rearranging the terms in the previous equation, bearing in mind the $\hat{O}(\mathbf{R})$ operates only on $\psi_n(\mathbf{R})$. Thus

$$<\hat{O}> = \sum_n \int \hat{O}(\mathbf{R})\psi_n(\mathbf{R})\psi_n^*(\mathbf{R}')d\mathbf{R} \tag{10.7}$$

and the rule is to put $\mathbf{R}' = \mathbf{R}$ after the operation. A statistical and time-dependent density matrix can then be usefully defined as follows:

$$\rho(\mathbf{R}, \mathbf{R}', t) = \sum_{nn'} \psi_n(\mathbf{R}, t) f(n, n') \psi^*_{n'}(\mathbf{R}', t) \tag{10.8}$$

whence Eq. (10.7) can be written

$$<\hat{O}> = \int \hat{O}(\mathbf{R}) \rho(\mathbf{R}, \mathbf{R}') \delta_{nn'} d\mathbf{R} \tag{10.9}$$

At thermodynamic equilibrium the density matrix must be diagonal and $f(n, n') = f_0(n)\ \delta_{nn'}$, where $f_0(n)$ is the Fermi–Dirac function, and the carrier density is given by

$$n_0(\mathbf{R}) = \rho(\mathbf{R}, \mathbf{R}') \delta_{nn'\mathbf{R},\mathbf{R}'} \tag{10.10}$$

The time dependence associated with the presence of a perturbation is obtainable from Schrodinger's equation. The total Hamiltonian is expressed as the sum of one-electron Hamiltonians $H(\mathbf{R}_i, t)$ and each one-electron wavefunction obeys a one-elecron equation of the form

$$i\hbar \frac{\partial \psi_n(\mathbf{R}, t)}{\partial t} = H(\mathbf{R}, t) \psi_n(\mathbf{R}, t) \tag{10.11}$$

Consequently

$$\begin{aligned} &\frac{\partial}{\partial t}(\psi_n(\mathbf{R}, t)\psi^*_{n'}(\mathbf{R}', t)) \\ &= \frac{1}{i\hbar} H(\mathbf{R}, t)\psi_n(\mathbf{R}, t)\psi^*_{n'}(\mathbf{R}, t) + \psi_n(\mathbf{R}, t)\left[\frac{1}{i\hbar} H(\mathbf{R}', t)\psi_{n'}(\mathbf{R}', t)\right]^* \end{aligned} \tag{10.12}$$

If we assign all the time dependence to the wavefunctions and keep the occupation probability, $f(n, n')$, time-independent, then this equation can be written

$$i\hbar \frac{\partial \rho_{nn'}(\mathbf{R}, \mathbf{R}', t)}{\partial t} = H(\mathbf{R}, t)\rho_{nn'}(\mathbf{R}, \mathbf{R}'t) - \rho_{nn'}(\mathbf{R}, \mathbf{R}'t) H(R', t) \tag{10.13}$$

in which $H(\mathbf{R}', t)$ must be taken to operate on the left. This is a variation of the Liouville equation.

Introducing a small perturbation, $\mathbf{H}_1$ causes a small perturbation in the density matrix ρ_1. Dropping the explicit depiction of dependent variables we obtain, after linearization

$$i\hbar \frac{\partial \rho_0}{\partial t} = H_0 \rho_0 - \rho_0 H_0 = 0 \tag{10.14a}$$

$$i\hbar \frac{\partial \rho_1}{\partial t} = H_1\rho_0 - \rho_0 H_1 + H_0\rho_1 - \rho_1 H_0 \tag{10.14b}$$

Let the unperturbed states be represented by $| a\rangle$; then $\rho_0 | a\rangle = f_0(E_a) | a\rangle$ and $H_0 | a\rangle = E_a | a\rangle$, E_a is the energy of the state. We suppose the perturbation to begin in the infinite past where the system is entirely unperturbed and to have the time dependence

$$H_1 = H_1(\mathbf{R})e^{-i\omega t + \delta t} \tag{10.15}$$

and we evaluate the effect at $t = 0$, with $\delta \to 0$. The perturbed density will have an identical time dependence and Eq. (10.14b) becomes

$$(\hbar\omega + i\hbar\delta)\rho_1 = H_1\rho_0 - \rho_0 H_1 + H_0\rho_1 - \rho_1 H_0 \tag{10.16}$$

The expectation values are then

$$\begin{aligned}(\hbar\omega + i\hbar\delta)\langle a'|\rho_1|a\rangle &= \langle a'|H_1\rho_0|a\rangle - \langle a'|\rho_0 H_1|a\rangle \\ &\quad + \langle a'|H_0\rho_1|a\rangle - \langle a'|\rho_1 H_0|a\rangle \\ &= [f(E_a) - f(E_{a'})]\langle a'|H_1|a\rangle + (E_{a'} - E_a)\langle a'|\rho_1|a\rangle\end{aligned} \tag{10.17}$$

whence

$$\langle a'|\rho_1|a\rangle = \prod\nolimits_{aa'} \langle a'|H_1|a\rangle \tag{10.18}$$

where

$$\prod\nolimits_{aa'} = \frac{f(E_{a'}) - f(E_a)}{E_{a'} - E_a - \hbar\omega - i\hbar\delta} \tag{10.19}$$

The perturbation of the density matrix reflects the perturbation of the wave-functions. If the perturbed states are depicted by $| \beta\rangle$, then

$$\rho(\mathbf{R}, \mathbf{R}', t) = \sum_{\beta,\beta'} |\beta\rangle f_{\beta\beta'} \langle\beta'| \tag{10.20}$$

and the perturbed states can be expanded in terms of the unperturbed states thus:

$$|\beta\rangle = \sum_a a_a|a\rangle, \quad a_a = \langle a|\beta\rangle \tag{10.21}$$

We can now relate the disturbance in the density matrix to the disturbance in the electron density via

$$n(\mathbf{R},t) = \rho(\mathbf{R},\mathbf{R}',t)\delta_{\beta\beta'}\delta_{\mathbf{RR}'} = \sum_{\beta} |\beta\rangle f_\beta \langle\beta| \tag{10.22}$$

Expanding the disturbed states, we obtain

$$\begin{aligned} n(\mathbf{R},t) &= \sum_{\beta a' a} |a'\rangle\langle a'|\beta\rangle f_\beta \langle\beta|a\rangle\langle a| \\ &= \sum_{aa'} |a'\rangle\langle a|\langle a'|\rho(\mathbf{R},\mathbf{R}',t)\delta_{\beta\beta'}\delta_{\mathbf{RR}'}|a\rangle \end{aligned} \tag{10.23}$$

Putting $n = n_0 + n_1$ gives with Eq. (10.18)

$$\begin{aligned} n_1 &= \sum_{a'a} |a'\rangle\langle a|\langle a'|\rho_1|a\rangle \\ &= \sum_{aa'} |a'\rangle\langle a| \prod\nolimits_{aa'} \langle a'|H_1|a\rangle \end{aligned} \tag{10.24}$$

We can put $H_1 = eV$, where V is the self-consistent potential of Eq. (10.1).

In the preceding equations, we have implicitly assumed that no correlated motion of the electrons occurs. Thus we are neglecting exchange and assuming a random phase or Hartree approximation.

10.3 The Dielectric Function

In order to get the screening potential we have to solve Poisson's equation

$$\nabla^2 V_s(\mathbf{R},t) = -\frac{en_1(\mathbf{R},t)}{\varepsilon_0} \tag{10.25}$$

with $n_1(\mathbf{R},t)$ given by Eq. (10.24). Complications occur in low-dimensional systems as a consequence of the differences in permittivity in adjacent regions; that means that Poisson's equation has to be solved in each distinct region and the usual electrostatic boundary conditions applied to connect the solutions. (Note that we assume that retardation effects are negligible in all of this.) These complications cannot be avoided when free-standing slabs are considered, but in purely solid systems involving the usual semiconductors the differences in permittivity are rather small and may be ignored in a first approximation. To begin with we will adopt this approximation since it simplifies matters substantially, but we will return to the problem later.

We therefore assume that whatever quantum-confinement effects are present, the screening occurs in an ambient medium of uniform dielectric properties.

Consequently we may expand $V_s(\mathbf{R},t)$ in a 3D Fourier series and similarly $n_1(\mathbf{R}, t)$, viz.:

$$V_s(\mathbf{R},t) = \frac{\Omega}{(2\pi)^3}\int V_{s\mathbf{Q}}(t)e^{i\mathbf{Q}\cdot\mathbf{R}}d\mathbf{Q} \tag{10.26a}$$

$$n_1(\mathbf{R},t) = \frac{\Omega}{(2\pi)^3}\int n_{\mathbf{Q}}(t)e^{i\mathbf{Q}\cdot\mathbf{R}}d\mathbf{Q} \tag{10.26b}$$

where the normalizing volume is Ω, and hence

$$V_{s\mathbf{Q}}(t) = \frac{en_Q}{\varepsilon_0 Q^2}(t) \tag{10.27}$$

and

$$n_{\mathbf{Q}}(t) = \frac{1}{\Omega}\int n_1(\mathbf{R},t)e^{-i\mathbf{Q}\cdot\mathbf{R}}d\mathbf{R} \tag{10.28}$$

The screening potential is given by

$$V_s(t) = \frac{e^2}{(2\pi)^3\varepsilon_0}\iint \frac{1}{\mathbf{Q}^2}\sum_{a_ia_j}|a_i\rangle\langle a_j|\prod\nolimits_{a_ia_j}\langle a_i|V(t)|a_j\rangle e^{i\mathbf{Q}\cdot(\mathbf{R}-\mathbf{R}')}d\mathbf{R}'d\mathbf{Q} \tag{10.29}$$

(For brevity the dependence of V_s on $\mathbf{R}$ is not explicitly shown.) The matrix element of this potential between two states a_g and a_h is then

$$\langle a_g|V_s(t)|a_h\rangle = \frac{e^2}{(2\pi)^3\varepsilon_0}\int \frac{1}{\mathbf{Q}^2}\sum_{a_ia_j}\prod\nolimits_{a_ia_j} G_{ij}^{gh}(Q)\langle a_i|V(t)|a_j\rangle d\mathbf{Q} \tag{10.30}$$

where

$$G_{ij}^{gh}(\mathbf{Q}) = \int \langle a_g|\left(|a_i\rangle\langle a_j|e^{i\mathbf{Q}\cdot(\mathbf{R}-\mathbf{R}')}\right)|a_h\rangle\, d\mathbf{R}' \tag{10.31}$$

The response of the system to a perturbation by a potential V_0 is thus characterized by

$$\langle a_g|V(t)|a_h\rangle = \langle a_g|V_0(t)|a_h\rangle + \langle a_g|V_s(t)|a_h\rangle \tag{10.32}$$

Further analysis depends upon the nature of the electron states, and this depends upon the degree of quantum confinement.

10.4 The 3D Dielectric Function

In bulk material the wavefunction of an electron in a conduction band is

$$\psi(\mathbf{R}) = \frac{1}{\Omega^{1/2}} e^{i\mathbf{K}\cdot\mathbf{R}} \tag{10.33}$$

The G factor is then

$$\begin{aligned} G_{ij}^{gh}(\mathbf{Q}) &= -\frac{1}{\Omega^2}\iint e^{-i\mathbf{K}_g\cdot\mathbf{R}} e^{i(\mathbf{K}_i-\mathbf{K}_j)\cdot\mathbf{R}'} e^{i\mathbf{Q}\cdot(\mathbf{R}-\mathbf{R}')} e^{i\mathbf{K}_h\cdot\mathbf{R}} d\mathbf{R}' d\mathbf{R} \\ &= \delta_{\mathbf{K}_g\mathbf{K}_h+\mathbf{Q}}\delta_{\mathbf{K}_i\mathbf{K}_j+\mathbf{Q}} \end{aligned} \tag{10.34}$$

and

$$\langle a_i|V(t)|a_j\rangle = \delta_{\mathbf{K}_i\mathbf{K}_j+\mathbf{Q}} V_{\mathbf{Q}}(t) \tag{10.35}$$

where $V_Q(t)$ is the 3D Fourier coefficient of $V(t)$. Incorporating these results in Eq. (10.32) leads to a dielectric function given by

$$\frac{\varepsilon(\mathbf{Q},\omega)}{\varepsilon_0} = 1 - \frac{e^2}{\varepsilon_0 Q^2\Omega}\sum_{\mathbf{K}} \frac{f(E_{\mathbf{K}+\mathbf{Q}}) - f(E_{\mathbf{K}})}{E_{\mathbf{K}+\mathbf{Q}} - E_{\mathbf{K}} - \hbar\omega - i\hbar\delta} \tag{10.36}$$

which is the Lindhard formula (Lindhard, 1954).

This expansion has to be evaluated with δ going to zero and this leads to a permittivity consisting of a real and an imaginary component, viz.

$$\varepsilon(\mathbf{Q},\omega) = \varepsilon_1(\mathbf{Q},\omega) + i\varepsilon_2(\mathbf{Q},\omega) \tag{10.37}$$

and these can be obtained using the formula

$$\lim_{\delta\to 0}\frac{1}{z-\delta} = P\left(\frac{1}{z}\right) + i\pi\delta(z) \tag{10.38}$$

where P stands for the principal part. Thus we obtain the well-known result

$$\frac{\varepsilon_1(\mathbf{Q},\omega)}{\varepsilon_0} = 1 - \frac{e^2}{\varepsilon_0 Q^2\Omega}\sum_{\mathbf{k}} \frac{f(E_{\mathbf{k}+\mathbf{Q}}) - f(E_{\mathbf{k}})}{E_{\mathbf{k}+\mathbf{Q}} - E_{\mathbf{k}} - \hbar\omega} \tag{10.39}$$

and

$$\frac{\varepsilon_2(\mathbf{Q},\omega)}{\varepsilon_0} = -\frac{\pi e^2}{\varepsilon_0 Q^2\Omega}\sum_{\mathbf{k}} [f(E_{\mathbf{k}+\mathbf{Q}}) - f(E_{\mathbf{k}})]\delta(E_{\mathbf{k}+\mathbf{Q}} - E_{\mathbf{k}} - \hbar\omega) \tag{10.40}$$

An often more convenient form of the real component can be obtained by exploiting the independence of $E_{\mathbf{K}}$ and $f(E_{\mathbf{K}})$ on the sign of $\mathbf{K}$ viz.:

$$\frac{\varepsilon_1(\mathbf{Q},\omega)}{\varepsilon_0} = 1 + \frac{e^2}{\varepsilon_0 Q^2 \Omega} \sum_{\mathbf{K}} f(E_{\mathbf{K}}) \left[\frac{1}{E_{\mathbf{K}+\mathbf{Q}} - E_{\mathbf{K}} + \hbar\omega} + \frac{1}{E_{\mathbf{K}+\mathbf{Q}} - E_{\mathbf{K}} - \hbar\omega} \right] \tag{10.41}$$

For a degenerate gas at $T = 0K$ the dielectric function becomes

$$\begin{aligned} \frac{\varepsilon_1(\mathbf{Q},\omega)}{\varepsilon_0} = 1 + \frac{e^2 N(E_F)}{2\varepsilon_0 Q^2} \times \Bigg[1 - \frac{1}{4\eta^3} \Bigg(& \left[\gamma^2 - (1-2\gamma)\eta^2 + \eta^4\right] ln \left| \frac{\eta + \eta^2 + \gamma}{\eta - \eta^2 - \gamma} \right| \\ & + \left[\gamma^2 - (1+2\gamma)\eta^2 + \eta^4\right] \ln \left| \frac{\eta + \eta^2 - \gamma}{\eta - \eta^2 + \gamma} \right| \Bigg) \Bigg] \end{aligned} \tag{10.42}$$

where $N(E_F)$ is the density of states at the Fermi surface, $\eta = Q/2k_F$; k_F is the Fermi wavevector; and $\gamma = \hbar\omega/4E_F$. For slowly varying or static potentials $\gamma \approx 0$ and

$$\frac{\varepsilon_1(Q,0)}{\varepsilon_0} = 1 + \frac{e^2 N(E_F)}{2\varepsilon_0 Q^2} \left[1 + \left(\frac{K_F}{Q} - \frac{Q}{4k_F} \right) \ln \left| \frac{1 + Q/2K_F}{1 - Q/2K_F} \right| \right] \tag{10.43}$$

If, in addition, the perturbation has a long wavelength so $Q \approx 0$ we obtain

$$\frac{\varepsilon_1(Q,0)}{\varepsilon_0} = 1 + \frac{Q_0^2}{Q^2}, \; Q_0^2 = \frac{e^2 N(E_F)}{\varepsilon_0} \tag{10.44}$$

Where Q_0 is the reciprocal screening length. This approximation to the dielectric function is often used to model the effect of screening of charged impurities and of acoustic phonons.

For certain values of frequency and wavevector the dielectric function vanishes ($\varepsilon = 0$), and when this occurs the electron gas exhibits collective oscillation whose quanta are plasmons. Because there always exists at finite temperatures an imaginary part of the dielectric function, plasma oscillations are always damped by being absorbed by an individual electron. However, at zero temperature such absorption processes are limited to regions of (ω, $\mathbf{k}$) in which conservation of energy and momentum is possible – the region of so-called single-particle excitation. Another region can be similarly defined for emission processes. Outside these regions a plasma wave can propagate without being damped. At finite temperatures the single-particle excitation regimes are no longer sharply defined and some damping always occurs outside their bound. Nevertheless, such damping can be weak enough for plasma modes to be well-defined, and to act as a further inelastic scattering agent for electrons. In 3D long-wavelength plasma oscillations occur at the well-known plasma frequency $\omega_p = (e^2 n/\varepsilon m^*)^{1/2}$, where n is the electron density and ε is the lattice permittivity.

10.5 The Quasi-2D Dielectric Function

The electrons are now assumed to be confined in a quantum well of width 'a' with a wavefunction

$$\psi_n(\mathbf{R}) = \frac{1}{\sigma^{1/2}} e^{i\mathbf{k}.\mathbf{r}} \psi_n(z) \tag{10.45}$$

where σ is the normalizing area, $\mathbf{k}$ is the in-plane wavevector and $\mathbf{R} = (\mathbf{r}, z)$. The perturbing potential can be taken to be expandable in a 2D Fourier series, thus

$$V(\mathbf{R}, t) = \frac{\sigma}{(2\pi)^2} \int V_q(z, t) e^{iq.r} d\mathbf{q} \tag{10.46}$$

where $\mathbf{q}$ is an in-plane wavevector, and the z-dependence remains implicit. However, our expression for the screening potential, Eq. (10.29), is in the form of a 3D Fourier expansion, but as there is now no implicit dependence on q_z, where $\mathbf{Q} = (\mathbf{q}, q_z)$, we can integrate over q_z to reduce the expansion to a 2D form. Using the standard integral

$$\int_{-\infty}^{\infty} \frac{e^{iq_z(z-z')}}{q^2 + q_z^2} dq_z = \frac{\pi}{q} e^{-q|z-z'|} \tag{10.47}$$

we obtain

$$V_s(t) = \frac{e^2}{8\pi^2 \varepsilon_0} \iint \frac{1}{q} \sum_{a_i a_j} |a_i\rangle\langle a_j| \Pi_{a_i a_j} \langle a_i|V(t)|a_j\rangle e^{i\mathbf{q}.(\mathbf{r}-\mathbf{r}')} e^{-q|z-z'|} d\mathbf{R}' d\mathbf{Q} \tag{10.48}$$

and the G factor becomes

$$G_{ij}^{gh}(\mathbf{q}) = \int \langle a_g \big| \left(|a_i\rangle\langle a_j| e^{i\mathbf{q}.(\mathbf{r}-\mathbf{r}')} e^{-q|z-z'|} \right) \big| a_h \rangle d\mathbf{R}' \tag{10.49}$$

Substitution of the wavefunctions and integration over $\mathbf{r}$ and $\mathbf{r}'$ lead to

$$G_{ij}^{gh}(\mathbf{q}) = \delta_{\mathbf{k}_g \mathbf{k}_h + \mathbf{q}} \delta_{\mathbf{k}_i \mathbf{k}_j + \mathbf{q}} F_{n_1 n}^{n_3 n_2}(q) \tag{10.50}$$

where F is a form factor given by

$$F_{n_2n_1}^{n_4n_3}(q) = \iint \psi_{n_4}(z)\psi_{n_3}(z)\psi_{n_2}(z')\psi_{n_1}(z')e^{-q|z-z'|}dzdz' \tag{10.51}$$

Furthermore

$$\langle a_i|V(t)a_j|\rangle = \delta_{\mathbf{k}_i,\mathbf{k}_j+\mathbf{q}}\langle n_2|V_\mathbf{q}(z,t)|n_1\rangle \tag{10.52}$$

The dielectric response is now described by

$$\langle n_4|V_\mathbf{q}(z,t)|n_3\rangle = \langle n_4|V_{0\mathbf{q}}(z,t)|n_3\rangle + \langle n_4|V_{s\mathbf{q}}(t)|n_3\rangle \tag{10.53}$$

where

$$\langle n_4|V_{sq}(t)n_3|\rangle = \frac{e^2}{2\varepsilon q\sigma}\sum_{n_2,n_2,\mathbf{k}}\frac{f(E_{n_2\mathbf{k}+q}) - f(E_{n_1,\mathbf{k}})}{E_{n_2\mathbf{k}+q} - E_{n_1\mathbf{k}} - \hbar\omega - i\hbar\delta}F_{n_2n_1}^{n_4n_3}(q)\langle n_2|V_q(z,t)|n_1\rangle \tag{10.54}$$

A response function can be defined by noting that

$$\langle n_4|V_q(z,t)n_3|\rangle = \sum_{n_2,n_1}\delta_{n_2,n_4}\delta_{n_1,n_3}\langle n_2|V_q(z,t)|n_1\rangle \tag{10.55}$$

so that

$$\langle n_4|V_{0q}(z,t)n_3|\rangle = \frac{1}{\varepsilon_0}\sum_{n_2,n_1}\varepsilon_{n_2,n_1}^{n_4,n_3}(\mathbf{q},\omega)\langle n_2|V_q(z,t)|n_1\rangle \tag{10.56}$$

where

$$\frac{\varepsilon_{n_2,n_1}^{n_4,n_3}(\mathbf{q},\omega)}{\varepsilon_0} = \delta_{n_2,n_4}\delta_{n_1,n_3} - \frac{e^2}{2\varepsilon_0 q\sigma}\sum_\mathbf{k}\frac{f(E_{n_2\mathbf{k}+q}) - f(E_{n_1,\mathbf{k}})}{E_{n_2\mathbf{k}+q} - E_{n_1,\mathbf{k}} - \hbar\omega - i\hbar\delta}F_{n_2n_1}^{n_4n_3}(q) \tag{10.57}$$

Eq. (10.56) shows that the relation between $\mathbf{V}_{0\mathbf{q}}$ and $\mathbf{V}_\mathbf{q}$ is not simple but depends upon the particular transition as well as the response of the system. In general, a straightforward screening factor cannot be defined without the operation of a matrix inversion (Price, 1981). There are, fortunately, cases where a screening factor can be obtained without labour, and it is worth examining the simple situation where only the ground state is occupied and the electron wavefunctions are assumed to be strictly confined to the region $0 \leq z \leq a$.

It is useful in calculating the effect of screening to note that as a consequence of the symmetry of the Coulomb potential in the well, the form factor vanishes if

$n_1 + n_2 + n_3 + n_4$ is odd. Thus, for example, $F_{11}^{12} F_{12}^{22}$ are zero. Also, the ordering and elevation of the suffices are irrelevant. Thus, for example, $F_{11}^{22} = F_{22}^{11}$, $F_{11}^{22} = F_{22}^{11}$

Taking the wavefunction to be

$$\psi_n(z) = \left(\frac{2}{a}\right)^{1/2} \sin kz, \quad ka = n\pi \tag{10.58}$$

we first compute the form factor, and obtain

$$F_{n_2,n_1}^{n_4,n_3}(q) = \frac{q}{a}\left\{\frac{\delta_{k_3-k_4,\pm(k_1-k_2)} - \delta_{k_3+k_4,\pm(k_1-k_2)}}{q^2 + (k_1-k_2)^2} - \frac{\delta_{k_3-k_4,(k_1+k_2)} - \delta_{k_3+k_4,(k_1+k_2)}}{q^2+(k_1+k_2)^2}\right.$$
$$-\frac{q}{a}[1 + \cos(k_3-k_4)a\cos(k_1-k_2)a - e^{-qa}(\cos(k_3-k_4)a + \cos(k_1-k_2)a)]$$
$$\left.\times\left[\frac{1}{q^2+(k_1-k_2)^2} - \frac{1}{q^2+(k_1-k_2)^2}\right]\left[\frac{1}{q^2+(k_3-k_4)^2} - \frac{1}{q^2+(k_3-k_4)^2}\right]\right\} \tag{10.59}$$

Now consider an unscreened interaction potential that is symmetric with respect to the plane at $z = a/2$. We assume that this property is not changed in the self-consistent potential. In this case *transitions* are allowed only between subbands of the same symmetry. In particular, intrasubband transitions are allowed, so $n_3 = n_4$ and $n_1 = n_2$. If, in addition, it is assumed that only the lowest subband is occupied, then also $n_3 = n_2 = 1$ and a straightforward screening function is obtained, viz.:

$$\frac{\varepsilon_{11}^{11}(q,\omega)}{\varepsilon_0} = 1 - \frac{e^2}{2\varepsilon_0 q\sigma} F_{11}^{11}(q) \sum_{\mathrm{k}} \frac{f(E_{1,k+q}) - f(E_{1,\mathrm{k}})}{E_{1,k+q} - E_{1,k} - \hbar\omega - i\hbar\delta} \tag{10.60}$$

where

$$F_{11}^{11}(\eta) = \frac{1}{2}\frac{[\eta(\eta^2+\pi^2)(3\eta^2+2\pi^2) - \pi^4(1-e^{-2\eta})]}{[\eta(\eta^2+\pi^2)]^2} \tag{10.61}$$

with $\eta = qa/2$. Note that $F_{11}^{11}(\eta)$ goes to unity as η goes to zero. Thus for a strictly 2D gas the form factor is unity (Fig. 10.1).

For an antisymmetric potential, transitions are allowed only between subbands of opposite symmetry. In this case screening arises solely from intersubband virtual transitions. When only the lowest subband is occupied, the subbands involved are bands 1 and 2 and the form factor is

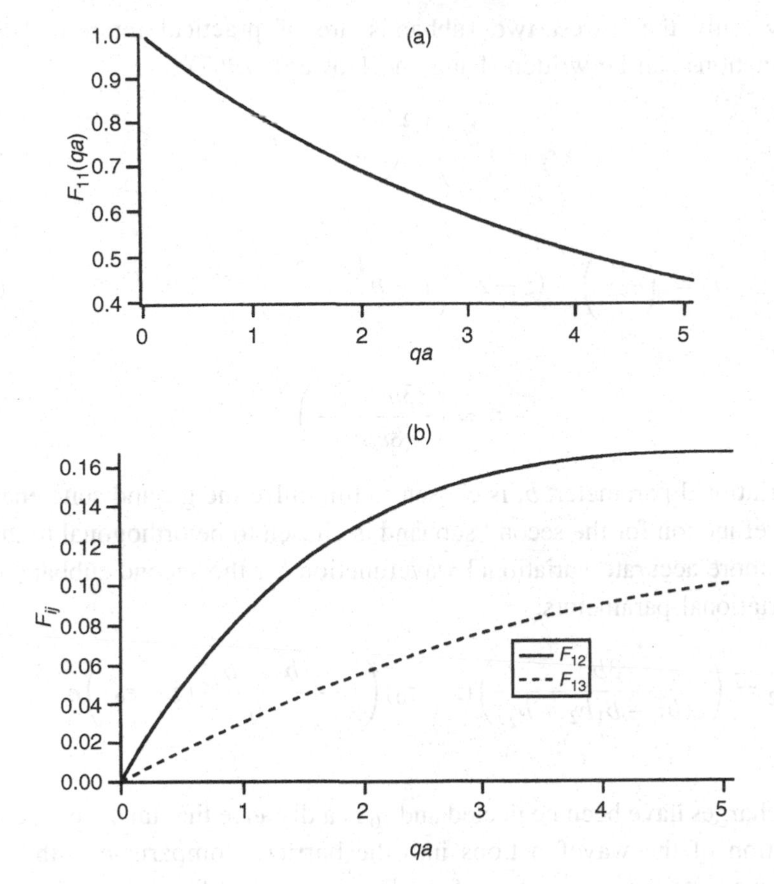

Fig. 10.1 Form factors for completely confined electrons in a quantum well: (a)$F_{11}^{11}(qa)$; (b)$F_{12}^{12}(qa)$; and $F_{13}^{13}(qa)$.

$$F_{21}^{21}(q) = \frac{4\eta}{(4\eta^2+\pi^2)(4\eta^2+9\pi^2)}\left[4\eta^2+5\pi^2-\frac{128\eta\pi^4(1+e^{-2\eta})}{(4\eta^2+\pi^2)(4\eta^2+9\pi^2)}\right] \tag{10.62}$$

which vanishes as $\eta \to 0$ (Fig. 10.1). This, coupled with the relatively large energy denominator in Eq. (10.19), makes the screening of intersubband transitions much weaker than that of intrasubband transitions. Symmetric and antisymmetric potentials are relevant to the case of confined optical modes.

Analytical expressions for the form functions can also be derived for wavefunctions associated with a modulation-doped heterostructure (see Section 2.2).

Usually, only the lowest two subbands are of practical interest. The model wavefunctions can be written (Fang and Howard, 1967)

$$\psi_1 = \left(\frac{b^3}{2}\right)^{1/2} (z - z_0) e^{-b(z-z_0)/2} \tag{10.63a}$$

$$\psi_2 = \left(\frac{3b^3}{2}\right)^{1/2} (z - z_0)\left(1 - b\frac{(z - z_0)}{3}\right) e^{-b(z-z_0)/2} \tag{10.63b}$$

$$\mathrm{b} = \left(\frac{33m^* e^2 N_s}{8\varepsilon_s \hbar^2}\right)^{1/3} \tag{10.63c}$$

The variational parameter, b, is chosen to minimize the ground state energy, and the wavefunction for the second subband is chosen to be orthogonal to that of the first. A more accurate variational wavefunction for the second subband employs two variational parameters:

$$\psi_2 = \left(\frac{3b^3}{2(b_1^2 - b_1 b_2 + b_2^2)}\right)(z - z_0)\left(1 - \frac{(b_1 + b_2)}{6}(z - z_0)\right) e^{-b(z-z_0)/2} \tag{10.63d}$$

Image charges have been neglected and z_0 is a distance that takes into account the penetration of the wavefunctions into the barrier. Comparison with numerical results suggests that z_0 is about 5 Å. The preceding forms are convenient for getting simple form functions. We quote, for brevity, the results only for the case of a single variational parameter:

$$\begin{aligned}
F_{11}^{11}(q) &= \frac{b}{8(q+b)^3}(8b^2 + 9qb + 3q^2) \\
F_{22}^{22}(q) &= \frac{b}{4(q+b)^5}(4b^4 + 5qb^3 + 9q^2b^2 + 5q^3b + q^4) \\
F_{22}^{11}(q) &= \frac{b}{16(q+b)^5}(16b^4 + 25qb^3 + 29q^2b^2 + 15q^3b + 3q^4)
\end{aligned} \tag{10.64}$$

These are depicted in Fig. 10.2. More sophisticated analytical forms have been used by Ando (1982), and by Takada and Uemura (1977).

In the case of Coulomb interactions, such as charged-impurity scattering, intersubband transition rates are weak and screening is predominantly due to intrasubband processes, as is scattering. Unconfined acoustic phonons produce comparable rates for inter- and intrasubband transitions, and, in general therefore,

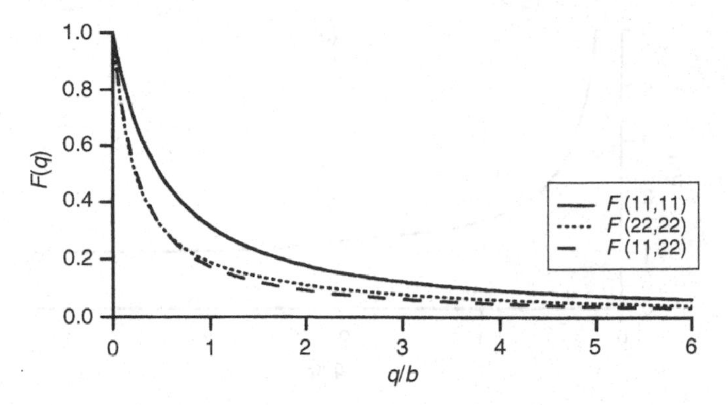

Fig. 10.2 Form factors for approximate heterojunction wavefunctions.

both types of process contribute to the screening, but in narrow wells the inter-subband components will be weaker because of the larger energy denominator, which reflects the large separation of subbands, and once again attention can usually be restricted to intrasubband processes within the lowest subband.

It is useful to obtain an explicit expression for the intrasubband dielectric response, which can be done for the special case of a degenerate gas at zero temperature when only the lowest state is occupied. The sum over **k** can be replaced by an integral and it is convenient to use the expanded form analogous to Eq. (10.41). We obtain, for the real part (Stern, 1967)

$$\mathrm{Re}\,\frac{\varepsilon_{11}^{11}}{\varepsilon_0} = 1 + \frac{e^2 m n_s}{\varepsilon_0 \hbar^2 q^2 k_F} F_{11}^{11}(q)\left[\frac{q}{k_F} - c_+(a_+^2 - 1)^{1/2} - c_-(a_-^2 - 1)^{1/2}\right] \quad (10.65a)$$

where n_s is the areal density of electrons, k_F is the Fermi wavevector, $c_\pm = sgn a_\pm$ if $a_\pm^2>1$ and zero otherwise, where $a_\pm = (q/2k_F) \pm (m^*\omega/\hbar q k_F)$, and for the imaginary part

$$\mathrm{Im}\,\frac{\varepsilon_{11}^{11}}{\varepsilon_0} = \frac{e^2 m n_s}{\varepsilon_0 \hbar^2 q^2 k_F} F_{11}^{11}(q)\left[d_-(1 - a_-^2)^{1/2} - d_+(1 - a_-^2)^{1/2}\right] \quad (10.65b)$$

Here, $d_\pm = 1$ provided that $a_\pm^2<1$, and zero otherwise (Fig. 10.3). The condition $a_\pm^2 \leq 11$ is the one satisfied for scattering with conservation of energy and momentum. Thus, if this condition for single-particle excitation is satisfied for a collective oscillation of the electron gas, the oscillation will be damped, a process known as Landau damping. Frequencies lying outside of the spectrum of single-particle excitation can propagate and may couple with phonons. Fig. 10.4 illustrates the excitation spectrum in a quasi-2D electron gas.

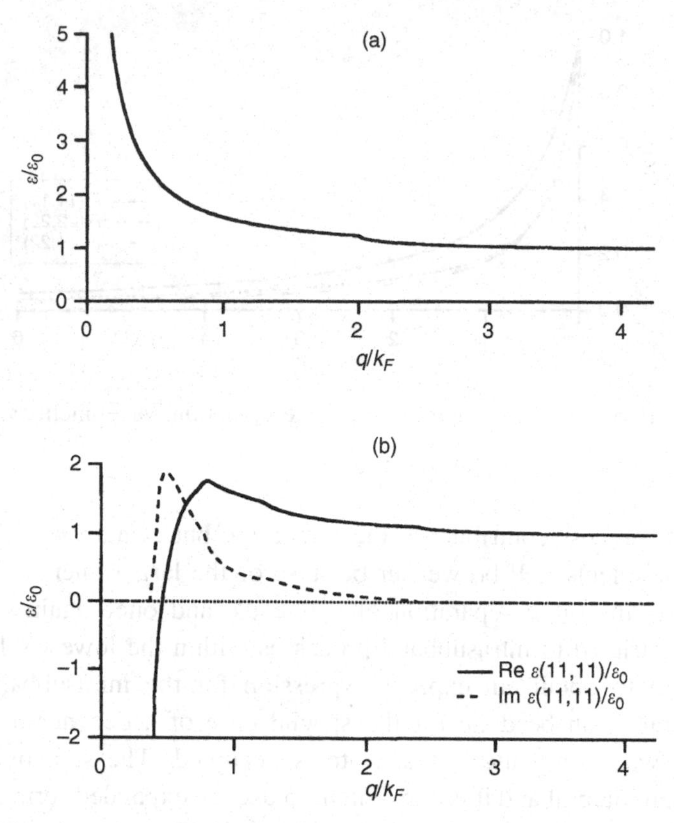

Fig. 10.3 The 2D dielectric functions at $T = 0$: (a) $\omega = 0$; (b) $\omega = \omega_{\rm LO}$(GaAs) (parameters as for GaAs with $\varepsilon_0 = \varepsilon_s$, the static permittivity).

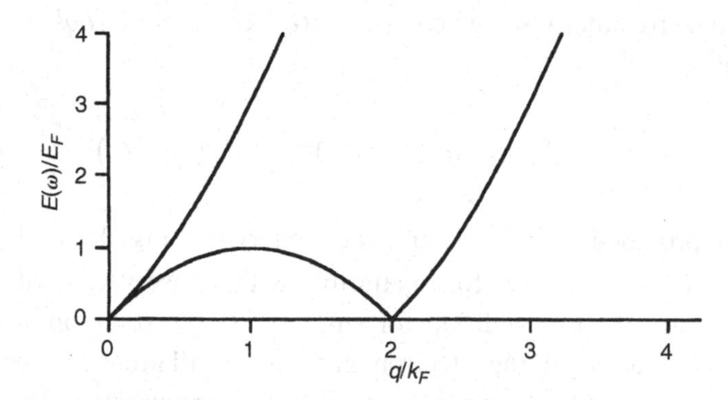

Fig. 10.4 The 2D single-particle excitation spectrum ($E(\omega) = \hbar\omega$).

In the long-wavelength limit ($q \to 0, F_{11}^{11}(q) = 1$, finite frequency)

$$\frac{\varepsilon_{11}^{11}}{\varepsilon_0} = 1 - \frac{\omega_{ps}^2}{\omega^2}, \quad \omega_{ps}^2 = \frac{e^2 n_s q}{2\varepsilon_0 m} \tag{10.66}$$

The plasma frequency is now proportional to the square root of the wavevector. In the static limit ($\omega \to 0$), and also in the Landau-damping regime ($a_{\pm}^2 < 1$),

$$\frac{\varepsilon_{11}^{11}}{\varepsilon_0} = 1 + \frac{q_s}{q}, \quad q_s = \frac{e^2 m^*}{2\pi\varepsilon_0 \hbar^2} F_{11}^{11}(q) \tag{10.67}$$

and we have used the degenerate condition $n_s = k_F^2/2\pi$. When there are $2N$ layers separated from each other by a barrier width b, the reciprocal screening length becomes

$$q_s(N) = q_L\left(1 + 2\frac{1 - e^{-qNb}}{1 - e^{-qb}} e^{-qb}\right) \tag{10.68}$$

so that when $N \to \infty$ and a, $b \to 0$ in such a way that $n_s/b \to n$ and Na remains finite, the bulk result is recovered.

10.6 The Quasi-1D Dielectric Function

In the case of quasi-1D structures, there is the question of what shape of cross-sectional area is appropriate. A common choice is to treat the case of a cylindrical quantum wire, and this will be our choice here when we come to specifying actual wavefunctions. In general, the wavefunction of the electrons takes the form

$$\psi_{nl}(\mathbf{R}) = \frac{1}{L^{1/2}} e^{ikx} \psi_{nl}(\mathbf{r}) \tag{10.69}$$

where L is the length of the normalizing cavity, x is the axial coordinate along the wire, $\mathbf{r}$ is the position vector in the cross-sectional plane, and nl are quantum numbers. The perturbing potential can be expanded in a 1D Fourier series

$$V(\mathbf{R}, t) = \frac{L}{2\pi} \int V_q(\mathbf{r}, t) e^{iqx} dq \tag{10.70}$$

where q is an axial wavevector. The 3D wavevector specifying the screening potential in Eq. (10.29) is now written as Q = (q, $\mathbf{q}_r$), where $\mathbf{q}_r$ is the wavevector in the plane perpendicular to the axis. Writing

$$\mathbf{Q}.(\mathbf{R} - \mathbf{R}') = q(x - x') + q_r|r - r'|\cos\theta \tag{10.71}$$

where θ is the angle between $\mathbf{q}_r$ and $\mathbf{r}$, $-\mathbf{r}'$, and exploiting the integrals

$$\iint \frac{e^{iq_r|\mathbf{r}-\mathbf{r}'|\cos\theta}}{q^2+q_r^2} q_r dq_r d\theta = 2\pi \int_0^\infty \frac{J_0(q_r|\mathbf{r}-\mathbf{r}'|)}{q^2+q_r^2} q_r dq_r = 2\pi K_0(q|\mathbf{r}-\mathbf{r}'|) \quad (10.72)$$

where $J_0(z)$ is the Bessel function and $K_0(z)$ is the modified Bessel function, we obtain the 1D Fourier expansion of $V_s(\mathbf{R}, t)$

$$V_s(t) = \frac{e_2}{(2\pi)^2 \varepsilon_0} \iint K_0(q|\mathbf{r}-\mathbf{r}'|) \sum_{a_i a_j} |a_i\rangle\langle a_j| \prod\nolimits_{a_i a_j} \langle a_i|V(t)|a_j\rangle e^{i\mathbf{q}.(x-x')} d\mathbf{R}' d\mathbf{q} \quad (10.73)$$

The G factor is then given by

$$G_{ij}^{gh}(\mathbf{q}) = \delta_{k_{k}k_i+q} \delta_{k_{i}k_j+q} F_{ij}^{gh}(q) \quad (10.74)$$

where

$$F_{ij}^{gh}(q) = \iint \psi_{n_4 l_4}^*(\mathbf{r}) \psi_{n_3 l_3}(\mathbf{r}) \psi_{n_2 l_2}(\mathbf{r}') \psi_{n_1 l_1}^*(\mathbf{r}') K_0(q|\mathbf{r}-\mathbf{r}'|) d\mathbf{r} d\mathbf{r}' \quad (10.75)$$

For a cylindrical wire in which the wavefunction disappears at the boundaries we have

$$\psi_{nl}(r,\theta) = \frac{J_n(k_{nl} r) e^{in\theta}}{(\pi a^2)^{1/2} J_{n+1}(k_{nl} a)} \quad (10.76)$$

where a is the radius of the wire and k_{nl} is a wavevector that is a solution of $J_n(k_{nl} a) = 0$. The energy of the state is given by

$$E_{nl} = \frac{\hbar^2}{2m*}(k^2 + k_{nl}^2) \quad (10.77)$$

The matrix element for the Fourier coefficient of the screening potential is therefore

$$\langle n_4 l_4|V_{sq}(\mathbf{r},t)|n_3 l_3\rangle = \frac{e^2}{2\pi\varepsilon_0 L} \sum_{n_2 l_2, n_1 l_1, k} \frac{f(E_{n_2 l_2 k+q}) - f(E_{n_1 l_1, k})}{E_{n_2 l_2 k+q} - E_{n_1 l_1 k} - \hbar\omega - i\hbar\delta} F_{ij}^{gh}(q) \langle n_2 l_2|V_q(\mathbf{r},t)|n_1 l_1\rangle \quad (10.78)$$

and the dielectric response function is then

$$\frac{\varepsilon_{ij}^{gh}(q,\omega)}{\varepsilon_0} = \delta_{gi}\delta_{hj} - \frac{e^2}{2\pi\varepsilon_0 L} \sum_k \frac{f(E_{jk+q}) - f(E_{ik})}{E_{jk+q} - E_{ik} - \hbar\omega - i\hbar\delta} F_{ij}^{gh}(q) \quad (10.79)$$

The suffices are $g = n_4 l_4$, $h = n_3 l_3$, $j = n_2 l_2$, $i = n_1 l_1$.

Even with the simple wavefunction of Eq. (10.69), the evaluation of $F(q)$ is not straightforward, to say the least.

The vector $\mathbf{r} - \mathbf{r}'$ is dependent on the angle θ' between $\mathbf{r}$ and $\mathbf{r}'$ through

$$|\mathbf{r} - \mathbf{r}'|^2 = r^2 + r'^2 - 2rr' \cos \theta' \quad (10.80)$$

and we can use the following expansion in terms of modified Bessel functions (Watson, 1958) for $r > r' > 0$:

$$K_0(q(r^2 + r'^2 - 2rr' \cos \theta)^{1/2}) = K_0(qr)I_0(qr') + 2\sum_{v=1}^{\infty} K_v(qr)I_v(qr') \cos v\theta' \quad (10.81)$$

For narrow wires where only the ground state is occupied the integration over θ' is straightforward:

$$\int_0^{2\pi} K_0(q|\mathbf{r} - \mathbf{r}'|)\, d\theta' = 2\pi \begin{cases} K_0(qr)I_0(qr')|_{r'<r} \\ K_0(qr')I_0(qr)|_{r<r} \end{cases} \quad (10.82)$$

At $T = 0$ the intrasubband response function becomes (Fig. 10.5)

$$\mathrm{Re}\frac{\varepsilon_{11}^{11}}{\varepsilon_0} = 1 + \frac{e^2 m^*}{2\pi^2 \varepsilon_0 \hbar^2 q} ln \left| \frac{(k_F + q/2)^2 - (m^*\omega/\hbar q)^2}{(k_F - q/2)^2 - (m^*\omega/\hbar q)^2} \right| F_{11}^{11}(q) \quad (10.83)$$

$$\begin{aligned} \mathrm{Im}\frac{\varepsilon_{11}^{11}}{\varepsilon_0} &= \frac{e^2 m^*}{4\pi \varepsilon_0 \hbar^2 q}[-c_+ + c_-]F_{11}^{11}(q) \\ c_+ &= 1, \left|\frac{m^*\omega}{\hbar q} + \frac{q}{2}\right| \le k_F, \quad c_- = 1, \left|\frac{m^*\omega}{\hbar q} - \frac{q}{2}\right| \le k_F \\ &= 0, \text{ otherwise} \qquad\qquad = 0, \text{ otherwise} \end{aligned} \quad (10.84)$$

Logarithmic singularities occur for

$$\left(k_F \pm \frac{q}{2}\right)^2 = \left(\frac{m^*\omega}{\hbar q}\right)^2 \quad (10.85)$$

which, for static perturbations, implies $q = \pm 2k_F$. For either sign the permittivity becomes infinite; that means that the electron gas polarizes to screen out the perturbation completely. This effect is at the core of the Peierls transition, whereby a lattice distortion with wavevector $2k_F$ occurs in a 1D metal. The singularities weaken and disappear as the temperature rises.

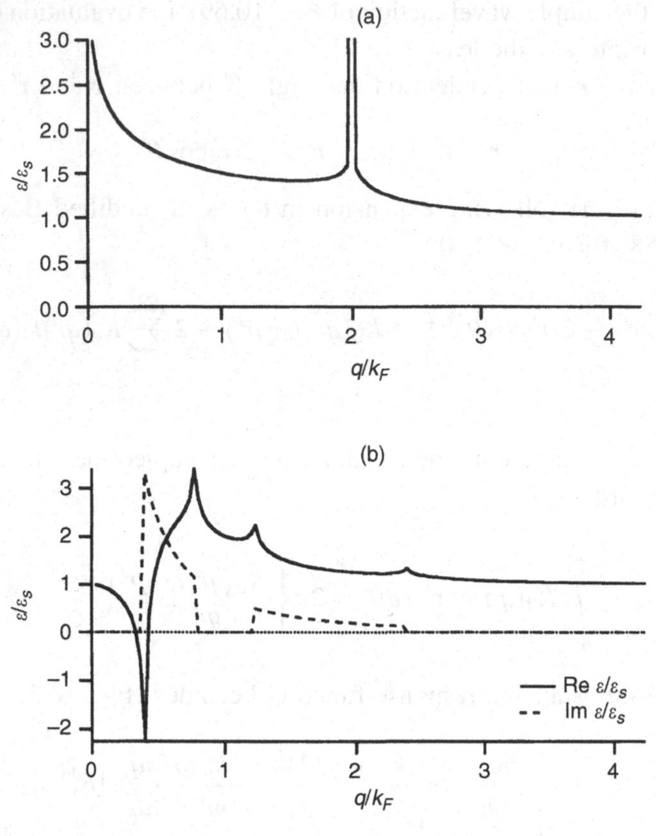

Fig. 10.5 The 1D dielectric function at $T = 0$: (a) $\omega = 0$; (b) $\omega = \omega_{\mathrm{LO}}$(GaAs) (parameters as for GaAs).

For finite values of qa the form function of Eq. (10.75) is obtainable in analytic form only by making approximations to the wavefunctions. The simplest is to assume that the wavefunction is constant over a range of radii and zero otherwise (Fig. 10.6). Thus, for the ground state we can follow Lee and Spector (1985) and take

$$\psi_{11}(r,\theta) \approx (\pi r_0^2)^{-1/2}, \quad 0 \leq r \leq r_0 \tag{10.86}$$

and use standard indefinite integrals of Bessel functions to obtain

$$F_{11}^{11}(q) = \frac{4}{(qr_0)^2}\left[\frac{1}{2} - K_1(qr_0)I_1(qr_0)\right] \tag{10.87}$$

Other analytic solutions for the form functions have been obtained by approximating the wavefunction for cylindrical wires more nearly to the Bessel function

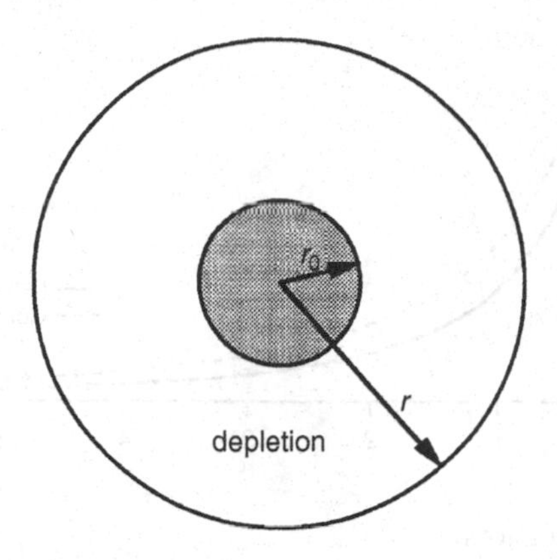

Fig. 10.6 Circular-cross-section wire.

using polynomials (Gold and Ghazali, 1990). Thus for the ground and first excited state

$$\psi_{01}(r,\theta) = \left(\frac{3}{\pi a^2}\right)^{1/2}\left(1 - \frac{r^2}{a^2}\right) \tag{10.88a}$$

$$\psi_{11}(r,\theta) = \left(\frac{12}{\pi a^2}\right)^{1/2}\left(\frac{r}{a} - \frac{r^3}{a^3}\right)e^{\pm i\theta} \tag{10.88b}$$

These reproduce the Bessel functions reasonably well and allow the use of standard integrals. The form factor then becomes

$$F_{11}^{11}(\eta) = \left(\frac{36}{\eta^2}\right)\left(\frac{1}{10} - \frac{2}{3\eta^2} + \frac{32}{3\eta^4} - \frac{64}{\eta^4}I_3(\eta)K_3(\eta)\right) \tag{10.88c}$$
$$\eta = qr_0$$

Fig. 10.7 depicts the form function. The effect of cross-sectional shape has been investigated by Bennett *et al.* (1994), who show that the intrasubband components are only marginally dependent.

At zero temperature with only the lowest subband occupied, the intrasubband plasmon dispersion is given by $\varepsilon_{11}^{11} = 0$ and the intersubband plasmon dispersion by $\varepsilon_{12}^{12} = 0$ (Li and Das Sarma, 1991). For the long-wavelength intrasubband mode the dependence on wavevector is of the form

$$\omega_{11}^2 \approx \frac{e^2 n_s q^2}{2\pi m^* \varepsilon}\left|\ln\left(\frac{qx}{2}\right)\right| \tag{10.89}$$

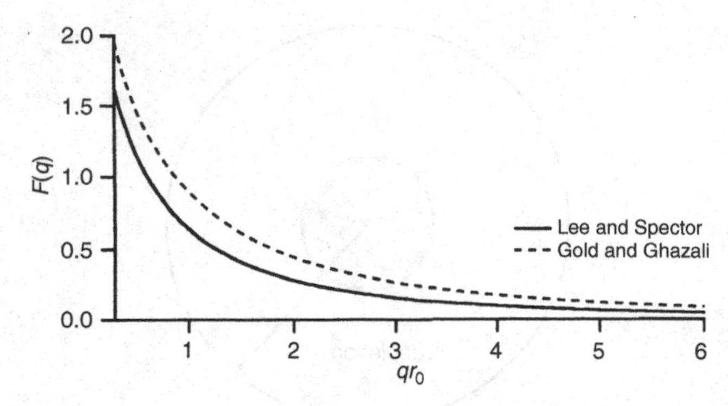

Fig. 10.7 1D Form factors

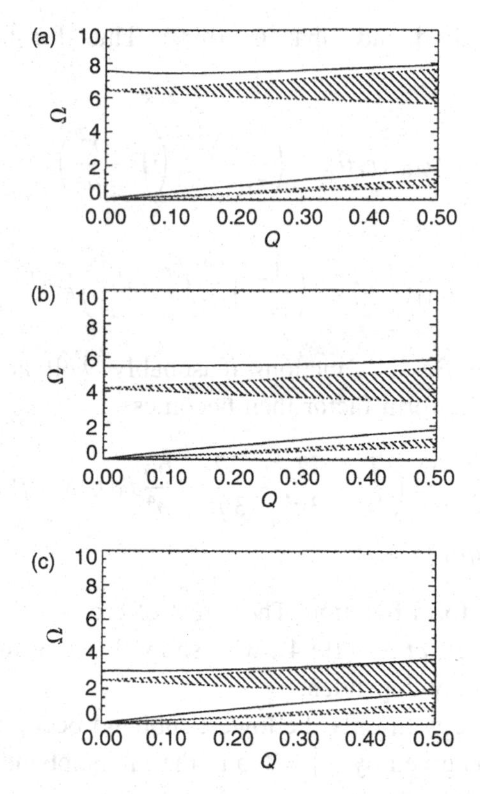

Fig. 10.8 1D Plasmon spectrum for various elliptical cross-sections: (a) circle; (b) 2:1 ellipse; (c) 10:1 ellipse ($\Omega = \hbar\omega/E_F$, $Q = q/k_F$). Shaded regions correspond to the single-particle continua (Bennett *et al.*, 1994).

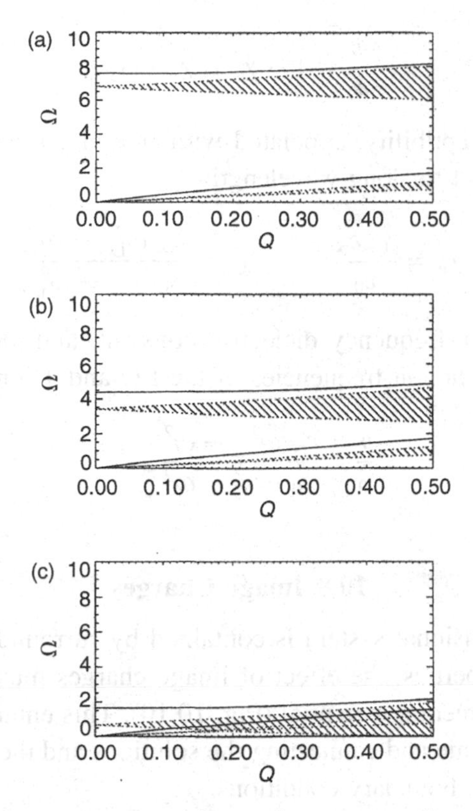

Fig. 10.9 1D Plasmon spectrum for various rectangular cross-sections: (a) square; (b) 2:1 rectangle; (c) 10:1 rectangle ($\Omega = \hbar\omega / E_F$, $Q = q/k_F$). Shaded regions correspond to the single-particle continua (Bennett *et al.*, 1994).

where x is a characteristic length dependent on cross-section shape (Figs. 10.8 and 10.9). The intersubband plasmon frequency is more sensitive to cross-section mainly because of the equivalent sensitivity of the subband energy separation. The single-particle continua, defined by the imaginary part of the dielectric function, are also shown in Figs. 10.8 and 10.9 and it is clear that the intersubband component is, once more, sensitive to the shape of the cross-section.

10.7 Lattice Screening

In addition to the screening produced by free electrons there is always a background screening due to the lattice. In a polar material lattice, screening occurs through the valence electronic polarization of the ions, quantified by a susceptibility χ_v, plus ionic displacement susceptibility χ_I. In general, the dielectric function is made up as follows:

$$\frac{\varepsilon}{\varepsilon_0} = 1 - \chi_e - \chi_v - \chi_I \tag{10.90}$$

where χ_e is the susceptibility associated with free electrons – the one we have been discussing – and for long wavelengths

$$\chi_v = \frac{\varepsilon_0 - \varepsilon_\infty}{\varepsilon_0}, \qquad \chi_1 = \frac{\varepsilon_\infty}{\varepsilon_0}\frac{\omega_{\mathrm{LO}}^2 - \omega_{\mathrm{TO}}^2}{\omega^2 - \omega_{\mathrm{TO}}^2} \tag{10.91}$$

where ε_∞ is the high-frequency dielectric constant, and ω_{LO} and ω_{TO} are the zone-centre optical-phonon frequencies of the LO and TO modes. Thus

$$\frac{\varepsilon}{\varepsilon_0} = \frac{\varepsilon_\infty}{\varepsilon_0}\frac{\omega_{\mathrm{LO}}^2 - \omega^2}{\omega_{\mathrm{TO}}^2 - \omega^2} - \chi_e \tag{10.92}$$

10.8 Image Charges

When the low-dimensional system is contained by surrounding media with different dielectric properties, the effect of image charges must be included in the calculation of the screening potential (Fig. 10.10). This entails solving Poisson's equation in each regime and connecting the solutions and the interfaces using the standard electrostatic boundary conditions.

The simplest case is that of a single interface. A charge e at z induces a polarization in the surrounding dielectric that is equivalent to an image charge e' at $-z$ with the dielectric replaced by the semiconductor. Standard electrostatics gives

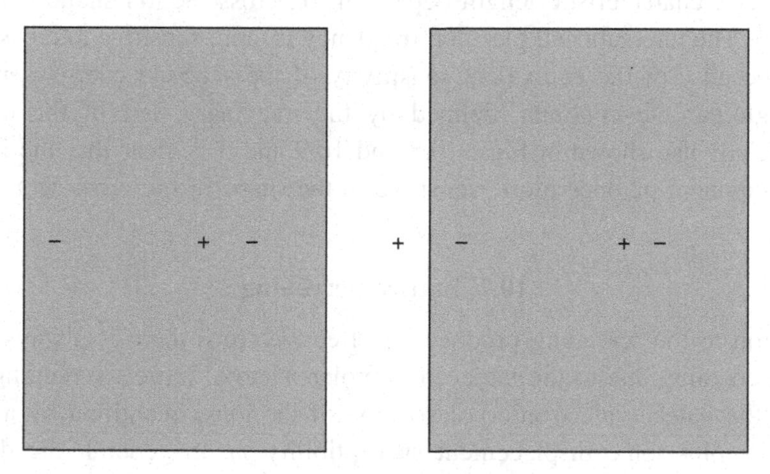

Fig. 10.10 Image charges.

$$e' = -\frac{\varepsilon_B - \varepsilon_W}{\varepsilon_B + \varepsilon_W} e \tag{10.93}$$

where ε_B, ε_w are the permittivities of the barrier (dielectric) and well (semiconductor). Thus in the form factor exp $-q[z - z']$ is replaced by

$$e^{-q|z-z'|} - \frac{\varepsilon_B - \varepsilon_W}{\varepsilon_B + \varepsilon_W} e^{-q|z+z'|} \tag{10.94}$$

When there are two plane parallel interfaces we can continue to use the image-charge method. Measuring now from the midplane with the interfaces at $z = \pm a/2$ the form factor exp $-q[z - z']$ is replaced by

$$\begin{aligned} & e^{-q|z-z'|} + \gamma(e^{-q|a-z-z'|} + e^{-q|a+z+z'|}) \\ & + \sum_{k=1}^{\infty} \left[\gamma^{2k}(e^{-q|2ka-z-z'|} + e^{-q|2ka-z+z'|}) \right. \\ & \left. + \gamma^{2k+1}(e^{-q|(2k+1)a-z-z'|} + e^{-q|(2k+1)a+z+z'|}) \right] \end{aligned} \tag{10.95}$$

where $\gamma = (\varepsilon_w - \varepsilon_B)/(\varepsilon_w + \varepsilon_B)$. In many cases only the first three terms in the preceding equation need be considered.

Alternatively, we return to Poisson's equation for the screening potential. This time, instead of expanding $V_S(\mathbf{R})$ and $n(\mathbf{R})$ in a 3D Fourier series, we expand only in a 2D Fourier series, e.g.

$$V_s(R) = \frac{\sigma}{(2\pi)^2} \int V_{sq}(z) e^{-i\mathbf{q}\cdot\mathbf{r}} d\mathbf{q} \tag{10.96}$$

where $\mathbf{q}$ is a wavevector in the plane, and obtain

$$\begin{aligned} \frac{d^2 V_{sq}(z)}{dz^2} - q^2 V_{sq}(z) &= -\frac{e}{\varepsilon_W} n_q(z), \quad 0 \le z \le a \\ &= 0, \quad \text{otherwise} \end{aligned} \tag{10.97}$$

The density is obtained from Eq. (10.24), its dependence on z determined by the wavefunctions. The solutions must be continuous at the interfaces and so must the electric displacements, i.e.

$$\varepsilon_W \frac{dV_1}{dz} = \varepsilon_B \frac{dV_2}{dz} \tag{10.98}$$

The form factor now contains terms dependent on the ratio $\varepsilon_W/\varepsilon_B$ (see, for example, Lee and Spector, 1983). A similar procedure is necessary for the quasi-1D case, and for cases involving multiple interfaces.

The quantum theory of image charges involves a calculation of the self-energy of a charge in the vicinity of a polarizable medium. For a metal it is straightforward to treat the problem using the dielectric function

$$\varepsilon = \varepsilon_0 \left(1 - \frac{\omega_p^2}{\omega^2}\right) \tag{10.99}$$

where ω_p is the plasma frequency. The Hamiltonian can be diagonalized and the classical result obtained. In the case of a dielectric the use of the usual dielectric function

$$\varepsilon = \varepsilon_\infty \frac{\omega^2 - \omega_{\mathrm{LO}}^2}{\omega^2 - \omega_{\mathrm{TO}}^2} \tag{10.100}$$

gives the wrong result. This is because the self-energy involves the virtual excitation of valence electrons as well as ionic polarization, which, in turn, involves the frequency dependence of the high-frequency permittivity. The latter must be incorporated in the form

$$\varepsilon = \varepsilon_0 \frac{\omega^2 - \omega_{CB}^2}{\omega^2 - \omega_{VB}^2} \tag{10.101}$$

where ω_{CB}, ω_{VB} are band-structure frequencies (Constantinou, 1995).

10.9 The Electron-Plasma/Coupled-Mode Interaction

Collective oscillations of the degenerate electron gas at $T = 0$ whose frequencies lie outside the spectrum of single-particle excitation are longitudinally polarized electromagnetic modes that can couple to an electron whose energy lies above the Fermi level. At finite temperatures the single-particle spectrum becomes diffuse, and collective modes are damped whatever their frequency, but where the damping is small there can be significant interaction with incident hot electrons. Besides this there can also be a significant coupling to the polar-optical modes of the lattice.

Electrons are scattered by electric-field fluctuations through the usual Hamiltonian

$$H = \frac{e}{Q} \sum_{\mathbf{Q}} iF(Q) e^{i\mathbf{Q}\cdot\mathbf{R}} \tag{10.102}$$

where the field is driven by a polarization $\mathbf{P}$ according to

$$\mathbf{F}(Q) = -\frac{\mathbf{p}(Q)}{\varepsilon(Q,\omega)} \tag{10.103}$$

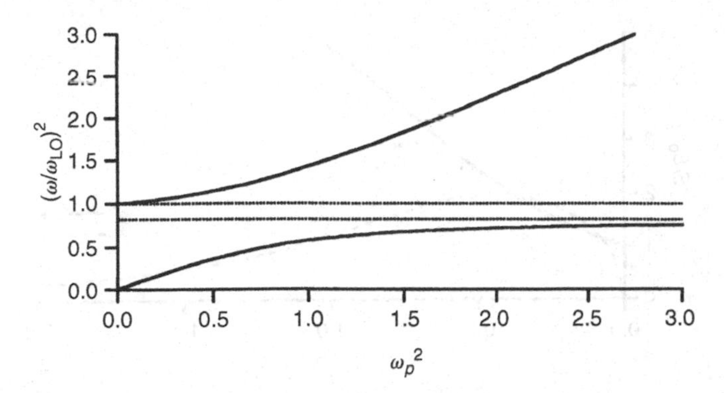

Fig. 10.11 Coupled-mode frequencies for long wavelengths (GaAs).

and $\varepsilon(Q, \omega)$ is the permittivity function. The mean power spectrum of the field fluctuations is given by the dissipation-fluctuation theorem to be

$$\langle F(Q)F^*(Q)\rangle = \frac{\hbar}{\pi}\{n(\omega)+1\}\mathrm{Im}\left(-\frac{1}{\varepsilon(Q,\omega)}\right) \tag{10.104}$$

where $n(\omega)$ is the Bose–Einstein number, and Im, once again, stands for the imaginary part. In general, therefore, the scattering rate to lowest order is

$$W(\mathbf{K}) = \frac{2\pi}{\hbar}\sum_{\mathbf{Q}}\int_{-\infty}^{\infty}\{1-f(\mathbf{K}+\mathbf{Q})\}\{n(\omega)+1\}\frac{e^2}{Q^2} \mathrm{Im}\left(-\frac{1}{\varepsilon(Q,\omega)}\right)\delta(E_{\mathbf{K+Q}}-E_{\mathbf{K}}+\hbar\omega)\frac{d(\hbar\omega)}{\pi} \tag{10.105}$$

The interactive modes are those defined by the maxima of $\mathrm{Im}\{-1/\varepsilon(Q, \omega)\}$.

In the long-wavelength limit, the total permittivity can be written

$$\varepsilon(Q,\omega) = \varepsilon_\infty\left(1-\frac{\omega_p^2}{\omega^2}-\frac{\omega_{\mathrm{LO}}^2-\omega_{\mathrm{TO}}^2}{\omega^2-\omega_{\mathrm{TO}}^2}\right) \tag{10.106}$$

where ω_p is the plasma frequency. The zeros of this function define two modes with frequencies (Fig. 10.11)

$$\omega_\pm^2 = \frac{1}{2}[\omega_{\mathrm{LO}}^2+\omega_p^2 \pm \{(\omega_{\mathrm{LO}}^2-\omega_p^2)^2+4\omega_p^2(\omega_{\mathrm{LO}}^2-\omega_{\mathrm{TO}}^2)\}^{1/2}] \tag{10.107}$$

and thus

$$\frac{1}{\varepsilon(Q,\omega)} = \frac{1}{\varepsilon_\infty}\frac{\omega^2(\omega^2-\omega_{\mathrm{TO}}^2)}{(\omega^2-\omega_+^2)(\omega^2-\omega_-^2)} \tag{10.108}$$

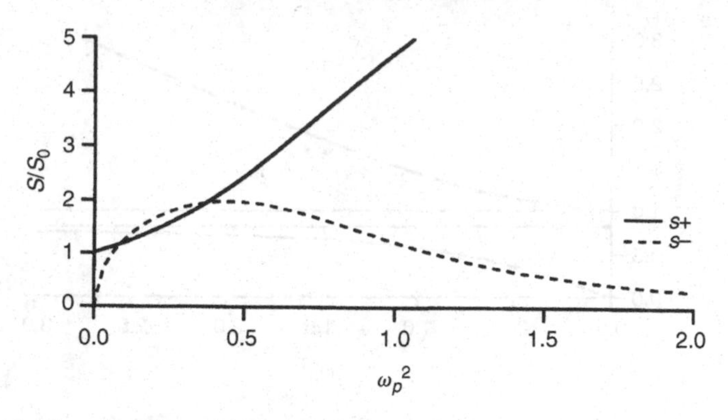

Fig. 10.12 Coupling-strength ratio for coupled modes (GaAs).

Noting that

$$\mathrm{Lim}_{\varepsilon\to 0}\frac{1}{x+i\varepsilon} = P\left(\frac{1}{x}\right) \mp i\pi\delta(x) \tag{10.109}$$

where P denotes principal value we obtain

$$\begin{aligned}\mathrm{Im}\left(-\frac{1}{\varepsilon(Q,\omega)}\right) = \frac{\pi}{2\varepsilon_\infty}\omega^2(\omega^2-\omega_{\mathrm{TO}}^2)&\left[\frac{\delta(\omega-\omega_+)-\delta(\omega-\omega_+)}{\omega_+(\omega_+^2-\omega_-^2)}\right.\\ &\left.+\frac{\delta(\omega-\omega_-)-\delta(\omega+\omega_-)}{\omega_-(\omega_-^2-\omega_+^2)}\right]\end{aligned} \tag{10.110}$$

For small electron densities ω_p is small (in all dimensions) and $\omega_+ \approx \omega_{\mathrm{LO}}$, $\omega_- \approx \omega_p$. The two modes are approximately the LO mode and the plasma mode, respectively. For the LO mode we recover the usual Fröhlich interaction strength

$$S_0 = \frac{1}{\varepsilon_\infty}\frac{(\omega_{\mathrm{LO}}^2-\omega_{\mathrm{TO}}^2)}{\omega_{\mathrm{LO}}} \tag{10.111}$$

The plasmon coupling strength is

$$S_P = \frac{\omega_P}{\varepsilon_S} \tag{10.112}$$

When the electron density is very high $\omega_+ \approx \omega_p$, $\omega_- \approx \omega_{\mathrm{TO}}$. In general the coupling-strength ratio is

$$\frac{S_\pm}{S_0} = \frac{\omega_\pm\omega_{\mathrm{LO}}(\omega_\pm^2-\omega_{\mathrm{TO}}^2)}{(\omega_\pm^2-\omega_\pm^2)(\omega_{\mathrm{LO}}^2-\omega_{\mathrm{TO}}^2)} \tag{10.113}$$

(Kim *et al.*, 1978), which approaches zero as $\omega_{-} \to \omega_{TO}$, corresponding to heavy screening. At intermediate densities the interaction with the phonon-like mode actually increases, denoting antiscreening, and the coupling strength with the plasmon-like mode is significant even at relatively low densities (Fig. 10.12).

The unscreening Fröhlich interaction is quite strong in spite of the rather small difference (in III-Vs) between ω_{LO} and ω_{TO}. The interaction with plasmons is basically much stronger because the frequency equivalent to ω_{TO} is zero. There is no restoring force for electrons other than the electrostatic one. Nevertheless, the overall scattering rate is not proportionately stronger in 3D and 2D because of Landau damping, which effectively restricts the sum for the plasma-like mode at the boundaries of the single-particle spectrum, thereby limiting the interaction to the relatively few long-wavelength modes. On the other hand, these are exactly the modes that couple most strongly. In 1D, however, even this limited interaction can be extremely effective.

In many cases the regime of ω and q of interest to the screening of the polar-optical interaction coincides with the Landau-damping regime. This means that the long-wavelength solutions for the coupled-mode frequencies do not apply, and, instead, solutions have to be found for the damped modes (Fig. 10.13). It is found that damping reduces the frequency of the high-frequency mode below ω_{LO} and toward ω_{TO}, whence it recovers to ω_{LO} as q increases. When the frequency is above ω_{LO} the interaction is antiscreened; below it is screened. A popular, if simplistic, approach is to regard one as balancing out the other and to neglect coupled-mode and screening effects entirely, at least as far as the interaction with the LO mode is concerned. The low-frequency mode enters a track that roughly parallels the undamped branch, the frequency reducing with reducing wavevector. Its interaction strength is therefore weak but, nevertheless, this

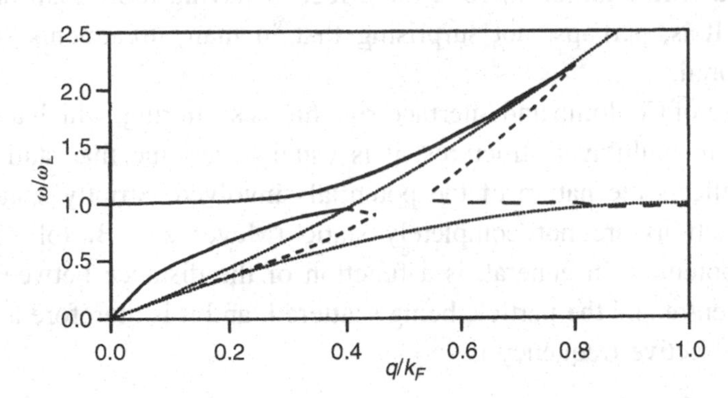

Fig. 10.13 Coupled-mode dispersion in a GaAs quantum well ($N_s = 10^{16}\text{m}^{-2}$, $F_{(q)} = 1$, $T = 0\text{K}$).

mode is of some importance. Its low energy makes its emission more accessible (Kim *et al.*, 1978), and at low temperatures in 3D and 2D systems, low-frequency plasmon emission may become an energy-loss mechanism that competes significantly with that associated with acoustic phonons (Das Sarma *et al.*, 1988). In 1D the interaction is more striking as a consequence of the divergence in the interaction density of states at the emission threshold (Yu-Kuang Hu and Das Sarma, 1993; Bennett *et al.*, 1995a). Indeed, this can overshadow the phonon-emission rate by an order of magnitude. Yu-Kuang Hu and Das Sarma predict the appearance of negative differential resistance (NDR) in a 1D hot-electron transistor as a consequence of this immensely strong threshold.

A point rarely addressed quantitatively is the question of plasmon lifetime. An LO phonon decays in a matter of a few picoseconds into two acoustic phonons. What happens to this process in the case of coupled modes is unclear. Plasmons emitted by a fast electron at finite temperature are eventually Landau-damped, but at low temperatures this process is slow. There is then a tendency for such plasmons to become hot, if copiously emitted, and hence readily reabsorbed by the initial electron. The coupling of optical phonons to plasmons has recently been shown to shorten the phonon lifetime (Dyson and Ridley, 2008).

10.10 Discussion

Many low-dimensional systems of interest have high carrier densities and the role of screening on the interactions that scatter carriers is one that cannot be ignored. The problem is quite complex, in general. Quite apart from the effects of quantum confinement, there are issues regarding spatial extent and dynamic effects that have to be decided. In terms of the Fourier components, this means determining the q-dependence and the ω-dependence of the dielectric function influencing the scattering rate. In addition there is the effect of having more than one subband populated. It is, perhaps, not surprising that in many treatments screening is simply ignored.

In the case of Coulomb and interface-roughness scattering, which are always of importance in multilayer structures, it is usual to assume that static screening correctly reflects the nature of the potentials involved. Strictly speaking, even these interactions are not completely static (Meyer and Bartoli, 1983). The scattering potential, in general, is a function of the distance between the static scattering centre and the particle being scattered, and it is therefore a function of time. The effective frequency is

$$\omega = \mathbf{q} \cdot \mathbf{v}_{\mathrm{CM}} \tag{10.114}$$

where $\mathbf{v}_{\mathrm{CM}}$ is the velocity of the centre-of-mass of the two-body system, viz:

$$\mathbf{v}_{\mathrm{CM}} = \frac{d\mathbf{R}}{dt}, \quad \mathbf{R} = \frac{m_1 r_1 + m_2 r_2}{m_1 + m_2} \tag{10.115}$$

When the scattering centre is a charged ion and the particle is an electron, then $m_1 >> m_2$, where m_1, m_2 are the respective masses, and $\mathbf{R} \approx \mathbf{r}_1$, which is independent of time, and therefore $\mathbf{v}_{\mathrm{CM}} \approx 0$. Taking a static approximation is valid in this case. But it would not be valid were the scattering centre to be another electron or a hole. The screening in this case would be dynamic. Interactions with phonons are always dynamic.

As regards spatial effects, it is a common approximation to assume the interaction to be long-wavelength-like and to evaluate the dielectric function in the limit $q = 0$. This would seem to be a good approximation for the long-range Coulomb interaction, but it turns out not to be so straightforward (Takimoto, 1959). Basically, this is because screening makes the interaction short-range. In general, the screening wavevector, $q_s(0)$, is replaced by

$$q_s(q) = q_s(0)\Gamma(q) \tag{10.116}$$

where $\Gamma(q)$ is a function of the wavevector obtained, for example, from the Lindhard expression. Thus, for a degenerate 3D gas at zero temperature, Eq. (10.43)

$$\Gamma(q) = \frac{1}{2}\left[1 + \frac{1}{2\eta}(1 - \eta^2)\ln\left|\frac{1+\eta}{1-\eta}\right|\right] \qquad \eta = \frac{q}{2k_F} \tag{10.117}$$

The Thomas–Fermi result is obtained in the limit $\eta = 0$, when $\Gamma(q) = 1$. For intense screening the interaction occurs only with the nearest ion and the electron therefore tends to be simply reflected back the way it came. In this case, $\eta = 1$, when $\Gamma(q) = 1/2$. In the equivalent case for a 2D gas, Eq. (10.65a), it turns out that $\Gamma(q) = 1$ for all q, and so the long-wavelength approximation is quite valid. However, in quasi-2D systems it must be remembered that although $\Gamma(q) = 1$ may be a reasonable approximation, the screening wavevector contains the q-dependent form factor $F(q)$.

The dynamic screening of the phonon interactions can be rather complicated. Its efficacy depends on the relation between the plasmon frequency ω_p, and the frequency of the phonon; screening is effective when $\omega << \omega_p$, and ineffective when $\omega >> \omega_p$. As ω approaches ω_p from below, screening weakens and eventually changes sign near $\omega \approx \omega_p$. This antiscreening effect maximizes when the screening electrons oscillate 180° out of phase and the effect eventually disappears with increasing co, being replaced by a trend toward static screening. In 3D the plasmon frequency is usually much higher than the relevant acoustic-phonon frequencies, and so a static-screening approximation is usually valid. In 2D things are different. The plasmon frequency is a function of q, Eq. (10.66), and the acoustic-phonon frequency can be comparable or even exceed ω_p (Fischetti and Laux, 1993). If the latter, the response tends to be unscreened. It is particularly important for piezoelectric scattering to establish a correct description of dynamic screening in 2D in view of the divergent nature of this interaction. Usually it is assumed that static screening is adequate, but normally without justification. In general, we may take it that using the static-screening approximation overestimates the effect of screening.

The static-screening approximation can be valid for optical modes only within the Landau-damping regime. In many practical cases it is necessary to consider seriously the coupling between plasmons and optical phonons, as discussed in the previous section, and this has led to a substantial literature concerning collective excitations (e.g., Ando *et al.*, 1982; Lee and Spector, 1983; Tselis and Quinn, 1984; Das Sarma and Mason, 1985; Lei, 1985; Bechstedt and Enderlein, 1985; Wendler and Pechstedt, 1986, 1987; and Das Sarma *et al.*, 1988). Dyson and Ridley (2008) have shown that the coupling reduces the phonon lifetime.

11
The Electron Distribution Function

Let thy tongue tang arguments of state;
put thyself into the trick of singularity.
Twelfth Night, *W. Shakespeare*

11.1 The Boltzmann Equation

Contact with experiment usually can be made only by considering the behaviour of large numbers of carriers and the effect of scattering on the distribution of particles in the available states. Here we assume that the states remain well-defined in terms of energy, wavevector and wavefunction so that the evolution of the distribution function in the presence of external forces can be described by the Boltzmann equation

$$\frac{\partial f(\mathbf{K})}{\partial \mathbf{t}} = \left(\frac{\partial f(\mathbf{K})}{\partial \mathbf{t}}\right)_{\text{vol}} - \nabla.(\mathrm{v}(\mathbf{K})f(\mathbf{K})) \tag{11.1}$$

where $f(\mathbf{K})$ is the probability that the state characterized by the wavevector $\mathbf{K}$ is occupied and $\mathrm{v}(\mathbf{K})$ is the group velocity. The volume rate is the sum of individual rates associated with applied fields, scattering, generation and recombination viz.:

$$\left(\frac{\partial f(\mathbf{K})}{\partial t}\right)_{\text{vol}} = \left(\frac{\partial f(\mathbf{K})}{\partial t}\right)_{\text{fields}} + \left(\frac{\partial f(\mathbf{K})}{\partial t}\right)_{\text{scat}} + \left(\frac{\partial f(\mathbf{K})}{\partial t}\right)_{\text{gen}} + \left(\frac{\partial f(\mathbf{K})}{\partial t}\right)_{\text{recomb}} \tag{11.2}$$

Solutions of the Boltzmann equation are usually obtained for particular situations numerically via the use of Monte Carlo techniques. Analytic solutions, inevitably approximate, are useful for providing general insights to the physics that underlie experimental situations, and we will focus on these in what follows.

A comprehensive treatment is beyond the scope of this book. In the first part of this chapter we will concentrate on solutions relating to two experimental situations, one involving optical excitation, the other involving an electric field. In both cases it will be assumed that polar optical-mode scattering is dominant and that the distribution is spatially homogeneous. For simplicity, we assume that the optical modes are all at the zone-centre frequency, and we will ignore screening.

11.2 Net Scattering Rate by Bulk Polar-Optical Phonons

Previously we have been dealing with the rate at which an electron is scattered by absorbing or, more likely in most situations, emitting an optical phonon. Here we must take into account the rates associated with the reverse processes, whereby electrons scatter from other states into the state of interest in order to get a net rate. In this way the distribution functions of the states in question become interlinked. Thus

$$\left(\frac{\partial f(\mathbf{K})}{\partial t}\right)_{scat} = \frac{\Omega}{(2\pi)^3}\mathbf{x}$$

$$\left[\sum_{\mathbf{Q},\mathbf{K}'} W(\mathbf{K}',\mathbf{K},\mathbf{Q})\{f(K')(1-f(\mathbf{K}))(n(\omega)+1)-f(\mathbf{K})(1-f(\mathbf{K}'))n(\omega)\}\right.$$
$$\delta(E(\mathbf{K}')-E(\mathbf{K})-\hbar\omega)$$
$$+\sum_{\mathbf{Q},\mathbf{K}'} W(\mathbf{K}'',\mathbf{K},\mathbf{Q})\{f(K'')(1-f(\mathbf{K}))(n(\omega)-f(\mathbf{K})(1-f(\mathbf{K}''))(n(\omega)+1)\}$$
$$\left.\delta(E(\mathbf{K}'')-E(\mathbf{K})+\hbar\omega)\right] \tag{11.3}$$

where $\mathbf{Q}$ is the phonon wavevector, $n(\omega)$ is the Bose–Einstein number, $\hbar\omega$ is the energy of the optical phonon (assumed independent of $\mathbf{Q}$) and Ω is the normalization volume.

For polar scattering (unscreened),

$$W(\mathbf{K}',\mathbf{K},\mathbf{Q}) = \frac{\pi e^2\omega}{\Omega Q^2}\left(\frac{1}{\varepsilon_\infty}-\frac{1}{\varepsilon_s}\right)|G(\mathbf{K}',\mathbf{K},\mathbf{Q})|^2 \tag{11.4}$$

where

$$G(\mathbf{K}',\mathbf{K},\mathbf{Q}) = \int \psi(\mathbf{K}',\mathbf{R})\phi(\mathbf{Q},\mathbf{R})\psi(\mathbf{K},\mathbf{R})d\mathbf{R} \tag{11.5}$$

and $\psi(\mathbf{K}', \mathbf{R})$ and $\psi(\mathbf{K}, \mathbf{R})$ are the final-state and initial-state wavefunctions and $\phi(\mathbf{Q}, \mathbf{R})$ is the phonon envelope function. We assume that the cell overlap integral is unity.

These equations are true for all dimensions, and for confined, interface and bulk modes. Dimensionality affects the integration over the final states and it affects the overlap integral, Eq. (11.5). The overlap integral determines the selection rule concerning crystal momentum and this becomes as follows:

$$G(\mathbf{K}', \mathbf{K}, \mathbf{Q}) = \begin{cases} \delta_{\mathbf{K}',\mathbf{K}+\mathbf{Q}} & & \text{3D} \\ \delta_{\mathbf{k}',\mathbf{k}+\mathbf{q}} & G(k_z', k_z, q_z) & \text{2D} \\ \delta_{k_z',k_z+q_z} & G(\mathbf{k}', \mathbf{k}, \mathbf{q}) & \text{1D} \end{cases} \tag{11.6}$$

where the Gs are form factors and $\mathbf{K} = (\mathbf{k}, v_z)$ for 2D and $(k_z, v_x\ v_y)$ for 1D, and $\mathbf{Q} = (\mathbf{q}, q_z)$. We then have

$$\sum_{\mathbf{Q},\mathbf{K}'} \Im(\mathbf{K}', \mathbf{K}, \mathbf{Q}) = \begin{cases} \int \Im(\mathbf{K}+\mathbf{Q}, \mathbf{K}, \mathbf{Q}) \frac{\Omega}{(2\pi)^3} Q^2 dQ d(-\cos\theta_Q) d\phi & \text{3D} \\ \int \Im(\mathbf{k}+\mathbf{q}, \mathbf{k}, \mathbf{q}) \sum_{v} \frac{\sigma}{(2\pi)^2} q dq d\theta_q \frac{a}{2\pi} dq_z & \text{2D} \\ \int \Im(\mathbf{k}+\mathbf{q}_z, \mathbf{k}_z, \mathbf{q}_z) \sum_{v_x, v_y} \frac{a}{2\pi} dq_z \frac{\sigma}{(2\pi)^2} q dq d\theta & \text{1D} \end{cases} \tag{11.7}$$

where σ is the normalization area in 2D and wire area in 1D, a is the normalization length in 1D and well-width in 2D, θ is the angle between phonon and electron wavevectors, ϕ is an azimuthal angle and v is a subband index for the final states.

Integration over phonon wavevectors directed along the direction(s) of electron confinement can be carried out straightaway for bulk modes to give, in 2D

$$\frac{a}{2\pi} \int \frac{|G(k_z', k_z, q_z)|^2}{q^2 + q_z^2} dq_z = \frac{a}{2q} F_{v,v'}(q)$$
$$F_{v,v'}(q) = \iint \psi_{v'}(z')\psi_v(z')\psi_{v'}(z)\psi_v(z) e^{-q|z-z'|} dz \tag{11.8}$$

and in 1D

$$\frac{\sigma}{(2\pi)^2} \int \frac{|G(k', k, q)|^2}{q^2 + q_z^2} q dq d\theta = \frac{\sigma}{2\pi} F_{v,v'}(q_z)$$
$$F_{v,v'}(q_z) = \iint \psi_{v'}(\mathbf{r}')\psi_v(\mathbf{r}')\psi_{v'}(\mathbf{r})\psi_v(\mathbf{r}) K_0(q_z|\mathbf{r} - \mathbf{r}'|) dr' d\mathbf{r} \tag{11.9}$$

where $K_0(x)$ is the modified Bessel function of zero order. The form factors $F_{vv'}(q)$ are familiar from previous calculations; cf. Eqs. (1.35), (10.51), (10.75).

The net rate can be written

$$\left(\frac{\partial f(\mathbf{K})}{\partial t}\right)_{scat} =$$

$$\sum_{v'} W_0 \left(\frac{\hbar^3\omega}{2m^*}\right)^{1/2} \mathbf{x}\left[\int (\{f(\mathbf{K}')(1-f(\mathbf{K}))(n(\omega)+1) - f(\mathbf{K})(1-f(\mathbf{K}'))n(\omega)\}\right))$$

$$\delta(E(\mathbf{K}') - E(\mathbf{K}) - \hbar\omega)d\boldsymbol{\kappa}'$$

$$+ \int (\{f(\mathbf{K}'')(1-f(\mathbf{K}))n(\omega) - f(\mathbf{K})(1-f(\mathbf{K}''))(n(\omega)+1)\}$$

$$\delta(E(\mathbf{K}'') - E(\mathbf{K}) + \hbar\omega)d\boldsymbol{\kappa}'')] \tag{11.10}$$

where

$$d\boldsymbol{\kappa} = \begin{cases} dQd(-\cos\theta_Q)\frac{d\phi}{2\pi}, & \mathbf{K}' = \mathbf{K} + \mathbf{Q} \quad \text{3D} \\ F_{v,v'}(q)dq\frac{d\theta_q}{2}, & \mathbf{K}' = (\mathbf{k} + \mathbf{q}, v_z) \quad \text{2D} \\ F_{v,v'}(q_z)dq_z, & \mathbf{K}' = (k_z + q_z, v_x, v_y) \quad \text{1D} \end{cases}$$

$$W_0 = \frac{e^2}{4\pi\hbar}\left(\frac{1}{\varepsilon_\infty} - \frac{1}{\varepsilon_s}\right)\left(\frac{2m^*\omega}{\hbar}\right)^{1/2} \tag{11.11}$$

For confined and interface modes the form factors will differ from Eqs. (11.8) and (11.9). In this case we have to use the overlap integrals of Section 1.4 or, for hybrid modes, those of Chapter 8 and settle for numerical integration.

11.3 Optical Excitation

We consider the case of weak monochromatic excitation of electrons across the direct forbidden gap of a polar semiconductor. Conservation of energy and momentum picks out a set of states in the conduction band that receives electrons. In the simplest case of isotropic band structure, to be assumed here, all of these states have the same energy E_e. An electron excited to one of these states either recombines or scatters by absorbing or emitting an optical phonon. (We assume that carrier–carrier scattering is negligible.) Usually the recombination rate is orders of magnitude smaller than the scattering rate and can be neglected. Under these circumstances the steady-state distribution function in the absence of fields is obtained from

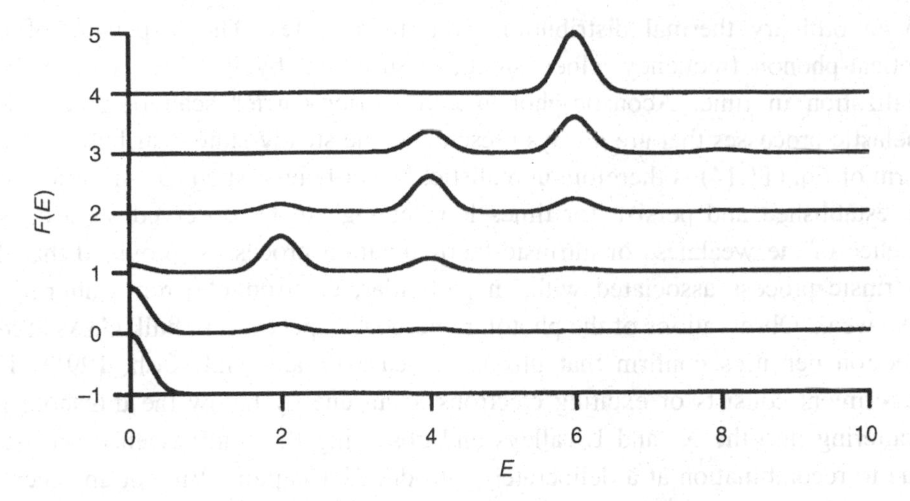

Fig. 11.1 Phonon cascade (schematic). Time increases downward.

$$\left(\frac{\partial f(\mathbf{K})}{\partial t}\right)_{\text{scat}} + \left(\frac{\partial f(\mathbf{K})}{\partial t}\right)_{\text{gen}} = 0 \tag{11.12}$$

and we can assume that

$$\left(\frac{\partial f(\mathbf{K})}{\partial t}\right)_{\text{gen}} = g\delta_{E,E_e}(1 - f(\mathbf{K})) \tag{11.13}$$

Regarding the conduction band to be empty to begin with, we obtain in the non-degenerate limit a distribution function of the general form

$$f(\mathbf{K}) = \mathrm{f}(E)\delta_{E,E\pm j\hbar\omega} \tag{11.14}$$

where j is an integer. States are populated by the optical excitation only if they can be reached by the emission or absorption of one or more optical phonons (Fig. 11.1). Moreover, in the non-degenerate case, it is clear from the form of the net scattering rate that

$$f(E + \hbar\omega) = f(E)e^{-\hbar\omega/kT} \tag{11.15}$$

which means that the occupied states have occupation probabilities related by Maxwell–Boltzmann statistics. Absolute values can be obtained by computing the recombination rate using this distribution and equating to the generation rate.

The model we have used is very simple; there are, of course, many processes that, in time, will broaden each delta-function-like peak and will eventually lead to an ordinary thermal distribution over all energies. The dispersion of the optical-phonon frequency, albeit small, is sufficient by itself to achieve thermalization in time. Acoustic-phonon and carrier–carrier scattering are other inelastic processes that are always present. A true steady-state distribution of the form of Eq. (11.14) is therefore unrealistic. Nevertheless, such a distribution will be established and persist, for times long enough to be observed, as a consequence of the weakness of intrinsic thermalization processes, provided that the extrinsic process associated with, in particular, electron–electron scattering is also weak. Observations of the photoluminescence spectrum in bulk GaAs at low electron densities confirm that this is the case (Tsang and Kash, 1991). The experiment consists of exciting electrons to an energy below the threshold for scattering into the X- and L-valleys and observing the luminescence spectrum due to recombination at a deliberately introduced acceptor. Without an acceptor level, luminescence could not occur at energies above the band edge since the wavevectors of cascading electrons and holes would not coincide, as they must for momentum conservation, given the smallness of the momentum of the photon. The peaks are shown in Fig. 11.2.

The width of the lines is determined by a number of factors, among them the rate of emission of an LO phonon. A significant fraction is due to band-structure effects and when these are taken into account a scattering time of 130 fs was deduced by Hackenberg and Fasol (1989). More directly, the scattering time can be deduced by using time-resolved Raman scattering to observe the LO phonons emitted during a cascade and this procedure allowed Tsang and Kash (1991) to deduce a time of 190 fs, which is close to the theoretical value of 200 fs. A third approach, by Mirlin *et al.* (1980), exploited the observation that under the excitation by linear polarized light the hot-electron luminescence was partially polarized, and that this polarization could be reduced in a magnetic field. The polarization is associated with the alignment of bound hole wavefunctions along the <111> directions and the effect of the magnetic field is to force the electrons to precess at the cyclotron frequency, $\omega_c = eB/m^+$. The reduction in polarization is then dependent on $\omega_c \tau$, from which τ was deduced to be 100 fs.

These experiments in bulk GaAs were carried out with electron densities typically below 3×10^{15} cm^{-3}. Increasing the carrier density to 2×10^{16} cm^{-3} reduces the peakiness of the spectrum and Tsang and Kash (1986) have estimated that phonon and carrier scattering rates become equal at a density of 8×10^{16} cm^{-3}, roughly in line with theoretical expectation.

The phenomenon of phonon cascade applies to all dimensions but it has been extensively studied only in bulk material. In virtually all experiments on 2D and 1D structures the electron densities have tended to be too high. A low 2D density of 2×10^{10} cm^{-2} in a 50 Å well corresponds to a bulk density of 4×10^{16} cm^{-3},

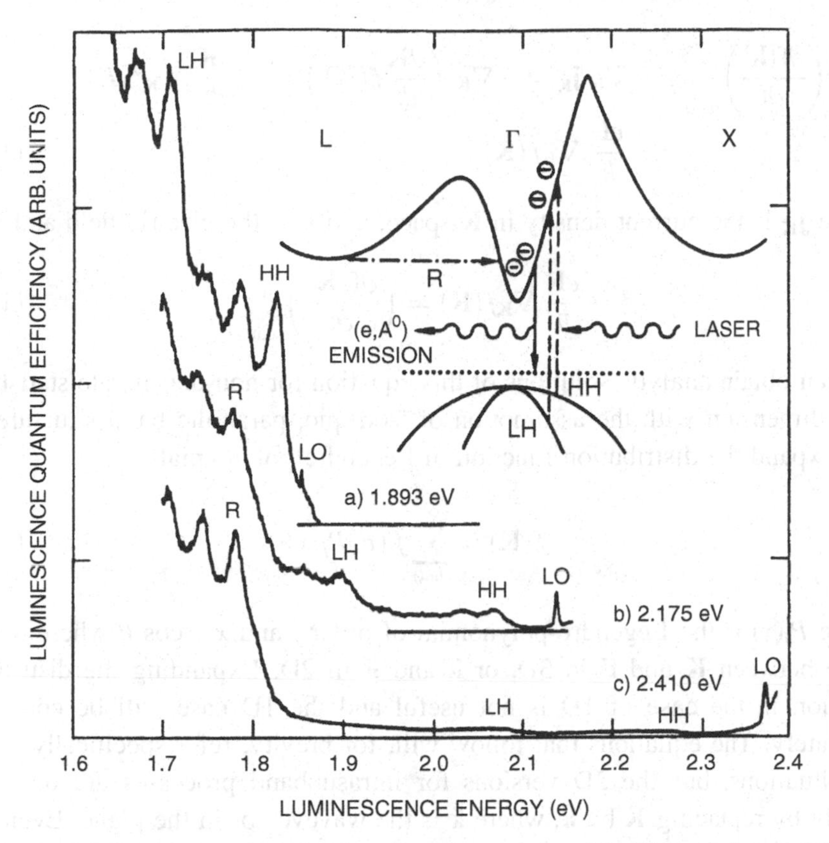

Fig. 11.2 High-energy photoluminescence spectrum at 2 K of GaAs doped with the neutral acceptor Mg and excited at three different phonon energies (Tsang and Kash, 1991).

which is already too high. Most experiments therefore measure average energy-relaxation times, which at lowest densities are around 100 fs as expected. At larger densities the energy-relaxation time rises above 1 ps as a consequence of the production of hot phonons, an effect first observed by Shah *et al.* (1983). In this regime of high carrier density, the electron distribution function is usually taken to be Maxwell–Boltzmann with an electron temperature determined by the details of energy balance.

11.4 Transport

We turn now to the effects produced by an electric field, limiting our attention once again to the steady state and spatial homogeneity. We retain, therefore, only the field and scattering volume rates. The field term can be expressed as follows:

$$\left(\frac{\partial \mathrm{f}(\mathbf{k})}{\partial \mathrm{t}}\right)_{field} = -\nabla_{\mathbf{K}}.\mathbf{j}_{\mathbf{K}} = -\nabla_{\mathbf{K}}.\left(\frac{d\mathbf{K}}{dt}f(\mathbf{K})\right) = -\frac{d\mathbf{K}}{dt}.\nabla_{\mathbf{K}}f(K)$$
$$= -\frac{e\mathbf{F}}{\hbar}.\nabla_{\mathbf{K}}f(\mathbf{K}) \tag{11.16}$$

where $\mathbf{j_K}$ is the current density in $\mathbf{K}$-space, and $\mathbf{F}$ is the electric field and hence

$$\frac{e\mathbf{F}}{\hbar}.\nabla_{\mathbf{K}}f(\mathbf{K}) = \left(\frac{\partial \mathrm{f}(\mathbf{K})}{\partial \mathrm{t}}\right)_{\mathrm{scat}} \tag{11.17}$$

We can obtain analytic solutions of this equation for non-degenerate statistics in each dimension with the assumption of isotropic, parabolic band structure. We first expand the distribution function in Legendre polynomials

$$f(\mathbf{K}) = \sum_{j=0}^{\infty} f_j(E)P_j(x) \tag{11.18}$$

where $P_j(x)$ is the Legendre polynomial of order j and $x = \cos\theta$ where θ is the angle between $\mathbf{K}$ and $\mathbf{F}$ in 3D, or $\mathbf{k}$ and $\mathbf{F}$ in 2D. Expanding the distribution function in the case of 1D is not useful and the 1D case will be considered separately. The equations that follow will, for brevity, refer specifically only to 3D situations, but the 2D versions for intrasubband processes are obtainable simply by replacing $\mathbf{K}$ by $\mathbf{k}$, where $\mathbf{k}$ is the wavevector in the plane. Eventually the difference in geometry enters and this will be indicated where it occurs. The components of $\nabla_{\mathbf{K}}f(\mathbf{K})$ are

$$\nabla_{\mathbf{K}}f(\mathbf{k}) = \left(\frac{\partial f(K)}{\partial K}, \frac{1}{K}\frac{\partial f(K)}{\partial \theta}\right) = \left(\hbar v(E)\frac{\partial f(K)}{\partial E}, \frac{1}{K}\frac{\partial f(K)}{\partial \theta}\right) \tag{11.19}$$

Thus

$$\frac{eF}{\hbar}.\nabla_{\mathbf{K}}f(K) = eFv(E).\sum_j P_j(x)\frac{\partial f_j(E)}{\partial E} + \frac{eF}{m^* v(E)}(1-x^2)\sum_j f_j(E)\frac{dP_j(x)}{dx} \tag{11.20}$$

where $v(E)$ is the magnitude of the velocity. Using the relations

$$\begin{aligned}(1-x^2)\frac{dP_j(x)}{dx} &= jP_{j-1}(x) - jxP_j(x)\\ xP_j(x) &= \frac{j+1}{2j+1}P_{j+1}(x) + \frac{j}{2j+1}P_{j-1}(x)\end{aligned} \tag{11.21}$$

we obtain

$$\frac{e\mathbf{F}}{\hbar}.\nabla_K f(K) = eF\sum_j \left[\left(\frac{j+1}{2j+1}\frac{f_j(E)}{m^*v(E)} + \frac{v(E)}{2j+1}\frac{\partial f_j(E)}{\partial E}\right)jP_{j-1}(x) + \left(v(E)\frac{\partial f_j(E)}{\partial E} - \frac{j}{m^*v(E)}f_j(E)\right)\frac{j+1}{2j+1}P_{j+1}(x)\right] \tag{11.22}$$

Retaining only the zero-order symmetric component of the distribution function, $f_0(E)$, and the first-order antisymmetric component, $f_1(E)$, we obtain

$$\frac{e\mathbf{F}}{\hbar}.\nabla_K f(K) = eF\left[\frac{1}{3}\left(\frac{2f_1(E)}{m^*v(E)}(P_0(x) - P_2(x)) + v(E)\frac{\partial f_1(E)}{\partial E}(P_0(x) + 2P_2(x))\right) + v(E)\frac{\partial f_0(E)}{\partial E}P_1(x)\right] \tag{11.23}$$

The scattering rate can likewise be expanded in Legendre polynomials, retaining only terms in $f_0(E)$ and $f_1(E)$. The second-order Legendre function, $P_2(x)$, can be averaged over all directions to convert it into a spherically symmetric form; then equating symmetric and antisymmetric coefficients we obtain, neglecting products such as $f_0(E)f_0(E')$ and assuming non-degeneracy

$$\frac{eF}{3}\left[\frac{2f_1(E)}{m^*v(E)}(1 - \langle P_2(x)\rangle) + v(E)\frac{\partial f_1(E)}{\partial E}(1 + 2\langle P_2(x)\rangle)\right] = W_0\left(\frac{\hbar^3\omega}{2m^*}\right)^{1/2}\left[\int [f_0(E')(n(\omega)+1) - f_0(E)n(\omega)]\delta(E' - E - \hbar\omega)d\mathbf{\kappa}' + \int [f_0(E'')(n(\omega) - f_0(n(\omega)+1)]\delta(E'' - E + \hbar\omega)d\mathbf{\kappa}''\right] \tag{11.24}$$

$$eFv(E)\frac{\partial f_0(E)}{\partial E} = W_0\left(\frac{\hbar^3\omega}{2m^*}\right)^{1/2}\left[\int [f_1(E')\frac{x'}{x}(n(\omega)+1) - f_1(E)n(\omega)]\delta(E' - E - \hbar\omega)d\mathbf{\kappa}' + \int [f_1(E'')\tfrac{x'}{x}n(\omega) - f_1(E)(n(\omega)+1)]\delta(E'' - E + \hbar\omega)d\mathbf{\kappa}''\right] \tag{11.25}$$

Here $x' = \cos\theta'$ where θ' is the angle between $\mathbf{K}'$ and $\mathbf{F}$, and similarly x''. Now, in general

$$\cos\theta' = \cos\alpha' \cos\theta + \sin\alpha' \sin\theta \sin\phi \tag{11.26}$$

where ϕ is the azimuthal angle in 3D that is to be integrated over, and α' is the angle between $\mathbf{K}'$ and $\mathbf{K}$. Nothing in the integrand depends upon ϕ except the second term in Eq. (11.26). Integration over ϕ between the limits 0 and 2π gives zero; thus, effectively, $x'/x = \cos\alpha'$ and $x''/x = \cos\alpha''$. In the 2D case ϕ is constant ($\pi/2$), but ultimately there is an integration over α' and α'' that will eliminate the sine product, so effectively the same result applies. It is also true for 1D.

Both energy and momentum are conserved, therefore, for a parabolic band

$$E' - E - \hbar\omega = \frac{\hbar^2 K'^2}{2m^*} - \frac{\hbar^2 K^2}{2m^*} - \hbar\omega = \frac{\hbar^2 Q^2}{2m^*} + \frac{\hbar^2 KQ}{m^*}\cos\theta_Q - \hbar\omega = 0 \tag{11.27a}$$

$$E'' - E + \hbar\omega = \frac{\hbar^2 K^2}{2m^*} - \frac{\hbar^2 K''^2}{2m^*} + \hbar\omega = \frac{\hbar^2 Q^2}{2m^*} - \frac{\hbar^2 KQ}{m^*}\cos\theta_Q + \hbar\omega = 0 \tag{11.27b}$$

(Q becomes q, the in-plane wavevector, in 2D, and q_z, the axial wavevector, in 1D.) In Eq. (11.27a) $Q_1 \leq Q \leq Q_2$, and in Eq. (11.27b) $Q_3 \leq Q \leq Q_4$, where

$$\begin{aligned} Q_1 &= K\left[\left(1+\left\{\frac{\hbar\omega}{E}\right\}\right)^{1/2} - 1\right], \quad Q_3 = K\left[1 - \left(1-\left\{\frac{\hbar\omega}{E}\right\}\right)^{1/2}\right] \\ Q_2 &= K\left[\left(1+\left\{\frac{\hbar\omega}{E}\right\}\right)^{1/2} + 1\right], \quad Q_4 = K\left[1 + \left(1-\left\{\frac{\hbar\omega}{E}\right\}\right)^{1/2}\right] \end{aligned} \tag{11.28}$$

and it is understood that the integral involving Q_3 and Q vanishes if $E < \hbar\omega >$. (In 3D, E is the total energy; in 2D it is the kinetic energy of motion in the plane; and in 1D it is the kinetic energy of motion along the wire.) In all dimensions

$$\begin{aligned} \frac{x'}{x} &= \cos\alpha' = \frac{1 - (Q^2/2K^2) + (\hbar\omega/2E)}{(1 + \{\hbar\omega/E\})^{1/2}} \\ \frac{x''}{x} &= \cos\alpha'' = \frac{1 - (Q^2/2K^2) - (\hbar\omega/2E)}{(1 - \{\hbar\omega/E\})^{1/2}} \end{aligned} \tag{11.29}$$

We are now ready to perform the integrations in Eqs. (11.24) and (11.25), and we must now distinguish 3D from 2D from 1D.

11.4.1 The 3D Case

The solution in 3D is well-known (Conwell, 1967; Nag, 1972, 1990) so we need not dwell on the details of the integration. In this case $<P_2(x')> = 0$, and the result is

$$\frac{eF}{3}\left[\frac{2f_1(E)}{m^*v(E)}+v(E)\frac{\partial f_1(E)}{\partial E}\right]=$$
$$W_0\left(\frac{\hbar\omega}{E}\right)^{1/2}\Big[\{f_0(E+\hbar\omega)(n(\omega)+1)-f_0(E)n(\omega)\}\sinh^{-1}(E/\hbar\omega)^{1/2}$$
$$+\{f_0(E-\hbar\omega)n(\omega)-f_0(E)(n(\omega)+1)\}\sinh^{-1}\{(E/\hbar\omega)-1\}^{1/2}\Big] \tag{11.30}$$

$$eFv(E)\frac{\partial f_0(E)}{\partial E}=$$
$$W_0\left(\frac{\hbar\omega}{E}\right)^{1/2}\left[\left\{f_1(E+\hbar\omega)\frac{1+(\hbar\omega/2E)}{\{1+(\hbar\omega/E)\}^{1/2}}(n(\omega)+1)-f_1(E)n(\omega)\right\}\sinh^{-1}\left(\frac{E}{\hbar\omega}\right)^{1/2}\right.$$
$$+\left\{f_1(E-\hbar\omega)\frac{1-(\hbar\omega/2E)}{\{1-(\hbar\omega/E)\}^{1/2}}n(\omega)-f_1(E)(n(\omega)+1)\right\}\sinh^{-1}\left\{\left(\frac{E}{\hbar\omega}\right)-1\right\}^{1/2}$$
$$\left.-\frac{1}{2}\{f_1(E+\hbar\omega)(n(\omega)+1)+(f_1(E-\hbar\omega)n(\omega)\}\right] \tag{11.31}$$

These coupled equations imply an infinite set relating the symmetric and antisymmetric components of the distribution function at energies $\xi \pm < j\hbar\,\omega$, where j is an integer and ξ is defined in the interval $0<\xi<\hbar\omega$. It is interesting to note that Eq. (11.30) can be rewritten as follows:

$$\frac{\partial[(\xi+j\hbar\omega)f_1(\xi+j\hbar\omega)]}{\partial\xi}=$$
$$\frac{3W_0}{eF}\left(\frac{\hbar\omega m^*}{2}\right)^{1/2}\Big[\{f_0(\xi+(j+1)\hbar\omega)(n(\omega)+1)-f_0(\xi+j\hbar\omega)n(\omega)\}$$
$$\sinh^{-1}\{(\xi/\hbar\omega)+j\}^{1/2}$$
$$+\{f_0(\xi-(j-1)\hbar\omega)n(\omega)-f_0(\xi+j\hbar\omega)(n(\omega)+1)\}$$
$$\sinh^{-1}\{(\xi/\hbar\omega)+j-1\}^{1/2}\Big] \tag{11.32}$$

If we add the equivalent equation for $j+1$ and then for $j-1$, and continue summing for all $j\geq 0$, we discover a sum-rule

$$\sum_{j=0}^{\infty}\frac{\partial[(\xi+j\hbar\omega)f_1(\xi+j\hbar\omega)]}{\partial\xi}=0 \tag{11.33}$$

and therefore

$$\sum_{j=0}^{\infty}[(\xi+j\hbar\omega)f_1(\xi+j\hbar\omega)] = g_1 \tag{11.34}$$

where g_1 is a constant with units of energy. This is directly related to the current density, **J**. Thus, if $N(\xi+j\hbar\omega)$ is the density of states, then

$$J = e\sum_j \int_0^{\hbar\omega} v(\xi+j\hbar\omega)f_1(\xi+j\hbar\omega)N(\xi+j\hbar\omega)d\xi = \frac{2em^*\omega}{\pi^2\hbar^2}g_1 \tag{11.35}$$

Each ladder founded on an energy $\xi < \hbar\omega$ contributes the same amount to the current.

11.4.2 The 2D Case

The integration for 2D is not as straightforward. Since the delta-function conserving energy involves cos θ_q it is convenient to make the substitution

$$\int_0^{2\pi} \ldots d\theta q \to 2\int_{-1}^{+1} \frac{d(\cos\theta_q)}{\sin\theta_q} \tag{11.36}$$

and integrate with respect to cos θ_q. On the left-hand side $\langle P_2(x)\rangle = 1/4$, and we obtain for intrasubband transitions in the first subband

$$\begin{aligned}
&\frac{eF}{2}\left[\frac{f_1(E)}{m^*v(E)}+v(E)\frac{\partial f_1(E)}{\partial E}\right] = \\
&W_0\left(\frac{\hbar}{2m^*\omega}\right)^{1/2}\frac{\hbar\omega}{E}[\{f_0(E+\hbar\omega)(n(\omega)+1)-f_0(E)n(\omega)\}H_a(E) \\
&+\{f_0(E-\hbar\omega)n(\omega)-f_0(E)(n(\omega)+1)\}H_e(E)]
\end{aligned} \tag{11.37a}$$

$$\begin{aligned}
&eFv(E)\frac{\partial f_0(E)}{\partial E} = W_0\left(\frac{\hbar}{2m^*\omega}\right)^{1/2} \\
&\frac{\hbar\omega}{E}\left[\begin{array}{l} f_1(E+\hbar\omega)(n(\omega)+1)H_a(a',E)-f_1(E)n(\omega)H_a(E) \\ +f_1(E-\hbar\omega)n(\omega)H_e(a'',E)-f_1(E)(n(\omega)+1)H_e(E)\end{array}\right]
\end{aligned} \tag{11.37b}$$

where

$$H_a(E) = \int_{q_1}^{q_2} \frac{F_{11}(q)}{(\sin\theta_q)_+} dq, \quad H_a(\alpha', E) = \int_{q_1}^{q_2} \frac{F_{11}(q)\cos\alpha'}{(\sin\theta_q)_+} dq$$
$$H_e(E) = \int_{q_3}^{q_4} \frac{F_{11}(q)}{(\sin\theta_q)_-} dq, \quad H_e(\alpha'', E) = \int_{q_3}^{q_4} \frac{F_{11}(q)\cos\alpha''}{(\sin\theta_q)_-} dq \tag{11.38}$$

Here q_1, q_2, q_3, q_4, cos α' and cos α'' are given by Eqs. (11.28) and (11.29), and

$$(\sin\theta_q)_\pm = \left[-\left(\frac{q}{k}\right)^4 + 2\left(\frac{q}{k}\right)^2 \left(2 \pm \frac{\hbar\omega}{E}\right) - \left(\frac{\hbar\omega}{E}\right)^2 \right]^{1/2} \tag{11.39}$$

We note that

$$q_1(\xi + j\hbar\omega) = q_3(\xi + \{j+1\}\hbar\omega)$$
$$q_1(\xi + j\hbar\omega) = q_4(\xi + \{j+1\}\hbar\omega) \tag{11.40}$$

and

$$(\sin\theta_q)_-^{j+1} = \frac{\xi + j\hbar\omega}{\xi + (j+1)\hbar\omega} (\sin\theta_q)_+^j \tag{11.41}$$

and so

$$H_e(\xi + (j+1)\hbar\omega) = \frac{\xi + (j+1)\hbar}{\xi + j\hbar\omega} H_a(\xi + j\hbar\omega) \tag{11.42}$$

The sum-rule in this case is

$$\sum_{j=0}^{\infty} \frac{\partial}{\partial\xi} [(\xi + j\hbar\omega)^{1/2} f_1(\xi + j\hbar\omega)] = 0 \tag{11.43}$$

hence

$$\sum_{j=0}^{\infty} [(\xi + j\hbar\omega)^{1/2} f_1(\xi + j\hbar\omega)] = g_1 \tag{11.44}$$

and

$$J = \frac{e(2m^*)^{1/2}\omega}{\pi\hbar} g_1 \tag{11.45}$$

11.4.3 The 1D Case

Conservation of energy and momentum restricts the wavevectors of the phonon to the extremal values in Eq. (11.28), which we denote q_1, q_2, q_3 and q_4, and all lie along the axis of the wire. There are thus four corresponding form-factors F_1, F_2, F_3 and F_4. Since $\cos\theta_q$ is now 1 or -1, we obtain $<P_2(x)> = 1$. Therefore

$$eFv(E)\frac{\partial f_1(E)}{\partial E} = W_0(\hbar\omega)^{1/2}\left[\frac{1}{(E+\hbar\omega)^{1/2}}\{f_0(E+\hbar\omega)(n(\omega)+1) - f_0(E)n(\omega)\}(F_1(E)+F_2(E)) + \frac{1}{(E-\hbar\omega)^{1/2}}\{f_0(E-\hbar\omega)n(\omega) - f_0(E)(n(\omega)+1)\}(F_3(E)+F_4(E))\right] \tag{11.46}$$

$$eFv(E)\frac{\partial f_0(E)}{\partial E} = W_0(\hbar\omega)^{1/2}\left[\frac{1}{(E+\hbar\omega)}\{f_0(E+\hbar\omega)(n(\omega)+1)(F_1(E)+F_2(E)) - f_0(E)n(\omega)(F_1(E)+F_2(E))\} + \frac{1}{(E-\hbar\omega)}\{f_0(E-\hbar\omega)n(\omega)(F_3(E)+F_4(E)) - f_0(E)(n(\omega)+1)(F_3(E)+F_4(E))\}\right] \tag{11.47}$$

We note that Eq. (11.40) applies and that therefore

$$\begin{aligned} F_1(\xi+j\hbar\omega) &= F_3(\xi+\{j+1\}\hbar\omega) \\ F_2(\xi+j\hbar\omega) &= F_4(\xi+\{j+1\}\hbar\omega) \end{aligned} \tag{11.48}$$

There is, once more, a sum-rule derived from Eq. (11.46), which is

$$\sum_{j=0}^{\infty}\frac{\partial}{\partial\xi}[f_1(\xi+j\hbar\omega)] = 0 \tag{11.49}$$

hence

$$\sum_{j=0}^{\infty} f_1(\xi + j\hbar\omega) = g_1 \tag{11.50}$$

and

$$J = \frac{2e\omega}{\pi} g_1 \tag{11.51}$$

which agrees with the result of Leburton (1992).

11.4.4 Discussion

We have shown that a simple sum-rule for the antisymmetric component of the distribution function can be derived in each dimension. Its general form is

$$\sum_{j=0}^{\infty} [v(\xi + j\hbar\omega)N(\xi + j\hbar\omega)f_1(\xi + j\hbar\omega)] = g \tag{11.52a}$$

where g is a constant related to the current density. This sum-rule is nothing more than a consequence of the conservation of particle number. If we define $N(\xi + j\hbar\omega)$ as the density of particles in unit energy range, we can write a continuity equation in energy space as follows:

$$\frac{\partial n(\xi + j\hbar\omega)}{\partial t} = \left(\frac{\partial n(\xi + j\hbar\omega)}{\partial t}\right)_{\text{vol}} - \frac{\partial J(\xi + \hbar\omega)}{\partial \xi} \tag{11.52b}$$

where $J\ (\xi + j\hbar\omega)$ is the particle energy current density, which is just the summand of Eq. (11.52a) mutiplied by the force eF. From the condition that the total particle number remain invariant, it is clear that a sum over all energy regions will entail that the divergence of Eq. (11.52b) vanish in the steady state, which is exactly the sum-rule.

This is all very well, but the fact of the matter is that these equations are insoluble as they stand. We are dealing here with an infinite set of coupled difference-differential equations where the difference part is second-order and the differential part is first-order. In the case of a second-order differential equation, a knowledge of value and slope at a point is sufficient to generate the solution for all positions. The equivalent requirement for a second-order difference equation is a knowledge of the value at two points, say at ξ and $\xi + \hbar\omega$. But ξ is a continuous variable lying between 0 and $\hbar\omega$ and so we need to know the distribution function entirely within any two intervals, for example, 0 and 2 $\hbar\omega$, before a full solution can be obtained. There is nothing in the system to suggest what this might be. Even with zero field, the best that can be done is to set the antisymmetric component to zero and obtain the relation

$$\begin{aligned} f_0(\xi + (j+1)\hbar\omega)(n(\omega)+1) &= f_0(\xi + j\hbar\omega)n(\omega) \\ f_0(\xi + (j-1)\hbar\omega)n(\omega) &= f_0(\xi + j\hbar\omega)(n(\omega)+1) \end{aligned} \tag{11.53a}$$

which tells us nothing about the distribution function in any interval $\xi + j\hbar\omega$ to $\xi + (j+1)\ \hbar\omega$. The general solution has the form

$$f(\varepsilon + j\hbar\omega) = \phi(E)e^{j\hbar\omega/k_BT} \tag{11.53a}$$

where $\phi(E)$ is an arbitrary function.

The problem is, basically, that there is no "communication" between states whose energy difference is less than $\hbar\omega$. Thus, the system is non-ergodic. It is therefore simply unrealistic to treat optical-phonon scattering on its own as we have. Remedies include taking into account dispersion in order to introduce a spread of frequencies; to add acoustic-phonon scattering, which restores communication between neighbouring states; or to assume the presence of a robust scattering mechanism such as frequent electron–electron collisions to randomize the distribution and maintain a drifted Maxwellian, or drifted Fermi–Dirac distribution. The latter has been the usual face-saving device. With this we assume that we know what the distribution function is except for its electron temperature and its drift, both of which are obtainable from energy and momentum balance equations. However, this way out of the problem is not available for 1D systems; in these electron–electron scattering is ineffective. When the electron density in 3D or 2D systems is high enough for the gas to be degenerate, the spherical part of the distribution function can be taken to be given by Fermi–Dirac statistics. Approximate solutions for the antisymmetric component can then be obtained using the concept of a momentum-relaxation time defined for electrons on the Fermi surface. An example is given in the appendix to this chapter.

The remarks made previously are, of course, applicable to deformation-potential optical-phonon scattering and intervalley scattering. They are also applicable to the case of confined optical modes. In the equations, confinement can be accommodated by suitably modifying the form factors.

We look at acoustic-phonon scattering next.

11.5 Acoustic-Phonon Scattering

For deformation-potential acoustic-phonon scattering (unscreened), Eq. (11.3) applies with

$$W(K', K, \mathbf{Q}) = \frac{\pi \Xi^2 Q^2}{\rho\omega\Omega} |G(K', K, \mathbf{Q})|^2 \tag{11.54}$$

where Ξ is the deformation potential (assuming a spherical valley and unity for the cell overlap integral), ρ is the mass density and $G(\mathbf{K}', \mathbf{K}, \mathbf{Q})$ is given by Eqs. (11.5) and (11.6). Where degeneracy is zero or weak the net scattering rate is linear in the distribution function (products like $f(\mathbf{K}')f(\mathbf{K})$ being neglected) and so

$$\begin{aligned}\frac{\partial f(K)}{\partial t} &= \\ &\frac{\pi\Xi^2}{\rho v_s\Omega}\sum_{\mathbf{Q},K'} Q\{[f(K')(n(\omega)+1)-f(K)n(\omega)] \\ &\delta(E(K') - E(K) - \hbar\omega)|G(K',K,\mathbf{Q})|^2 \\ &+[f(K)n(\omega)-f(K)(n(\omega)+1)] \\ &\delta(E(K) - E(K)+\hbar\omega)|G(K,K,\mathbf{Q})|^2\}\end{aligned} \tag{11.55}$$

where v_s is the velocity of sound. Momentum in the unconfined directions is conserved, Eq. (11.6). For brevity, we limit attention to intrasubband processes in 2D and 1D. The sum over phonon states can be replaced by an integral in the usual way, and we can write

$$\sum_{\mathbf{Q},\mathbf{k}'} \Im(\mathbf{k}',\mathbf{k},\mathbf{Q}) = \begin{cases} \int\cdots Q^3 dQ d(-\cos\theta_Q)d\phi \; 3\text{D} \\ \int\cdots (q^2+q_z^2)^{1/2} q dq d\theta dq_z \; 2\text{D} \\ \int\cdots (q^2+q_z^2)^{1/2} q dq d\theta dq_z \; 1\text{D} \end{cases} \tag{11.56}$$

where $Q^2 = q^2 + q_z^2$. In 2D q is the in-plane wavevector and q_z is the wavevector parallel to the direction of electron confinement. In 1D q_z is along the wire and q is in the plane of electron confinement.

Acoustic-phonon scattering turns out to be more difficult to handle in low dimensions than optical-phonon scattering because of the dependence of frequency on the wavevector. It is therefore more convenient to examine the three cases separately. In each case we expand the distribution function into symmetric and antisymmetric components as usual. In 3D and 2D it is convenient to exploit the small magnitude of the acoustic-phonon energy and expand $f(E')$ in terms of $f(E)$ via a Taylor expansion, retaining only the first three terms.

11.5.1 The 3D Case

The 3D case is straightforward and is familiar from studies of bulk semiconductors. With $f(\mathbf{K}) = f_0(E) + f_1(E)P_1(\cos\theta)$, we obtain, assuming equipartition and $K > m^* v_s/\hbar$,

$$\frac{\partial f_0(E)}{\partial t} = \frac{\Xi^2 m^*}{4\pi\rho v_s \hbar^2 K}\left[\int_0^{2K+2m^* v_s/\hbar} \{f_0(E) + (k_BT + \hbar\omega)f_0'(E) + \frac{1}{2}k_BT\hbar\omega f_0''(E)\}Q^2 dQ \right.$$
$$\left. + \int_0^{2K-2m^* v_s/\hbar} \{-f_0(E) - k_BTf_0'(E) + \frac{1}{2}k_BT\hbar\omega f_0''(E)\}Q^2 dQ\right] \quad (11.57)$$

$$\frac{\partial f_1(E)}{\partial t} = \frac{\Xi^2 m^*}{4\pi p\, v_s \hbar^2 K}\int_0^{2K} f_1(E)\left(\frac{2k_BT}{\hbar\omega}+1\right)(\cos a - 1)Q^2 dQ$$
$$\cos a = 1 - \frac{Q^2}{2K^2} \quad (11.58)$$

Here, the limiting wavevectors are derived from Eq. (11.28) with $\hbar\omega / E <<1$, and the prime(s) on $f_0(E)$ denotes differentiation with respect to energy. Differentials of $f_1(E)$ have been ignored as being relatively small quantities. Performing the integrations gives

$$\frac{\partial f_0(E)}{\partial t} = \frac{1}{\tau_0(E)}[2f_0(E) + (2k_BT + E)f_0'(E) + k_BTEf_0'']$$
$$= \frac{1}{\tau_0(E)}\frac{k_BT}{E}\frac{d}{dE}\left[E^2\left(f_0'(E) + \frac{f_0(E)}{k_BT}\right)\right] \quad (11.59)$$

$$\frac{\partial f_1(E)}{\partial t} = -\frac{1}{\tau_m(E)}f_1(E) \quad (11.60)$$

The times $\tau_0(E)$ and $\tau_m(E)$ are, respectively, the energy- and momentum-relaxation times, given by

$$\frac{1}{\tau_0(E)} = \frac{\Xi^2(2m^*)^{5/2}E^{1/2}}{2\pi p\, \hbar^4} = \frac{2m^* v_s^2}{k_BT}\frac{1}{\tau_m(E)} \quad (11.61a)$$

$$\frac{1}{\tau_m(E)} = \frac{\Xi^2(2m^*)^{3/2}k_BTE^{1/2}}{2\pi\rho\, v_s^2\, \hbar^4} \quad (11.61b)$$

Note that at thermodynamic equilibrium $f_0(E)$ has the required Maxwell–Boltzmann form to give zero net rate. Hot-electron solutions are discussed in the book by Esther Conwell (1967) and Ridley (1999).

11.5.2 The 2D Case

The confinement of electrons introduces a new element, namely the relaxation of momentum conservation in the direction of confinement. This means that the acoustic-phonon frequency is no longer determined, as it is in 3D, solely by momentum conservation, but instead it is shaped both by momentum conservation in the plane and by the form factor. In principle, all frequencies are allowed to participate, but, in practice, the form factor sets limits on the wavelength of the order of the width of the quantum well. As long as the confinement is not extreme, these limits allow us to continue to regard the interaction of acoustic phonons with confined electrons as quasi-elastic. We will also ignore the effects of acoustic mismatches at the interfaces and regard the acoustic-phonon spectrum to be entirely bulklike.

The equivalent equations for the 2D case, with equipartition and quasi-elastic processes ($\hbar\omega << E$), are

$$\frac{\partial f_0(E)}{\partial t} = \frac{\Xi^2 m^*}{4\pi^2 \rho v_s \hbar^2 k} \mathbf{x}$$
$$\left[\int_{-\infty}^{\infty} \int_0^{2k+2m^*\omega/\hbar q} \{f_0(E) + (k_B T + \hbar\omega) f_0'(E) + \frac{1}{2} k_B T \hbar\omega f_0''(E)\} \right.$$
$$\frac{|G_{11}(q_z)|^2}{(\sin\theta_q)_+} (q^2 + q_z^2)^{1/2} dq dq_z \tag{11.62}$$
$$+ \int_{-\infty}^{\infty} \int_0^{2k-2m^*\omega/\hbar q} \{-f_0(E) - k_B T f_0'(E) + \frac{1}{2} k_B T \hbar\omega f_0''(E)\}$$
$$\left. \frac{|G_{11}(q_z)|^2}{(\sin\theta_q)_- (q^2 + q_z^2)^{1/2} dq dq_z} \right]$$

$$\frac{\partial f_1(E)}{\partial t} = \frac{\Xi^2 m^*}{4\pi^2 \rho v_s \hbar^2 k} \int_{-\infty}^{\infty} \int_0^{2k} f_1(E) \left(\frac{2k_B T}{\hbar\omega} + 1 \right) (\cos a - 1)$$
$$\frac{|G_{11}(q_z)|^2}{(\sin\theta_q)} (q^2 + q_z^2)^{1/2} dq dq_z \tag{11.63}$$

Here

$$G_{11}(q_z) = \int \psi_1^2(z) e^{iq_L z} dz \tag{11.64}$$

and

$$(\sin\theta_q)_\pm = \left[1 - \left(\frac{m^*\omega}{\hbar kq} \mp \frac{q}{2k}\right)^2\right]^{1/2} \tag{11.65a}$$

$$\sin\theta_q = \left[1 - \left(\frac{q}{2k}\right)^2\right]^{1/2} \tag{11.65b}$$

$$\cos a = 1 - \frac{q^2}{2k^2} \tag{11.65c}$$

In general the integrals must be evaluated numerically (e.g., Bockelmann and Bastard, 1990; Vickers, 1992).

Note that the limits on the in-plane wavevector, q, set by in-plane momentum conservation depend on the perpendicular wavevector, q_z, through the frequency. In the quasi-elastic approximation we assume that $m^*\omega/\hbar q << k$ for the vast majority of possible interactions.

We note that if the electron wavefunctions are fully confined within a quantum well of width "a", then

$$\int_{-\infty}^{\infty} |G_{11}(q_z)|^2\, dq_z = \frac{3\pi}{a} \quad \text{and} \quad \int_{-\infty}^{\infty} |G_{11}(q_z)|^2\, d_z^2 dq_z = \frac{4\pi^3}{a^3} \tag{11.66}$$

and Eq. (11.62) reduces to

$$\frac{\partial f_0(E)}{\partial t} = \frac{1}{\tau_0(E)} k_B T \frac{d}{dE}\left[\left(E + \frac{2E_1}{3}\right)\left(f\prime_0(E) + \frac{f_0(E)}{k_B T}\right)\right] \tag{11.67a}$$

where $E_1 = \pi^2\hbar^2/2m^* a^2$ is the energy of the ground state, and

$$\frac{\partial f_1(E)}{\partial t} \approx \frac{-f_1(E)}{\tau_m}, \quad \frac{1}{\tau_m} = \frac{3\Xi^2 m^* k_B T}{2\rho v_s^2 \hbar^3 a} \tag{11.67b}$$

$$\frac{1}{\tau_0(E)} = \frac{2m^* v_s^2}{k_B T}\frac{1}{\tau_m(E)} = \frac{3\Xi^2 m^{*2}}{\rho\hbar^3 a} \tag{11.67c}$$

Cf. Eq. (1.29).

11.5.3 The 1D Case

In the 1D case, instead of expanding f(E′) in a Taylor expansion, we expand the δ-function viz.:

$$\delta(E' - E \pm \hbar\omega) = \delta(E' - E) \mp \hbar\omega\delta'(E' - E) + \frac{1}{2}(\hbar\omega)^2\delta''(E' - E)\ldots \quad (11.68)$$

where the primes denote differentiation with respect to energy. In an integration over the energy of final states we can exploit the formula

$$\int_{-\infty}^{\infty} f(x)\frac{d^n}{dx^n}\delta(x)dx = (-1)^n\frac{d^n}{dx^n}f(x) \quad (11.69)$$

We could have used this approach for the 2D case, as the reader can, no doubt, verify. Replacing dq_z in Eq. (11.56) with dk_z' and using

$$dk_z' = \frac{1}{2}\frac{(2m*)^{1/2}}{\hbar E'^{1/2}}dE' \quad (11.70)$$

we obtain, assuming once more that $n(\omega) \approx k_BT/\hbar\omega$ and $k > m^+v_s/\hbar$,

$$\frac{\partial f_0(E)}{\partial t} = \frac{\Xi^2(2m*)^{1/2}}{16\pi^2\rho v_s\hbar E^{1/2}}\mathbf{x}$$
$$\left[\int\{[f_0(E) + (k_BT + \hbar\omega)f_0'(E)\right.$$
$$+\frac{1}{2}k_BT\hbar\omega f_0''(E)](q^2+q_z^2)^{1/2}(\delta_{q_z2k+2m*v_s/\hbar}) + [-f_0(E) - k_BTf_0'(E)$$
$$\left.+\frac{1}{2}k_BT\hbar\omega f_0''(E)](q^2+q_z^2)^{1/2}(\delta_{q_z,0} + \delta_{q_z2k-2m*v_s/\hbar})\}|G_{11}(q)|^2qdqd\theta\right]$$
$$= \frac{1}{\tau_0(E)}k_BTE\frac{d}{dE}\left[\frac{E+\frac{2E_1}{3}}{E}\left(f_0'(E) + \frac{f_0(E)}{k_BT}\right)\right] \quad (11.71)$$

where $E_1 = \pi^2\hbar^2/m^*a^2$ is the ground-state energy of a wire of square cross-section. Furthermore

$$\frac{\partial f_1(E)}{\partial t} = \frac{\Xi^2(2m*)^{1/2}}{8\pi^2\rho v_s\hbar E^{1/2}}$$
$$\int f_1(E)\left(\frac{2k_BT}{\hbar v_s(q^2+(2k)^2)^{1/2}} + 1\right)|G_{11}(\mathbf{q})|^2(q^2+(2k)^2)^{1/2}qdqd\theta \quad (11.72)$$
$$= -\frac{1}{\tau_m(E)}f_1(E), \quad \frac{1}{\tau_m(E)} = \frac{9}{4}\frac{\Xi^2(2m*)^{1/2}k_BT}{\rho\hbar^2v_s^2a^2}E^{-1/2}$$

The energy-relaxation time is given by

$$\frac{1}{\tau_0(E)} = \frac{2m * v_s^2}{k_B T}\frac{1}{\tau_m(E)} = \frac{9}{4}\frac{\Xi^2(2m*)^{3/2}}{\rho\hbar^2 a^2}E^{-1/2} \tag{11.73}$$

Note that forward scattering ($q_z = 0$) contributes to the symmetric rate but not to the asymmetric rate; i.e. there is energy relaxation but no momentum relaxation. It is also useful to note that the relationship between energy and momentum relaxation rates is independent of dimensionality.

A fuller treatment of acoustic-phonon scattering in quantum wires which takes into account inelastic scattering and non-equipartition can be found in Ridley and Zakhleniuk (1996).

11.5.4 Piezoelectric Scattering

In the case of piezoelectric scattering Eq. (11.54) is replaced by

$$W(\mathbf{K}', \mathbf{K}, \mathbf{Q}) = \frac{e^2 \pi K_{av}^2 c_{av}}{\varepsilon \rho \omega \Omega} |G(\mathbf{K}', \mathbf{K}, \mathbf{Q})|^2 \tag{11.74}$$

where K_{av} is the electromechanical coupling coefficient suitably averaged over direction and mode polarization; c_{av} is the elastic constant, similarly averaged; and ε is the static permittivity. We continue to assume that the cell overlap integral is unity. The *general* expressions for the rates derived remain valid except for the change in coupling parameters, which involves dividing all integrands by Q^2.

Piezoelectric scattering can usually be ignored in all dimensions at room temperatures. At low temperatures it can become stronger than deformation-potential scattering, but only if the carrier density is too small to screen the interaction effectively. The piezoelectric interaction is stronger for small wavevectors and these are the most susceptible to screening. In most experimental situations involving 2D and 1D systems the carrier density is high enough to screen out the piezoelectric polarization, leaving the deformation-potential interaction as the dominant energy-relaxation mechanism.

11.6 Discussion

In general, one must determine the distribution function by including both optical- and acoustic-phonon contributions to the energy relaxation and adding all additional elastic-scattering mechanisms to the optical- and acoustic-phonon contributions to the momentum relaxation. The relative Boltzmann equations

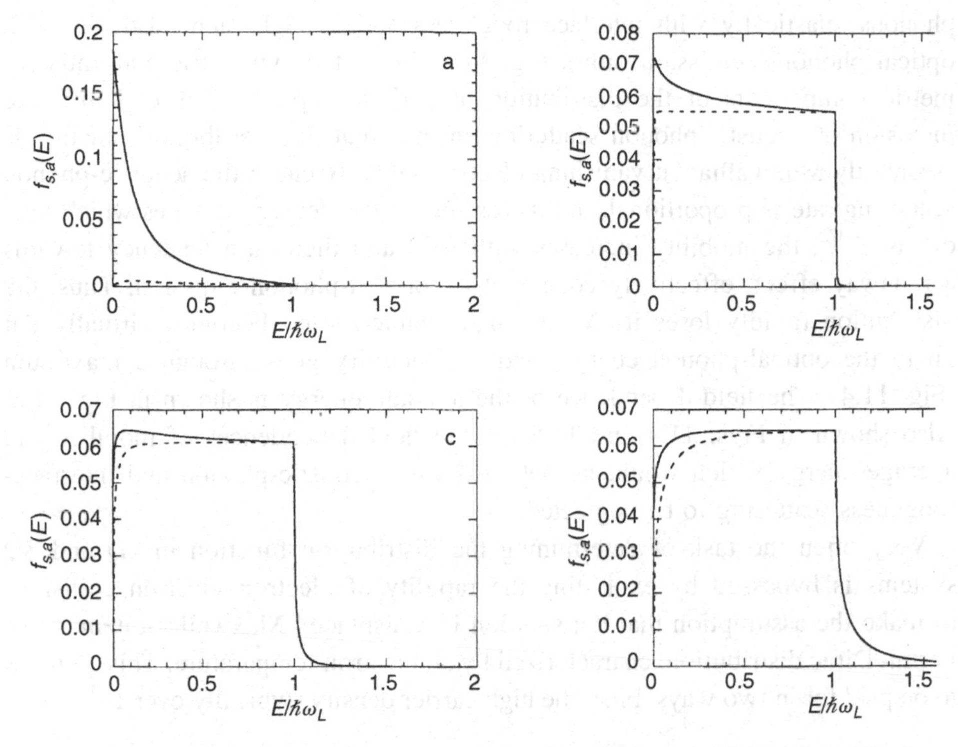

Fig. 11.3 The symmetric ($f_s(E)$, solid line) and antisymmetric ($f_a(E)$, dotted line) distribution function in a GaAs wire of circular cross-section, radius 100 Å, and linear electron density of 10^5 cm^{-1}, at 50K-scattering via interface roughness ($\Lambda = 20$ Å and $\Delta = 3$ Å), acoustic phonons and optical phonons. The applied electric fields are (a) F → 0, (b) F = 200 Vcm^{-1}, (c) F = 500 Vcm^{-1}, (d) F = 1000 Vcm^{-1}.

describing transport in an electric field are therefore

$$\left.\begin{array}{l} 3D \ \dfrac{2eF}{3m*v(E)}\dfrac{\partial(Ef_1(E))}{\partial E} \\ 2D \ \dfrac{eF}{(2m^*)^{1/2}}\dfrac{\partial(E^{1/2}f_1(E))}{\partial E} \\ 1D \ eFv(E)\dfrac{\partial f_1(E)}{\partial E} \end{array}\right\} = \left(\frac{\partial f_0(E)}{\partial t}\right)_{\text{opt}} + \left(\frac{\partial f_0(E)}{\partial t}\right)_{\text{ac}} \tag{11.75}$$

$$\begin{aligned} &3D, 2D, 1D\, eFv(E)\frac{\partial f_0(E)}{\partial E} \\ &= \left(\frac{\partial f_1(E)}{\partial t}\right)_{\text{opt}} + \left(\frac{\partial f_1(E)}{\partial t}\right)_{\text{ac}} + \left(\frac{\partial f_1(E)}{\partial t}\right)_{\text{elastic scat}} \end{aligned} \tag{11.76}$$

An analysis of hot-electron transport in the lowest subband of a cylindrical quantum wire at low lattice temperatures has been carried out by Zakhleniuk *et al.* (1996). The electrons were assumed to interact quasi-elastically with acoustic phonons, elastically with interface roughness and strongly non-elastically with optical phonons (emission only). Fig. 11.3 shows the symmetric and antisymmetric components of the distribution at various applied electric fields. The inclusion of acoustic-phonon scattering ensures that the distribution function is essentially Maxwellian at vanishing electric fields. Because the acoustic-phonon scattering rate is proportional, in this regime, to the density of states which falls off as $E^{-1/2}$, the mobility increases with field and there is a tendency towards a runaway effect, effectively countered by optical-phonon emission. Thus, the distribution rapidly loses its Maxwellian character and becomes virtually flat up to the optical-phonon energy, and the mobility goes through a maximum (Fig. 11.4). The field-dependence of the average energy is shown in Fig. 11.5. Also shown in Figs. 11.4 and 11.5 arc the field dependences of mobility and average energy which would be obtained were acoustic-phonon and interface-roughness scattering to be neglected.

Very often the task of determining the distribution function in 2D and 3D systems is bypassed by exploiting the rapidity of electron–electron collisions to make the assumption that the solution is a displaced Maxwell–Boltzmann or Fermi–Dirac distribution, characterized by an electron temperature. This gain has to be paid for in two ways. First, the high carrier density (typically over 10^{16} cm^{-3})

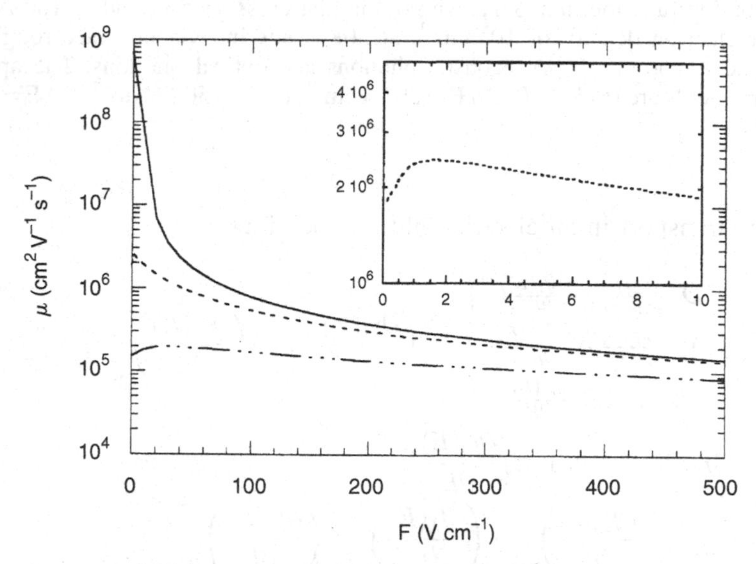

Fig. 11.4 Mobility as a function of field: radius = 50 Å (chained line), = 100 Å (dotted line). In each case the solid line is the mobility when the interaction is only with optical phonons (the high-field limit of the other lines). The inset is the mobility at low fields for a radius of 100 Å.

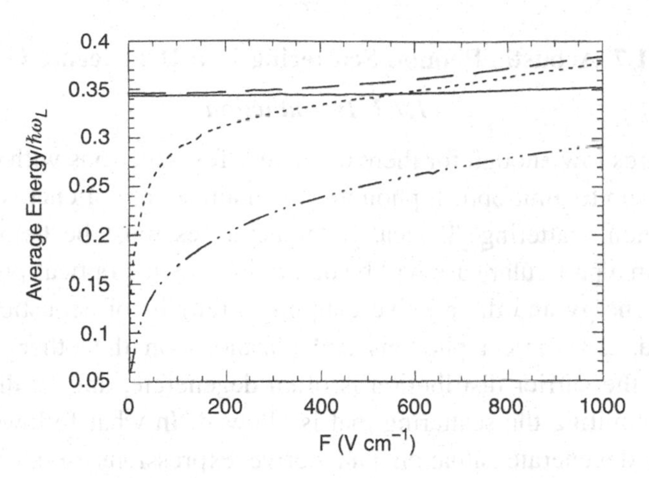

Fig. 11.5 Average energy. (Details as in Figs. 11.3 and 11.4.)

necessary for the assumption to be valid means that the whole question of screening has to be addressed. This can be quite complicated in low-dimensional systems with several subbands occupied, as we saw in Chapter 10. Second, the high density implies a high rate of production of phonons, whose lifetime is, of course, finite, being a few picoseconds for optical phonons at room temperature, somewhat longer for transverse acoustic (TA) modes and significantly longer for longitudinal acoustic (LA) modes. In multilayered structures grown on a bulk substrate there is no problem for acoustic modes, which do not suffer confinement (though zone folding can introduce interesting effects). Optical phonons, on the other hand, *are* confined. As a result their population can easily become hot in the sense that the occupation of those phonon states deeply involved in the scattering process becomes enhanced over its thermodynamic-equilibrium value. The main effect of this is to increase the probability of absorption and hence to weaken the rate of energy relaxation. The phonon gas will also suffer drag and share to some extent the drift of the electrons. Hot-phonon effects are commonly observed in quasi-2D systems (Shah, 1986; Gupta *et al.*, 1992). Theoretical description must add to the Boltzmann equations of Eqs. (11.75) and (11.76) a coupled set describing the drift and relaxation of the optical-phonon population, which in turn requires a description of processes affecting the lifetime and momentum relaxation of optical phonons. Hot phonons are discussed in review articles by Shah (1986), Lyon (1986), Kocevar (1987), Esipov and Levinson (1987) and Ridley (1991). Monte Carlo simulations have been carried out by Lugli *et al.* (1989). Phonon scattering processes were summarized in Section 9.6.

11.7 Acoustic-Phonon Scattering in a Degenerate Gas

11.7.1 Introduction

At temperatures low enough for there to be very few electrons with energies high enough for them to emit optical phonons, the main source of energy relaxation is acoustic-phonon scattering. Typically, temperatures must be below 30 K; the temperature in a particular case will be determined by the optical-phonon energy, the plasmon energy and the relative coupling strengths of acoustic phonons, on the one hand, and optical phonons and plasmons on the other. At such low temperatures the carrier distribution is often degenerate, and Pauli exclusion is important in limiting the scattering that is allowed. In what follows we concentrate on this degenerate situation and derive expressions of momentum- and energy-relaxation rates for polar and non-polar acoustic-phonon scattering, first in the low-temperature regime, where Pauli exclusion leads to rapid variations with temperature, and then in a high-temperature regime, where the system remains degenerate but the effect of exclusion is small. In each case we obtain rates for 3D, 2D and 1D systems. The 1D case is particularly interesting in the low-temperature regime in that the anomalous situation arises of a scattering mechanism that, to zero order, allows energy relaxation but not momentum relaxation – just the opposite of the situation for impurity scattering.

An advantage of working with a degenerate system is that electron–electron interactions in 2D and 3D can be assumed to maintain a distribution characterized by an electron temperature. In 1D this assumption is invalid unless higher order electron–electron interactions are postulated. The assumption means that systems can be studied experimentally by measuring steady-state properties such as mobility and Shubnikov–de Haas oscillations as a function of lattice temperature and electric field. There is also the practical advantage of the low-energy relaxation rate, which is associated with acoustic-phonon scattering, since this allows direct observation of relaxation processes using transient techniques, with voltage pulses that need not be shorter than 100 ps.

11.7.2 Energy- and Momentum-Relaxation Rates

When the system is degenerate we must return to Eq. (11.3), which describes the scattering rate taking into account the occupation of the final states. We can assume that a Fermi–Dirac distribution is always maintained, characterized by an electron temperature T_e, and therefore we may proceed directly to the computation of energy- and momentum-relaxation rates. The power input per unit volume is given by

$$P = \int E \frac{\partial f_0(E)}{\partial t} N(E)dE = \int (E - E_F) \frac{\partial f_0(E)}{\partial t} N(E)dE \tag{11.77}$$

where f_0(E) is the spherically symmetric part of the distribution function, E_F is the Fermi level and $N(E)$ is the density of states. The second equation follows provided the total number of electrons remains constant. We assume that the Fermi energy is always much larger than an acoustic-phonon energy so that scattering is concentrated around the Fermi surface. In this case we can define a momentum-relaxation time constant τ_m (E) thus

$$\frac{\partial f_1(E)}{\partial t} = -\frac{f_1(E)}{\tau_m(E)} \tag{11.78}$$

where $f_1(E)$ is (as before) the lowest-order antisymmetric component of the distribution function. The solution of the Boltzmann equation in the presence of an electric field is then

$$f_1(E) = -eFv(E)\tau_m(E)\frac{\partial f_0(E)}{\partial E} = \frac{eFv(E)\tau_m(E)}{k_B T} f_0(E)(1 - f_0(E)) \tag{11.79}$$

where F is the field and $v(E)$ is the group velocity associated with the state with energy E. The identity

$$\frac{\partial f_0(E)}{\partial E} = -\frac{1}{k_B T} f_0(E)(1 - f_0(E)) \tag{11.80}$$

follows from the form of the Fermi–Dirac function

$$f_0(E) = \frac{1}{1 + \exp[(E - E_F)/k_B T]} \tag{11.81}$$

The rates of change of $f_0(E)$ and $f_1(E)$ are as follows:

$$\begin{aligned}
\left(\frac{\partial f_0(K)}{\partial t}\right)_{scat} &= \frac{\Omega}{(2\pi)^3} x \\
&\sum_{\mathbf{Q}} \Big[\int (W(\mathbf{K}', \mathbf{K}, \mathbf{Q})\{f_0(E')(1 - f_0(E))(n(\omega) + 1) - f_0(E)(1 - f_0(E'))n(\omega)\} \\
&\delta(E' - E - \hbar\omega)d\mathbf{K}') \\
&+ \int (W(\mathbf{K}'', \mathbf{K}, \mathbf{Q})\{f_0(E'')(1 - f_0(E))(n(\omega) - f_0(E)(1 - f_0(E''))n(\omega) + 1)\} \\
&\delta(E'' - E + \hbar\omega)d\mathbf{K}'') \Big]
\end{aligned} \tag{11.82}$$

$$\left(\frac{\partial f_1(\mathbf{K})}{\partial t}\right)_{\text{scat}} = \frac{\Omega}{(2\pi)^3}\mathbf{x}$$

$$\sum_{\mathbf{Q}}\left[\int\left(W(\mathbf{K}',\mathbf{K},\mathbf{Q})\left\{\begin{array}{l}\{f_1(E')(1-f_0(E))\cos a\prime - f_0(E')f_1(E)\}(n(\omega)+1)\\ -\{f_1(E)(1-f_0(E'))-f_0(E)f_1(E\prime)\cos a'\}n(\omega)\end{array}\right\}\right.\right.$$

$$\left.\delta(E'-E-\hbar\omega)d\mathbf{K}'\right)$$

$$+\int\left(W(\mathbf{K}'',\mathbf{K},\mathbf{Q})\left\{\begin{array}{l}\{f_1(E'')(1-f_0(E))\cos a'' - f_0(E'')f_1(E)\}(n(\omega)\\ -\{f_1(E)(1-f_0(E''))-f_0(E)f_1(E\prime\prime)\cos a''\}n(\omega)+1)\end{array}\right\}\right.$$

$$\left.\left.\delta(E'-E-\hbar\omega)d\mathbf{K}'\right)\right] \qquad (11.83)$$

where a' is the angle between $\mathbf{K}$ and $\mathbf{K}'$, and terms involving the product $f_1(E)$ $f_1(E')$ and $f_1(E)f_1(E'')$ have been neglected. We have leaned heavily on our previous experience of handling net scattering rates in the early sections of this chapter in order to write down these equations. A general expression for acoustic-phonon scattering is

$$W(\mathbf{K},'\mathbf{K},\mathbf{Q}) = \frac{\pi C^2(Q)|G(\mathbf{K}',\mathbf{K},\mathbf{Q})|^2}{\rho\ \omega\Omega S(Q)} \qquad (11.84)$$

where $C^2(q) = \Xi^2 q^2$ for the non-polar interaction and $C^2(q) = e^2K_{\text{av}}^2 c_{av}/\varepsilon_s$ for the piezoelectric interaction. $S(Q)$ is the screening function, and the other symbols have their previous meanings.

These integrals can be evaluated approximately by exploiting the smallness of the phonon energy relative to the Fermi energy. In the 2D and 1D cases we assume that only the lowest subband is occupied.

We begin by noting that

$$n(\omega) = \frac{e^{-\hbar\omega/k_BT_L}}{2\sinh(\hbar\omega/2k_BT_L)}, \qquad n(\omega)+1 = \frac{e^{\hbar\omega/k_BT_L}}{2\sinh(\hbar\omega/2k_BT_L)} \qquad (11.85)$$

where T_L is the lattice temperature and we are going to assume that there are no hot-phonon effects, thereby implying low fields. We wish specifically to avoid assuming equipartition since this cannot hold at low temperatures for phonons that scatter electrons right across the Fermi surface.

Integrating over the final-state energies can be done immediately thanks to the delta-functions and $E' \to E + \hbar\omega$, and $E'' \to E - \hbar\omega$. Conservation of energy and momentum serves to define the angles a' and a'' as before. The expression for the power input can be integrated over energy E using the integral

$$\int_{-\infty}^{\infty} \frac{xe^{-x}}{(e^{\hbar\omega/k_BT_e}+e^{-x})(1+e^{-x})}dx = \frac{1}{2}\frac{(\hbar\omega/k_BT_e)^2}{e^{\hbar\omega/k_BT_e}-1} \tag{11.86}$$

We obtain, ignoring the small differences between the two integrals involving $\pm \hbar\omega$,

$$P = \frac{\hbar N(E_F)}{16\pi^2\rho}\int \frac{C^2(Q)|G(K',K,\mathbf{Q})|^2}{S(Q)}\frac{\sinh[(\hbar\omega/2k_BT_L)-(\hbar\omega/2k_BT_e)]}{\sinh(\hbar\omega/2k_BT_L)\sinh(\hbar\omega/2k_BT_e)}\hbar\omega\frac{d\mathbf{K}'}{dE'} \tag{11.87}$$

When $T_e > T_L$, this is the power generated by the electric field, which is equal to the power dissipated by the phonons. We will need to examine the term $d\mathbf{K}'/dE'$ in more detail later. In deriving this expression we have exploited the fact that terms such as $f_0(E \pm \hbar\omega)(1-f_0(E))$ are sharply peaked at the Fermi level and vary more rapidly than any other, so that other terms may be evaluated at the Fermi level.

To obtain the momentum-relaxation rate we must first substitute for $f_1(E)$, $f_1(E')$ and $f_1(E'')$ using Eq. (11.79), and evaluate the integrals by integrating over the final state energies using

$$\int_{-\infty}^{\infty} \frac{e^{-x}}{(e^{\hbar\omega/k_BT_e}+e^{-x})(1-e^{-x})}dx = \frac{\hbar\omega/k_BT_e}{e^{\hbar\omega/k_BT_e}-1} \tag{11.88}$$

An integration over E then allows us to use the delta-function-like properties of $f_0(E \pm \hbar\omega)(1 - f_0(E))$ and similar terms to evaluate $\tau_m(E_F)$, viz.:

$$\frac{1}{\tau_m(E_F)} = \frac{\hbar}{16\pi^2\rho k_BT_e}$$
$$\int \frac{C^2(Q)|G(\mathbf{K}',\mathbf{k},\mathbf{Q})|^2}{S(Q)}\frac{\cosh[(\hbar\omega/2k_BT_L)-(\hbar\omega/2k_BT_e)]}{\sinh(\hbar\omega/2k_BT_L)\sinh(\hbar\omega/2k_BT_e)}(1-\cos a')\frac{d\mathbf{K}'}{dE'} \tag{11.89}$$

In general the integrals describing energy and momentum relaxation must be evaluated numerically (see Vickers, 1992). For parabolic bands $dK'/dE' = m^*/\hbar^2k_F$, and thus

$$\int \ldots \frac{d\mathbf{K}'}{dE'} = \int \ldots \frac{m^*}{\hbar^2k_F}\mathbf{x}\begin{cases} k_F^2 d(-\cos a')d\phi & \text{3D} \\ k_F da' dq_z & \text{2D} \\ \mathbf{dq} & \text{1D}\end{cases} \tag{11.90}$$

where k_F is the Fermi wavevector. The overlap integral is given in Eq. (11.6), and $\cos\alpha'$ by Eq. (11.29) in the limit $\hbar\omega \to 0$.

11.7.3 Low-Temperature Approximation

In the limit of low temperatures the rates reduce sharply with increasing phonon energy. Energy and momentum conservation allows the phonon wavevector parallel to unconfined directions to be as high as $2k_F$, with $\alpha' = \pi$, but if $2\ \hbar v_s k_F / k_B T_e >> 1$ there will be few phonons to be absorbed, and emission will be inhibited because the final electron state is occupied. Small phonon energies will therefore be favoured, a condition that can be exploited in working out the 2D and 1D cases.

With these considerations in mind we make the low-temperature approximation

$$\sinh\left(\frac{\hbar\omega}{k_B T_L}\right)\sinh\left(\frac{\hbar\omega}{k_B T_e}\right) \approx \frac{1}{4}\exp\left[\frac{\hbar\omega}{2k_B}\left(\frac{1}{T_e}+\frac{1}{T_L}\right)\right] \tag{11.91}$$

with $\omega(=v_s Q)$ unrestricted. For brevity we consider only the unscreened case.

In 3D the resultant integrals are straightforward and we obtain

$$\begin{aligned} P &= \frac{6\Xi^2 m^{*2}}{\pi^3 \rho \hbar^7 v_s^4}\left[(k_B T_e)^5 - (k_B T_L)^5\right] \\ \frac{1}{\tau_m} &= \frac{15\Xi^2 m^*}{\pi\rho\,\hbar^7 v_s^6 k_F^3}\left[\frac{(k_B T_e)^6 + (k_B T_L)^6}{k_B T_e}\right] \text{non-polar} \end{aligned} \tag{11.92}$$

$$\begin{aligned} P &= \frac{e^2 K_{\text{av}}^2 m^* 2}{2\pi^3 \hbar^5 \varepsilon_s}\left[(k_B T_e)^3 - (k_B T_L)^3\right] \\ \frac{1}{\tau_m} &= \frac{3e^2 K_{\text{av}}^2 m^*}{4\pi\hbar^5 \varepsilon_s v_s^2 k_F^3}\left[\frac{(k_B T_e)^4 + (k_B T_L)^4}{k_B T_e}\right] \text{piezoelectric} \end{aligned} \tag{11.93}$$

The power expressions are essentially those given by Kogan (1963). The non-polar rate for momentum relaxation exhibits the well-known Gruneisen–Bloch temperature dependence for metals when $T_e = T_L$.

In 2D we have to worry about the dependence of the integrand on q_z, which is unrestricted in magnitude. Fortunately, this lack of restriction fits in well with our low-temperature approximation, which removes any restriction on the total wavevector **Q**. We can also exploit the weighting of the integrand toward small Q by approximating the overlap integral by unity. Thus with $1 - \cos\alpha' = \mathrm{q}^2/2\mathrm{k}_\mathrm{F}^2$, $\mathrm{d}\alpha' = dq/kF$, and transforming to coordinates $q = Q\sin\phi$, $q_z = Q\cos\phi$, we obtain

$$
\begin{aligned}
P &= \frac{6\Xi^2 m^{*}2}{\pi^2 \rho \hbar^7 v_s^4 k_F a}\left[(k_B T_e)^5 - (k_B T_L)^5\right] \\
\frac{1}{\tau_m} &= \frac{15\Xi^2 m^{*}2}{2\pi\rho\hbar^7 v_s^6 k_F^3}\left[\frac{(k_B T_e)^6 + (k_B T_L)^6}{k_B T_e}\right] \text{non-polar}
\end{aligned}
\tag{11.94}
$$

$$
\begin{aligned}
P &= \frac{e^2 K_{av}^2 m{*}^2}{2\pi^2 \hbar^5 \varepsilon_s k_F a}\left[(k_B T_e)^3 - (k_B T_L)^3\right] \\
\frac{1}{\tau_m} &= \frac{3e^2 K_{av}^2 m{*}}{8\pi\hbar^5 \varepsilon_s v_s^2 k_F^3}\left[\frac{(k_B T_e)^4 + (k_B T_L)^4}{k_B T_e}\right] \text{piezoelectric}
\end{aligned}
\tag{11.95}
$$

The non-polar momentum-relaxation rate is essentially that given by Milsom and Butcher (1986). The power dissipation is unchanged from that in 3D except for the change in density of states (a = well-width). The relaxation rate is reduced by a factor of 2.

In 1D the only scattering events permitted are those in which the scattering angle is either zero or π, but **q**, the wavevector transverse to the wire, is unrestricted. In this limit, once again replacing the overlap integral by unity, we obtain

$$
\begin{aligned}
P &= \frac{6\Xi^2 m^{*2}}{\pi^2 \rho \hbar^7 v_s^4 k_F^2 ab}\left[(k_B T_e)^5 - (k_B T_L)^5\right] \\
\frac{1}{\tau_m} &\approx 0 \text{ non-polar}
\end{aligned}
\tag{11.96}
$$

$$
\begin{aligned}
P &= \frac{e^2 k_{av}^2 m^{*2}}{2\pi^2 \hbar^5 \varepsilon_s k_F^2 ab}\left[(k_B T_e)^3 - (k_B T_L)^3\right] \\
\frac{1}{\tau_m} &\approx 0 \text{ piezoelectric}
\end{aligned}
\tag{11.97}
$$

where a, b are the dimensions of the wire, assumed to be rectangular. A finite relaxation rate can be obtained by focussing on the scattering at zero scattering angle taking into account its inelastic nature. For example, in an absorption process

$$
\frac{\hbar^2 k'2}{2m^*} = \frac{\hbar^2 k^2}{2m^*} + \hbar\omega \tag{11.98}
$$

and as both k' and k lie along the axis of the wire

$$\frac{\hbar^2}{2m^*}(k' + k)(k' - k) = \hbar\omega \tag{11.99}$$

The change in momentum is $k' - k = q_z$, where

$$q_z = \frac{v_s}{v_F} Q \tag{11.100}$$

or, since $q_z = Q\cos\phi$,

$$\cos\phi = \frac{v_s}{v_F} \tag{11.101}$$

Thus, a mode travelling at this angle to the axis of the wire can induce a change in momentum, but no other. Basically in 1D there is energy relaxation but only infinitesimal momentum relaxation – purely inelastic in contradistinction to the more familiar scattering mechanisms, which are purely elastic.

Screening will affect all exponents of temperature. In the extreme case of dominant static screening exponents increase by 2 (Price, 1982). However, as we discussed in Chapter 10, the screening of acoustic modes can be complicated by dynamic effects.

11.7.4 The Electron Temperature

The electron temperature is found by equating the field power to P, viz.:

$$P = \int e\mathbf{F} \cdot v(\mathbf{K}) f_1(E) \cos\theta \frac{d\mathbf{K}}{4\pi^3} \tag{11.102}$$

The power loss per electron is often studied experimentally by measuring Shubnikov–de Haas magneto resistance oscillations (Hirakawa and Sakaki, 1986; Mannion *et al.*, 1987; and Daniels *et al.*, 1989). Screening is expected to make piezoelectric scattering negligible compared with non-polar scattering. Consequently, a study of energy relaxation can yield a value of the deformation potential, which in bulk GaAs is 7 eV. Reported figures are 11 eV (Hirakawa and Sakaki, 1986) and 7.5 eV (Daniels *et al.*, 1989). Vickers (1992) has pointed out the danger of using approximate theoretical models in order to extract quantitative information.

11.7.5 High-Temperature Approximation

At temperatures just above where the low-temperature approximation breaks down we may assume equipartition for the phonon number and a continued

Fermi–Dirac distribution. In this case we can write Eqs. (11.87) and (11.89) as follows:

$$P = \frac{\hbar N(E_F)}{8\pi^2\rho}(k_BT_e - k_BT_L)\int \frac{C^2(Q)|G(\mathbf{K}',\mathbf{K}\cdot\mathbf{Q})|^2}{S(Q)}\frac{d\mathbf{K}'}{dE'}$$
$$\frac{1}{\tau_m} = \frac{k_BT_L}{4\pi^2\rho\hbar}\int \frac{C^2(Q)|G(\mathbf{K}',\mathbf{K}\cdot\mathbf{Q})|^2}{S(Q)}(1-\cos\alpha')\frac{d\mathbf{K}'}{dE'} \tag{11.103}$$

In 3D we obtain, for the unscreened case

$$P = \frac{3\Xi^2 m^{*2} n k_F}{\pi\rho\hbar^3}(k_BT_e - k_BT_L)$$
$$\frac{1}{\tau_m} = \frac{\Xi^2 m^* k_BT_L k_F}{\pi\rho v_s^2\hbar^3} \text{ non-polar} \tag{11.104}$$

$$P = \frac{3e^2K_{av}^2 m^{*2} v_s^2 n}{2\pi\hbar^3\varepsilon_s k_F}(k_BT_e - k_BT_L)$$
$$\frac{1}{\tau_m} = \frac{e^2K_{av}^2 m^* k_BT_L}{2\pi\hbar^3\varepsilon_s k_F} \text{ piezoelectric} \tag{11.105}$$

where n is the electron density. In 3D the Fermi wavevector is temperature-dependent according to

$$k_F = k_F(0)\left[1 - \frac{\pi^2}{24}\left(\frac{k_BT}{E_F(0)}\right)^2\right] \tag{11.106}$$

For weakly degenerative systems k_F must be replaced as follows:

$$k_F \to \frac{(2m * k_BT)^{1/2}}{\hbar}\frac{F_1(\eta)}{F_{1/2}(\eta)}, \qquad k_F^{-1} = \frac{\hbar}{(2m * k_BT)^{1/2}}\frac{F_0(\eta)}{F_{1/2}(\eta)} \tag{11.107}$$

where the $F_j(\eta)$ are the Fermi integrals

$$F_j(\eta) = \int_0^\infty \frac{x^j}{e^{(x-\eta)}+1}dx \tag{11.108}$$

where $\eta = E_F/k_BT$. The energy-loss rate is shown in Fig. 11.6.

For intrasubband processes in 2D the overlap integral becomes $G_{11}(q_z)$ (momentum conservation taken as read), and we are presented with integrals of the form

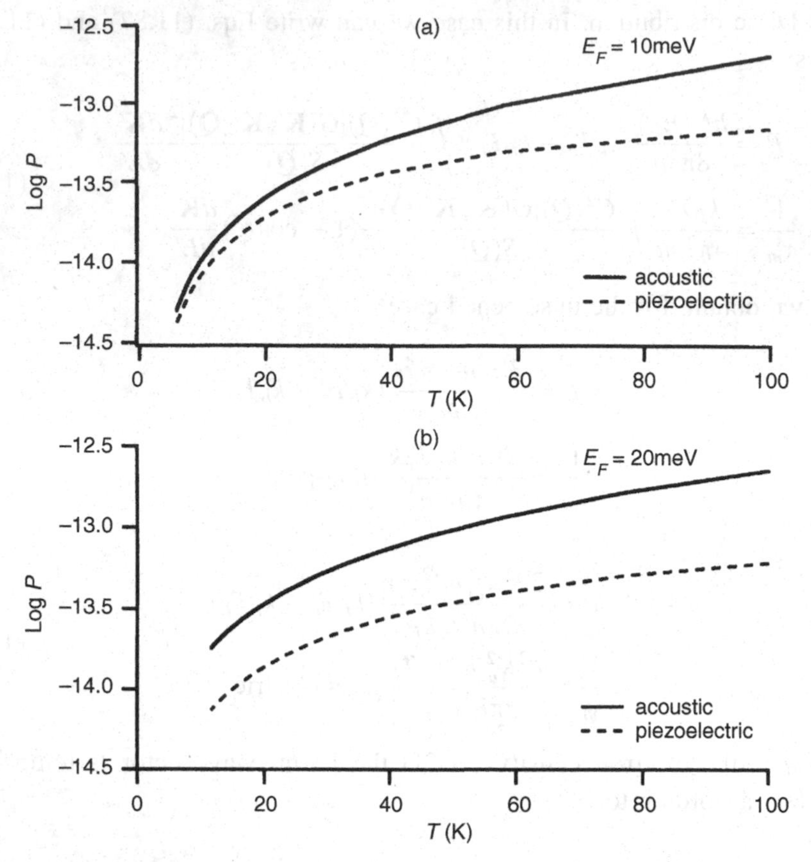

Fig. 11.6 Acoustic-phonon energy-relaxation rate in bulk GaAs at 2 K. The Fermi level is (a) 20 meV, (b) 10 meV. The lower curve is for piezoelectric, the upper for deformation-potential scattering.

$$I_\nu = \int_{-\infty}^{\infty} (q^2 + q_z^2)^\nu |G_{11}(q_z)|^2 \, dq_z \qquad (11.109)$$

where $\nu = 0$, 1 or – 1. For fully confined electrons

$$I_0 = \frac{3\pi}{a}, \quad I_1 = \frac{3\pi q^2}{a} + \frac{4\pi^3}{a^3}, \quad I_{-1} = \frac{\pi}{q} F_{11}(q) \qquad (11.110)$$

where a is the well-width and $F_{11}\ j(q)$ is the form factor familiar from the discussion of the optical-phonon scattering (Section 1.3.4). When these expressions are valid

$$P = \frac{\Xi^2 m^{*2}(3k_F^2 a^2 + 2\pi^2)}{2\pi\rho\hbar^3 a^4}(k_B T_e - k_B T_L)$$
$$\frac{1}{\tau_m} = \frac{3\Xi^2 m^* k_B T_L}{2\rho v_s^2 \hbar^3 a} \text{non-polar} \quad (11.111)$$

$$P = \frac{3e^2 K_{av}^2 m^* 2v_s^2}{4\pi\hbar^3 \varepsilon_s a^2}(k_B T_e - k_B T_L)$$
$$\frac{1}{\tau_m} = \frac{e^2 K_{av}^2 m^* k_B T_L}{4\hbar^3 \varepsilon_s k_F} \int_0^{2k_F} \frac{q F_{11}(q)}{2k_F^2(1 - q^2/4k_F^2)^{1/2}} dq \quad \text{piezoelectric} \quad (11.112)$$

In 2D k_F does not vary with temperature.

For quantum wires with rectangular cross-section

$$P = \frac{3\Xi^2 m *^2 \left[3k_F^2 ab + 2\pi^2 \left(a^2 + b^2\right)/ab\right]}{2\pi\rho\,\hbar^3 k_F^2 (ab)^3}(k_B T_e - k_B T_L)$$
$$\frac{1}{\tau_m} = \frac{9\Xi^2 m * k_B T_L}{2\rho v_s^2 \hbar^3 k_F ab} \quad \text{non-polar} \quad (11.113)$$

$$P = \frac{9e^2 K_{av}^2 m^{*2} v_s^2}{8\pi\hbar^3 \varepsilon_s k_F^2 (ab)^2}(k_B T_e - k_B T_L)$$
$$\frac{1}{\tau_m} = \frac{e^2 K_{av}^2 m * k_B T_L}{\hbar^3 \varepsilon_s k_F} \mathrm{F}_{11}^{1D}(2\mathrm{k_F}) \quad \text{piezoelectric} \quad (11.114)$$

We have assumed a circular cross-section in the expression for the relaxation rate for piezoelectric scattering, and

$$F_{11}^{1D} = \iint \psi^2(\mathbf{r}\prime)\psi^2(\mathbf{r}) K_0(2k_F|\mathbf{r} - \mathbf{r}\prime| d\mathbf{r}\mathbf{r}, \quad (11.115)$$

It should be remarked that the experimental situation is often such that the most easily accessible condition is one in which neither the high-temperature nor the low-temperature approximations are valid. Toward high temperatures (typically 10 K and above) other energy-relaxation mechanisms tend to intrude, particularly the mechanisms associated with plasmons and optical phonons. Toward low temperatures the energy-loss rate weakens and it becomes difficult to measure accurately, especially as its dependence on temperature is extremely rapid. Shubnikov–de Haas measurements aimed at measuring the energy-loss rate of hot electrons are carried out typically at between 2 K and 20 K. In this regime an extrapolation between the two approximations can be effected using the formula

$$P = (C_{np} + C_p)(k_B T_e - k_B T_L)F(T_e T_L)$$
$$F(T_e T_L) = \frac{\sinh(x_L - x_e)}{\sinh x_L \sinh x_e}\frac{x_L x_e}{x_L - x_e} \tag{11.116}$$

where C_{np} and C_p are the high-temperature parameters for the non-polar and polar interactions in Eqs. (11.111), and (11.112). The other parameters are

$$x_L = \frac{\langle\hbar\omega\rangle}{2k_B T_L}, \quad x_e = \frac{\langle\hbar\omega\rangle}{2k_B T_e} \quad \text{and} \quad \langle\hbar\omega\rangle = 2^{1/2}\hbar v_s k_F \tag{11.117}$$

Whereas the deformation interaction strengthens toward high wavevectors the reverse is true of the piezoelectric interaction. The latter is therefore more sensitive to screening, and in many cases it is weak and can be ignored.

12

Spin Relaxation

Hence you long-legged spinners, hence!
A Midsummer Night's Dream, *William Shakespeare*

12.1 Introduction

In recent years there has been considerable interest in the idea of novel devices that exploit the spin of the electron rather than its charge. The activity is now known as spintronics, with applications in areas that include electronics (e.g. spin transistors, spin filters), photonics (e.g. all-optical switching) and quantum computing. While there are problems associated with the injection and manipulation of spin-polarized populations of carriers still to be satisfactorily solved, the common concern of all applications is the lifetime of the spin-polarization in semiconductor structures, and it is the rate of spin relaxation that is the topic of this chapter.

Before getting into that, it may be helpful to recall what magnetic effects are associated with the spin of an electron and its orbital motion. A simple classical-physics model regarding the orbital motion, which will be adequate for our immediate purpose, is to consider the electron to be travelling in a circular orbit of radius r with velocity v. The magnetic dipole moment is $\vec{\mu}_\ell = i\mathbf{A}$, where i is the current $(ev/2\pi r)$ and $\mathbf{A}$ is the vector area, so $\mu_\ell = (ev/2\pi r)\pi r^2 = evr/2 = (e/2m)L$, where $L = mvr$ is the angular momentum. In terms of the Bohr magneton $\mu_B = e\hbar/2m$ the dipole moment is $\mu_\ell = (g_\ell \mu_B/\hbar)L$, where g_ℓ is the gyromagnetic factor, equal to unity in this case. It turns out that this relationship holds for all orbital motion. The magnetic moment of the electron is similarly related to its intrinsic angular momentum S: $\mu_s = (g_s \mu_B/\hbar)S$, but in this case $g_s = 2$. The total angular momentum of an electron in an atomic orbit is $\mathbf{J} = \mathbf{L} + \mathbf{S}$. Quantum theory allows only quantized values for the magnitude of each of these angular momenta and only quantized values for their projections along a given direction, say the z direction. Thus:

$$\begin{aligned} L &= \sqrt{\ell(\ell+1)}\hbar \ (\ell = 0...n-1); \ L_z = m_\ell \hbar \ (m_\ell = -\ell...0...+\ell) \\ S &= \sqrt{s(s+1)}\hbar \ (s = 1/2) \ S_z = m_s \hbar \ (m_s = -s...0...+s) \\ J &= \sqrt{j(j+1)}\hbar \ (j = l \pm 1/2) \ J_z = m_j \hbar \ (m_j = -j...0...+j) \end{aligned} \tag{12.1}$$

The direct way of getting a spin-polarized electron population in a semiconductor is to inject from a ferromagnetic contact in a magnetic field. The necessity of having a magnetic field and, commonly, a low temperature is an obvious drawback for practical devices; and another difficulty is the mismatch between conduction in the ferromagnetic and that in the semiconductor. Manipulation of spin using optical methods does not have these problems.

We recall that the spin of a photon is unity, which means that absorption of a photon by an electron takes place with selection rules:

$$\Delta\ell = \pm 1 \text{ and } \Delta m_j = \begin{cases} 0 \text{ linear polarization} \\ \pm 1 \text{ circular polarization} \end{cases} \tag{12.2}$$

The transition from the valence band of the semiconductor ($\ell = 1$) to the conduction band ($\ell = 0$) is therefore always allowed. (In this case $\Delta\ell = -1$.) Linearly polarized light cannot change the spin balance, but circularly polarized light can. Consider the action of right circularly polarized light on the heavy-hole state $j = 3/2$, $m_j = -3/2$. The transition is to the conduction-band state $j = 1/2$, $m_j = -1/2$. The corresponding transition from the light-hole state $j = 3/2$, $m_j = -1/2$ is to the conduction-band state $j = 1/2$, $m_j = +1/2$. Thus "spin-down" states are generated in the conduction band from the heavy-hole state, "spin-up" states from the light-hole state. Because of the form of the wavefunctions in the valence band the intensity of the heavy-hole transition is three times stronger than that of the light-hole transition; consequently, the absorption of right circularly polarized light produces a population of electrons in the conduction band that is spin-polarized with "spin-down" dominant. Left circularly polarized light also produces a polarized distribution via transitions from the $m_j = +3/2$ and $+1/2$ heavy-and light-hole states, but this time with "spin-up" dominating.

In whatever way a spin-polarized population is produced, its exploitation will depend on its lifetime. There are a number of processes that randomize the direction of spin, eventually, in the absence of a magnetic field, producing a population in which the directions of spin are entirely random. In what follows we briefly review the physics of these processes for zinc blende materials.

12.2 The Elliot–Yafet Process

This mechanism is simply a consequence of the familiar scattering processes undergone by electrons in a semiconductor, taking account of the more accurate wavefunction of the electron in the conduction band given by **k.p** theory (Elliot, 1954; Yafet, 1963). Only at $k=0$ are the degenerate spin states unambiguously "up" or "down"; elsewhere they are mixed. In Chapter 2, **k.p** theory was used to describe the effect of strain on the valence bands. In the next chapter, we will extend **k.p** theory to the wurtzite band structure and obtain explicit expressions for the mixed spin states of the conduction band. Here we will quote the relevant result for cubic materials (Eq. (12.6)) and focus on the statistics that lead to a net relaxation rate.

In the presence of a scattering potential $V_{\mathbf{k},\mathbf{k}'}$ the scattering rate is given by Fermi's Golden Rule:

$$W_{\mathbf{k},\mathbf{k}'} = \frac{2\pi}{\hbar}\int |\langle f|V_{\mathbf{k},\mathbf{k}'}|i\rangle|^2 \delta(E_f - E_i)\mathbf{dN}_f \tag{12.3}$$

where i denotes the initial state and f denotes the final state. The initial eigenfunction can be split into opposite spin components:

$$\begin{aligned} u_c(\uparrow) &= \psi_1(\uparrow) + \psi_1(\downarrow) \\ u_c(\downarrow) &= \psi_2(\downarrow) + \psi_2(\uparrow) \end{aligned} \tag{12.4}$$

We assume that the scattering potential does not act directly on spin and that it is slowly varying over a unit cell. Let $F_{\mathbf{k}}(\mathbf{r})$ be envelope function multiplying u_c. Then:

$$\langle f|V_{\mathbf{k},\mathbf{k}'}(\mathbf{r})|i\rangle = \langle F^*_{\mathbf{k}'}(\mathbf{r})|V_{\mathbf{k},\mathbf{k}'}(\mathbf{r})|F_{\mathbf{k}}(\mathbf{r})\rangle\left[\langle u^*_{c\mathbf{k}'}(\uparrow) \mid u_{c\mathbf{k}}(\uparrow)\rangle + \langle u^*_{c\mathbf{k}'}(\downarrow) \mid u_{c\mathbf{k}}(\uparrow)\rangle\right] \tag{12.5}$$

which is the sum associated with a spin-conserving transition and a spin-flip transition. The spin-flip part is given by (Zawadski and Szymanska, 1971; Chazalviel, 1975):

$$\begin{aligned} \langle u^*_{c\mathbf{k}'}(\downarrow) \mid u_{c\mathbf{k}}(\uparrow)\rangle &= \frac{E}{E^*_g}(k_+k'_z - k_zk'_+) = \frac{E}{E^*_g}\frac{k'}{k}e^{i\phi'}\sin\psi_{\mathbf{k}\mathbf{k}'} \\ \frac{1}{E^*_g} &= \left(1 - \frac{m^*}{m}\right)\frac{(2E_g + \Delta_0)\Delta_0}{E_g(E_g + \Delta_0)(3E_g + 2\Delta_0)} \end{aligned} \tag{12.6}$$

where m^* is the effective mass, m is the free-electron mass, E_g is the bandgap, Δ_0 is the spin-orbit splitting, $\mathbf{k}'$ is the wavevector associated with the final state, $\mathbf{k}$ is the wavevector associated with the initial state (energy E), $k_{\pm} = k_x \pm ik_y$. In cubic material $\mathbf{k}$ can be taken to be along the z direction, so the matrix element can be expressed in terms of $\psi_{\mathbf{kk}'}$, the angle between $\mathbf{k}$ and $\mathbf{k}'$, and $\tan\varphi = k'_y/k'_x$.

All the usual scattering processes will contribute to the relaxation of spin. In practical devices only those dominant at room temperature need be considered.

In bulk material the Fröhlich interaction with polar-optical phonons is, in pure material, likely to be the dominant scattering mechanism at room temperature. It is described by:

$$\left|\langle F^*_{\mathbf{k}'}(\mathbf{r})|V_{\mathbf{k},\mathbf{k}'}(\mathbf{r})|F_{\mathbf{k}}(\mathbf{r})\rangle\right|^2 = \left[\frac{e^2\hbar\omega\{n(\omega)+1/2\pm 1/2\}}{2V\varepsilon_p q^2}\right]\delta_{\mathbf{k}',\mathbf{k}\mp\mathbf{q}}$$

$$\frac{1}{\varepsilon_p} = \frac{1}{\varepsilon_\infty} - \frac{1}{\varepsilon_s} \tag{12.7}$$

where ω is the LO frequency (taken to be independent of $\mathbf{q}$), $\mathbf{q}$ is the phonon wavevector, V is the cavity volume, ε_∞ is the high-frequency permittivity, ε_s is the low-frequency permittivity, and the Kronecker delta function indicates the conservation of crystal momentum assuming normal processes. Since crystal momentum is conserved, the angle between $\mathbf{k}$ and $\mathbf{k}'$ can be expressed in terms of the angle θ between $\mathbf{k}$ and $\mathbf{q}$ according to:

$$\sin^2\psi_{\mathbf{kk}'} = (q/k')^2\sin^2\theta \tag{12.8}$$

We can find the spin-flip rate associated with out-scattering in terms of standard integrals and obtain:

$$W_k(\uparrow\downarrow)$$
$$= \frac{e^2 m^*\omega}{4\pi\varepsilon_p\hbar^2 k}\frac{E^2}{E_g^{*2}}\left[\begin{array}{l} n(\omega)\left\{\left(1+\frac{\hbar\omega}{2E}\right)\sqrt{1+\frac{\hbar\omega}{E}} - \frac{1}{2}\left(\frac{\hbar\omega}{E}\right)^2\sinh^{-1}\sqrt{\frac{E}{\hbar\omega}}\right\} + \\ \Theta(E-\hbar\omega)\times \\ \{n(\omega)+1\}\left\{\left(1-\frac{\hbar\omega}{2E}\right)\sqrt{1-\frac{\hbar\omega}{E}} - \frac{1}{2}\left(\frac{\hbar\omega}{E}\right)^2\sinh^{-1}\sqrt{\frac{E}{\hbar\omega}-1}\right\} \end{array}\right] \tag{12.9}$$

where $\Theta(x) = 0$ if $x < 0$, $= 1$ otherwise. Eq. (12.9) will be the spin-relaxation rate when the occupation of states with opposite spin is negligible, so no in-scattering. More usually, the imbalance between populations is rather small and, consequently, in-scattering has to be taken into account.

The Boltzmann equation describing spin relaxation in the absence of fields and concentration gradients may be written:

$$\frac{df(E)}{dt} = \left(\frac{df(E)}{dt}\right)_{col\uparrow\uparrow} + \left(\frac{df(E)}{dt}\right)_{col\uparrow\downarrow} + \left(\frac{df(E)}{dt}\right)_{rec} + G(E) \tag{12.10}$$

where $f(E)$ is the net occupation probability of the state with spin-up and energy E and $G(E)$ is the generation rate. A corresponding equation describes the rates for spin-down states. On the right, terms 1 to 3 are the spin-conserving collision rate, the spin-flip rate and the rate at which particles disappear, for example by recombination in the optically generated case. The spin-conserving collisions are usually by far the most frequent so we can assume that $f(E)$ is isotropic and given by the thermal equilibrium distribution function corresponding to the density of electrons with spin-up.

The spin-flip rate couples the two distributions at energies determined by the absorption and emission of optical phonons. Thus:

$$\left(\frac{df(E)}{dt}\right)_{col\uparrow\downarrow} = (-W_{out} + W_{in})_{abs} + (-W_{out} + W_{in})_{em} \tag{12.11}$$

where:

$$\begin{aligned}
(W_{in} - W_{out})_{abs} &= \int_{q_1}^{q_2} W(\mathbf{k},\mathbf{k}')\begin{bmatrix}(n+1)f(\downarrow, E+\hbar\omega)\{1-f(\uparrow,E)\}- \\ nf(\uparrow,E)\{1-f(\downarrow,E+\hbar\omega)\}\end{bmatrix}\delta(E)\mathbf{dq} \\
(W_{in} - W_{out})_{em} &= \int_{q_3}^{q_4} W(\mathbf{k},\mathbf{k}'')\begin{bmatrix}nf(\downarrow, E-\hbar\omega)\{1-f(\uparrow,E)\}- \\ (n+1)f(\uparrow,E)\{1-f(\downarrow,E-\hbar\omega)\}\end{bmatrix}\delta(E)\mathbf{dq}
\end{aligned} \tag{12.12}$$

Here, $\mathbf{q}$ is the phonon wavevector, n is the phonon occupation number (henceforth we drop the explicit dependence on ω), $\delta(E)$ is the delta function conserving energy, and q_i are the momentum and energy-conserving limits on $\mathbf{q}$. Eq. (12.12) takes into account that the distribution may be degenerate.

Let f_0(E) be single-spin occupation probability when the spin populations are equal, and let $f(\uparrow, E) = f_0(E) + \Delta f(E)$, $f(\downarrow, E) = f_0(E) - \Delta f(E)$. If quadratic deviations are neglected and the phonon occupation can be expressed by the Bose–Einstein factor, Eq. (12.12) can be written:

$$\begin{aligned}
(W_{in} - W_{out})_{abs} &= -\int_{q_1}^{q_2} W(\mathbf{k},\mathbf{k}')\begin{bmatrix} n\frac{f_0(E)}{f_0(E+\hbar\omega)}\Delta f(E+\hbar\omega)+ \\ (n+1)\frac{f_0(E+\hbar\omega)}{f_0(E)}\Delta f(E)\end{bmatrix}\delta(E)\mathbf{dq} \\
(W_{in} - W_{out})_{em} &= -\int_{q_3}^{q_4} W(\mathbf{k},\mathbf{k}'')\begin{bmatrix}(n+1)\frac{f_0(E)}{f_0(E-\hbar\omega)}\Delta f(E-\hbar\omega)+ \\ n\frac{f_0(E-\hbar\omega)}{f_0(E)}\Delta f(E)\end{bmatrix}\delta(E)\mathbf{dq}
\end{aligned} \tag{12.13}$$

Because of our assumption that spin-conserving collisions are much more frequent than spin-relaxing collisions, we can take the distribution functions to be thermalized. Assuming a spin imbalance, we will have different Fermi levels for the spin-up and spin-down states. This difference will be related to the difference of population density, ΔN, thus:

$$\Delta f(E) = \frac{df(E)}{dN}\Delta N \tag{12.14}$$

and this, plus corresponding expressions for $\Delta f(E \pm \hbar\omega)$ can be substituted into Eq. (12.13). Weighting both sides of Eq. (12.11) with the distribution function and (single spin) density of states and integrating over energy leads to the net spin-relaxation rate:

$$\begin{aligned}
&\left(\frac{df(E)}{dt}\right)_{col\uparrow\downarrow} = 2\left(\frac{d\Delta f(E)}{dt}\right)_{col\uparrow\downarrow}, \\
&\int \left(\frac{d\Delta f(E)}{dt}\right)_{col\uparrow\downarrow} f(E)N(E)dE = \frac{dN}{dt}, \\
&\frac{dN}{dt} = \int f(E)N(E)dE\big((-W_{out} + W_{in})_{abs} + (-W_{out} + W_{in})_{em}\big) \\
&= -\frac{1}{4}W_0 \Delta N \int W(E)f(E)N(E)dE
\end{aligned} \tag{12.15}$$

where:

$$\begin{aligned}
W_0 &= \frac{e^2}{4\pi\hbar\varepsilon_p}\left(\frac{2m^*\omega}{\hbar}\right)^{1/2}, \\
W(E) &= \left(\frac{\hbar\omega}{E_g^*}\right)^2 \left[\begin{array}{l} \left(n\frac{f_0(E)}{f_0(E+\hbar\omega)}\frac{df_0(E+\hbar\omega)}{dN} + (n+1)\frac{f_0(E+\hbar\omega)}{f_0(E)}\frac{df_0(E)}{dN}\right)A(E) + \\ \left((\mathrm{n}+1)\frac{f_0(E)}{f_0(E-\hbar\omega)}\frac{df_0(E-\hbar\omega)}{dN} + n\frac{f_0(E-\hbar\omega)}{f_0(E)}\frac{df_0(E)}{dN}\right)B(E) \end{array}\right]
\end{aligned} \tag{12.16}$$

The coefficients $A(E)$ and $B(E)$ that arise from integrations over the phonon wavevector are:

$$\begin{aligned}
A(E) &= \left(\frac{E}{\hbar\omega}\right)^{3/2}\left[\left(1+\frac{\hbar\omega}{2E}\right)\sqrt{1+\frac{\hbar\omega}{E}} - \frac{1}{2}\left(\frac{\hbar\omega}{E}\right)^2 \sinh^{-1}\left(\frac{E}{\hbar\omega}\right)^{1/2}\right] \\
B(E) &= \theta(E-\hbar\omega)\left(\frac{E}{\hbar\omega}\right)^{3/2}\left[\left(1-\frac{\hbar\omega}{2E}\right)\sqrt{1-\frac{\hbar\omega}{E}} - \frac{1}{2}\left(\frac{\hbar\omega}{E}\right)^2 \sinh^{-1}\left(\frac{E}{\hbar\omega}-1\right)^{1/2}\right]
\end{aligned} \tag{12.17}$$

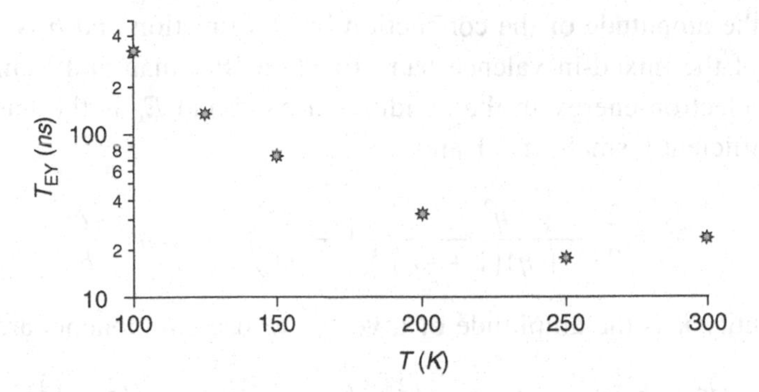

Fig. 12.1 Temperature dependence of the spin-relaxation time constant in GaAs for the E-Y mechanism.

In the non-degenerate case the distribution function can be written:

$$f_0(E) = \frac{N}{N_c} e^{-E/k_BT} \tag{12.18}$$

where N_c is the effective density of states in the conduction band and the energy zero has been taken to be the conduction-band edge. Consequently:

$$\frac{1}{T_{\mathrm{EY}}} = W_0 \left(\frac{\hbar\omega}{E_g^*}\right)^2 \frac{1}{\pi^{1/2}(k_BT)^{3/2}} \int e^{-E/k_BT}[nA(E) + (n+1)B(E)]E^{1/2}dE \tag{12.19}$$

Fig. 12.1 shows the magnitude of T_{EY} and its temperature dependence in GaAs (Dyson and Ridley, 2004). At near room temperature the EY process is much weaker than the D'yakonov–Perel process (D'yakonov and Perel, 1972), which we consider next.

12.3 The D'yakonov–Perel Process

The 8×8 Hamiltonian that describes the conduction- and valence-band energies in **k.p** theory that was used in the EY mechanism can be extended by including the interaction with more remote bands. Those with symmetry Γ_{15} yield a second-order matrix element that has the effect of lifting the spin degeneracy of the conduction band in all but the (100) and (111) directions. The Hamiltonian describing the conduction band is then of the form:

$$\begin{aligned} H &= H_0 \pm \hbar\Omega(\mathbf{k}) \\ \Omega(\mathbf{k}) &= \sqrt{2}|abB|\kappa/\hbar k \end{aligned} \tag{12.20}$$

Here, a is the amplitude of the conduction-band s-function and b is the relevant amplitude of the mixed-in valence-band function (Fishman and Lampel, 1977). If E is the electron energy in the conduction band and E_g is the bandgap, then for E/E_g sufficiently small, $a = 1$ and:

$$b^2 = \frac{2}{9}\frac{\eta^2}{(1+\eta)\left(1+\frac{2}{3}\eta\right)}\left(1-\frac{m^*}{m}\right)\frac{E}{E_g}, \quad \eta = \frac{\Delta}{E_g} \tag{12.21}$$

The quantity κ is the amplitude of a vector whose components are:

$$\kappa_x = k_x(k_y^2 - k_z^2), \qquad \kappa_y = k_y(k_z^2 - k_x^2), \qquad \kappa_z = k_z(k_x^2 - k_y^2) \tag{12.22}$$

Splitting occurs in all but the (100) and (111) directions.

In the D'yakonov–Perel mechanism κ is interpreted as a vector defining the direction of an effective magnetic field and Ω is then the precession frequency of the spin. Precession itself does not relax spin; collisions are required that change the direction of κ. The operative time is the momentum-relaxation time τ_p, as distinct from the scattering time, associated with the scattering mechanism. When $\Omega\tau_p >> 1$ the spin-relaxation time is just equal to the momentum-relaxation time associated with the dominant scattering mechanism. Typically, however, the spin-splitting of the conduction band is very small and the scattering time is very short and so $\Omega\tau_p << 1$.

This situation is conveniently described using spin-density-matrix methods. Thus, if $\rho(\mathbf{k})$ is the spin density and we take a spin imbalance to be produced optically with a generation rate G, the rate equation can be written:

$$\frac{d\rho(\mathbf{k})}{dt} = -\frac{\rho(\mathbf{k})}{\tau} - i\frac{1}{\hbar}[H_1(\mathbf{k}), \rho(\mathbf{k})] - W_{scatt} + G \tag{12.23}$$

where τ is the lifetime, $H_1(\mathbf{k}) = (1/2)\hbar(\sigma.\Omega(\mathbf{k}))$ is the interaction Hamiltonian, σ is the spin operator, W_{scatt} is some scattering rate and [,] is the commutator. If we resolve the density according to $\rho(\mathbf{k}) = \bar{\rho} + \rho_1(\mathbf{k})$, where the bar denotes an average over all directions of $\mathbf{k}$, then, since we suppose that the interaction strength is weak, we can take $\rho_1(\mathbf{k}) << \bar{\rho}$. Assuming further that the distribution is thermalized and in a steady state we can write:

$$\begin{aligned} \frac{\bar{\rho}}{\tau} + i\frac{1}{\hbar}\overline{[H_1(\mathbf{k}), \rho_1(\mathbf{k})]} - \bar{G} = 0 \\ \frac{\rho_1(\mathbf{k})}{\tau} + i\frac{1}{\hbar}[H_1(\mathbf{k}), \bar{\rho}] + W_{scatt} - G + \bar{G} = 0 \end{aligned} \tag{12.24}$$

The second term in the first equation is the spin-relaxation rate $1/T$:

$$\frac{1}{T} = i\frac{1}{\hbar}\overline{[H_1(\mathbf{k}), \rho_1(\mathbf{k})]} \qquad (12.25)$$

In order to evaluate this rate, we need an expression for the directional part, which we must obtain from the second equation. The latter can be simplified by neglecting the relatively slow generation and recombination rates compared with the scattering rate.

It is common to limit discussion to those scattering mechanisms involving elastic collisions, but this is not possible when the interaction is with optical phonons, except in the unrealistic high-energy limit when $E >> \hbar\omega$. In most cases (and temperatures) of practical interest E is less than or of the order of the phonon energy, thus we must take into account separately the rôle of absorption and emission processes. Taking into account occupation probabilities and degeneracy, as we did in the case of the EY mechanism, we then have:

$$i\frac{1}{\hbar}[H_1(\mathbf{k}), \bar{\rho}] + \sum\nolimits_{\mathbf{k}'} W^{abs}_{\mathbf{k}\mathbf{k}'}\left\{(n+1)\frac{f_0(E+\hbar\omega)}{f_0(E)}\rho_1(\mathbf{k}) - n\frac{f_0(E)}{f_0(E+\hbar\omega)}\rho_1(\mathbf{k}')\right\}$$
$$+\sum\nolimits_{\mathbf{k}''} W^{em}_{\mathbf{k}\mathbf{k}''}\left\{n\frac{f_0(E)}{f_0(E-\hbar\omega)}\rho_1(\mathbf{k}) - (n+1)\frac{f_0(E-\hbar\omega)}{f_0(E)}\rho_1(\mathbf{k}'')\right\} = 0$$
$$(12.26)$$

Connection is thereby made to states $E - \hbar\Omega$ and $E + \hbar\Omega$ and from these states to more remote states. Once again, it is not possible to define an overall momentum-relaxation time, but it is possible to define an energy-dependent time-constant $\tau^*(E)$ associated with each of the directionally dependent terms in analogy with the corresponding solution of the Boltzmann equation for optical-mode scattering. Thus we assume solutions of the form:

$$\rho_1(\mathbf{k}) = -i\frac{\tau^*(E)}{\hbar}[H_1(\mathbf{k}), \rho] \qquad (12.27)$$

The spin density can be written $\bar{\rho} = (1/2)(\mathbf{s}.\sigma)$, where $s_i = (1/2)Tr(\bar{\rho}.\sigma)$. We also have $H_1(\mathbf{k}) = (1/2)\hbar(\sigma.\Omega(\mathbf{k}))$. Using the properties of the spin operator we can evaluate the commutator in Eq. (12.23) and obtain the rate of decay of a spin component:

$$\frac{ds_x}{dt} = -\tau^*(E)\{s_x(\Omega_y^2 + \Omega_z^2) - s_y\Omega_x\Omega_y - s_z\Omega_x\Omega_z\} \qquad (12.28)$$

As Pikus and Titkov (1984) have pointed out, the rate is a tensor quantity that reduces to a scalar when an average over direction is taken since $\overline{\Omega_i\Omega_j} = 0, i \neq j$, and $\overline{\Omega_x^2} = \overline{\Omega_y^2} = \overline{\Omega_z^2} = 2|abB|^2\overline{\kappa_i^2}/(\hbar^2 k^2)$. The average over any of the squared vector components of κ yields a factor 4/105. Therefore:

$$
\begin{aligned}
\frac{1}{T} &= 2\tau^*(E)\overline{\Omega^2} \\
&= \frac{128}{945}\tau^*(E)\frac{\Delta^2 B^2}{\{1+\eta\}\{1+(2/3)\eta\}}\frac{m^{*2}}{\hbar^6}\left(1-\frac{m^*}{m}\right)\left(\frac{E}{E_g}\right)^3
\end{aligned}
\tag{12.29}
$$

The matrix element B has been estimated by Fishman and Lampel (1977) to be $10\hbar^2/2m$.

A very simple solution can be obtained at low temperatures when the interaction with optical phonons is essentially via absorption and the emission and "back-scattering" term can be ignored. In this extreme case, $\tau^* \approx \tau_p$, where (Callen, 1949):

$$
\begin{aligned}
\frac{1}{\tau_p(E)} &= \frac{1}{2}W_0\left(\frac{\hbar\omega}{E}\right)^{1/2} n\left[\left(1+\frac{\hbar\omega}{E}\right)^{1/2} - \frac{\hbar\omega}{E}\sinh^{-1}\left(\frac{E}{\hbar\omega}\right)^{1/2}\right] \\
W_0 &= \frac{e^2}{4\pi\hbar}\left(\frac{2m^*\omega}{\hbar}\right)^{1/2}\left(\frac{1}{\varepsilon_\infty}-\frac{1}{\varepsilon_s}\right)
\end{aligned}
\tag{12.30}
$$

In general, however, all the $\tau^*(E)$ can be obtained using the ladder technique.

In order to apply this method to the DP mechanism, we need to know how the interaction Hamiltonian of the scattered state is related to that for the initial state.

The interaction term can be expressed as a series of spherical harmonics:

$$
H_1(\mathbf{k}) \equiv f(k,a,\phi) = \sum_{m=-3}^{3} A_m k^3 Y_3^m(a,\phi) \equiv k^3 g(a,\phi) \tag{12.31}
$$

The order of the spherical harmonic has been determined by the power of k appearing in κ. Eq. (12.26) becomes:

$$
\begin{aligned}
& g(a,\phi) \\
& -\sum\nolimits_{\mathbf{k}'} W^{abs}_{\mathbf{kk}'}\left\{(n+1)\frac{f_0(E)}{f_0(E+\hbar\omega)}\tau^*(E)g(a,\phi) - n\frac{f_0(E+\hbar\omega)}{f_0(E)}\lambda'\tau^*(E')g(a',\phi')\right\} \\
& -\sum\nolimits_{\mathbf{k}''} W^{em}_{\mathbf{kk}''}\left\{n\frac{f_0(E)}{f_0(E-\hbar\omega)}\tau^*(E)g(a,\phi) - (n+1)\frac{f_0(E-\hbar\omega)}{f_0(E)}\lambda''\tau^*(E'')g(a'',\phi'')\right\} \\
& = 0 \\
& \lambda' = \left(\frac{k'}{k}\right)^3, \lambda'' = \left(\frac{k''}{k}\right)^3
\end{aligned}
\tag{12.32}
$$

The angle a (a') is the angle between $\mathbf{k}$ ($\mathbf{k}'$) and the polar axis. The collision interaction does not depend on the azimuthal angle so we may replace the angular functions by their azimuthal averages, e.g.:

$$\int_0^{2\pi} g(a,\phi)\frac{d\phi}{2\pi} = A_0 P_3(\cos a) \tag{12.33}$$

where $P_3(\cos a)$ is the Legendre function $\{P_3(\cos a) = (5\cos^3 a - 3\cos a)/2\}$. If β is the angle between $\mathbf{k}$ and $\mathbf{k}'$, then $\cos a' = \cos a \cos \beta' - \sin a \sin \beta' \cos(\phi - \phi')$ and the addition theorem allows us to express $P_3(\cos a')$ after averaging over the azimuthal angle as follows:

$$< P_3 \cos a') > \; = P_3(\cos a) P_3(\cos \beta') \tag{12.34}$$

Eq. (29) then becomes:

$$\begin{aligned} &\sum\nolimits_{\mathbf{k}'} W^{abs}_{\mathbf{kk}'}\left\{(n+1)\frac{f_0(E)}{f_0(E+\hbar\omega)}\tau^*(E) - n\frac{f_0(E+\hbar\omega)}{f_0(E)}\lambda'\tau^*(E')P_3(\cos\beta')\right\} + \\ &\sum\nolimits_{\mathbf{k}''} W^{em}_{\mathbf{kk}''}\left\{n\frac{f_0(E)}{f_0(E-\hbar\omega)}\tau^*(E) - (n+1)\frac{f_0(E-\hbar\omega)}{f_0(E)}\lambda''\tau^*(E'')P_3(\cos\beta'')\right\} \\ &= 1 \end{aligned} \tag{12.35}$$

(Note that we recover the usual equation for elastic collisions when $E = E' = E''$ and $\lambda = \lambda' = \lambda'' = 1$.) The effective time-constants are obtained from the coupled equations:

$$A'(E)\tau^*(E+\hbar\omega) + B'(E)\tau^*(E-\hbar\omega) + C'(E)\tau^*(E) = 1 \tag{12.36}$$

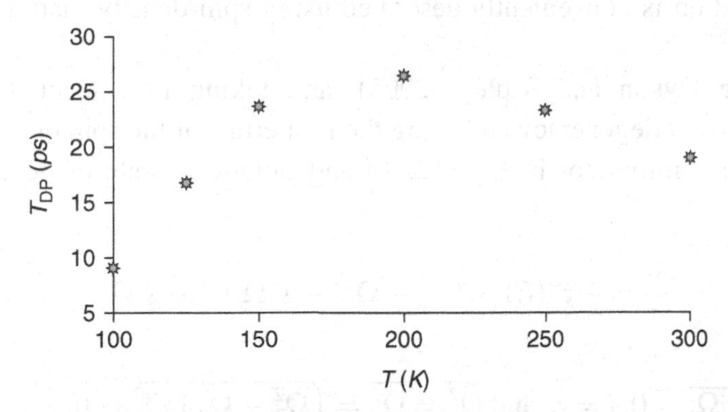

Fig. 12.2 Temperature dependence of the D-P time constant in GaAs.

with $E = \xi + p\hbar\omega$, $0 < \xi < \hbar\omega$ and $p = 0, 1, 2$, etc. and $A'(E)$, $B'(E)$ and $C'(E)$ are the relevant collision integrals (see Appendix 1):

Fig. 12.2 shows the temperature dependence in GaAs (Dyson and Ridley, 2004). For p sufficiently large ($E >> \hbar\omega$) a time-constant can be defined and then used to solve for all the lower energy time-constants in a step-wise procedure. Well-nigh exact results are achievable.

12.3.1 The DP Mechanism in a Quantum Well

Taking the confinement direction to be along z we can replace $<k_z^2>$ with K_i^2 by taking the expectation value:

$\Omega_x = \gamma k_x(k_y^2 - K_i^2)$, $\Omega_y = \gamma k_y(K_i^2 - k_y^2)$, $<\Omega_z> = 0$ where i is the subband index. Thus we have:

$$\begin{aligned} \Omega_x &= -\Omega_1 \cos a - \Omega_3 \cos 3a \\ \Omega_y &= \Omega_1 \sin a - \Omega_3 \sin 3a \\ \Omega_1 &= \gamma k\left(\mathrm{K}_i^2 - \frac{k^2}{4}\right), \quad \Omega_3 = \gamma\frac{k^3}{4}, \quad \tan a = \frac{k_y}{k_x} \end{aligned} \tag{12.37}$$

Scattering takes $\mathbf{k}$ to $\mathbf{k}'$ and K_i to K_j. For simplicity we consider only intrasubband transitions. The tensor components of the scattering times are obtained by first considering $\mathbf{k}$ to be along the x-direction. The scattered component is:

$$\Omega'_x = -\Omega'_1 \cos\theta - \Omega'_3 \cos 3\theta \tag{12.38}$$

where θ is the scattering angle. Similarly, the y-component is:

$$\Omega'_y = \Omega'_1 \cos\theta + \Omega'_3 \cos 3\theta \tag{12.39}$$

We now have two time-constants, one associated with θ the other with 3θ.

This situation is conveniently described using spin-density matrix methods as before.

Following Dyson and Ridley (2005), and taking into account occupation probabilities and degeneracy and using the properties of the spin operator, we can evaluate the commutator in Eq. (12.26) and obtain the rate of decay of a spin component:

$$\frac{ds_x}{dt} = -\tau^*(E)\{s_x(\Omega_y^2 + \Omega_z^2) - s_y\Omega_x\Omega_y - s_z\Omega_x\Omega_z\} \tag{12.40}$$

$$\overline{\Omega_i\Omega_j} = 0, i \neq j, \text{ and } \overline{\Omega_{\mathrm{x}}^2} = \overline{\Omega_{\mathrm{y}}^2} = \left(\overline{\Omega_1^2 + \Omega_3^2}\right), \overline{\Omega_{\mathrm{z}}^2} = 0 \tag{12.41}$$

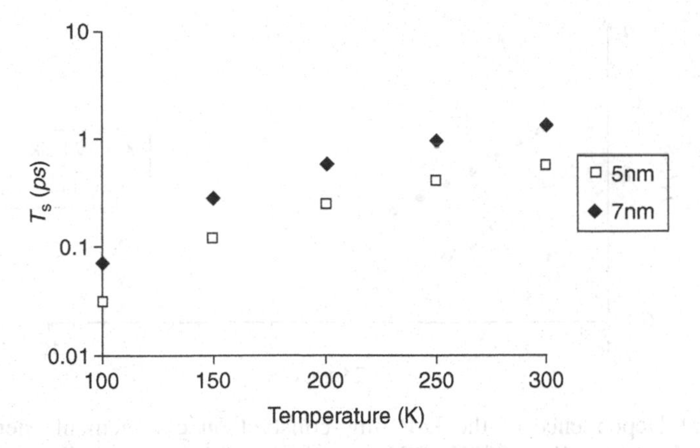

Fig. 12.3 Temperature dependence of the D-P time-constant in GaAs for two well-widths.

$$\frac{1}{T_z} = \frac{2}{T_{x,y}} = 2(\Omega_1^2\tau_1 + \Omega_3^2\tau_3) \tag{12.42}$$

In order to apply this method to the DP mechanism, we need to know how the interaction Hamiltonian of the scattered state is related to that for the initial state:

$$\sum\nolimits_{\mathbf{k}'} W^{abs}_{\mathbf{kk}'}\left\{(n+1)\frac{f_0(E+\hbar\omega)}{f_0(E)}\tau_p^*(E) - n\frac{f_0(E)}{f_0(E+\hbar\omega)}\lambda'_p\tau_p^*(E')\cos p\beta'\right)\right\}+$$
$$\sum\nolimits_{\mathbf{k}''} W^{em}_{\mathbf{kk}''}\left\{n\frac{f_0(E-\hbar\omega)}{f_0(E)}\tau_p^*(E) - (n+1)\frac{f_0(E)}{f_0(E-\hbar\omega)}\lambda''_p\tau_p^*(E)\cos p\beta''\right)\right\} = 1 \tag{12.43}$$

where $\lambda'_p = \Omega_p/\Omega_p$, $\cos\beta' = \frac{1-(q^2/2k^2)+(\hbar\omega/2E)}{\sqrt{1+(\hbar\omega/E)}}$ $\cos\beta'' = \frac{1-(q^2/2k^2)-(\hbar\omega/2E)}{\sqrt{1-(\hbar\omega/E)}}$ with $p = 1$ or 3.

The effective time-constants are obtained from the coupled equations:

$$A'(E)\tau^*(E+\hbar\omega) + B'(E)\tau^*(E-\hbar\omega) + C'\tau^*(E) = 1 \tag{12.44}$$

with $E = \xi + m\hbar\omega$, $0 < \xi < \hbar\omega$ and $m = 0, 1, 2$, etc. and $A'(E)$, $B'(E)$ and $C'(E)$ are the relevant collision integrals (see Appendix 2). For m sufficiently large ($E >> \hbar\omega$) a time-constant can be defined and then used to solve for all the lower energy time-constants. Matrix inversion techniques are employed to solve the set of m equations.

Fig. 12.3 shows the temperature dependence in a GaAs quantum well and Fig. 12.4 shows the dependence on subband energy at 300K.

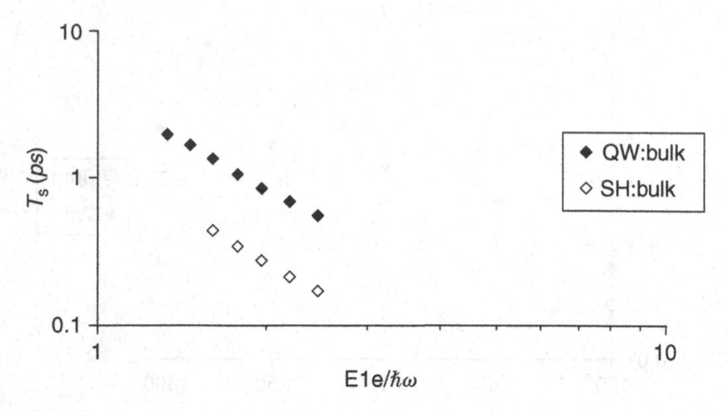

Fig. 12.4 Dependence of the D-P time-constant on confinement energy in a GaAs quantum well at 300K. (SH = single heterojunction, bulk means bulk phonons were assumed.)

12.3.2 Quantum Wires

Taking the confinement directions to be along x and y results in scattering along the wire either forward or backward in z. Again we limit ourselves to intrasubband scattering due to the large energy separation to the next subband. The minimum energy gap for the widest wire is $\sim 3kT$. The expectation values are $<k_x{}^2> = K_i^2$ and $< k_y{}^2> = \Lambda_i^2$ with i and j the subband indices:

$$\overline{\Omega_i \Omega_j} = 0, i \neq j, \text{ and } \overline{\Omega_x^2} = \overline{\Omega_y^2} = 0, \ \overline{\Omega_z^2} = \left[\gamma k_z (K_i - \Lambda_j)\right]^2 \tag{12.45}$$

$$\frac{1}{T_z} = 0, \ \frac{1}{T_{x,y}} = \tau^*(E)\overline{\Omega_z^2} \tag{12.46}$$

It is immediately apparent from Eq. (12.46) that for electrons injected with their spins aligned along the axis of the wire, there will be no DP relaxation. Spins aligned in the x–y plane will relax. In order to calculate the energy dependent time constant, $\tau^*(E)$, we need to know how the interaction Hamiltonian of the scattered state is related to that for the initial state:

$$\sum\nolimits_{\mathbf{k}'} W_{\mathbf{kk}'}^{abs} \left\{ (n+1)\frac{f_0(E+\hbar\omega)}{f_0(E)}\tau^*(E) - n\frac{f_0(E)}{f_0(E+\hbar\omega)}\frac{k_z'\left(K_{i'}^2 - \Lambda_{j'}^2\right)}{k_z\left(K_i^2 - \Lambda_j^2\right)}\tau^*(E') \right\} +$$
$$\sum\nolimits_{\mathbf{k}''} W_{\mathbf{kk}''}^{em} \left\{ n\frac{f_0(E-\hbar\omega)}{f_0(E)}\tau^*(E) - (n+1)\frac{f_0(E)}{f_0(E-\hbar\omega)}\frac{k_z''\left(K_{i''}^2 - \Lambda_{j''}^2\right)}{k_z\left(K_i^2 - \Lambda_j^2\right)}\tau^*(E'') \right\} = 1 \tag{12.47}$$

where $k_z'(k_z'')$ is the scattered wavevector for absorption (emission).

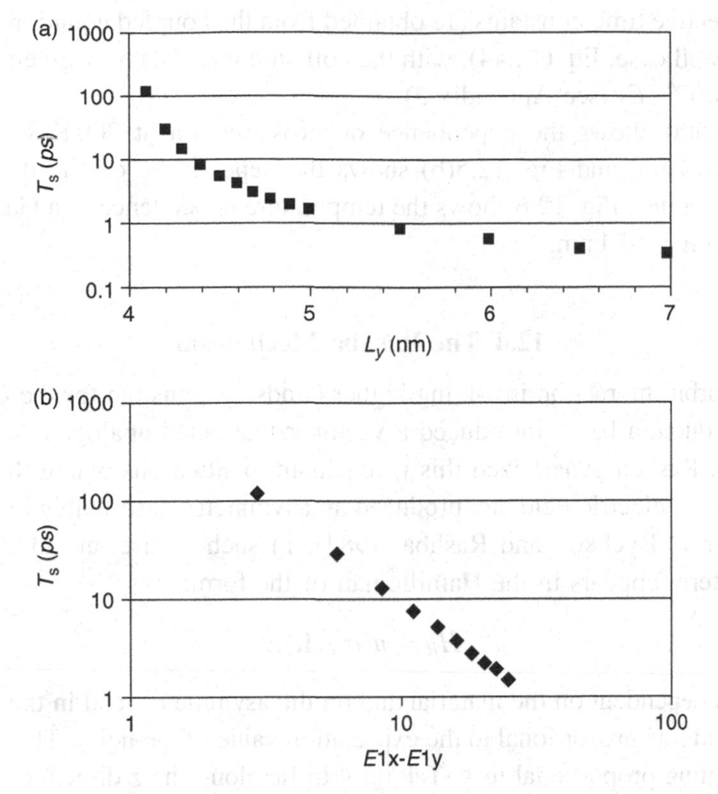

Fig. 12.5 Variation of spin-relaxation time on the cross-section of a GaAs quantum wire at 300K with $L_x = 4$ nm. (a) the variation with L_y; (b) the variation with difference.

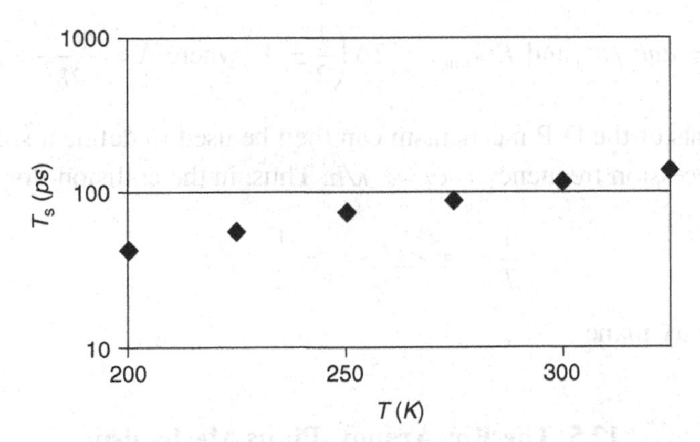

Fig. 12.6 Temperature dependence of the spin-relaxation time in a GaAs quantum wire of cross-section 4×4.1 nm.

The effective time-constants are obtained from the coupled equations as for the quantum-well case, Eq. (12.44), with the collision integrals now given by $A''(E)$, $B''(E)$ and $C''(E)$ (see Appendix 3).

Fig. 12.5(a) shows the dependence on cross-section at 300 K with L_x held constant at 4 nm, and Fig. 12.5(b) shows the dependence on the difference of subband energies. Fig. 12.6 shows the temperature dependence in a GaAs wire of cross-section 4×4.1 nm.

12.4 The Rashba Mechanism

The spin-orbit interaction involving higher bands, responsible for the k^3 splitting of the conduction band, introduced a vector κ that acted analogously to a magnetic field. Rashba generalized this mechanism to situations where the presence of strain or an electric field etc. produced an asymmetry describable in terms of a unit vector **v** (Bychkov and Rashba, 1984). In such a case, an additional perturbation term appears in the Hamiltonian of the form:

$$H_R = a(\sigma \times \mathbf{k}).\mathbf{v} \tag{12.48}$$

where a is dependent on the material and on the asymmetry (and in the case of an electric field it is proportional to the expectation value of the field). This introduces a spin-splitting proportional to k. Taking **v** to lie along the z direction, we get:

$$E = \frac{\hbar^2 k^2}{2m^*} \pm a\left|\sigma_x k_y - \sigma_y k_x\right| = \frac{\hbar^2 k^2}{2m^*} \pm ak \tag{12.49}$$

which results in a minimum at:

$$k_{\min} = am^*/\hbar^2, \text{and } E(k_{\min}) = 2\Delta\left(\frac{1}{2} \pm 1\right) \text{where } \Delta = \frac{m^* a^2}{2\hbar^2} \tag{12.50}$$

The arguments of the D-P mechanism can then be used to define a spin-relaxation rate. The precession frequency is $\omega = a\,k/\hbar$. Thus, in the collision-dominated case:

$$\frac{1}{T} = a^2 \langle k^2 \rangle \tau_p = \frac{1}{2} a^2 k^2 \tau_p \tag{12.51}$$

with k in the xy plane.

12.5 The Bir–Aranov–Pikus Mechanism

The BAP mechanism for spin flip exploits the exchange interaction with holes (Bir, Aranov and Pikus, 1976). This interaction is basically the coulomb interaction between electron and hole with exchange of electron and hole coordinates. Spin comes into the picture because the wavefunctions depend on the total spin of the two-particle system and different energies are obtained for different total spins. In the case of two electrons, for example, with spins s_1 and s_2, the total spin $S = s_1 \pm s_2$, that is since $s = 1/2$, $S = 1$ or 0. Since there are three ways of combining one-electron wavefunctions to give $S = 1$ ($S_z = 1, 0, -1$) this is a triplet state, whereas there is only one way to give $S = 0$, so this state is a singlet state. The energies that emerge from the exchange interaction can be written E_1 for $S = 1$ and E_0 for $S = 0$ and the difference $E_0 - E_1$ is referred to as the exchange splitting. Since going from $S = 1$ to $S = 0$ (or vice versa) involves a spin flip the relevance for spin relaxation becomes clearer. However, spin flips in electron–electron collisions through exchange do not produce a net spin relaxation, whereas electron–hole collisions can produce a net spin relaxation in the electron gas.

To calculate the rate it is convenient to make use of the analogy with magnetic effects, rather than calculate the exchange/coulomb matrix element directly. We therefore use the so-called Heisenberg Hamiltonian:

$$H_{ex} = -J_{12}\mathbf{S}_1.\mathbf{S}_2 \tag{12.52}$$

where $\mathbf{S}_1$, $\mathbf{S}_2$ are the spin operators for the electron and hole respectively and J_{12} is the exchange energy. The total spin operator is $\mathbf{S}$ such that:

$$\begin{aligned} \mathbf{S}^2 &= (\mathbf{S}_1 + \mathbf{S}_2)^2 = \mathbf{S}_1^2 + \mathbf{S}_2^2 + 2\mathbf{S}_1.\mathbf{S}_2 \\ \mathbf{S}_1.\mathbf{S}_2 &= \frac{1}{2}(S(S+1) - S_1(S_1+1) - S_2(S_2+1)) \end{aligned} \tag{12.53}$$

For the electron $S_1 = 1/2$ and for the hole (in the uppermost valence band) $S_2 = 3/2$, and so $S = 2$ if the electron and hole spins are aligned and $S = 1$ if they are oppositely aligned. The eigenvalues for the spin product are then 3/4 when $S = 2$ and $-5/4$ when $S = 1$. If the exchange energy is E_1 for $S = 1$ and E_2 for $S = 2$, then the Hamiltonian can be expressed as follows:

$$H_{ex} = \frac{1}{8}(3E_1 + 5E_2) - \frac{1}{2}(E_1 - E_2)\mathbf{S}_1.\mathbf{S}_2 \tag{12.54}$$

The first term on the right is just a constant energy that can be eliminated by redefining the zero of energy and so J_{12} can be expressed in terms of the splitting:

$$J_{12} = \frac{1}{2}(E_1 - E_2) \tag{12.55}$$

Because of the nature of the exchange mechanism, the electron and hole must be in the same location for exchange to occur, i.e.:

$$H_{ex} = -\frac{1}{2}(E_1 - E_2)\mathbf{S}_1.\mathbf{S}_2 V_{ex}\delta(\mathbf{r}_1 - \mathbf{r}_2) \tag{12.56}$$

where V_{ex} is some appropriate volume. We can relate the splitting energy to that observed in the exciton Δ_{ex} and we can take the volume to be of order of the volume of the exciton ground state, πa_B^3, where a_B is the Bohr radius.

The matrix element for the transition can then be written:

$$M(S) = -\frac{1}{2}\Delta_{ex}\pi a_B^3 \int \mathbf{dr_1 dr_2}\times$$
$$\langle \psi_e^*(\mathbf{k}', s', \mathbf{r}_1)\psi_h^*(\mathbf{p}', j', \mathbf{r}_2)|\mathbf{S}_1.\mathbf{S}_2\delta(\mathbf{r}_1 - \mathbf{r}_2)|\psi_e(\mathbf{k}, s, \mathbf{r}_1)\psi_h(\mathbf{p}, j, \mathbf{r}_2)\rangle \tag{12.57}$$
$$\psi_e(\mathbf{k}, s, \mathbf{r}_1) = u_c(\mathbf{k}, s, \mathbf{r}_1)F_c(\mathbf{k}, \mathbf{r}_1)$$
$$\psi_h(\mathbf{p}, j, \mathbf{r}_2) = u_v(\mathbf{p}, j, \mathbf{r}_2)F_v(\mathbf{p}, \mathbf{r}_2)$$

F_c and F_v are envelope functions which are normalized plane-wave functions in bulk material enhanced by the Sommerfeld factor:

$$C = \frac{2\pi\eta}{1 - e^{-2\pi\eta}}, \eta = \left(\frac{m^*v^2}{2E_B}\right)^{-1/2} \tag{12.58}$$

where v is the carrier velocity and E_B is the exciton Bohr energy. The spin operators yield the eigenvalues a_S where a_2=3/4 and a_1=−5/4, as we have seen. The rest of the integration gives the factorized electron and hole overlap integrals:

$$I_e(\mathbf{k}', \mathbf{k}) = \int_{cell} \mathbf{dr} u_c^*(\mathbf{k}', \mathbf{r})u_c(\mathbf{k}, \mathbf{r})$$
$$I_h(\mathbf{p}', \mathbf{p}) = \int_{cell} \mathbf{dr} u_v^*(\mathbf{p}', \mathbf{r})u_v(\mathbf{p}, \mathbf{r})$$
$$G(\mathbf{k}', \mathbf{k}, \mathbf{p}', \mathbf{p}) = \int \mathbf{dr_1 dr_2} F_c(\mathbf{k}', \mathbf{r}_1)F_c(\mathbf{k}, \mathbf{r}_1)F_v(\mathbf{p}', \mathbf{r}_2)F_v(\mathbf{p}, \mathbf{r}_2)\delta(\mathbf{r}_1 - \mathbf{r}_2) \tag{12.59}$$

The matrix element is then

$$M(S) = -\frac{1}{2}\Delta_{ex}\pi a_B^3 a_S I_e(\mathbf{k}', \mathbf{k})I_h(\mathbf{p}', \mathbf{p})G(\mathbf{k}', \mathbf{k}, \mathbf{p}', \mathbf{p}) \tag{12.60}$$

The spin-relaxation rate is given by:

$$\frac{1}{T}(S,\mathbf{k}) = \frac{2\pi}{\hbar}\int |M(S)|^2 f_{\mathbf{p}}(1-f_{\mathbf{p}'})\delta\{E(\mathbf{k}')+E(\mathbf{p}')-E(\mathbf{k}) - E(\mathbf{p})\}\mathbf{dk'dpdp'}\frac{\mathbf{V^3}}{(\mathbf{2\pi})^9} \quad (12.61)$$

where f_p is the hole distribution function and we have assumed that the electron gas is non-degenerate. The overlap integrals for the electron and the heavy or light holes are

$$|I_e(\mathbf{k}',\mathbf{k})|^2 \approx 1, \qquad |I_h(\mathbf{p}',\mathbf{p})|^2 = \frac{1}{4}(1+3\cos^2\beta) \quad (12.62)$$

where β is the angle between $\mathbf{p}$ and $\mathbf{p}'$. In bulk material:

$$G(\mathbf{k}',\mathbf{k},\mathbf{p}',\mathbf{p}) = \frac{1}{V}\delta_{\mathbf{k}'+\mathbf{p}',\mathbf{k}+\mathbf{p}}C_eC_h \quad (12.63)$$

12.6 Hyperfine Coupling

A purely magnetic interaction between electrons in the conduction band and nuclei with magnetic moments can flip the spin. The mechanism is basically the classical interaction between two magnetic dipoles, the nuclear dipole moment being μ_N and the electron dipole moment being $g_0\mu_B$ ($\mu_B = e\hbar/2m$). At a distance $\mathbf{r}$ from the nucleus the vector potential is given by $\mathbf{A} = (\mu_0/4\pi)\frac{\mu\times\mathbf{r}}{r^3}$, where μ_0 is the vacuum permeability, and the magnetic field is $\mathbf{B} = \nabla\times\mathbf{A}$. The full Hamiltonian is $H = \frac{1}{2m}(\mathbf{p}+e\mathbf{A})^2 + g_0\mu_B\mathbf{s}.\nabla\times\mathbf{A}$. Noting that p.A and $\mathbf{A.p} = (\mu_0/4\pi)\frac{\mu_N\times\mathbf{r}}{r^3}.\mathbf{p} = (\mu_0/4\pi)\frac{\mu_N.\mathbf{r}\times\mathbf{p}}{r^3} = (\mu_0/4\pi)\frac{\mu_N.\hbar\mathbf{L}}{r^3}$ where $\mathbf{L}$ is the angular momentum, we can write the interaction Hamiltonian as follows:

$$H = g_0\mu_B\left((\mu_0/4\pi)\frac{\mu_N.\mathbf{L}}{r^3} + \mathbf{s}.\nabla\times\mathbf{A}\right) \quad (12.64)$$

In the expression for $\mathbf{A}$, $\frac{\mu_N\times\mathbf{r}}{r^3} = \nabla\times\left(\frac{\mu_N}{r}\right)$ and $\nabla\times\nabla\times\left(\frac{\mu_N}{r}\right) = \left[\nabla\nabla.\left(\frac{\mu_N}{r}\right) - \nabla^2\left(\frac{\mu_N}{r}\right)\right]$. The Hamiltonian is then:

$$H = (\mu_0/4\pi)g_0\mu_B\left(\frac{\mu_N.\mathbf{L}}{r^3} + \left[\mathbf{s}.\nabla\mu_N.\nabla\left(\frac{1}{r}\right) - \mathbf{s}.\mu_N\nabla^2\left(\frac{1}{r}\right)\right]\right) \quad (12.65)$$

There is still some way to go. It is worth spelling out the manipulations in full:

$$\left(\mu_{\mathbf{N}}.\nabla\left(\frac{1}{r}\right)\right) = -\left(\mu_{Nx}\frac{x}{r^3}+\mu_{Ny}\frac{y}{r^3}+\mu_{Nz}\frac{z}{r^3}\right) = -\phi$$

$$\begin{aligned} \mathbf{s}.\nabla\left(\mu_{\mathbf{N}}.\nabla\left(\frac{1}{r}\right)\right) &= -s_x\frac{\partial\phi}{\partial x} - s_y\frac{\partial\phi}{\partial y} - s_z\frac{\partial\phi}{\partial z} \\ &= -s_x\left[\mu_{Nx}\left(\frac{1}{r^3}-\frac{3x^2}{r^5}\right) - \mu_{Ny}\frac{3xy}{r^5} - \mu_{Nz}\frac{3xz}{r^5}\right] \\ &\quad - s_y\left[-\mu_{Nx}\frac{3xy}{r^5} + \mu_{Ny}\left(\frac{1}{r^3}-\frac{3y^2}{r^5}\right) - \mu_{Nz}\frac{3yz}{r^5}\right] \\ &\quad - s_z\left[-\mu_{Nx}\frac{3xz}{r^5} - \mu_{Ny}\frac{3yz}{r^5} + \mu_{Nz}\left(\frac{1}{r^3}-\frac{3z^2}{r^5}\right)\right] \\ &= \frac{3(\mathbf{s}.\mathbf{r})(\mu_{\mathbf{N}}.\mathbf{r})}{r^5} - \frac{\mathbf{s}.\mu_{\mathbf{N}}}{r^3} \end{aligned} \tag{12.66}$$

The last term in Eq. (12.66) is actually zero for all $r \neq 0$. However, rather than eliminating this term, it is important to retain it for a reason that will become apparent. Thus $\nabla^2\left(\frac{1}{r}\right) = -\frac{3}{r^3} + \frac{3(x^2+y^2+z^2)}{r^5}$ and the Hamiltonian becomes

$$H = (\mu_0/4\pi)g_0\mu_B\left(\frac{\mu_N.\mathbf{L}}{r^3} + \left[\frac{3\mathbf{s}.\mathbf{r}\mu_{\mathbf{N}}.\mathbf{r}}{r^5} - \mathbf{s}.\mu_N\frac{x^2+y^2+z^2}{r^5}\right] - \frac{2}{3}\mathbf{s}.\mu_N\nabla^2\left(\frac{1}{r}\right)\right) \tag{12.67}$$

One third of the last term has been used to eliminate the last term in Eq. (12.66) and so the quantity in the square brackets contains nothing but quadratic factors which are expressible as spherical harmonics $Y_\ell^m(\theta,\phi) = e^{im\phi}P_\ell^m(\cos\theta)$ with l=2. The electron wavefunction can be expressed in terms of spherical harmonics and the spin Hamiltonian matrix element, $\int\langle\ell'|H_s(\ell=2)|\ell\rangle r^2 dr d\Omega$, is zero unless $l+l' \geq 2$. Since the dependence on r in the spherical harmonic is r^l, the power of r in the matrix element is $l+l'+2-3$ and, therefore, the matrix element remains finite as r goes to zero. The last term in Eq. (12.67) is zero for $r \neq 0$ and $\nabla^2\left(\frac{1}{r}\right) \xrightarrow{r=0} \delta(\mathbf{r})$. Finally, the Hamiltonian that describes the magnetic interaction between the electron and the nucleus is

$$H = -(\mu_0/4\pi)g_0\mu_B\gamma\hbar\mathbf{J}.\left(\frac{\mathbf{L}}{r^3} + \left[\frac{3\mathbf{r}\mathbf{s}.\mathbf{r}}{r^5} - \frac{\mathbf{s}}{r^3}\right] - \frac{2}{3}\mathbf{s}\delta(\mathbf{r})\right) \tag{12.68}$$

where $\mu_N = -\gamma\hbar\mathbf{J}$, γ is the gyromagnetic ratio and $\mathbf{J}$ is the angular momentum vector.

For an s-electron the basic element of the Hamiltonian is of the form:

$$H = \Delta_{HF}\mathbf{J}.\mathbf{S}V_N\delta(\mathbf{r}) \tag{12.69}$$

where **J** is the total angular momentum of the nucleus which is at the origin of coordinates and V_N is the nuclear volume. The coupling energy is of the form:

$$\Delta_{HF} \approx \left(\frac{\mu_0}{4\pi V_N}\right)\frac{2}{3}\gamma\hbar g_0\mu_B \tag{12.70}$$

The matrix element can be written

$$M = \int \left\langle J, m_j'; \mathbf{k}', \downarrow \middle| H \middle| J, m_j; \mathbf{k} \uparrow \right\rangle \mathbf{dr} \tag{12.71}$$

where $|J, m\rangle$ represents the nuclear state and $|\mathbf{k}, \uparrow\rangle = (1/V)^{1/2}e^{i\mathbf{k}.\mathbf{r}}u_c(\mathbf{r})|\uparrow\rangle$ represents the electron state (for brevity, considering only the bulk situation). We have assumed that the nucleus is much more massive than the electron so that the collision can be taken to be elastic, $k' = k$. The matrix element depends on the electron wavefunction at $\mathbf{r} = 0$ which is zero except for an s-state:

$$M = \Delta_{HF}|\mathbf{J}.\mathbf{S}||u_c(0)|V_{cell}^{1/2}V_N/V \tag{12.72}$$

and the spin-flip rate is

$$\frac{1}{T}(S) = \frac{2\pi}{\hbar}N_N\Delta_{HF}^2 a_S^2|u_c(0)|^2 V_{cell}V_N^2 N(E) \tag{12.73}$$

where S is the total spin, a_S is the eigenvalue of the **J.S** interaction, N_N is the density of nuclei, and $N(E)$ is the single-spin density of states in the conduction band. Since the cell-periodic wavefunction of the conduction band has s-like symmetry, it will have a non-vanishing magnitude at the nucleus, which would not be the case for holes with their p-like symmetry.

Evaluation of a_S is not straightforward. The total angular momentum of the nucleus in the **LS** coupling scheme is $\mathbf{J} = \sum_n (\mathbf{L}_n + \mathbf{S}_n)$ where L is the orbital angular momentum and the sum is over all nucleons. In the shell model, a nucleus with one or more filled shells has $\mathbf{J} = 0$ since the components of **L** and **S**, characterized by the quantum numbers m_L and m_s sum to zero. The simplest case, therefore, is that of a nucleus with closed shells plus one nucleon, in which case the angular momentum will be determined entirely by that nucleon. Putting $g_j\mu_n = \gamma\hbar$, where g_j is the nuclear g-factor and μn is the nuclear magneton, we obtain the g-factor of the form

$$g_j\mu_N\mathbf{J} = (g_L\mathbf{L} + g_s\mathbf{S})\mu_N \tag{12.74}$$

The expression for g_j can be written

$$\begin{aligned} g_j\mathbf{J} &= \frac{1}{2}(g_L + g_s)(\mathbf{L}+\mathbf{S}) + \frac{1}{2}(g_L - g_s)(\mathbf{L}-\mathbf{S}), \\ g_jJ(J+1) &= \frac{1}{2}(g_L+g_s)J(J+1) + \frac{1}{2}(g_L - g_s)[L(L+1) - S(S+1)] \end{aligned} \tag{12.75}$$

Noting that $J = L \pm s$, we get

$$g_j = g_L \pm \frac{g_s - g_L}{2L+1} \tag{12.76}$$

For the proton $g_L = 1$ and $g_s = 5.585$; for the neutron $g_L = 0$ and $g_s = -3.826$. We still need to know the L value for the new shell into which the proton or neutron enters in order to know the total spin of the interaction so that spin eigenvalue, a_S, can be evaluated, as in the previous section.

Appendix 1

$$A'(E) = -W_o\left(\frac{\hbar\omega}{E}\right)^{1/2} n\frac{f_0(E)}{f_0(E+\hbar\omega)}\left(\frac{E+\hbar\omega}{E}\right)^{3/2}\frac{1}{2}\int_{q^a_-}^{q^a_+} P_3(\cos\beta')\frac{dq}{q},$$

$$q^a_\pm = k(\sqrt{1+(\hbar\omega/E)} \pm 1),\ \cos\beta' = \frac{1-(q^2/2k^2)+(\hbar\omega/2E)}{\sqrt{1+(\hbar\omega/E)}},$$

$$\frac{1}{2}\int_{q^a_-}^{q^a_+} P_3(\cos\beta')\frac{dq}{q} = (1+(\hbar\omega/E))^{-3/2}\times$$

$$\left[\begin{array}{l}(1+(\hbar\omega/2E))\left(1+(\hbar\omega/E)+(5/8)(\hbar\omega/E)^2\right)\sinh^{-1}(E/\hbar\omega)^{1/2} - \\ (11/12)(1+(\hbar\omega/E))^{1/2}\left(1+(\hbar\omega/E)+(15/44)(\hbar\omega/E)^2\right)\end{array}\right] \tag{12.A1}$$

For $B'(E)$:

$$B'(E) = -W_o\left(\frac{\hbar\omega}{E}\right)^{1/2}[n+1]\frac{f_0(E)}{f_0(E-\hbar\omega)}\left(\frac{E-\hbar\omega}{E}\right)^{3/2}\frac{1}{2}\int_{q_-^e}^{q_+^e} P_3(\cos\beta'')\frac{dq}{q},$$

$$q_\pm^e = k(1 \pm \sqrt{1-(\hbar\omega/E)}),\ \cos\beta'' = \frac{1-(q^2/2k^2)-(\hbar\omega/2E)}{\sqrt{1-(\hbar\omega/E)}},$$

$$\frac{1}{2}\int_{q_-^e}^{q_+^e} P_3(\cos\beta'')\frac{dq}{q} = (1-(\hbar\omega/E))^{-3/2}\times$$

$$\left[\begin{array}{c}(1-(\hbar\omega/2E))\left(1-(\hbar\omega/E)+(5/8)(\hbar\omega/E)^2\right)\cosh^{-1}(E/\hbar\omega)^{1/2}- \\ (11/12)(1-(\hbar\omega/E))^{1/2}\left(1-(\hbar\omega/E)+(15/44)(\hbar\omega/E)^2\right)\end{array}\right] \tag{12.A2}$$

For $E < \hbar\omega_o$, $B'(E) = 0$. Finally, for $C'(E)$ we have:

$$C'(E) = W_o\left(\frac{\hbar\omega}{E}\right)^{1/2}\frac{1}{2}\left[(n+1)\frac{f_0(E+\hbar\omega)}{f_0(E)}\int_{q_-^a}^{q_+^a}\frac{dq}{q} + n\frac{f_0(E-\hbar\omega)}{f_0(E)}\int_{q_-^e}^{q_+^e}\frac{dq}{q}\right]$$

$$= W_o\left(\frac{\hbar\omega}{E}\right)^{1/2}\left[\begin{array}{c}(n+1)\frac{f_0(E+\hbar\omega)}{f_0(E)}\sinh^{-1}(E/\hbar\omega)^{1/2}+ \\ n\frac{f_0(E-\hbar\omega)}{f_0(E)}\cosh^{-1}(E/\hbar\omega)^{1/2}\end{array}\right] \tag{12.A3}$$

and the second term in $C(E)$ is zero if $E < \hbar\omega$. The basic rate is:

$$\begin{aligned} W_o &= (e^2/4\pi\hbar\varepsilon_p)(2m^*\omega/\hbar)^{1/2} \\ \varepsilon_p^{-1} &= (\varepsilon_\infty^{-1} - \varepsilon_s^{-1}) \end{aligned} \tag{12.A4}$$

Appendix 2

The scattering rate for the 2D polar-optical-phonon interaction can be expressed as an integral over phonon states with wavevector **q** in the usual way. Taking account of the conservation of energy and crystal momentum, we obtain the integrals that derive from Eq. (12.12) as follows:

$$A'(E) = -\frac{W_o}{2}\left(\frac{\hbar\omega}{E}\right)^{1/2} n\frac{f_0(E)}{f_0(E+\hbar\omega)}\frac{\Omega_p'}{\Omega_p}\int_{q_-^a}^{q_+^a}\frac{F(q)\cos p\beta'}{q\left[1-\left(\left(\frac{m\omega}{\hbar kq}\right)+\frac{q}{2k}\right)^2\right]^{1/2}}dq,$$

$$q_\pm^a = k(\sqrt{1+(\hbar\omega/E)}\pm 1),\ \cos\beta' = \frac{1-(q^2/2k^2)+(\hbar\omega/2E)}{\sqrt{1+(\hbar\omega/E)}}$$

$$\mathrm{QW}F_{11}(q) = \frac{1}{2}\frac{\mu(\mu^2+\pi^2)(3\mu^2+2\pi^2)-\pi^4(1-e^{-2\mu})}{[\mu(\mu^2+\pi^2)]^2},\ \mu = \frac{qa}{2}$$

$$\mathrm{SH}F_{11}(q) = \frac{b}{8(q+b)^3}(8b^2+9qb+3q^2)b = \left(\frac{33m^*e^2N_s}{8\varepsilon_s\hbar^2}\right)^{1/3} \tag{12.A5}$$

For $B'(E)$:

$$B'(E) = W_o\left(\frac{\hbar\omega}{E}\right)^{1/2}\frac{1}{2}\left[\begin{array}{l}(n+1)\frac{f_0(E+\hbar\omega)}{f_0(E)}\int_{q_-^a}^{q_+^a}\frac{F_{11}(q)dq}{q\left[1-\left(\left(\frac{m\omega}{\hbar kq}\right)+\frac{q}{2k}\right)^2\right]^{1/2}}\\ +n\frac{f_0(E-\hbar\omega)}{f_0(E)}\int_{q_-^e}^{q_+^e}\frac{F_{11}(q)dq}{q\left[1-\left(\left(\frac{m\omega}{\hbar kq}\right)-\frac{q}{2k}\right)^2\right]^{1/2}}\end{array}\right] \tag{12.A6}$$

and the second term in $B'(E)$ is zero if $E < \hbar\omega$. Finally, for $C'(E)$ we have:

$$C'(E) = -\frac{W_o}{2}\left(\frac{\hbar\omega}{E}\right)^{1/2}[n+1]\frac{f_0(E)}{f_0(E-\hbar\omega)}\frac{\Omega_p''}{\Omega_p}\int_{q_-^e}^{q_+^e}\frac{F_{11}(q)\cos p\beta}{q\left[1-\left(\left(\frac{m\omega}{\hbar kq}\right)-\frac{q}{2k}\right)^2\right]^{1/2}}dq,$$

$$q_\pm^e = k(1\pm\sqrt{1-(\hbar\omega/E)}),\ \cos\beta = \frac{1-(q^2/2k^2)-(\hbar\omega/2E)}{\sqrt{1-(\hbar\omega/E)}} \tag{12.A7}$$

For $E < \hbar\omega$, $C'(E) = 0$. The basic rate is

$$\begin{aligned} W_o &= (e^2/4\pi\hbar\varepsilon_p)(2m^*\omega/\hbar)^{1/2} \\ \varepsilon_p^{-1} &= (\varepsilon_\infty^{-1}-\varepsilon_s^{-1}) \end{aligned} \tag{12.A8}$$

and p takes values of 1 and 3.

Appendix 3

Similarly for the 1D polar-optical-phonon interaction:

For $A''(E)$:

$$A''(E) = -W_0 n \frac{f_0(E)}{f_0(E+\hbar\omega)} \left(\frac{E+\hbar\omega}{E}\right)^{1/2} \int \frac{1}{2}\left(\frac{\hbar\omega}{E+\hbar\omega}\right)^{1/2} \left[\frac{qF(q,n,m)}{q^2+Q_{a+}^2} + \frac{qF(q,n,m)}{q^2+Q_{a-}^2}\right] dq \tag{12.A14}$$

For $B''(E)$:

$$\begin{aligned} B''(E) = W_0(n+1)\frac{f_0(E+\hbar\omega)}{f_0(E)} \int \frac{1}{2}\left(\frac{\hbar\omega}{E+\hbar\omega}\right)^{1/2} \left[\frac{qF(q,n,m)}{q^2+Q_{a+}^2} + \frac{qF(q,n,m)}{q^2+Q_{a-}^2}\right] dq + \\ W_0 n \frac{f_0(E-\hbar\omega)}{f_0(E)} \int \frac{1}{2}\left(\frac{\hbar\omega}{E-\hbar\omega}\right)^{1/2} \left[\frac{qF(q,n,m)}{q^2+Q_{e+}^2} + \frac{qF(q,n,m)}{q^2+Q_{e-}^2}\right] dq \end{aligned} \tag{12.A15}$$

and the second term in $B(E)$ is zero if $E < \hbar\omega$. Finally, for $C''(E)$ we have:

$$C''(E) = -W_0(n+1)\frac{f_0(E)}{f_0(E-\hbar\omega)} \left(\frac{E-\hbar\omega}{E}\right)^{1/2} \int \frac{1}{2}\left(\frac{\hbar\omega}{E+\hbar\omega}\right)^{1/2} \left[\frac{qF(q,n,m)}{q^2+Q_{e+}^2} + \frac{qF(q,n,m)}{q^2+Q_{e-}^2}\right] dq \tag{12.A16}$$

For $E < \hbar\omega_o$, $C''(E) = 0$ where

$$\begin{aligned} F(q,n,m) &= G_{n,n'}^2(q_x) G_{m,m'}^2(q_y) \\ G_{r,r'}^2(q_a) &= (\pi^2 r r')^2 \frac{(q_a L_a/2)^2 \sin^2(q_a L_a/2 + (\pi/2)(r+r'))}{\left[(q_a L_a/2)^2 - (\pi^2/4)(r-r')^2\right]^2 \left[(q_a L_a/2)^2 - (\pi^2/4)(r+r')^2\right]^2} \end{aligned} \tag{12.A17}$$

with

$$\begin{aligned} Q_{e\pm} &= Q_0(\sqrt{E} \mp \sqrt{E-\hbar\omega}) \\ Q_{a\pm} &= Q_0(-\sqrt{E} \pm \sqrt{E+\hbar\omega}) \end{aligned} \tag{12.A18}$$

for forward and backward scattering.

13

Electrons and Phonons in the Wurtzite Lattice

What immortal hand or eye
Could frame thy fearful symmetry?
Songs of Experience, *William Blake*

13.1 The Wurtzite Lattice

Wurtzite is zinc oxide. Its lattice is made up of tetrahedrally bonded atoms, as it is in zinc blende, but it differs in the stacking sequence: in zinc blende the structure has face-centred cubic symmetry, in wurtzite the structure has hexagonal symmetry. The wurtzite lattice is the preferred thermodynamically stable phase of a number of binary semiconductors, among them the nitrides, AlN, GaN, and InN, which have become of considerable technical importance in recent years. Among the applications are high-power microwave transistors, light-emitting diodes, and highly efficient solar cells. If the paradigm of the cubic III-V semiconductors is GaAs, the paradigm of the hexagonal nitrides is GaN. Being familiar with the properties of GaAs, we need to know how the lower symmetry of GaN affects the band structure of electrons and holes, the phonon spectrum and the electron–phonon interaction. We focus on these topics in this chapter. Fuller accounts of the properties of the nitrides can be found in a number of excellent reviews (e.g. Ambacher, 1998; Ambacher *et al.*, 2002; Piprek (ed.) *Nitride Semiconductor Devices*, 2007).

The unit cell of the wurtzite lattice is depicted in Fig. 13.1. It consists of stacked hexagonal planes so that were the atoms to be spherical with the same radius, the lattice would be a close-packed hexagonal structure. The direction of stacking defines a polar axis at right angles to the base. The lattice is characterized by two lattice constants, c, the height of the hexagon, and a, the length of an edge in the base. GaAs also has a polar axis, along the [111] direction, because it lacks a centre of inversion. Because all the bonds are identical in the cubic case, there is no electric polarization unless the crystal is strained. This would be the case for the close-packed hexagonal structure since the bonds would also be identical. If this is

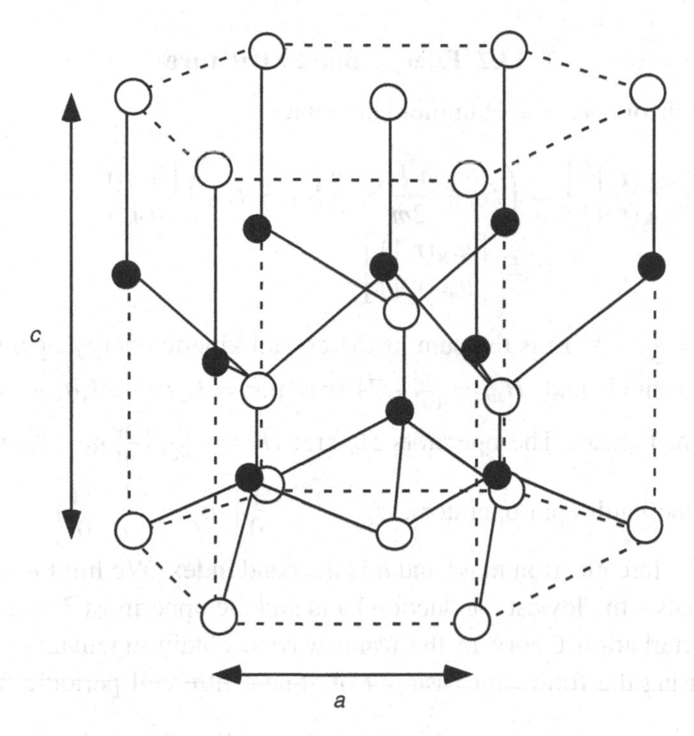

Fig. 13.1 The wurtzite unit cell

the case and all the bonds are of the same length, then it is a matter of simple trigonometry to show that the ratio $c/a = \sqrt{8/3} = 1.63299$. For the nitrides, the observed ratios are 1.6010 (AlN), 1.6259 (GaN) and 1.6116 (InN). The result of these departures from ideality is the appearance of spontaneous electric polarization in each case. Its origin can be understood by the inequivalence of the bonds connecting nearest neighbours, some having a more ionic character than others. A member of the class of materials exhibiting this form of spontaneous polarization is known as a pyroelectric, insofar as the magnitude of the polarization (but not its direction) is a function of temperature, which can be understood by the expansion or contraction of the lattice. In fact, the nitrides show only a weak temperature dependence. But the pyroelectrics are to be distinguished from the other class of materials exhibiting spontaneous polarization, namely the ferroelectrics whose polarization can be inverted by applying a strong enough electric field. In both GaAs and GaN, elastic strain can induce a piezoelectric polarization, but the appearance of spontaneous polarization in the absence of strain in GaN marks a fundamental difference between the two semiconductors. We will return to the technological role of spontaneous polarization when we consider heterostructures. We can expect, however, that the lower symmetry of the wurtzite lattice will affect the band structure and the phonon spectrum.

13.2 Energy Band Structure

Kane's **k.p** theory has the Hamiltonian equation:

$$H\begin{bmatrix} u_{n\mathbf{k}}(\mathbf{r}\uparrow) \\ u_{n\mathbf{k}}(\mathbf{r}\downarrow) \end{bmatrix} = \left(H_0 + \frac{\hbar^2k^2}{2m} + \frac{\hbar}{m}\mathbf{k}.\mathbf{p} + H_{so}\right)\begin{bmatrix} u_{n\mathbf{k}}(\mathbf{r}\uparrow) \\ u_{n\mathbf{k}}(\mathbf{r}\downarrow) \end{bmatrix}$$
$$= E_n\begin{bmatrix} u_{n\mathbf{k}}(\mathbf{r}\uparrow) \\ u_{n\mathbf{k}}(\mathbf{r}\downarrow) \end{bmatrix} \quad (13.1)$$

where; $H_0 = \frac{p^2}{2m} + V(\mathbf{r})$ is the sum of the crystal kinetic energy operator and the periodic potential and $H_{so} = \frac{\hbar^2}{4m^2c^2}\nabla V(\mathbf{r})\times\mathbf{p}.\sigma = H_x\sigma_x + H_y\sigma_y + H_z\sigma_z$ is the spin-orbit interaction. The operators H_α are: $H_x = \frac{\hbar^2}{4m^2c^2}\left(\frac{\partial V}{\partial y}p_z - \frac{\partial V}{\partial z}p_y\right)$ etc. and the σ_α are the Pauli spin operators: $\sigma_x = \begin{bmatrix} 0 & 1 \\ 1 & 0 \end{bmatrix}, \sigma_y = \begin{bmatrix} 0 & -i \\ i & 0 \end{bmatrix}, \sigma_z = \begin{bmatrix} 1 & 0 \\ 0 & 1 \end{bmatrix}$.

Here m is the free electron mass and n is the band index. We limit the calculation to just 4 bands – the lowest conduction band and the uppermost 3 valence bands – and use perturbation theory in the usual way to obtain eigenvalues and eigenfunctions using the following basis set of zone-centre cell-periodic functions:

$$u_{c1} = |iS\uparrow\rangle, \quad u_1 = \left|-\frac{X+iY}{\sqrt{2}}\uparrow\right\rangle, \quad u_2 = \left|\frac{X-iY}{\sqrt{2}}\uparrow\right\rangle, \quad u_3 = |Z\uparrow\rangle,$$
$$u_{c2} = |iS\downarrow\rangle, \quad u_4 = \left|\frac{X-iY}{\sqrt{2}}\downarrow\right\rangle, \quad u_5 = \left|-\frac{X+iY}{\sqrt{2}}\downarrow\right\rangle, \quad u_6 = |Z\downarrow\rangle$$

We take the spin quantization along the z axis. In the case of wurtzite we take the z axis to lie along the c-direction. Unlike the case of the cubic lattice matrix elements involving X and Y, i.e. the basal plane functions, will be different from the matrix elements involving Z. We quantify this by introducing the crystal-field splitting Δ_1, as follows:

$$\langle Z|H_0|Z\rangle = E_v, \langle X|H_0|X\rangle = \langle Y|H_0|Y\rangle = E_v + \Delta_1, \langle S|H_0|S\rangle = E_c \quad (13.2)$$

If Δ_1 is positive, E_v is the lowest energy. A similar distinction must be made in the case of the momentum matrix elements that connect the conduction and valence bands:

$$\langle S|p_z|Z\rangle = \frac{m}{\hbar}P_1, \langle S|p_z|X\rangle = \langle S|p_z|Y\rangle = \frac{m}{\hbar}P_2 \quad (13.3)$$

and for the spin-orbit splitting:

$$\langle X|H_z|Y\rangle = -i\Delta_2, \langle Y|H_x|Z\rangle = \langle Z|H_y|X\rangle = -i\Delta_3 \quad (13.4)$$

Noting that $H_\alpha = -H_\alpha^*$ and $\langle Y|H_z|X\rangle = \langle X|H_z^* Y\rangle = -\langle X|H_z|Y\rangle$, we obtain the diagonal terms:

$$\langle u_1|H_z|u_1\rangle = i\Delta_2, \langle u_2|H_z|u_2\rangle = -i\Delta_2 \tag{13.5}$$

The off-diagonal terms can be similarly obtained and the resultant 8×8 matrix for $H_0 - \frac{\hbar^2 k^2}{2m}$ is:

$$\begin{vmatrix}
E_c & -\frac{k_+}{\sqrt{2}}P_2 & \frac{k_-}{\sqrt{2}}P_2 & k_zP_1 & 0 & 0 & 0 & 0 \\
-\frac{k_-}{\sqrt{2}}P_2 & E_v+\Delta_1+\Delta_2 & 0 & 0 & 0 & 0 & 0 & 0 \\
\frac{k_+}{\sqrt{2}}P_2 & 0 & E_v+\Delta_1-\Delta_2 & 0 & 0 & 0 & 0 & \sqrt{2}\Delta_3 \\
k_zP_1 & 0 & 0 & E_v & 0 & 0 & \sqrt{2}\Delta_3 & 0 \\
0 & 0 & 0 & 0 & E_c & \frac{k_-}{\sqrt{2}}P_2 & -\frac{k_+}{\sqrt{2}}P_2 & k_zP_1 \\
0 & 0 & 0 & 0 & \frac{k_+}{\sqrt{2}}P_2 & E_v+\Delta_1+\Delta_2 & 0 & 0 \\
0 & 0 & 0 & \sqrt{2}\Delta_3 & -\frac{k_-}{\sqrt{2}}P_2 & 0 & E_v+\Delta_1-\Delta_2 & 0 \\
0 & 0 & \sqrt{2}\Delta_3 & 0 & k_zP_1 & 0 & 0 & E_v
\end{vmatrix} \tag{13.6}$$

where $E_{v1} = \Delta_1 + \Delta_2$, $E_{v2} = \Delta_1 - \Delta_2$, and $k_\pm = k_x \pm ik_{\_y}$. Solving the secular determinant with $E_v = 0$, gives the band-edge energies (Fig. 13.2):

$$\begin{aligned}
E_c &= E_1 + E_g \\
E_1 &= \Delta_1 + \Delta_2 \\
E_2 &= \frac{\Delta_1 - \Delta_2}{2} + \sqrt{\left(\frac{\Delta_1 - \Delta_2}{2}\right)^2 + 2\Delta_3^2} \\
E_3 &= \frac{\Delta_1 - \Delta_2}{2} - \sqrt{\left(\frac{\Delta_1 - \Delta_2}{2}\right)^2 + 2\Delta_3^2}
\end{aligned} \tag{13.7}$$

where E_g is the bandgap. The result for the cubic case is obtained by the substitutions $\Delta_1 = 0$, $\Delta_2 = \Delta_3 = \Delta_0/3$, where Δ_0 is the spin-orbit splitting.

Away from the band-edge, the secular determinant takes the form:

$$\begin{aligned}
&-(E_c - E')(E_{v1} - E')[(E_{v2} - E')(E_v - E') - 2\Delta_3^2] \\
&+ [(E_v + \Delta_1 - E')(E_v - E') - \Delta_3^2]P_2^2(k_x^2 + k_y^2) \\
&+ (E_{v1} - E')(E_{v2} - E')P_1^2 k_z^2 = 0
\end{aligned} \tag{13.8}$$

$$E' = E - \hbar^2 k^2/2m$$

For brevity, we consider only the conduction band. This has the form:

$$E = E_c + \frac{\hbar^2(k_x^2 + k_y^2)}{2m_\perp^*} + \frac{\hbar^2 k_z^2}{2m_z^*} \tag{13.9}$$

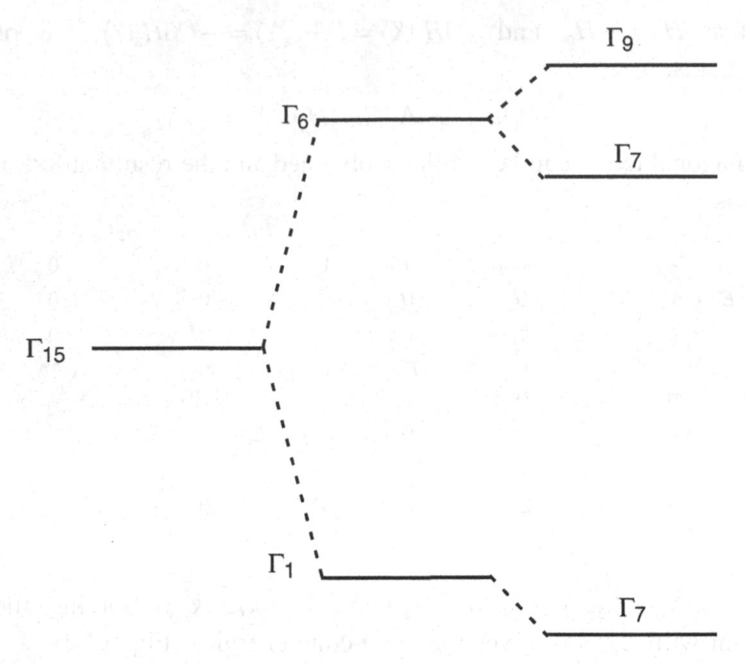

Fig. 13.2 Valence band-edges.

where $m^*_\perp$ is the effective mass of electrons moving in the basal plane and m^*_z is the effective mass component along the c axis. They are related to the momentum matrix elements as follows:

$$\begin{aligned}\frac{1}{m^*_\perp} &= \frac{1}{m} + \frac{(E_g+\Delta_2)(E_g+\Delta_1+\Delta_2)-\Delta_3^2}{(E_g+2\Delta_2)(E_g+\Delta_1+\Delta_2)-2\Delta_3^2}\frac{2P_2^2}{\hbar^2 E_g}\\ \frac{1}{m^*_z} &= \frac{1}{m} + \frac{E_g+2\Delta_2}{(E_g+2\Delta_2)(E_g+\Delta_1+\Delta_2)-2\Delta_3^2}\frac{2P_1^2}{\hbar^2}\end{aligned} \tag{13.10}$$

The splitting energies are each of order 10 meV (Chuang and Chang, 1996), whereas the bandgaps for GaN (3.4 eV) and AlN (6.2 eV) are much greater, even for InN (0.7 eV). Moreover, the momentum matrix elements in the III-V binaries do not vary dramatically, so it is reasonable to expect that, roughly, $P_1 \approx P_2$, in which case the anisotropy in the effective mass is not expected to be large. Measurements confirm this expectation.

13.3 Eigenfunctions

The **k.p** interaction mixes conduction-band and valence-band states, and as we pointed out in connection with spin relaxation, this mixing is responsible for the

Elliot–Yafet mechanism. Eigenfunctions are obtained from the solutions of the simultaneous equations implied by Eq. (13.1) operating on the basis functions. Here we will focus on the conduction-band functions.

In general, the conduction-band cell-periodic function involving $|iS\uparrow\rangle$ is the linear combination:

$$\begin{aligned} &au_{c1} + bu_1 + cu_2 + du_3 + eu_4 + fu_5 + gu_6 \\ &a^2 + b^2 + c^2 + d^2 + e^2 + f^2 + g^2 = 1 \end{aligned} \tag{13.11}$$

In terms of the ratios with respect to a (e.g. $r_1 = b/a$, $r_2 = c/a$, etc):

$$a^2 = \frac{1}{1 + \sum_{n=1}^{6} r_n} \tag{13.12}$$

The ratios can be obtained from Eq. (13.6) with E_v' subtracted from the diagonal terms. From the second row, since E_{v1} is replaced by $-E_g$, we have $r_1 = -k_- P_2/(\sqrt{2}E_g)$, and from the fifth row $r_4 = 0$ since u_{c2}. From the other rows we get:

$$\begin{aligned} &\frac{k_+ P_2}{\sqrt{2}} - (E_g + 2\Delta_2) r_2 + \sqrt{2}\Delta_3 r_6 = 0 \\ &k_z P_1 - (E_g + \Delta_1 + \Delta_2) r_3 + \sqrt{2}\Delta_3 r_5 = 0 \\ &r_5 = \frac{\sqrt{2}\Delta_3}{E_g + 2\Delta_2} r_3 \\ &r_6 = \frac{\sqrt{2}\Delta_3}{E_g + \Delta_1 + \Delta_2} r_2 \end{aligned} \tag{13.13}$$

The result is:

$$\begin{aligned} &a = \frac{E_a E_b - 2\Delta_3^2}{D}, \quad b = \frac{-k_- P_2 (E_a E_b - 2\Delta_3^2)}{\sqrt{2} E_g D}, \quad c = \frac{k_+ P_2 E_a}{\sqrt{2} D} \\ &d = \frac{k_z P_1 E_b}{D}, \quad e = 0, \quad f = \frac{k_z P_1 \sqrt{2}\Delta_3}{D}, \quad g = \frac{k_+ P_2 \Delta_3}{D} \\ &D^2 = [E_a E_b - 2\Delta_3^2]^2 \left(1 + \frac{k_\perp^2 P_2^2}{2E_g^2}\right) + [E_a^2 + 2\Delta_3^2] \frac{k_\perp^2 P_2^2}{2} + [E_b^2 + 2\Delta_3^2] k_z^2 P_1^2 \\ &E_a = E_g + \Delta_1 + \Delta_2, E_b = E_g + 2\Delta_2 \end{aligned} \tag{13.14}$$

For the opposite spin:

$$a = \frac{E_a E_b - 2\Delta_3^2}{D}, \quad b = 0, \quad c = \frac{\sqrt{2}\Delta_3 k_z P_1}{D}, \quad d = -\frac{k_- P_2 \Delta_3}{D},$$
$$e = \frac{k_+ P_2 [E_a E_b - 2\Delta_3^2]}{\sqrt{2} E_g}, \quad f = -\frac{k_- P_2 E_a}{\sqrt{2} D}, \quad g = \frac{k_z P_1 E_b}{D}, \tag{13.15}$$

The matrix elements for the Elliot–Yafet spin-flip process are therefore:

$$\langle \mathbf{k}' \downarrow | \, \mathbf{k} \uparrow \rangle = (k_+ k'_z - k_z k'_+)(2E_g + \Delta_1 + 3\Delta_2)\frac{P_1 P_2 \Delta_3}{D^2}$$
$$\langle \mathbf{k}' \uparrow | \, \mathbf{k} \downarrow \rangle = (k_z k'_- - k_- k'_z)(2E_g + \Delta_1 + 3\Delta_2)\frac{P_1 P_2 \Delta_3}{D^2} \tag{13.16}$$

Although these matrix elements are unequal it should be noted that the squared moduli are equal. The result for the cubic case is recovered with $P_1 = P_2$, $\Delta_1 = 0$, $\Delta_2 = \Delta_3 = \Delta_0/3$ and the use of Eq. (13.10).

In the conduction band and in the E_1 uppermost valence bands, spins are mixed only for non-zero k, but in the lower valence bands spins are mixed even at $k = 0$. Thus in the E_2 level the two components are mixed according to $au_2 + bu_6$ and $au_5 + bu_3$. It easy to see that $a = \frac{E_2 - E_v}{\sqrt{(E_2 - E_v) + 2\Delta_3}}$ and $b = \frac{\sqrt{2}\Delta_3}{\sqrt{(E_2 - E_v)^2 + 2\Delta_3^2}}$. For the E_3 level the mixture is $-au_6 + bu_2$, and $-au_3 + bu_5$. In the cubic case $a = 1/\sqrt{3}$ and $b = \sqrt{2/3}$ so the eigenfunctions at $\mathbf{k} = 0$ are:

$$\text{Heavy-hole band}: \left\{ \begin{array}{l} \left|\frac{3}{2}, \frac{3}{2}\right\rangle \uparrow = -\frac{1}{\sqrt{2}} |X + iY\rangle \uparrow \\ \left|\frac{3}{2}, -\frac{3}{2}\right\rangle \downarrow = \frac{1}{\sqrt{2}} |X - iY\rangle \downarrow \end{array} \right\}$$
$$\text{Light-hole band}: \left\{ \begin{array}{l} \left|\frac{3}{2}, \frac{1}{2}\right\rangle \uparrow = \left(-\frac{1}{\sqrt{6}} |X + iY\rangle \downarrow + \sqrt{\frac{2}{3}} |Z\rangle \uparrow\right) \\ \left|\frac{3}{2}, -\frac{1}{2}\right\rangle \downarrow = \left(\frac{1}{\sqrt{6}} |X - iY\rangle \uparrow + \sqrt{\frac{2}{3}} |Z\rangle \downarrow\right) \end{array} \right\} \tag{13.17}$$
$$\text{Split} - \text{offband}: \left\{ \begin{array}{l} \left|\frac{1}{2}, \frac{1}{2}\right\rangle \uparrow = \left(-\frac{1}{\sqrt{3}} |X + iY\rangle \downarrow - \frac{1}{\sqrt{3}} |Z\rangle \uparrow\right) \\ \left|\frac{1}{2}, -\frac{1}{2}\right\rangle \downarrow = \left(-\frac{1}{\sqrt{3}} |X - iY\rangle \uparrow + \frac{1}{\sqrt{3}} |Z\rangle \downarrow\right) \end{array} \right\}$$

The valence-band structures in the wurtzite nitrides have been studied by a number of authors (e.g. Susuki *et al.*, 1995; Sirenko *et al.*, 1996, 1997; Kim *et al.*, 1997). Fig. 13.3 shows the general form in GaN.

The results of an empirical pseudopotential calculation of the band structure is shown in Fig. 13.4.

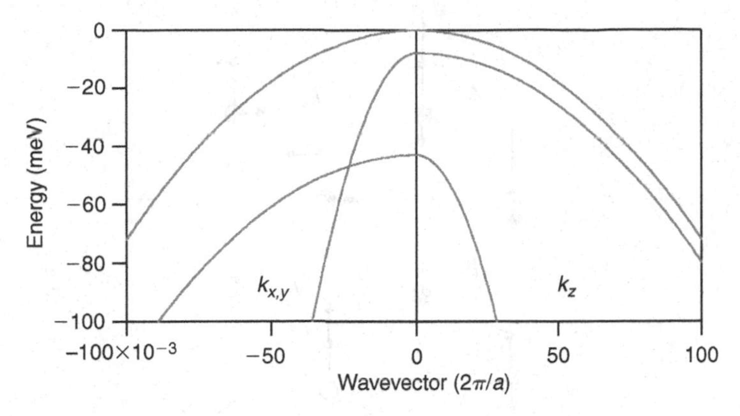

Fig. 13.3 Valence band structure of GaN.

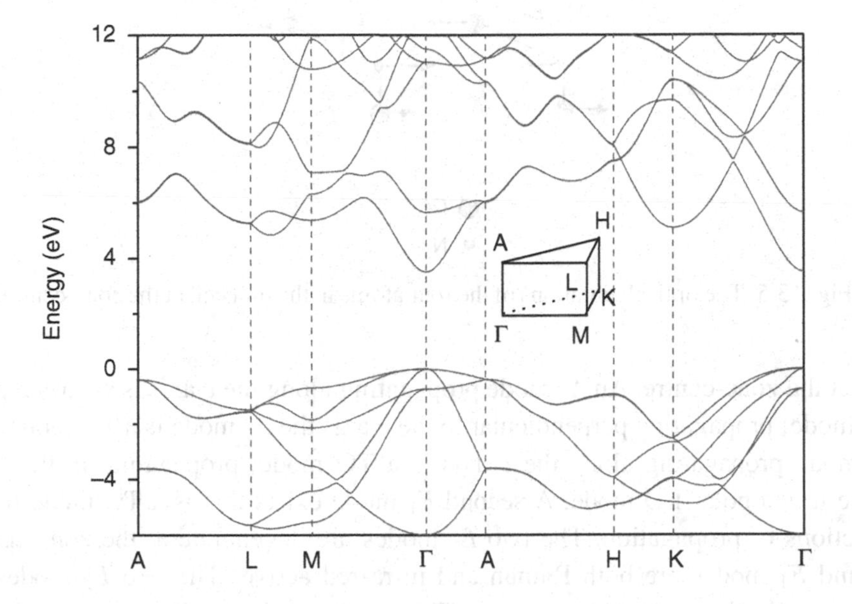

Fig. 13.4 Band structure of GaN from an empirical pseudopotential model by Bulutay *et al.* (2000). The dashed lines show a comparison with results derived using the form factors of Yeo *et al.* (1998).

13.4 Optical Phonons

In the primitive unit cell of the wurtzite lattice there are two molecules and the point group of the wurtzite lattice is C_{6v} (class 6 mm). Group theory predicts the symmetries of Γ-point optical-phonon modes to be A_1+E_1+$2E_2$+$2B_1$ (see Appendix). Fig. 13.5 depicts the optical vibrations of the four atoms in the unit

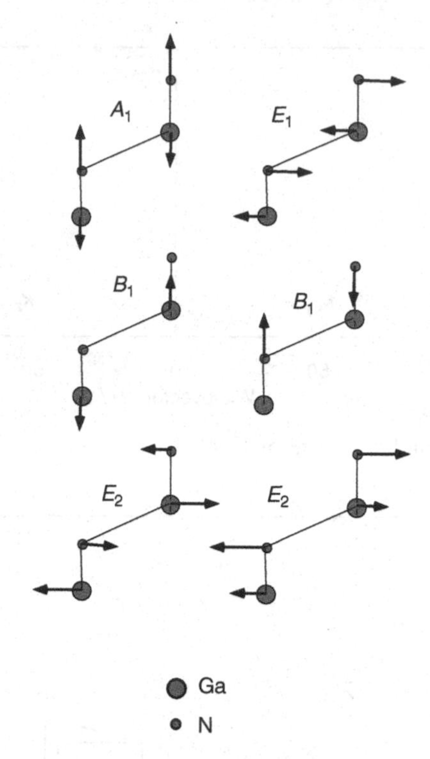

Fig. 13.5 The optical vibrations of the four atoms in the unit cell at the zone-centre.

cell at the zone-centre. An A_1 mode propagating along the c axis is clearly a polar LO mode; propagating perpendicular to the c axis the A_1 mode is a TO mode. The E_1 mode propagating along the c axis is a TO mode; propagating in the basal plane it is a polar LO mode. A second E_1 mode exists that is a TO mode for all directions of propagation. The two E_1 modes are degenerate at the zone-centre. A_1 and E_1 modes are both Raman and infra-red active. The two E_2 modes are Raman active but not infra-red active. The two B_1 modes are Raman and infra-red inactive (often denoted "silent modes"). In both types, the higher frequency mode corresponds to the nitrogen atom having the largest amplitude and the lower frequency mode corresponds to the heavier metallic atom having the largest amplitude. The second E_1, the E_2 and the B_1 modes exhibit no anisotropy with propagation directions at an angle θ to the c axis, but this is not the case for the polar modes. Their frequencies have to satisfy Maxwell's equations, which, in the unretarded limit, reduce to the Fresnel equation:

$$\frac{\omega_{L\perp}^2 - \omega^2}{\omega_{T\perp}^2 - \omega^2}\varepsilon_{\infty\perp}\sin^2\theta + \frac{\omega_{L\|}^2 - \omega^2}{\omega_{T\|}^2 - \omega^2}\varepsilon_{\infty\|}\cos^2\theta = 0 \tag{13.18}$$

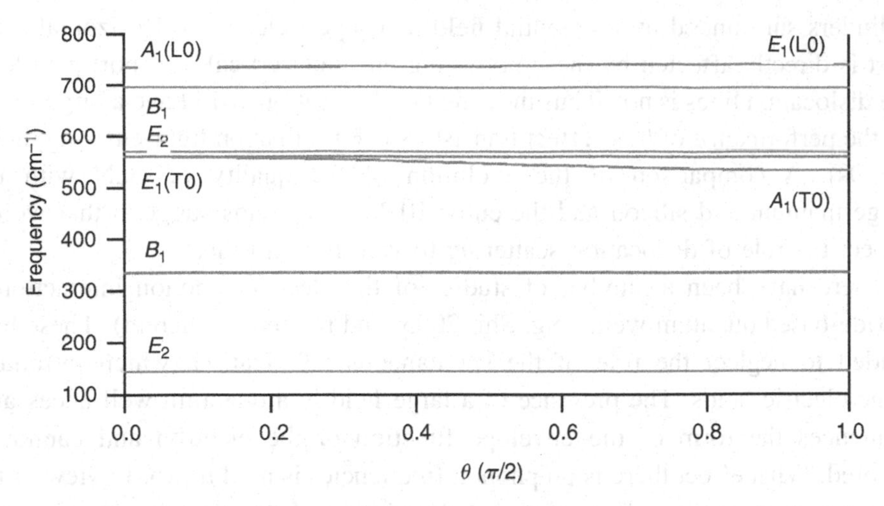

Fig. 13.6 Anisotropy of the optical modes in GaN

where $\varepsilon_{\infty\perp}$ and $\varepsilon_{\infty\parallel}$ are the components of the high frequency permittivity, which in wurtzite is a tensor quantity. It is often an acceptable approximation to regard the difference between the two components as negligible. If, in addition, the large splitting between LO and TO frequencies is exploited, Eq. (13.18) reduces to (Loudon, 1964):

$$\begin{aligned} \omega^2 &= \omega_{L\perp}^2 \sin^2\theta + \omega_{L\parallel}^2 \cos^2\theta \\ \omega^2 &= \omega_{T\parallel}^2 \sin^2\theta + \omega_{T\perp}^2 \cos^2\theta \end{aligned} \tag{13.19}$$

Here $\omega_{L\perp}=\omega[E_1(LO)]$, $\omega_{L\parallel}=\omega[A_1(LO)]$, $\omega_{T\parallel}=\omega[E_1(TO)]$, $\omega_{T\perp}=\omega[A_1(TO)]$. The anisotropy of the modes is depicted for GaN in Fig. 13.6. The anisotropy is rather weak and can often be neglected in order to simplify the description of the electron–phonon interaction, which reduces to the cubic form.

The transport properties of GaN are studied typically with material grown on the non-ideal substrates sapphire or SiC by molecular-beam epitaxy or by vapour deposition. The lack of lattice matching to the substrate results in a high dislocation density (10^7 to 10^{11} cm^{-2}) that can be moderated by various growth techniques. Theoretical estimates of the electron mobility based on the dominance of polar-optical-phonon scattering at room temperature give some 2,000 cm^2/Vs, but measured Hall mobilities are typically half of this or less. The difference is accounted for by charged-impurity scattering and by the scattering of threading dislocations (Weimann and Eastman, 1998). When dislocation scattering is not negligible a distinction has to be made between horizontal and vertical transport. Dislocation lines are negatively charged and therefore like cylinders surrounded by a potential field that repels electrons. Horizontal transport is directly affected by these no-go regions, but vertical transport parallel to the dislocation lines is not. Thus the effect of dislocations will have a larger effect on the performance of field-effect transistors (FETs) than on light-emitting diodes (LEDs). A comparison of the evolution of the quality of GaN with that of germanium and silicon and the cubic III-V compounds suggests that we can expect the role of dislocation scattering to vanish with time.

There have been a number of studies of the electron–phonon interaction in nitride-based quantum wells (e.g. Shi, 2003, and references therein). These have tended to neglect the role of the spontaneous polarization, which introduces large electric fields. The presence of a large field in a quantum well necessarily influences the form of the envelope function of the electron and cannot be ignored. What effect there is on phonon frequencies is not known. In view of this it seems premature to offer a detailed description of the electron–phonon interaction in nitride heterostructures. In a later section we will limit the description of low-dimensional effects to the energy band-structure of the AlN/GaN superlattice, since it offers some advantages as a cascade laser.

13.5 Spontaneous Polarization

The GaN lattice consists of alternate layers of Ga and N (not equally spaced). In normal epitaxial growth the Ga layer is uppermost, and this defines the direction of spontaneous polarization. This means that the uppermost layer carries a negative surface charge density and the lowermost layer a positive surface charge density, which introduces a substantial electric field directed upwards parallel to the c axis. Thermodynamic equilibrium can be achieved only by neutralizing this field to give a spatially independent Fermi level, and this means neutralizing the layer charges at the top and bottom. In the absence of external electric circuits, and in pure material, this can be done only through the movement of the electrons away from the upper surface and towards the lower surface. In addition to the electrons in the valence band there are electrons that occupy surface states whose quasi-two-dimensional energy bands span the valence band and the forbidden gap. The density of surface atoms is of order $10^{20}\,\mathrm{m}^{-2}$. If only a tenth have complex band states in the forbidden gap from which electrons can be readily removed or to which electrons can be attached, this corresponds to a substantial reservoir charge density of order $1\,\mathrm{Cm}^{-2}$. The spontaneous polarization charge densities of AlN, GaN, and InN are 0.0898, 0.0339, and $0.0413\,\mathrm{Cm}^{-2}$ respectively (Bernadini *et al.*, 2001; Bernadini and Fiorentini, 2002). It would need only a small population change of electrons in the surface states to neutralize any one of these polarization charge densities. At thermodynamic equilibrium there will be at each surface a surface charge density equal and opposite in sign to the spontaneous polarization layer density, and therefore zero field in the bulk of the crystal.

A surface state on a clean surface has a wavefunction that decays exponentially into the vacuum and also into the bulk with decaying oscillations. Below what is termed the branch point the states at 0K are filled and, donor-like, they can contribute electrons; above the branch point the states are empty and, therefore, acceptor-like. What determines the branch point is the Fermi level and this, in turn, is determined by the middle of the average energy gap in the semiconductor. The high density of surface states ensures that the Fermi level remains virtually pinned at the branch point, thereby determining surface barriers. It is of some importance, therefore, to determine the branch point of each semiconductor. These are summarized in Monch's book on *Electronic Properties of Semiconductor Interfaces*. Here we quote the results for AlN, GaN, and InN, which are at energies 2.97, 2.37, and 1.51 eV above the respective valence bands (Monch, 1996). The situation in InN is interesting in that it confirms the reality of the branch-point Fermi level. The forbidden gap in InN is only 0.7 eV (Wu *et al.*, 2002), which implies that the branch point is 0.81 eV above the conduction-band edge, which further implies the existence of a degenerate electron gas at the surface. The existence of this gas has been amply confirmed (Lu *et al.*, 2003). A recent investigation of the properties of Si-doped InN shows that the branch point is 1.83 eV above the valence band rather than 1.51 eV (King *et al.*, 2008).

Appendix 1 Symmetry

The symbols that describe the symmetry of the optical phonons are defined by the C_{cv} character table:

$C6_v$	E	C_2	$2C_3$	$2C_6$	$3\sigma_d$	$3\sigma_v$	
A_1	1	1	1	1	1	1	z, z^2, x^2+y^2
A_2	1	1	1	1	−1	−1	
B_1	1	−1	1	−1	−1	1	
B_2	1	−1	1	−1	1	−1	
E_1	2	−2	−1	1	0	0	(x,y) (xz, yz)
E_2	2	2	−1	−1	0	0	(xy, x^2-y^2)

E is the identity operation; C_n n-fold rotation; σ_d reflection in the vertical dihedral plane; σ_v reflection in the vertical plane that includes the c axis.

The permittivity tensor:

$$\begin{vmatrix} \varepsilon_{11} & 0 & 0 \\ 0 & \varepsilon_{11} & 0 \\ 0 & 0 & \varepsilon_{33} \end{vmatrix}$$

The piezoelectric tensor:

$$\begin{vmatrix} 0 & 0 & 0 & 0 & e_{15} & 0 \\ 0 & 0 & 0 & e_{15} & 0 & 0 \\ e_{31} & e_{31} & e_{33} & 0 & 0 & 0 \end{vmatrix}$$

Elastic constants:

$$\begin{vmatrix} c_{11} & c_{12} & c_{13} & 0 & 0 & 0 \\ c_{12} & c_{11} & c_{13} & 0 & 0 & 0 \\ c_{13} & c_{13} & c_{33} & 0 & 0 & 0 \\ 0 & 0 & 0 & c_{44} & 0 & 0 \\ 0 & 0 & 0 & 0 & c_{44} & 0 \\ 0 & 0 & 0 & 0 & 0 & \frac{1}{2}(c_{11} - c_{120} \end{vmatrix}$$

14

Nitride Heterostructures

It was a miracle of rare device,
Kubla Khan, *Samuel Taylor Coleridge*

14.1 Single Heterostructures

In bulk material, neutralization of the enormous electric fields produced by the spontaneous polarization can be achieved by a relatively modest adjustment of the population of surface states. This is no longer possible in the case of inhomogeneous material exhibiting inhomogeneous spontaneous polarization, such as the AlGaN/GaN heterostructure. A layer of AlGaN grown epitaxially on GaN has not only a spontaneous polarization larger than that of GaN but it has, in addition, a piezoelectric polarization associated with the elastic strain inevitably present as a consequence of the lattice mismatch. No straightforward adjustment of the population of the surface states can eliminate the resultant fields. As a result, nitride heterostructures differ uniquely from heterostructures of cubic semiconductors in exhibiting large, built-in electric fields, and these result in the spontaneous formation of a quasi-2D electron gas at the AlGaN/GaN interface. In the cubic case such a gas must be produced by doping, thereby, ineluctably, introducing charged-impurity scattering even when the system is modulation doped. Doping is unnecessary in nitride structures so, in principle, enormous low-temperature mobilities are possible and will, no doubt, be achieved as growth techniques evolve.

An estimate of the density of the polarization-induced electron population at the AlGaN/GaN interface can be obtained assuming the existence of a common branch-point Fermi energy and simple electrostatics. For simplicity we will assume that the materials are undoped and the structure exists with its upper and lower surfaces in the vacuum. We thereby decouple the structure from external influences. Fig.14.1 depicts the energy profile of the heterojunction.

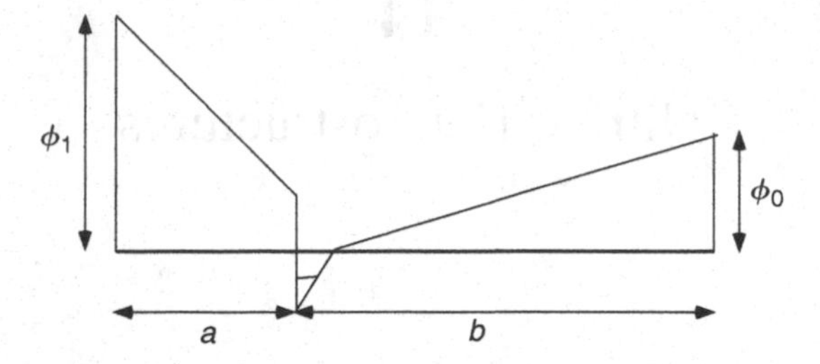

Fig. 14.1 Energy profile of idealized heterostructure.

Define the following areal charge densities:

$-\sigma_1$ = polarization charge density on the AlGaN surface
$+\sigma_1$ = polarization charge density on the AlGaN interface
$-\sigma_0$ = polarization charge density on the GaN interface
$+\sigma_0$ = polarization charge density on the GaN surface
$+\sigma_{s1}$ = surface charge density on the upper surface
$-\sigma_{s0}$ = surface charge density on the lower surface
$-\sigma_n$ = induced electron density at the AlGaN/GaN interface

If E_1 is the electric field in AlGaN and E_0 is the field in GaN and assuming zero field in the vacuum we get:

$$\begin{aligned} \varepsilon_1 E_1 &= \sigma_{s1} - \sigma_1 \\ \varepsilon_0 E_0 &= \varepsilon_1 E_1 + \sigma_1 - \sigma_0 - \sigma_n = \sigma_{s1} - \sigma_0 - \sigma_n \\ 0 &= \varepsilon_0 E_0 + \sigma_0 - \sigma_{s0} = \sigma_{s1} - \sigma_n - \sigma_{s0} \end{aligned} \tag{14.1}$$

The induced electron density is thus just the difference between the upper and lower surface charge densities.

We can obtain these densities from the condition that the voltage difference across the structure is zero. The electron gas at the AlGaN/GaN interface will be a quasi-2D gas in a quantum well. For simplicity we take the gas to be degenerate and the well-depth to be the sum of the lowest subband level, W, and the difference between the Fermi energy and the subband edge. The latter is given by σ_n/D, where D is the $2D$ density of states. Using the usual sign convention we obtain respectively the sum over the AlGaN layer and over the GaN as follows:

$$\begin{aligned} -\phi_1 - E_1 a + \Delta W_c - W - \sigma_n/D = 0 \\ -E_0 b + \phi_0 = 0 \end{aligned} \tag{14.2}$$

where ϕ_1 and ϕ_0 are the surface barrier heights, i.e. the differences between the branch point energy and the conduction-band edges; and ΔW_c is the discontinuity between the AlGaN and GaN conduction-band edges. Elimination of

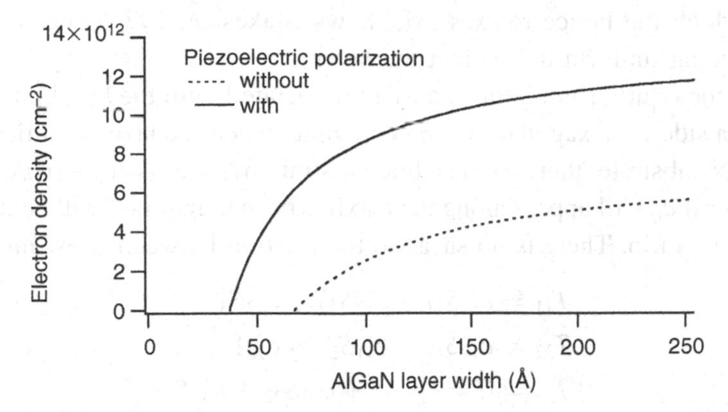

Fig. 14.2 Induced electron density as a function of AlGaN layer width showing the contribution that piezoelectric polarization makes.

the surface charge densities from Eqs. (14.1) and (14.2) gives the induced electron density:

$$\sigma_n = \frac{\sigma_1 - \sigma_0 - \frac{\varepsilon_1}{a} W - \phi_0\left(\frac{\varepsilon_1}{a} + \frac{\varepsilon_0}{b}\right)}{1 + \frac{\varepsilon_1}{aD}} \tag{14.3}$$

and we have assumed that $\phi_1 = \phi_0 + \Delta W_c$ and we have ignored complications associated with band bending in the quantum well. More rigorously, it would be necessary to obtain a self-consistent solution of the Poisson and Schrödinger equations. In most cases it is sufficient to ignore entirely the contributions from the quantum-well structure and neglect the terms involving W and D and work with the simpler expression:

$$\sigma_n = \sigma_1 - \sigma_0 - \phi_0\left(\frac{\varepsilon_1}{a} + \frac{\varepsilon_0}{b}\right) \tag{14.4}$$

We illustrate the dependence on the width of the AlGaN layer in Fig. 14.2 with and without the piezoelectric polarization and taking $b \gg a$ and $\phi_0 = 1\text{eV}$.

14.2 Piezoelectric Polarization

In cubic III-V semiconductors there is only one piezoelectric coefficient, e_{14}, which relates the polarization to the shear strain. In hexagonal crystals there are three, e_{31}, e_{33} and e_{15} (see Appendix, p. 369). Growing AlGaN on GaN produces elastic strains in both materials, but when, as is usually the case for the single heterojunction, the GaN layer is much thicker than the AlGaN layer, the GaN layer is unstrained and all the strain occurs in the AlGaN layer. We consider the more general case later. As long as the thickness of the AlGaN layer is below the limit at which the lattice relaxes (Matthews-Blakeslee, 1974), the strain can be taken as being uniform throughout the layer.

If a_0 is the equilibrium lattice constant associated with the basal plane (i.e. the length of a side of hexagonal) and a is its value when the layer is lattice-matched to the GaN substrate, there exists a biaxial strain $S_{11} = S_{22} = (a - a_0)/a_0$. A strain $S_{33} = (c - c_0)/c_0$ will appear along the c axis whose magnitude will be determined by Poisson's ratio. There is no shear so the relation between stress and strain is:

$$\begin{aligned} T_{11} &= c_{11}S_{11} + c_{12}S_{22} + c_{13}S_{33} \\ T_{22} &= c_{12}S_{11} + c_{11}S_{22} + c_{13}S_{33} \\ T_{33} = 0 &= c_{13}S_{11} + c_{13}S_{22} + c_{33}S_{33} \end{aligned} \tag{14.5}$$

From the third equation we get the relation between the axial and basal strains:

$$S_{33} = -2\frac{c_{13}}{c_{33}}S_{11} \tag{14.6}$$

and from the first two equations we get the stress:

$$T_{11} = T_{22} = \left(c_{11} + c_{12} - 2\frac{c_{13}^2}{c_{33}}\right)S_{11} \equiv CS_{11} \tag{14.7}$$

where C is a uniaxial elastic constant.

The piezoelectric polarization is directed parallel to the c axis and is given by:

$$\sigma_{pz} = 2e_{31}S_{11} + e_{33}S_{33} = 2\left(e_{31} - e_{33}\frac{c_{13}}{c_{33}}\right)S_{11} \tag{14.8}$$

The lattice constants of the alloy obey Vegard's Law (Ambacher *et al.*, 2002):

$$\begin{aligned} a_0 &= (3.1986 - 0.0891x)\text{Å} \\ c_0 &= (5.2262 - 0.2323x)\text{Å} \end{aligned} \tag{14.9}$$

Similar linear extrapolations give the piezoelectric and elastic coefficients:

$$\begin{aligned} e_{31} &= (-0.34 - 0.19x)\text{Cm}^{-2} \\ e_{33} &= (0.67 + 0.83x)\text{Cm}^{-2} \\ c_{13} &= (110 - 10x)\text{GPa} \\ c_{33} &= (390 - 0x)\text{GPa} \end{aligned} \tag{14.10}$$

To a good approximation the piezoelectric polarization in AlGaN on GaN is given by the quadratic equation:

$$\sigma_{pz} = [-0.0525x + 0.0282x(1 - x)]\text{Cm}^{-2} \tag{14.11}$$

Piezoelectric polarization makes a significant contribution, as Fig. 14.2 shows. Formulae for other nitride heterosystems can be found in Ambacher *et al.* (2002). Non-linearity can be taken into account by using the following formulae:

$$\begin{aligned}\sigma_{pz}^{AlN} &= -1.808S + 5.624S^2 \quad S<0 \\ &= -1.808S - 7.888S^2 \quad S>0 \\ \sigma_{pz}^{GaN} &= -0.918S + 9.541S^2\end{aligned} \tag{14.12}$$

The subband structure of the well at the interface is most simply obtained using the Fang–Howard wavefunctions (Section 2.2) with an interface charge density $\sigma_1 - \sigma_0$.

Before considering the problem of describing polarization in nitride multilayers, we need to make three further points concerning the single heterojunction. The presence of spontaneous polarization has been shown to give rise to a quasi-2D electron gas in structures where the heterojunction has a net positive layer charge. It should, therefore, be possible to produce a quasi-2D hole gas at a heterojunction which has a net negative charge such as may occur in a structure where GaN was grown on AlGaN (Fig. 14.3). The formula equivalent to Eq. (14.4) is:

$$\sigma_p = \sigma_1 - \sigma_0 - \phi_0^*\left(\frac{\varepsilon_0}{a} + \frac{\varepsilon_1}{b}\right) \tag{14.13}$$

where $\phi_0^* = E_g - \phi_0$, and E_g is the bandgap of GaN. Since ϕ_0^* is significantly larger than ϕ_0 a much thicker layer is required which may exceed the Matthews–Blakeslee limit. Obtaining an induced hole gas is more difficult than obtaining an electron gas, which, no doubt, accounts for the lack of good evidence for its achievement.

The second point concerns the optical properties of a nitride quantum well. Quantum wells in cubic semiconductors exhibit a blue-shift in absorption and photoluminescence due to the confinement of electrons and holes, which increases the effective energy gap. This blue-shift is not always observed in typical nitride structures in spite of the confinement of carriers in nitride quantum wells increasing the effective gap. The explanation is found in the action of the polarization-induced electric field, which introduces a red-shift associated with the Franz–Keldysh effect. The red-shift can be eliminated in structures grown on a substrate such that the c axis is horizontal. The polarization fields are then directed horizontally and are neutralized by surface charges and play no part in the confinement of carriers.

The final point refers to the effect of replacing the vacuum of the thought experiment with real material. What is the effect of the substrate on which the structure is grown, and what is the effect of adding a passivation layer on the upper surface? The latter problem was addressed by Shealy *et al.* (2003). The theoretical question was: What was the charge at the interface of the Si_xN_y? Measurements

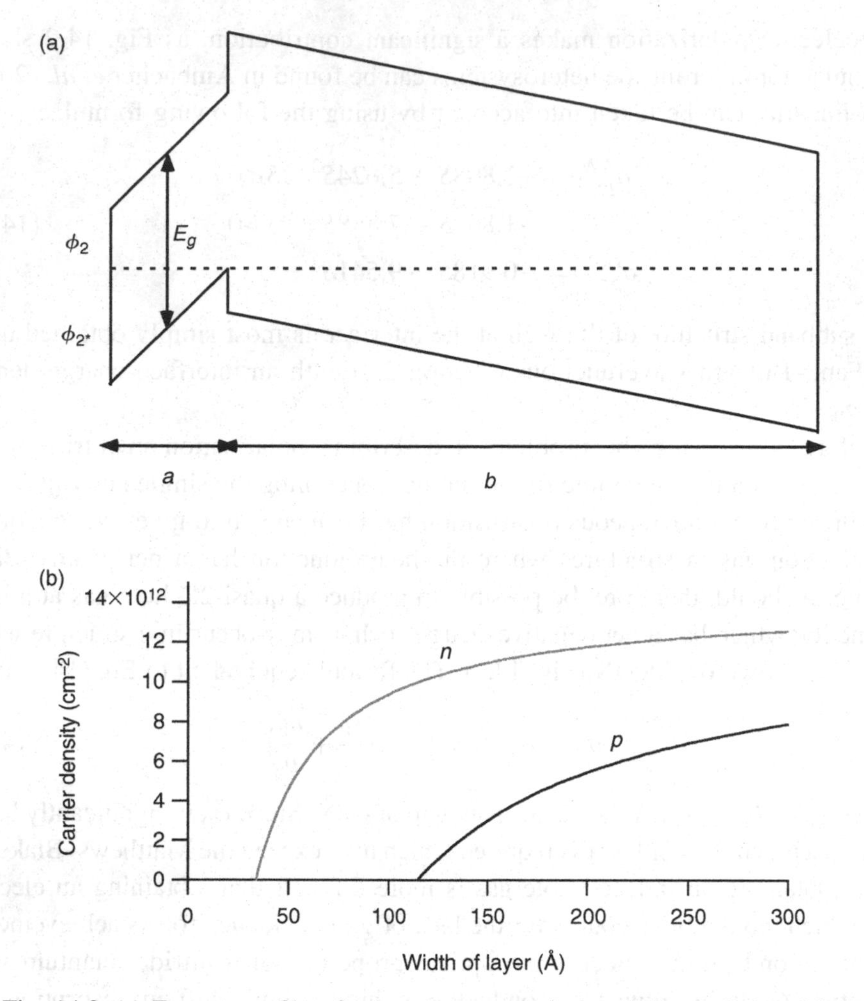

Fig. 14.3 (a) Energy profile of a heterostructure designed for an induced hole gas; (b) Comparison of electron and hole densities for given barrier width.

showed that the polarization charge of the AlGaN at the nitride interface was neutralized by charges that developed in the dielectric/AlGaN interface states and a surface charge appeared on the Si_xN_y surface.

14.3 Polarization Model of Passivated HFET with Field Plate

An application of the simple electrostatic model can be made to describe the induced electron profiles in a passivated HFET with a field plate. A structure is illustrated in Fig. 14.4 depicting a field plate and also a narrow AlN layer which helps to confine the electrons. The field plate is introduced as an extension of

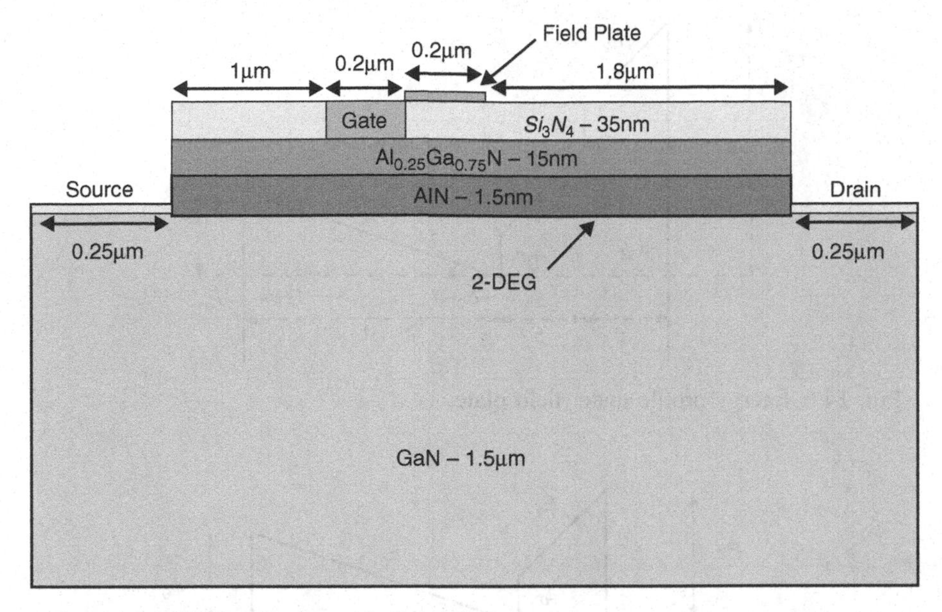

Fig. 14.4 HFET structure with field plate.

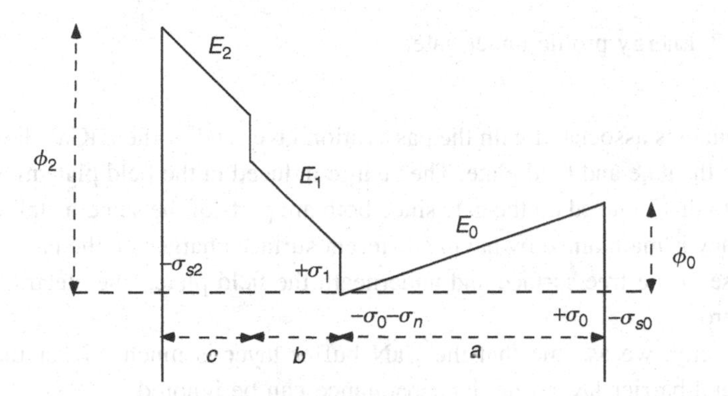

Fig. 14.5 Energy profile with passivation layer.

the gate across the passivation layer in order to modify the field in the gate-drain region and so reduce the applied-voltage threshold of impact ionization. For brevity we ignore the effect of the AlN layer. Figs. 14.5–14.7 depict the three profiles of interest, namely (1) Free surface, (2) Field plate, (3) Gate. We make the following simplifying assumptions:

1. The polarization charge on the top surface of the AlGaN barrier (negative) has been neutralized by charge contributed by the passivation layer. A negative surface charge therefore appears on the surface of the passivating layer.

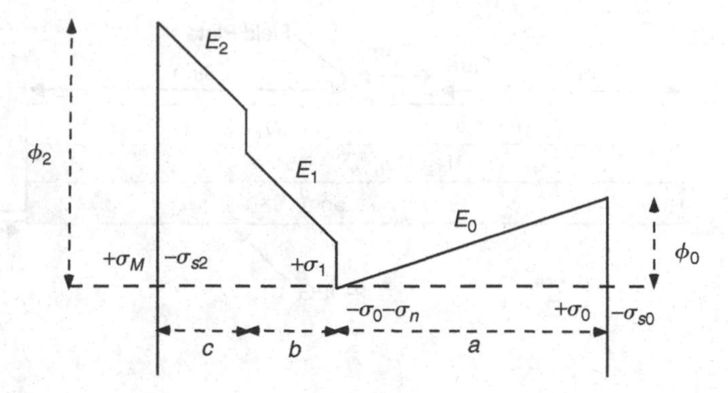

Fig. 14.6 Energy profile under field plate.

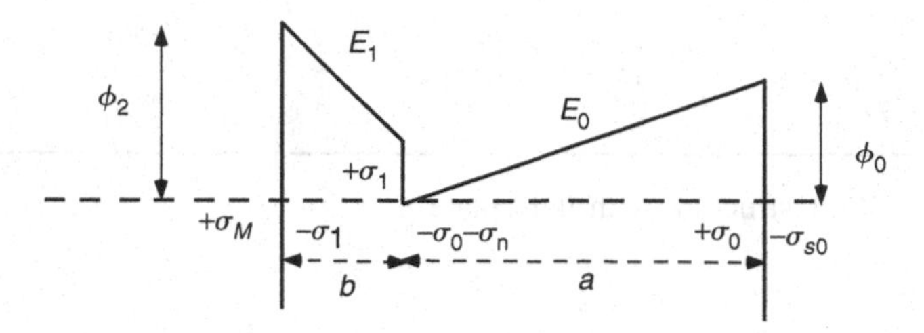

Fig. 14.7 Energy profile under gate.

2. Surface charges associated with the passivation layer and of the AlGaN layer induce a charge on the gate and field plate. The charge induced in the field plate must of course be equal to that induced on the gate since both are parts of the same metallic electrode. Consistency is maintained by having different surface charges on the passivation layer in the case of the free surface and underneath the field plate. The electric field in the gate is zero.
3. For simplicity, we assume that the GaN buffer layer is much thicker than the passivation and barrier layers, i.e. its capacitance can be ignored.
4. For simplicity, we also assume that the surface/Schottky barrier are related to the Schottky barrier of GaN simply via the conduction-band discontinuities: for the passivation layer $\phi_2 = \phi_0 + \Delta E_2 + \Delta E_1$, and for the gate $\phi_2 = \phi_0 + \Delta E_1$, where ϕ_0 is the Schottky barrier for GaN. The assumptions regarding barrier heights are rough approximations made for simplicity; real interfaces and free surfaces call for a more rigorous assessment.
5. For simplicity, we ignore 2D and quantization effects.
6. We assume that there is no free space charge other than the induced electron gas.

These assumptions allow us to isolate as far as possible the classical electrostatic elements of the problem and to illustrate the principal features of the density profile in the channel.

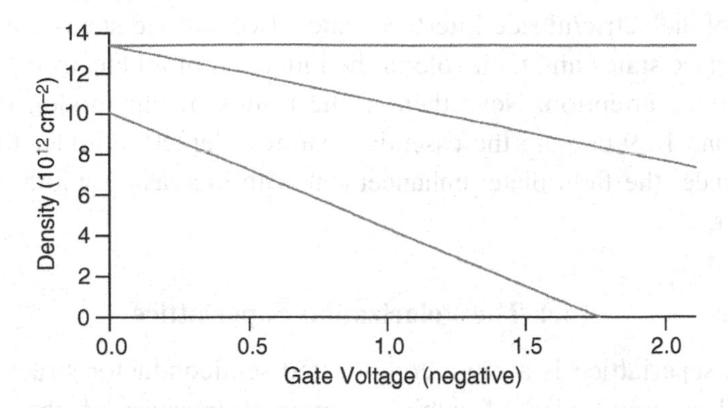

Fig. 14.8 Induced electron density. (Eq. (14.14) with $b=c=10\,\text{nm}$, $a \gg b,c$; $\psi=1.17\,\text{eV}$, $\varepsilon_0 \approx \varepsilon_1 \approx \varepsilon_2 = 10.3\varepsilon_{00}$; zero gate-drain field.)

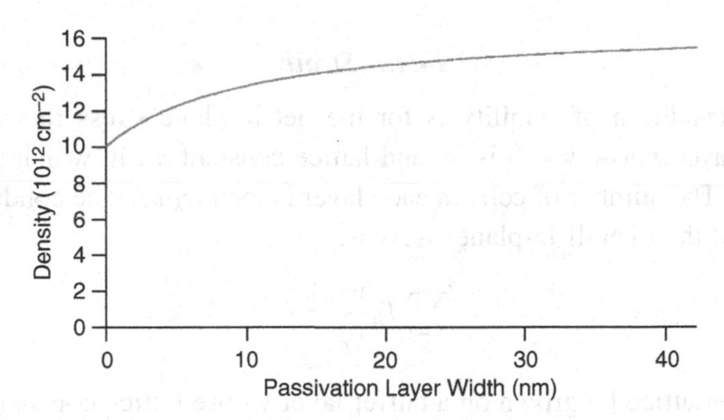

Fig. 14.9 Variation with passivation layer width ($V=0$, $b=10\,\text{nm}$).

With a gate bias equal to $-V$, the induced electron charges in the three regions are given by:

$$\sigma_n = \sigma_1 - \sigma_0 - \begin{cases} \phi_0 C_{cb} & \text{Free} \\ (\phi_0 + V)C_{cb} & \text{Field plate} \\ (\phi_0 + V)C_b & \text{Gate} \end{cases} \tag{14.14}$$

$$C_{cb} = \left(\frac{c}{\varepsilon_2} + \frac{b}{\varepsilon_1}\right)^{-1} \quad C_b = \frac{\varepsilon_1}{b}$$

$$\sigma_0 = 0.034\text{C/m}$$
$$\sigma_1^{sp} = [0.09x + 0.034(1-x) - 0.021x(1-x)]\text{C/m}$$
$$\sigma_1^{pz} = [0.0525x - 0.0282x(1-x)]\text{C/m}$$
$$\phi_0 = 1.17eV$$

The study of dielectric/nitride interface states, free surface states, and the metal/nitride interface states and their role in the induction of an electron gas is clearly worthy of more attention. Nevertheless, the results of our model, illustrated in Figs. 14.8 and 14.9 capture the essential features: depletion under the gate; less depletion under the field plate; enhancement with increasing width of the passivation layer.

14.4 The Polarization Superlattice

The nitride superlattice is a more complicated semiconductor structure than the more familiar superlattice of cubic compounds because of the presence of polarization, both spontaneous and piezoelectric. Elastic strain is, however, a common feature, given the mismatch of the lattice constants.

14.4.1 Strain

A general condition of stability is for the net in-plane stress to vanish. Let k identify a layer whose width is w_k and lattice constant a_k, in which the in-plane stress is T_k. The number of cells in each layer is then w_k/a_k. The condition for the vanishing of the overall in-plane stress is:

$$\sum_k T_k \frac{w_k}{a_k} = 0 \tag{14.15}$$

Let the superlattice be grown on a buffer layer whose lattice constant is a. Then the strain in layer k is $S_k = (a - a_k)/a_k$ and in layer j $S_j = (a - a_j)/a_j$. Eliminating a in favour of S_k we have $S_j = (S_k a_k + a_k - a_j)/a_j$. Recalling Eq. (14.15), we have:

$$\sum_j T_j \frac{w_j}{a_j} = \sum_j C_j S_j \frac{w_j}{a_j} = C_k \frac{w_k}{a_k} S_k + \sum_{j \neq k} C_j \frac{w_j}{a_j} \frac{S_k a_k + a_k - a_j}{a_j}$$

and therefore:

$$S_k = \frac{\sum\limits_{j \neq k} C_j w_j \frac{a_j - a_k}{a_j^2}}{a_k \sum\limits_j C_j \frac{w_j}{a_j^2}} \tag{14.16}$$

Since $S_k = (a - a_k)/a_k$, the condition for zero in-plane stress implies that:

$$a = \frac{\sum\limits_j C_j \frac{w_j}{a_j}}{\sum\limits_j C_j \frac{w_j}{a_j^2}} \tag{14.17}$$

14.4.2 Deformation Potentials

A biaxial strain can be split into a dilatational strain plus a uniaxial strain. In a cubic material, the uniaxial strain splits the light and heavy-hole bands (see Section 2.3). In wurtzite, there are three bandgaps to consider plus the crystal-field splitting. In the absence of splitting, the energy gap varies with strain according to:

$$\Delta E_g = -[d_1 S_3 + d_2(S_1 + S_2)] \tag{14.18}$$

The crystal-field splitting (but not the spin-orbit splitting) is:

$$\Delta_1 = \Delta_1(0) - [d_3 S_3 + d_4(S_1 + S_2)] \tag{14.19}$$

Shifts in the energy gap of adjacent layers affect the conduction-band and valence-band offsets, which are important in determining the confinement barriers, and these are taken to be proportional to the difference in energy gaps, e.g.

$$\Delta E_c = 0.69(E_g(0) + \Delta E_g).$$

14.4.3 Fields

We turn now to the polarizations and fields. The electric displacements between the kth and $(k+1)$th layers and between the $(k+1)$th and $(k+2)$th layers are as follows:

$$\begin{aligned} \varepsilon_{k+1}F_{k+1} &= \varepsilon_k F_k + \sigma_k - \sigma_{k+1} \\ \varepsilon_{k+2}F_{k+2} &= \varepsilon_{k+1}F_{k+1} + \sigma_{k+1} - \sigma_{k+2} = \varepsilon_k F_k + \sigma_k - \sigma_{k+2} \end{aligned} \tag{14.20}$$

It follows that the field in any layer is related to the field in a particular layer according to:

$$F_j = \frac{1}{\varepsilon_j}(\varepsilon_k F_k + \sigma_k - \sigma_j) \tag{14.21}$$

(F is field and we retain E for energy.) In the absence of an applied field the volts dropped across the superlattice must be zero:

$$\sum_j w_j F_j = w_k F_k + \sum_{j \neq k} w_j F_j = 0 \tag{14.22}$$

Substitution from Eq. (14.21) gives an expression for the field in any layer:

$$F_k = \frac{\sum\limits_{j \neq k} (\sigma_j - \sigma_k)(w_j/\varepsilon_j)}{\varepsilon_k \sum\limits_j (w_j/\varepsilon_j)} \tag{14.23}$$

We now have the tools to describe any nitride superlattice. Eq. (14.16) gives us the strain in each layer from which the piezoelectric polarization and the bandgap shifts can be deduced. Eq. (14.17) specifies the lattice constant of the substrate required to ensure that the superlattice is overall stress free. Eq. (14.23) gives us the field in each layer which is a constant throughout the layer in the absence of space charge. The sub-band structure can then be calculated using Schrödinger's equation with, assuming a constant effective mass for the carrier, wavefunctions described by Airy functions (see Appendix.)

14.5 The AlN/GaN Superlattice

As an example we consider the AlN/GaN superlattice (Ridley *et al.*, 2003). The data used are those culled from Ambacher *et al.* (1999, 2002) and are displayed in Table 14.1. The strain and field in each layer and the lattice constant of the buffer (and hence the value of x in $Al_xGa_{1-x}N$) are obtained from the equations above specialized to two types of layer. Explicit results are quoted for the superlattice consisting of one monolayer of AlN and seven monolayers of GaN. Fig. 14.10 shows the E-k diagram for the conduction band along the c direction (Bulutay *et al.*, 2000) and a comparison with the parabolic and **k.p** approximations. The energy dependence of the mass in AlN requires knowledge of the complex band structure, which is not available. In view of this and the comparatively weak dependence of the Airy functions on mass, energy-independent effective masses are assumed throughout.

The eigenfunctions are of the form:

$$\psi(z) = AAi(\chi) + BBi(\chi), \chi = \mu(Fz + \Delta E_c - E) \tag{14.24}$$

where the energies are in electron volts and $\Delta E_c = 0$ in the GaN well. Thus:

$$\begin{aligned} \psi(z) &= AAi(z) + BBi(z) \quad 0 \le z \le w_2 \qquad \text{for GaN} \\ &= CAi(z) + DBi(z) \quad -w_1 \le z \le 0 \qquad \text{for AlN} \\ &= [CAi(z) + DBi(z)]e^{ikd} \quad w_2 \le z \le d \qquad \text{for AlN} \\ d &= w_1 + w_2 \end{aligned} \tag{14.25}$$

where $\boldsymbol{k}$ is the superlattice wavevector. The coefficients are obtained by matching amplitude and mass-modified slope at $z=0$ and $z=w_2$. (Note that $\frac{dAi(z)}{dz} = \mu F \frac{dAi(\chi)}{d\chi}$.) Solutions are obtained numerically. The subband structure for AlN_1/GaN_7 is shown in Fig. 14.11 and in Table 14.2 and the wavefunction in Fig. 14.12. For comparison Fig. 14.13 shows the wavefunctions for the symmetrical AlN_6/GaN_6 superlattice.

Table 14.1. *Superlattice data* $(AlN)_1(GaN)_7$

	AlN	GaN
Lattice parameters:		
Monolayer	2.490Å	2.594Å
Layer thickness (a, b)	2.490Å	18.158Å
Lattice constant	3.103Å	3.189Å
Dielectric constant	10.31	10.28
c_{11}	410 GPa	370 GPa
c_{12}	140	145
c_{13}	100	110
c_{33}	390	390
Buffer layer		
C	499 GPa	453 GPa
a (buffer) / x	3.178Å / 0.128	3.178Å / 0.128
Strain (ε)	+0.02417	−0.003449
Bandgap		
Hydrostatic strain (ε_0)	0.03595	−0.004952
ΔE_g	−0.313 eV	+0.06413 eV
E_g	5.817 eV	3.484 eV
CB offset ΔE_c	1.47 eV	1.47 eV
Polarization		
Piezo	$-0.0483\ \mathrm{Cm^{-2}}$	$+0.003280\ \mathrm{Cm^{-2}}$
Spontaneous	−0.090	−0.034
Total	−0.1383	−0.0307
Field	−10.37 MV/cm	+1.422 MV/cm
Volts per layer	+0.258 V	−0.258 V
Airy functions		
Effective-mass ratio	0.48	0.228
μ	$2.271\ (\mathrm{eV})^{-1}$	$6.664\ (\mathrm{eV})^{-1}$
$X_1(0)$, $X(0)$	2.271(1.47-E)	−6.664E
$X_1(b)$, $X(b)$	$2.271(0.1037a+1.47-E)$	6.664(0.01422b−E)
γ	−1.181	−1.181

$$C = c_{11} + c_{12} - 2c_{13}^2/c_{33},\ a(buffer) = \left(\frac{ad_1}{a_{01}} + \frac{bd_2}{a_{02}}\right) / \left(\frac{ad_1}{a_{01}^2} + \frac{bd_2}{a_{02}^2}\right)$$

$x = (3.189 - a(buffer))/0.086$, $\varepsilon = (a(buffer) - a_0)/a_0$, $\varepsilon_0 = 2(1 - c_{13}/c_{33})\varepsilon$, $\Delta E_g = -12.95\varepsilon_0\,\mathrm{eV}$, $\Delta E_c = 0.63(E_{g1} - E_{g2})$, $P^{pz}(\mathrm{AlN}) = -1.808\varepsilon - 7.888\varepsilon^2\,\mathrm{Cm^{-2}}$
$P^{pz}(\mathrm{GaN}) = -0.918\varepsilon + 9.541\varepsilon^2\ \mathrm{Cm^{-2}}$

$$F_1 = -\frac{(\sigma_1 - \sigma_2)b}{\varepsilon_2 a + \varepsilon_1 b},\ F_2 = -F_1\frac{a}{b},\ \mathrm{V} = -\mathrm{F_1 a} = \mathrm{F_2 b}$$

$$\mu = \left(\frac{2em*}{\hbar^2 F^2}\right)^{1/3},\ \gamma = -\left(\frac{m_2^*}{m_1^*}\right)^{2/3}\left(\frac{F_1}{F_2}\right)^{1/3}$$

Table 14.2. *Subband energies (eV)*

Subband	Kd=0	Kd=π	Width	Gap
1	0.258	0.438	0.180	
2	1.418	0.820	0.598	0.382

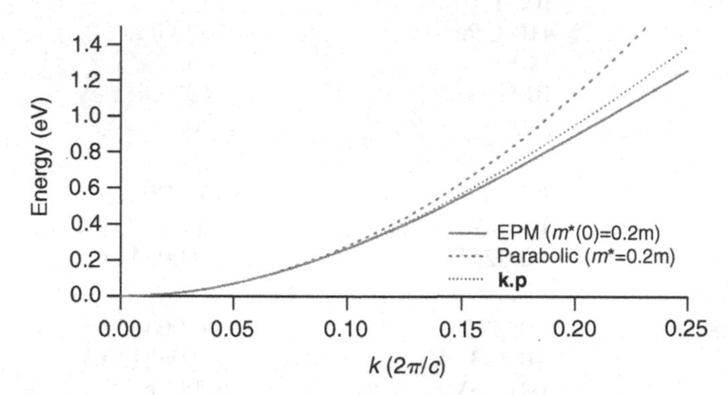

Fig. 14.10 E-k diagram for the GaN conduction band along the *c* direction with $m^*(0) = 0.2$ m without polaron correction. Parabolic and **k.p** approximations are shown for comparison.

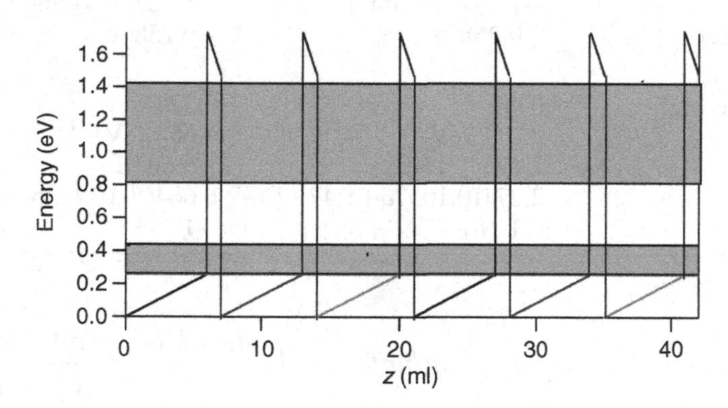

Fig. 14.11 Subband structure in the AlN_1GaN_7 superlattice.

E-k diagrams for AlN_n/GaN_7 with $n = 1,2,3$ are shown in Fig. 14.14. They depict the rapid narrowing of the subband width with increasing barrier width. Optical applications shift interest to the separation of the subbands. Fig. 14.15 shows how this varies with the number of AlN monolayers in the AlN_nGaN_7 structure.

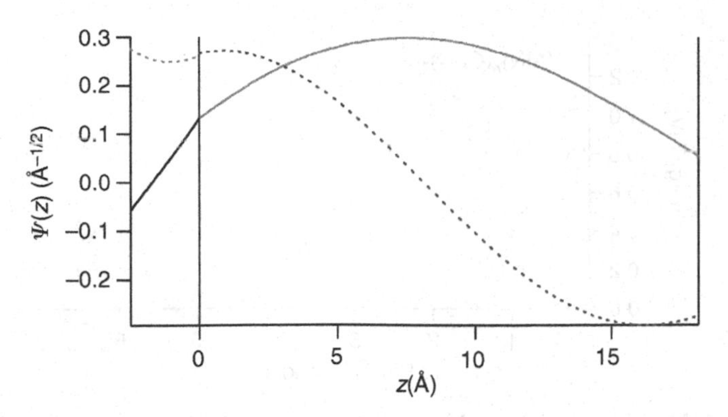

Fig. 14.12 Normalized wavefunctions for AlN_1GaN_7 superlattice. The continuous line is for the uppermost state in subband 1 and the dotted line is for the lowest state in subband 2. In both cases $kd = \pi$.

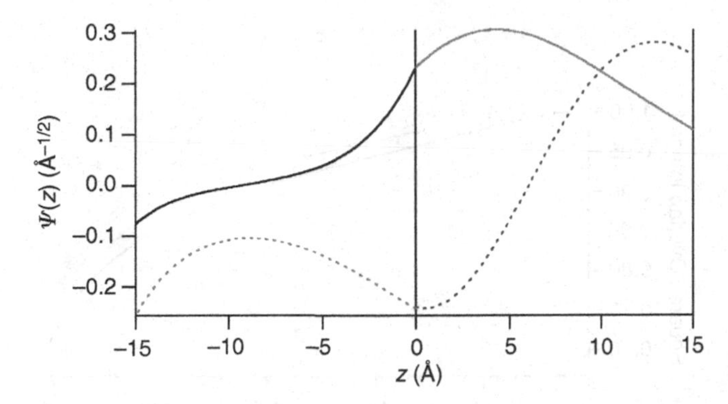

Fig. 14.13 Normalized wavefunctions for AlN_6GaN_6 superlattice. The continuous line is for the uppermost state in subband 1 and the dotted line is for the lowest state in subband 2. In both cases $kd = \pi$.

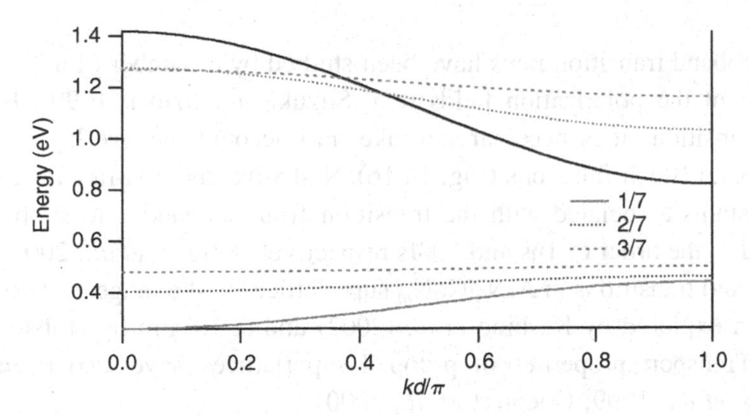

Fig. 14.14 E-k diagrams for AlN_nGaN_7 for $n = 1,2,3$.

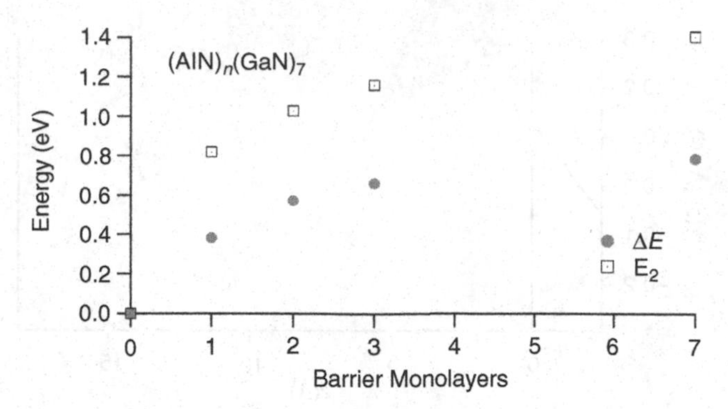

Fig. 14.15 Variation of subband energies with barrier thickness for the AlN_nGaN_7 superlattice. ΔE is the subband separation and E_2 is the energy at the bottom of subband 2.

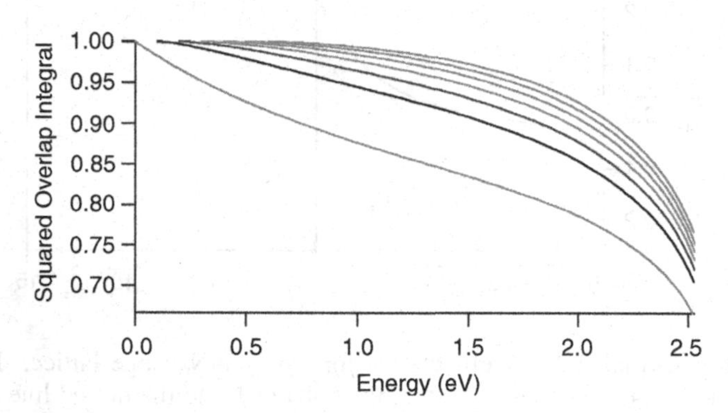

Fig. 14.16 Squared overlap integral of Block functions between energy states 0, 0.1, 0.2, 0.3, 0.4, 0.6 eV and higher energy states.

Intersubband transition rates have been studied by a number of authors taking into account the polarization fields (e.g. Suzuki and Iizuka, 1999). For large-energy transitions it is necessary to take into account the overlap of the conduction-band Bloch functions (Fig. 14.16). Radiative and non-radiative emission time-constants associated with the transition from subband 2 to subband 1 are estimated to the order of 1ns and 100fs respectively (Ridley *et al.*, 2003). Optical intersubband transitions in AlN_n/GaN_m superlattices in the range 1.08 to 1.61 μm have been explored by Kashino *et al.* (2002) and at 1.8 μm by Hofstetter *et al.* (2007). Transport properties of p-doped superlattices have also been studied (Kozodoy *et al.*, 1999; Goepfert *et al.*, 2000).

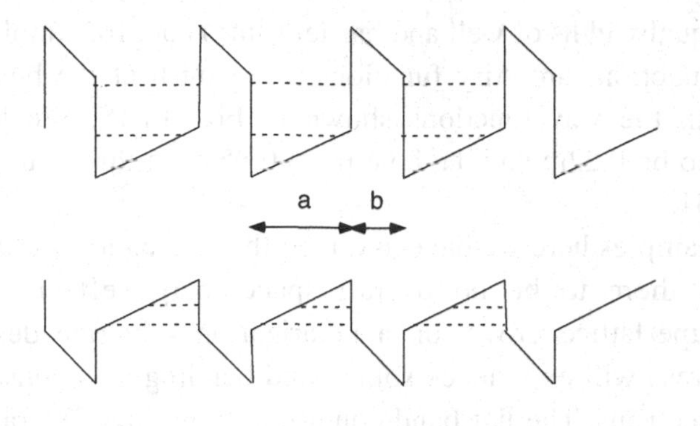

Fig. 14.17 An AlGaN/GaN superlattice.

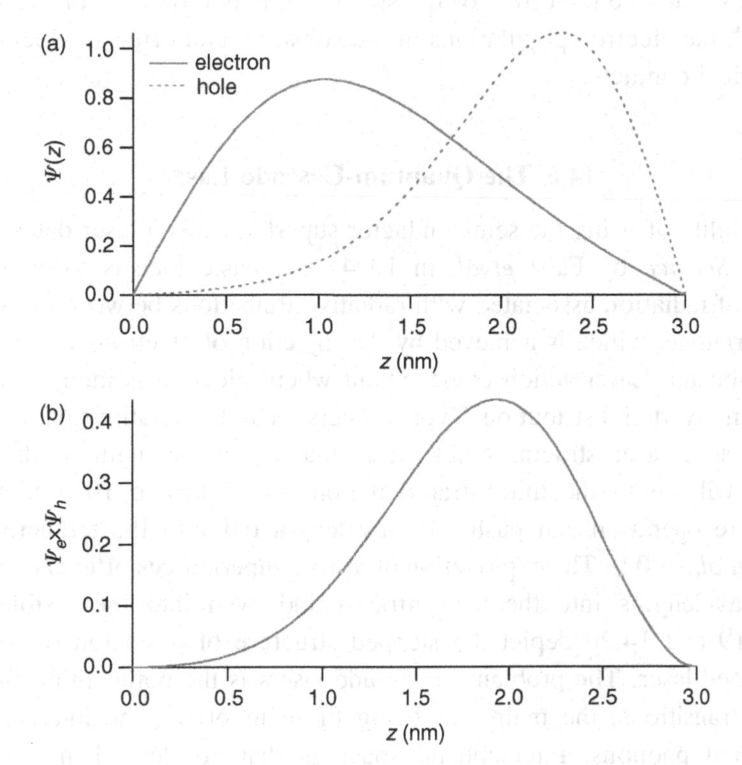

Fig. 14.18 (a) Normalized electron and hole wavefunctions in their respective wells. (b) Overlap. The overlap integral is 0.551.

As regards cross-gap optical properties, it is necessary to take into account the fact that, unlike the case in polarization-free superlattices, the wavefunctions of the electrons and holes do not coincide in space. A rough example can illustrate this. Fig. 14.17 depicts an $Al_{0.65}Ga_{0.35}N$/GaN in which the GaN is compressively strained. For equal widths of well and barrier (3nm) and, for simplicity, assuming that the wavefunctions are Airy functions that vanish at the boundaries of the well, we obtain the wavefunctions shown in Fig. 14.18. We have taken the electron mass to be 0.22m and the hole mass 0.85m. In this example the overlap integral is 0.551.

In all our examples here we have assumed that the bands are flat. That is, we have assumed there to be no overall space-charge effects. In general, a polarization superlattice grown on a polarization substrate designed to give overall zero stress will experience some band bending as a consequence of the induction of electrons. The flat-band condition, important for transport through the superlattice, can be engineered only at the expense of the stress-free condition. In a fully comprehensive polarization superlattice it is necessary, therefore, to include the effect of the spontaneous polarization of the substrate along with the electron populations in the substrate and effects associated with the electrical contacts.

14.6 The Quantum-Cascade Laser

The possibility of using the semiconductor superlattice as a laser dates from the article in *Science* by Faist *et al.* in 1994. The basic idea is to engineer the emission of radiation associated with radiative transitions between the subbands of a superlattice, which is achieved by the injection of electrons into the higher energy subband. Laser action comes about when injection is intense enough to produce an inverted distribution. Typical intersubband separations in GaAs-based and InP-based laser structures determine that the wavelength of the emitted radiation will be in the mid-infrared region, say 3 μm to 10 μm, and low-temperature operation can push the wavelength out into the terahertz regime (Kœhler *et al.*, 2002). The exploitation of nitride superlattices offers an expansion of the wavelengths into the near-infrared and even into the visible region. Figs. 14.19 and 14.20 depict the stepped structure of a section of a possible nitride-based laser. The problem in cascade lasers is the competition from non-radiative transitions, the main sort being those involving the interaction with polar-optical phonons. Intersubband spacings that are less than the optical-phonon energy are less affected by non-radiative processes. The larger phonon energies of the nitrides are therefore an advantage. The phonon energy in GaAs is 0.036 eV which corresponds to a wavelength of 34 μm (8.8 THz); in GaN the figures are 0.091eV and 13.4 μm (22 THz).

Further notes on quantum-cascade laser can be found in Section 15.6.

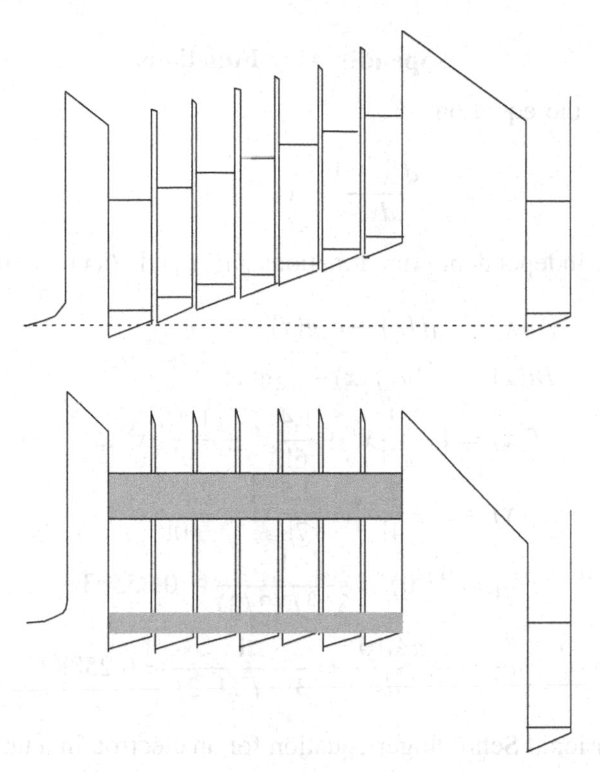

Fig. 14.19 An AlN/GaN stepped structure for a quantum-cascade laser. Upper figure is for zero applied field, the lower figure is for an applied field that allows electrons in the lower miniband to tunnel into the second subband of the quantum well.

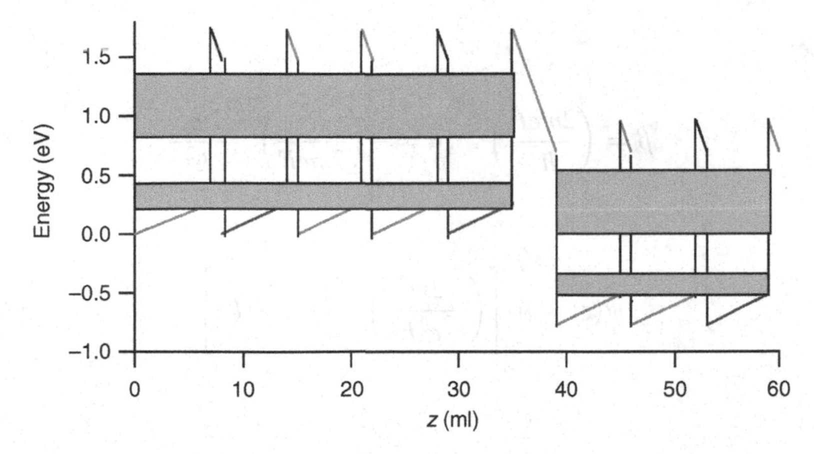

Fig. 14.20 Quantum-cascade structure.

Appendix Airy Functions

The solution of the equation:

$$\frac{d^2\psi(x)}{dx^2} - x\psi(x) = 0 \tag{A14.1}$$

are the linearly independent Airy functions $Ai(x)$ and $Bi(x)$ where:

$$\begin{aligned}
Ai(x) &= c_1 f(x) - c_2 g(x) \\
Bi(x) &= \sqrt{3}[c_1 f(x) + c_2 g(x)] \\
f(x) &= 1 + \frac{1}{3!}x^3 + \frac{1.4}{6!}x^6 + \frac{1.4.7}{9!}x^9 \ldots. \\
g(x) &= x + \frac{2}{4!}x^4 + \frac{2.5}{7!}x^7 + \frac{2.5.8}{10!}x^{10} \ldots \\
c_1 &= Ai(0) = \frac{1}{3^{2/3}\Gamma(2/3)} \approx 0.35503 \\
c_2 &= -\frac{dAi(0)}{dx} = \frac{1}{3^{1/3}\Gamma(1/3)} \approx 0.25882
\end{aligned} \tag{A14.2}$$

The one-dimensional Schrödinger equation for an electron in a uniform field F is:

$$-\frac{\hbar^2}{2m}\frac{d^2\psi(z)}{dz^2} + [eFz - E]\psi(z) = 0 \tag{A14.3}$$

Put $\psi(z) = Ai(\alpha + \beta z)$. We get, with $x = \alpha + \beta z$:

$$\frac{d^2Ai(x)}{dx^2} - \frac{2m}{\hbar^2\beta^2}(eFz - E)Ai(x) = 0 \tag{A14.4}$$

whence:

$$\beta = \left(\frac{2meF}{\hbar^2}\right)^{1/3}, \alpha = -\left(\frac{\hbar^2}{2meF}\right)^{2/3}\frac{2mE}{\hbar^2} \tag{A14.5}$$

Thus:

$$\psi(z) = Ai\left[\left(\frac{2m}{\hbar^2 e^2 F^2}\right)^{1/3}(eFz - E)\right] \tag{A14.6}$$

15
Terahertz Sources

What radiance of glory,
What bliss beyond compare.
Jerusalem the Golden, *James Mason Neale*

15.1 Introduction

In recent years there has been growing interest in sources of radiation whose frequencies occupied the region of the electromagnetic spectrum between, roughly, 300GHz and 30THz. The numerous applications include their use in medicine, molecular spectroscopy, communication, and security. The frequency range is covered by free-electron lasers, but there is an obvious need for more portable and adaptable sources, and this need has focussed attention on the properties of semiconductors and semiconductor multilayers. Existing semiconductor microwave generators such as the Gunn diode run out of power above 100GHz and IMPATTs above 300GHz. A new generation of devices that can extend the wavelength range beyond the familiar millimetre wave regime and into the sub-millimetre regime is required, and there has been substantial progress in recent years, much of it associated with the exploitation of the femtosecond Ti-sapphire laser and various non-linear optical processes. Semiconductors have also been used as THz detectors, notably, Ge and Si bolometers operating at liquid helium temperatures. The photon energy at 1THz is 4meV, and photon energies of room temperature radiation are around 25meV, so the problem of detection of THz radiation is that it is always against a competing, pervading background. Mention in Section 14.6 has already been made of cascade lasers operating in the near infrared. Recently, a continuous-wave laser operating at 9μm at room temperature has been reported (Beck *et al.*, 2002), but those that operate in the THz range require low temperatures. It is evident that nitride structures may allow room temperature operation. In this chapter we discuss other physical mechanisms in semiconductor structures that can yield THz radiation. This includes descriptions of ballistic and non-ballistic transport in a superlattice miniband; ballistic and non-ballistic transport in the conduction band of a large-bandgap semiconductor at very high electric fields; Bloch oscillations and Wannier–Stark states. We also look at the transport properties of the bulk nitrides in connection with the next generation of millimetre microwave and THz generators. Discussion of optically-induced THz generation is limited to surface excitation by femtosecond pulses, and to cw operation as exemplified in photoconductive mixing in a pin structure and in the quantum-cascade laser.

15.2 Bloch Oscillations

In the recent literature on THz generation, the phenomenon of Bloch oscillations has been suggested as a possible source. The concept of Bloch oscillations has had an interesting history, which we will briefly review here.

The kinetics of an electron in a periodic potential was revealed when Bloch (1928) showed that a ballistic electron, confined to the conduction band of a crystal, would exhibit oscillations under the influence of a constant applied electric field as a consequence of Bragg reflections at the Brillouin zone boundary. In the absence of collisions, the crystal momentum of the electron increases with time in the presence of an electric field according to the acceleration law, $d\hbar\mathbf{k}/dt = e\mathbf{F}$. The extent of the Brillouin zone in a major crystallographic direction is $2\pi/a$, where a is the lattice constant, so the Bloch angular frequency is $\omega_B = eaF/\hbar$. With $a \sim 5$ Å, the frequency would be in the THz range for a field of 10^5 V/cm, hence of interest to the THz community. Unfortunately, in bulk material, the scattering rate is also in the THz regime, which calls for fields of 1 MV/cm and more to retain the ballistic nature of the motion. But such high fields are beyond what most semiconductors can sustain without breakdown, so it is not surprising that Bloch oscillations have never been observed in bulk material. Nevertheless, Zener (1934) used Bloch's model to calculate the rate of tunnelling between bands as an approach to understanding the electric breakdown of dielectrics, and the formalism of Bloch's and Zener's approach was amplified by Houston (1940) who derived a general expression for the acceleration of an electron including its behaviour near a zone boundary.

But it was not *a priori* obvious that the concept of the conduction band in an applied field remained valid. On the one hand, the form of the conduction band rested on the existence of a periodic potential, but the scalar potential, eFx, that characterized the field was obviously not periodic, nor was it bounded, but Wannier (1960, 1962) showed that the concept indeed remained valid, but that the energy states became arranged in the form of a Stark ladder, whose steps were separated by the energy eFa. The existence of these Wannier–Stark (W-S) states was severely questioned by Zak (1968), and this led to some controversy (Wannier, 1969; Zak, 1969; Rabinovitch and Zak, 1971). The problem here was focussed on the role of other conduction bands, each of which has its own W-S ladder, into which electrons may tunnel, thus localization of the W-S sort could, at best, be only temporary. The time-independent interaction of these bands means that the rungs of the ladder of one band never coincide with those of another, as Bychkov and Dykhne (1965) pointed out. Changing the field means that ladders slide past one another so avoiding the possibility of resonances. The description of transport under these conditions becomes very complicated, as Avron (1982) has indicated, and Zak (1996) has recently described W-S ladders in terms of quasi-energy states. Problems associated with the coordinate representation, necessitated by the form of the scalar potential, were circumvented by Kreiger and Iafrate (1986) by representing the field by a vector potential. The existence of Bloch oscillations and W-S states was shown to follow. The observation of sufficiently long-lived W-S ladders in the minibands of superlattices by Menendez *et al.* (1988) and Voisin *et al.* (1988) has proved their reality beyond reasonable doubt. Nevertheless, their detection in bulk material has still to be reported. The problem of the electron dynamics in the multiband case continues to receive attention (e.g. Leo and Mackinnon, 1989; Rotvig *et al.*, 1996), and the situation in degenerate bands has begun to be studied (Foreman, 2000; the inhibiting effect of W-S quantization on the impact-ionization rate was pointed out by Ridley, 1998).

Spatial localization in a single non-degenerate conduction band was commonly discussed under the implicit assumption that the crystal was of infinite extent. The validity of the effective-mass approximation in finite crystals was investigated by Rabinovitch (1971) and Rabinovitch and Zak (1971), in which it was shown that the boundary conditions could change the situation drastically. This topic was further pursued by Fukuyama *et al.* (1973) and later by others (Churchill and Holmstrom, 1982; Davidson *et al.*, 1997; Onipko and Malysheva, 2001). In a 1D one-band model, a W-S ladder is formed in an infinite crystal with periodic boundary conditions or with rigid-wall boundary conditions. But in a finite crystal, length L, with finite conduction-band width, E_B, the energy spectrum reverts to that for free Bloch electrons when $eFL < E_b$. Moreover, in the case of 3D the spectrum is discrete along the field direction only if the field is directed along a principal crystallographic direction, and it is continuous with respect to motion perpendicular to the field direction. If the field does not lie along a principal crystallographic direction, the spectrum in the direction of the field becomes continuous, since the period of the motion becomes infinite. Onipko and Malysheva (2001) have pointed out that in addition to the presence of Bloch

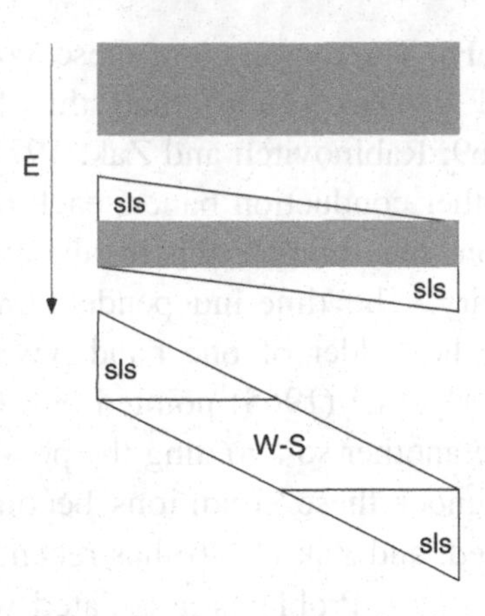

Fig. 15.1 Surface localized states and Wannier–Stark localized states.

extended states and W-S states there exist surface localized states extending spatially between one crystal boundary and a band edge (Fig. 15.1).

The transport of electrons in high electric fields has been a topic of intense interest for the best part of a century. Initially motivated by acquiring an understanding of electric breakdown in insulators (e.g. Von Hippel, 1937; Callen, 1949; Kane, 1959; Fröhlich, 1937), investigations expanded to include hot-electron effects, including negative differential resistance (NDR), in semiconductors (see Conwell, 1967). Pushing the hot-electron theme to its limits in the context of a narrow conduction band Bychkov and Dykhne (1965) predicted that in the limit of high fields the current would inevitably become inversely proportional to the field. Their argument was the simple one that once the electrons were hot enough to become uniformly spread throughout the band their average energy was no longer dependent on the field and, consequently, from the energy-balance equation, **j.F** was a constant, hence $j \sim 1/F$. Given that a current exists under these circumstances, a negative differential resistance appeared to be a fundamental feature of high-field transport. Incorporating W-S quantization gave results that supported this conclusion (Bryskin and Firsov, 1972; Levinson and Vasevitchyute, 1972). But it was unclear that the properties of any known bulk materials would correspond to the conditions under which the predicted NDR could be observed.

However, the concept of an artificial superlattice, made of alternate layers of different semiconductors, offered the possibility of realizing in the narrow miniband structure the right conditions. Esaki and Tsu (1970) showed that NDR was indeed to be expected as a consequence of electrons being excited into the negative mass regions of the miniband. Their analysis assumed the existence of momentum-relaxing collisions but no energy-relaxing collisions. In the presence

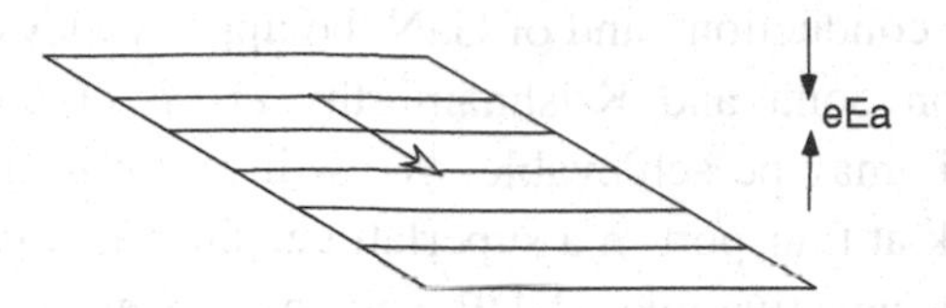

Fig. 15.2 Transport via hopping down the W-S ladder. (Here E is field.)

of a NDR a convective instability is expected (Ridley, 1963), and Lei *et al.* (1995) showed that a proper account of convective instability had to take into account energy relaxation as well as momentum relaxation. It is clear that when the electron is in a localized state, some form of energy relaxation is absolutely necessary in order for there to be a current. We return to this point later. The necessity for energy relaxation is most obvious for electrons in W-S states. A current in this case can only be generated by a hopping mechanism down the W-S ladder (Fig. 15.2), which can occur most efficiently via the emission of optical phonons, as discussed by Hacker (1969). The hopping mechanism was also discussed by Bryskin and Firsov (1972) and by Sawaki and Nishinaga (1977).

The focus of experimental work has naturally been on these effects in superlattices, but this has necessitated the ambient of low temperatures. It is still necessary to consider the situation in bulk material and to examine the possibility of achieving a robust NDR at high fields and at room temperature. Of course, there is the well-known electron transfer mechanism that is the source of the Gunn effect, which, in InP allows the generation of frequencies up to around 100 GHz, the limit being the rate of return of electrons from the upper heavy-mass band and the rate at which energy is relaxed. Other sources of NDR that have been studied are mixed scattering and non-parabolicity (e.g. Harris and Ridley, 1973). A high-field NDR that depends solely on intraband processes has the technological attraction of a significantly higher frequency performance, the limits being intraband scattering rates. Sokolov *et al.* (2004, 2005) have analysed the Gunn effect in GaN and predict frequency outputs between 140 GHz and 1.6 THz. But if a bulk semiconductor is to exhibit negative-mass NDR and W-S ladders, it must be one that can accommodate high electric fields without electrical breakdown, and GaN, with a band-gap of 3.4 eV, is a prime candidate.

15.3 Negative-Mass NDR

Negative-mass NDR goes back to Krömer's idea of exploiting the valence-band structure of Krömer (1958), an idea that is still relevant given the valence-band structure in quantum wells (see Section 2.3, also Cao *et al.*, 2001; Gribnikov *et al.*, 2001). In the conduction band of GaN the upper valleys have energies at or above the inflection point and Krishnamurthy *et al.* (1997) suggested that a negative-mass NDR may be achievable. We will return to this idea shortly, but first we should look at transport in a superlattice, since it was the first multilayer structure to offer a negative-mass NDR and, moreover, it was experimentally more accessible.

In describing electron transport through a superlattice, it is necessary to delineate several different regimes. The first thing to do is to define the condition in which it makes sense to talk about minibands. If τ is the scattering time-constant for the electrons, the uncertainty in energy will be at least $\hbar/\tau$. If E_B is the miniband width, the condition for there being a recognizable miniband is $E_B >> \hbar/\tau$. When this is not the case, the superlattice is better described as being multiple quantum wells, with transport occurring via sequential tunnelling. If the wells are identical, the subbands coincide in energy, and tunnelling can occur. In the presence of a field, however, the coincidence is weakened and ultimately destroyed with increasing field. This gives rise to a NDR and consequent domain formation. The description of sequential-tunnelling transport becomes complicated (see Grahn, 1995; Wacker, 1998) with current oscillations typically at MHz frequencies.

When $E_B >> \hbar/\tau$ we can speak about conventional transport in a conduction band whose shape is determined by the period of the superlattice. The simplest description is of the time-independent transport in an infinitely long diode. We adopt for clarity, generality, and simplicity a model band structure, and a model scattering mechanism, in particular, a tight-binding-like cosine band structure of the form:

$$E = (E_B/2)(1 - \cos ka) \tag{15.1}$$

where k is the wavevector and a is the superlattice period. Since the period and bandwidth vary from superlattice to superlattice, we choose to illustrate transport in the conduction band of GaN since the cosine structure as in Eq. (15.1) is a good approximation (Fig. 15.3). (But see the Appendix.) The band structure of Eq. (15.1) exhibits an inflection point where the mass becomes negative. A description of electron transport with this band structure is straightforward only if collisions are ignored and the electrons are regarded as moving ballistically according to the quantum-mechanical acceleration law. We consider this case in the next section. The challenge here is to incorporate the effects of scattering in an acceptable way. We look at three approaches to the problem.

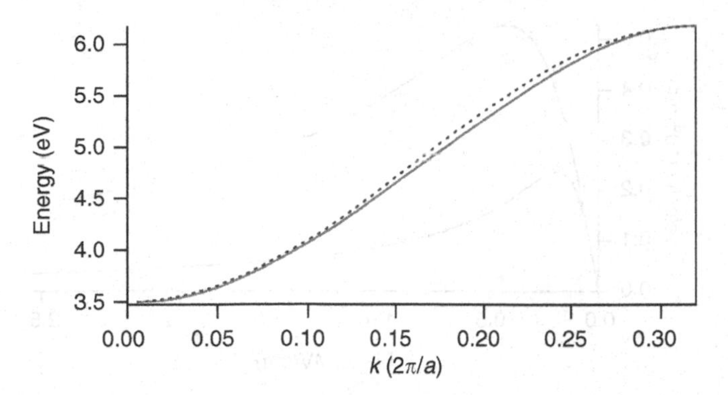

Fig. 15.3 Energy versus wavevector in the conduction band of GaN along the *c*-axis (Bulutay *et al.*, 2000). The dotted line is the cosine approximation.

15.3.1 The Esaki–Tsu approach

We assume that the electron gas is uniform and begin with the acceleration law:

$$\frac{dk}{dt} = \frac{eF}{\hbar} \tag{15.2}$$

The group velocity is given by $dE/d\hbar k$ and so:

$$\frac{d\text{v}}{dt} = \frac{1}{\hbar}\frac{d^2E}{dk^2}\frac{dk}{dt} = \frac{1}{\hbar}\frac{d^2E}{dk^2}\frac{eF}{\hbar} \tag{15.3}$$

In the absence of collision, integrating would give us the group velocity as a function of time, which, of course, we could have obtained directly from Eq. (15.1). However, we need to bring in collisions. We assume a simple momentum-relaxation time and weight the velocity with the probability of avoiding a momentum-relaxing collision:

$$\langle \text{v} \rangle = \frac{eF}{\hbar}\int_0^t e^{-t/\tau_m}\frac{1}{\hbar}\frac{d^2E}{dk^2}dt = \frac{eF}{\hbar}\int_0^t e^{-t/\tau_m}(E_B a^2/2)\cos(ka)dt \tag{15.4}$$

Putting $k = eFt/\hbar$ and integrating gives:

$$\langle \text{v} \rangle = \frac{E_B a}{2\hbar}\frac{\omega_B\tau_m}{1+(\omega_B\tau_m)^2}[1+(\omega_B\tau_m\sin\omega_B t - \cos\omega_B t)e^{-t/\tau_m}] \tag{15.5}$$

Esaki and Tsu then take the solution for $t \to \infty$, and obtain:

$$\langle \text{v} \rangle_\infty = \frac{E_B a}{2\hbar}\frac{\omega_B\tau_m}{1+(\omega_B\tau_m)^2} \tag{15.6}$$

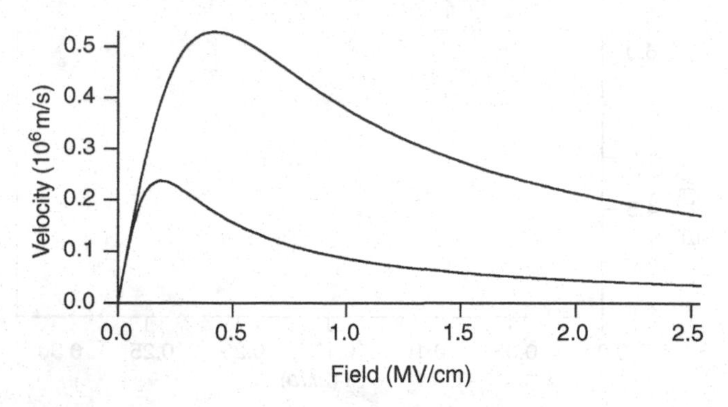

Fig. 15.4 Steady state average velocity with momentum and energy relaxation assuming a constant effective mass. Comparison is made with the Esaki–Tsu prediction (upper curve).

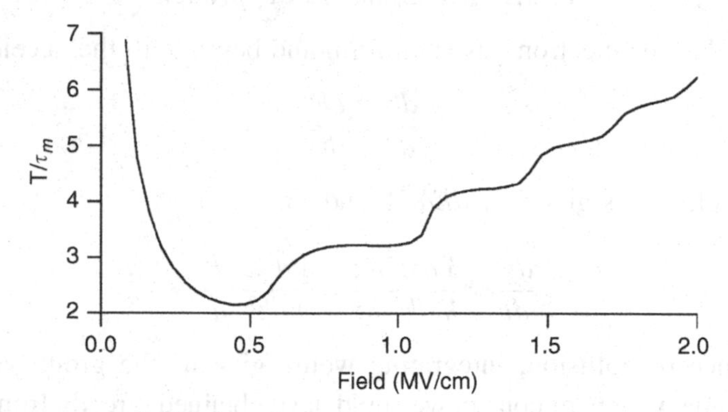

Fig. 15.5 Transit time for 400Å diode.

The NDR is shown in Fig. 15.4. In short diodes the transit time can be too short for a steady state to be established (Fig. 15.5) and, as a consequence, the NDR has time-dependent elements (Fig. 15.6).

15.3.2. Lucky Drift

It will be noticed that the Esaki–Tsu model does not include energy relaxation. This implies that the electrons will diffuse upwards in energy, eventually reaching the upper band-edge and suffering Bragg reflection and consequent localization. Including the probability of avoiding an energy-relaxing collision leads essentially to the lucky-drift model for impact ionization (Ridley, 1983). Rather than follow this approach further, we turn to the case where energy relaxation produces a steady state.

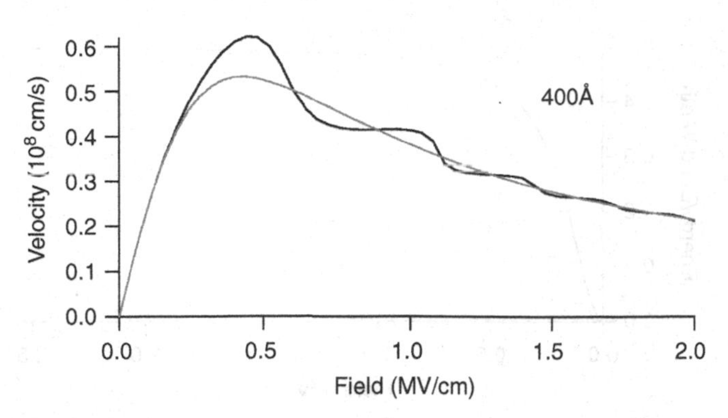

Fig. 15.6 Effect of transit time in a 400 Å diode.

15.3.3 The Hydrodynamic Model

We continue to assume a uniform electron concentration and write down the balance equations for dynamic quantities that are averages over the electron distribution function:

$$\begin{aligned} \frac{d\langle \mathrm{v}\rangle}{dt} &= eF\left\langle \frac{1}{m^*}\right\rangle - \frac{\langle \mathrm{v}\rangle}{\tau_m} \\ \frac{d\langle E\rangle}{dt} &= eF\langle \mathrm{v}\rangle - \frac{\langle E\rangle}{\tau_E} \end{aligned} \tag{15.7}$$

Here τ_m are τ_E the average momentum and energy-relaxation times respectively, and we have taken the initial energy to be zero. These equations are essentially those of the hydrodynamic model of transport, recently used to describe convective instability in a superlattice (Lei *et al.*, 1995). Solving the equations for the steady state gives the average energy and velocity as functions of field:

$$\begin{aligned} \langle \mathrm{v}\rangle &= \mathrm{v}_m \frac{\omega_B \tau_m}{1+\omega_B^2 \tau_m \tau_E} \\ \langle E\rangle &= \frac{E_B}{2}\frac{\omega_B^2 \tau_m \tau_E}{1+\omega_B^2 \tau_m \tau_E} \end{aligned} \tag{15.8}$$

where $\mathrm{v}_m = E_B a/\hbar$, $\omega_B = eFa/\hbar$. The variation of the drift velocity with field is depicted in Fig. 15.4 where it is compared with the Esaki–Tsu result. The average energy is shown in Fig. 15.7.

In short samples the upper limit of the time is the transit time, T, which is obtained from $L = \int_0^T \mathrm{v}(t)dt$, where L is the length of the diode. The finite length of the diode or superlattice becomes an important factor. The concept of a transit time is, of course, useful only up to the point of Bragg reflection. In the absence

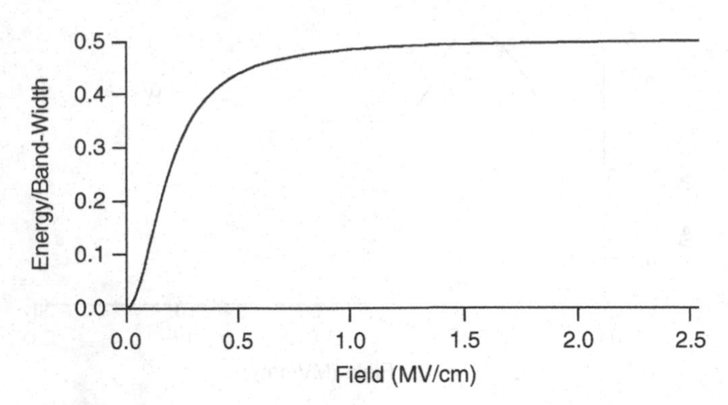

Fig. 15.7 Average energy.

of energy loss, the critical field for this is just $F = E_B/eL$. If the emission of optical phonons is taken into account, the critical field is modified by the average number of phonons emitted in the transit time, thus $F = (E_B + \hbar\omega T/\tau)/eL$, where τ is the time-constant for spontaneous emission (~10 fs in GaN). Beyond this critical field the electron executes oscillations and becomes localized. In times of order $E_B\tau/2\hbar\omega \approx 150$ fs the electrons will become thermalized with some occupation of the upper conduction-band valleys, and transport will be describable by conventional theory. In the absence of intervalley transfer, Bragg reflection ultimately forces the system into some constant energy state which will exhibit the Bychkov–Dykhne NDR.

We have said nothing about impact ionization, yet this phenomenon is bound to set a limit on the length of the bulk diode at high fields. If impact ionization across the forbidden gap is to be avoided, the applied voltage must not greatly exceed the bandgap E_g. Thus, ideally, the length should satisfy $eFL \leq E_g$, which implies that for fields of order 1 MV/cm and a gap of 3.4 eV the length should not exceed 340 Å. Transit times then become short enough for ballistic motion to become dominant.

15.4 Ballistic Transport

So far we have been dealing with the case in which diffusive transport is dominant. We now want to look at four cases where the collision rate is weak enough for the electrons to move ballistically in the applied field. These are:

(1) Optical-phonon-determined transit-time oscillations.
(2) Transit-time oscillations in a short diode.
(3) Negative-mass NDR
(4) Bloch oscillations

15.4.1 Optical-Phonon-Determined Transit-Time Oscillations

At low temperatures absorption of optical phonons can be ignored and the scattering rate for electrons with energy less than $\hbar\omega_{LO}$ is determined by acoustic phonons, impurities, and other defects. Compared to the emission rate for optical phonons, the low-energy rate is typically small. In a sufficiently strong field, low-energy electrons can move ballistically until their energy equals or exceeds the optical-phonon energy when they rapidly emit a phonon and return to near zero-energy. They are then accelerated once more up to the phonon-emission threshold and once again return. The electrons therefore execute oscillations in k space and in real space. Initially these oscillations will be coherent, but, ultimately, scattering and penetration above the optical-phonon emission threshold will damp the oscillations. Coherence is improved in quantum wells.

The abrupt emission threshold in quantum wells and its potential for initiating a NDR was pointed out by Ridley (1982, 1984) and, when electron–electron scattering is dominant, the streaming Maxwellian distribution leading to a squeezed distribution was analysed by Ridley *et al.* (2000). The large phonon energy of the nitrides make these materials particularly suitable for study. Starikov *et al.* (2001, 2002) have predicted that, up to liquid nitrogen temperatures, the bulk nitrides (InN, GaN, AlN) are potentially capable of generating radiation from 0.1 to 5THz.

15.4.2 Transit-Time Oscillations in a Short Diode

The classical theory of ballistic transport in vacuum diodes (Llewellyn, 1941) can be applied to ballistic electrons in semiconductors provided the energy is low enough for the effective mass to be assumed constant and the diode is free of carriers before injection. Electrons are taken to be injected into a carrier-free region from the cathode at $x=0$ with zero velocity, become accelerated by the applied field, and are collected at the anode at $x=L$. Since they introduce space charge, the current density is the sum of drift and displacement:

$$j = \rho \mathrm{v} + \varepsilon \frac{\partial F}{\partial t} \tag{15.9}$$

In one dimension Poisson's equation gives $\rho = \varepsilon \frac{\partial F}{\partial x}$ whence:

$$j = \varepsilon \left(\mathrm{v} \frac{\partial F}{\partial x} + \frac{\partial F}{\partial t} \right) = \varepsilon \frac{dF}{dt} = \frac{\varepsilon m}{e} \frac{d^2 \mathrm{v}}{dt^2} \tag{15.10}$$

The current density is taken to be separable into *dc* and *rf* components:

$$j = j_{dc} + j_{rf}e^{i\omega t} \tag{15.11}$$

The applied voltage is:

$$V = \int_0^L F dx = \frac{m}{e}\int_0^L \frac{dv}{dt} dx \tag{15.12}$$

After some manipulation these equations allow us to write down the real and imaginary parts of the impedance:

$$\begin{aligned} R &= \frac{ej_{dc}}{\varepsilon^2 m}\frac{2(1-\cos\omega T)-\omega T\sin\omega T}{\omega^4} \\ X &= -\frac{L}{\varepsilon\omega}-\frac{ej_{dc}}{\varepsilon^2 m}\frac{\omega T(1+\cos\omega T)-2\sin\omega T}{\omega^4} \end{aligned} \tag{15.13}$$

where T is the transit time given, in the absence of space charge, by:

$$T = \left(\frac{2mL}{eF}\right)^{1/2} \tag{15.14}$$

(Note that for the space-charge-limited case the transit time is just 3/2 times the space-charge-free result.) The real part is depicted in Fig. 15.8. Regions of NDR appear above $\omega T \approx 2\pi$. With L=1μm F=*0.1MV/cm*, $T \approx$*1ps*, and so the frequency will be in the THz regime. Thus, microscopic vacuum diodes have a role to play. However, the NDR is extremely weak. A scheme for enhancing the NDR by using a specially graded heterostructure which enhances a population inversion caused by the more rapid transit of lower energy electrons has been suggested by Kozlov *et al.* (2004). Here, the motion is non-ballistic with optical-phonon emission creating the low energy population in a region of stronger field.

15.4.3 Negative-Mass NDR

But in a real semiconductor the mass is no longer constant and Eq. (15.10) becomes:

$$j = \varepsilon\left(v\frac{\partial F}{\partial x}+\frac{\partial F}{\partial t}\right) = \varepsilon\frac{dF}{dt} = \frac{\varepsilon}{e\hbar}\frac{d^2k}{dt^2} \tag{15.15}$$

Adoption of the cosine band structure of Eq. (15.1) allows us to obtain analytic expressions for the real and imaginary parts of the impedance (Ridley *et al.*,

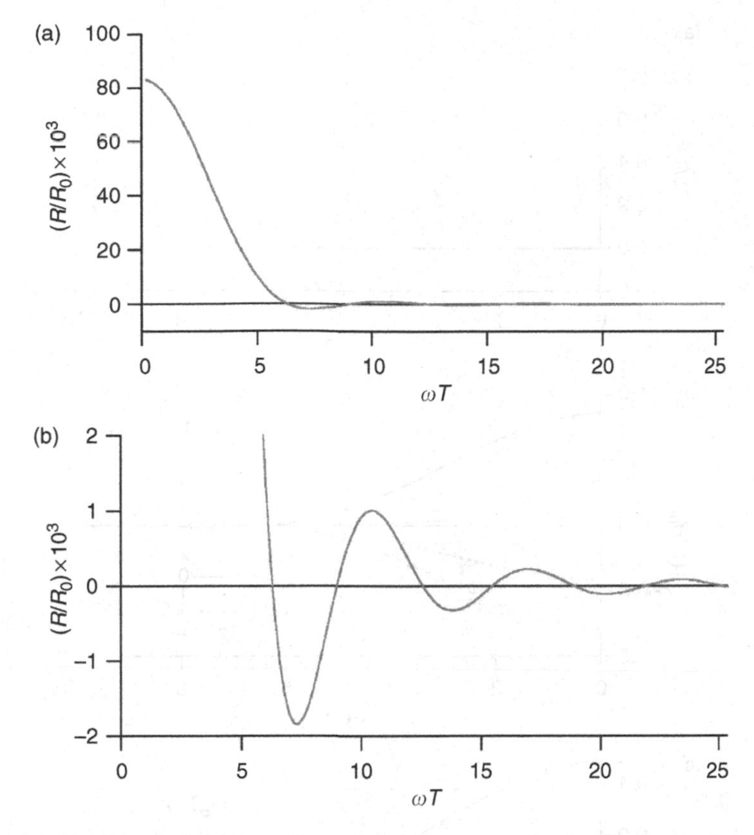

Fig. 15.8 Real part of the impedance for the vacuum diode. The lower figure is a magnified section showing the weak transit-time NDR regions.

2005). Allowing for the possibility of injection at high energies (e.g. over a barrier) leads to:

$$R = \frac{ej_{dc}}{\varepsilon^2 m_0^*} \frac{1}{\omega^2\omega_B(\omega^2 - \omega_B^2)} \begin{bmatrix} \omega_B(\cos y + \cos a)(1 - \cos\omega T) \\ -\omega(\sin y - \sin a)\sin\omega T \end{bmatrix}$$
$$X = -\frac{L}{\varepsilon\omega} - \frac{ej_{dc}}{\varepsilon^2 m_0^*} \frac{1}{\omega^2\omega_B(\omega^2 - \omega_B^2)} \begin{bmatrix} \omega_B(\sin y - \sin a)(1 + \cos\omega T) \\ -\omega(\cos y + \cos a)\sin\omega T \end{bmatrix} \quad (15.16)$$

where $\omega_B = eFa/\hbar$, $y = \omega_B T + a$ and $a = k_0 a$; m_0^* is the effective mass at the band-edge and k_0 is the injection wavevector. The transit time is:

$$T = \frac{1}{\omega_B}\left[\cos^{-1}\left(\cos a - m_0^*\omega_B aL/\hbar\right) - a\right] \quad (15.18)$$

Eqs. (15.16) and (15.17) reduce to Eqs. (15.13) and (15.14) in the limit. The real part of the impedance is depicted in Fig. 15.9. For cold injection ($k_0 = 0$)

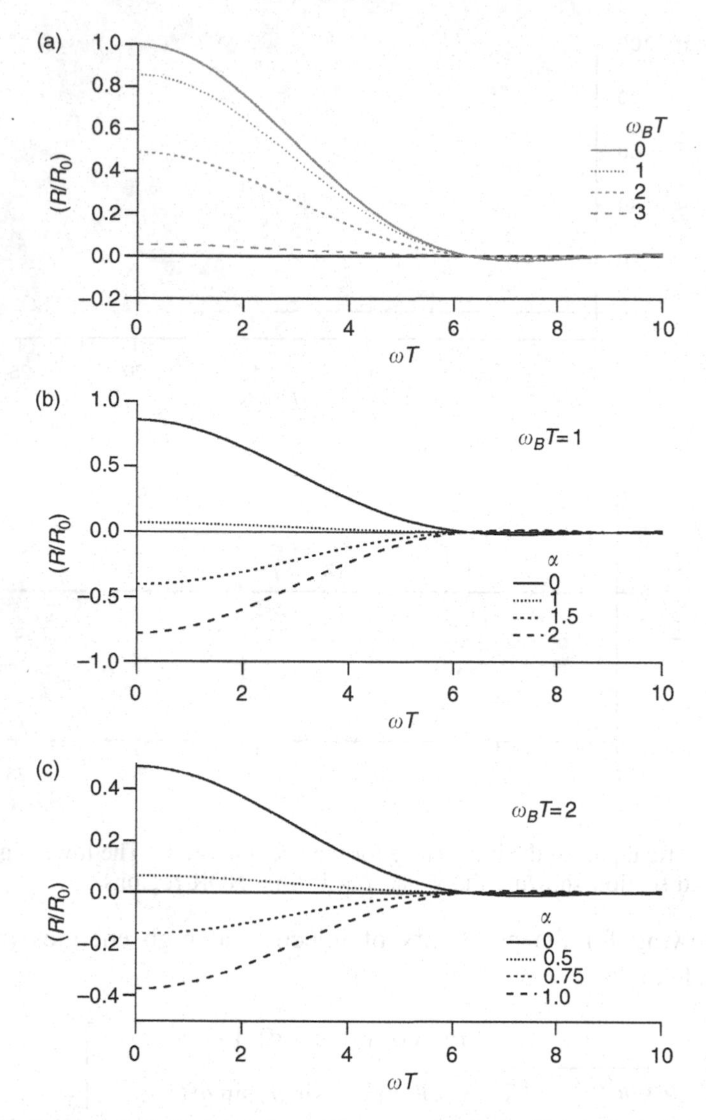

Fig. 15.9 Real part of the impedance for a crystal diode illustrating the effect of field and hot injection.

transit-time NDR appears as usual. For an injection energy that is sufficiently high, a substantial negative-mass NDR appears over a broad range of lower frequencies. Qualitatively similar behaviour is exhibited under space-charge-limited conditions (Dyson and Ridley, 2006). A structure consisting of an AlGaN/GaN junction with injection from the AlGaN barrier into a short GaN layer is able to exploit the large electric field associated with the spontaneous polarization (Dyson *et al.*, 2006). However, the phonon-emission time in GaN is very short (~10fs) which implies that for truly ballistic motion the diode has to be extremely short. The polar nature of the interaction favouring small-angle collisions means that the momentum-relaxation time (~30 fs) is the relevant time-constant at high energies rather than the scattering time. Nevertheless, even for quasi-ballistic motion the diode cannot be very long, unless the field is high enough for quasi-Bloch oscillations, that is $\omega_B\tau_m \geq 2\pi$, corresponding to a field greater than about 3MV/cm. Fields of this order associated with the spontaneous polarization are present in nitride heterostructures. This property, plus the large bandgap and the inhibition of impact ionization introduced by W-S state localization, make the AlGaN/GaN system very promising for exhibiting Bloch oscillations and W-S states in bulk material.

15.4.4 Bloch Oscillations

The implicit assumption in our discussion of ballistic transport is that the motion is unimpeded between cathode and anode. The implication is that the field is small enough to satisfy the condition $eFL<E_B$. When this is not the case, motion towards the anode is impeded by Bragg reflection at the upper band-edge (Fig. 15.10). Reflection introduces a localization (Fig. 15.1) with an increasing probability of scattering and energy relaxation. When $eFL<E_B$, ballistic motion can produce Bloch oscillations if $eFa\tau/\hbar>2\pi$. Fig. 15.11 shows the Bloch factor $\omega_B=eFa/\hbar$ in GaN as a function of field. A field over 1 MV/cm is required for fuzzy Bloch oscillations ($eFa\tau/\hbar>1$), or somewhat less than this for quasi-Bloch oscillations if the momentum-relaxation time is used instead of the scattering time. True Bloch oscillations become possible at fields around 10 MV/cm, and then the energy becomes quantized into W-S states. Scattering between W-S states then determines the current. The calculation of the current due to hopping has been the aim of a number of authors (e.g. Hacker, 1969; Bryksin and Firsov, 1972; Sawaki and Nishinaga, 1977). Their results describe a current that exhibits a series of rises and falls with increasing field which are related to the increasing gaps between adjacent states. The essential steps of the calculation are set out below for the case of a single electron interacting with polar-optical phonons.

The Schrödinger equation is:

$$(H_0 - eFx)\psi(\mathbf{r}) = E\psi(\mathbf{r}) \tag{15.18}$$

where F is the electric field directed along the x axis. The wavefunction can be assumed to be of the form:

$$\psi(\mathbf{r}) = Au_{\mathbf{k}_\perp}(\mathbf{r}_\perp)e^{i\mathbf{k}_\perp\cdot\mathbf{r}_\perp}\sum_{k_x} u_{k_x}(x)c_{k_x}e^{ik_x x} \tag{15.19}$$

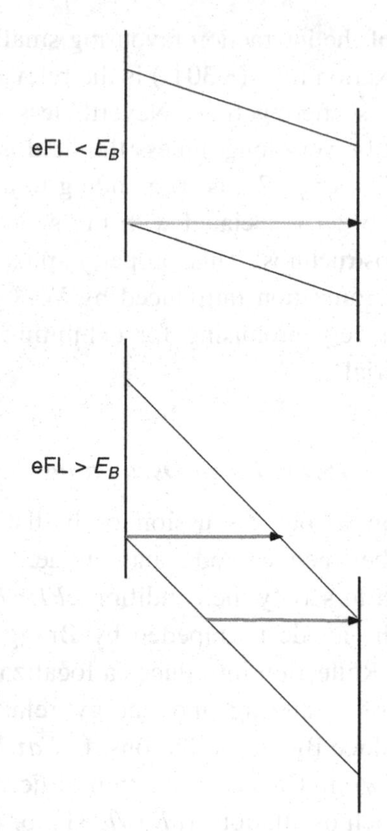

Fig. 15.10 Field ranges.

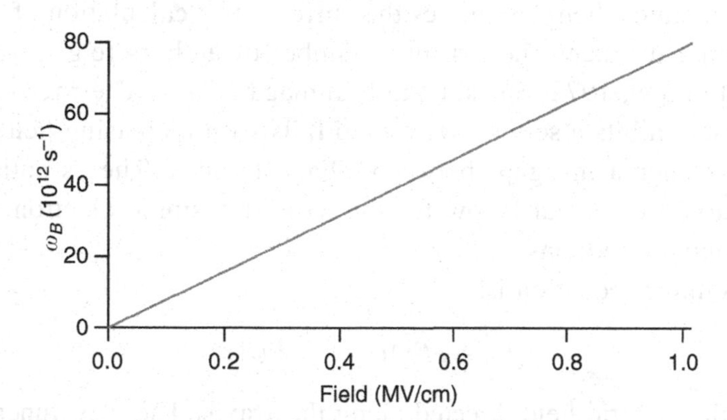

Fig. 15.11 Bloch-oscillation factor ($\omega_B = eFa/\hbar$).

For simplicity we will ignore the **k** dependence of the periodic Bloch function and take the energy of the electron in the band to be:

$$E = E_{\perp} + E_x = \frac{\hbar^2 k_{\perp}^2}{2m^*} + (E_B/2)(1 - \cos k_x a) \tag{15.20}$$

Insertion of Eq. (15. 19) into Eq. (15.18) and carrying out the usual manipulations yields:

$$c_{k'_x} = -\frac{\int \sum_{k_x} c_{k_x} e^{-ik'_x x} eFx e^{ik_x x} dx}{E_{k_x} - E_{k'_x}} = -\frac{ieF}{E_{k_x} - E_{k'_x}} \frac{dc_{k_x}}{dk'_x} \delta_{k_x k'_x} \tag{15.21}$$

and so:

$$c_{k'_x} = \exp\left(-\int \frac{E_{k_x} - E_{k'_x}}{ieF} dk'_x\right) \tag{15.22}$$

The initial energy can be written $E_{k_x} = E_O - eFna$, where n is an integer and a is the lattice constant, and $E_{k'_x} = (E_B/2)(1 - \cos k'_x a)$. Thus:

$$\psi(x) = \sum_{k_x} c_{k_x} e^{ik_x x} = AJ_m(z) \tag{15.23}$$

where $J_m(z)$ is the Bessel function of order $m = n' - n$ (since $E_0 - E'_0 = (n' - n)eFa$) and $z = E_O/eFa$. A is a normalizing factor, equal to unity since $\sum_{m=-\infty}^{\infty} J_m^2(z) = 1$. (Note that this assumes an infinite crystal!)

The scattering matrix element in the interaction with a travelling wave is:

$$M = \delta_{\mathbf{k}'_{\perp}, \mathbf{k}_{\perp}} \sum_{n=-\infty}^{\infty} e^{iq_x x} J_{m+n}(z) J_n(z) \tag{15.24}$$

Using the sum-rule for Bessel functions gives:

$$|M|^2 = J_m^2(u) = J_m^2(2z \sin\{q_x a/2\}) \tag{15.25}$$

The transition polar optical phonon rate is then:

$$W(n, \mathbf{k}_{\perp}) = \frac{e^2 \omega}{8\pi^2 \varepsilon_p} \sum_m \iiint \frac{J_m^2(u)}{q_x^2 + q_{\perp}^2} [(n(\omega) + 1/2 \pm 1/2)\delta(E' - E \pm \hbar\omega)] \times k'_{\perp} dk'_{\perp} d\theta dq_x \tag{15.26}$$

where

$$(E' - E \pm \hbar\omega) = \frac{\hbar^2}{2m^*}\left(k'_{\perp}2 - k_{\perp}^2\right) - meFa \pm \hbar\omega \tag{15.27}$$

Integration over $k'_{\perp}$ gives:

$$W(n, \mathbf{k}_{\perp}) = \frac{e^2\omega m*}{8\pi^2\varepsilon_p\hbar^2}[(n(\omega) + 1/2 \pm 1/2)]\sum_m \iint \frac{J_m^2(u)}{q_x^2 + q_{\perp}^2} d\theta dq_x \tag{15.28}$$

Now: $q_{\perp}^2 = k'_{\perp}2 + k_{\perp}^2 \mp 2k'_{\perp}k_{\perp}\cos\theta$, so:

$$\int_0^{2\pi} \frac{d\theta}{q_x^2 + q_{\perp}^2} = 2\int_0^{\pi} \frac{d\theta}{a \mp b\cos\theta} = \frac{2\pi}{\sqrt{a^2 - b^2}} \tag{15.29}$$

$$a = q_x^2 + k'^2_{\perp} + k_{\perp}^2, b = 2k'_{\perp}k_{\perp}$$

With $\Delta = meFa \mp \hbar\omega$,

$$W(n, \mathbf{k}_{\perp}) = \frac{e^2\omega}{8\pi\varepsilon_p}[(n(\omega) + 1/2 \pm 1/2)]\sum_m \int \frac{J_m^2(u)}{\sqrt{E_x^2 + 2E_x(2E_{\perp} \pm \Delta) + \Delta^2}} dq_x \tag{15.30}$$

where $E_x = \frac{\hbar^2 q_x^2}{2m*}$. Eq. (15.30) with the upper sign is Hacker's result for emission.

The limits on m can be found from the conservation of energy:

$$meFa = E'_{\perp} - E_{\perp} \pm \hbar\omega \text{ with } E'_{\perp}(\text{min}) = 0, E'_{\perp}(\text{max}) = E_B.$$

The rate can be obtained numerically. Fig. 15.12a illustrates possible transitions.

Let $x = q_x a/2$. The spontaneous rate is then:

$$W(n, \mathbf{k}_{\perp}) = \frac{e^2\omega m^* a}{8\pi\varepsilon_p\hbar^2}\sum_m \int \frac{J_m^2(E_B \sin x/eFa)}{\sqrt{x^4 + rx^2 + s}} dx \tag{15.31}$$

$$r = \frac{m^* a^2}{\hbar^2}(2E_{\perp} + meFa - \hbar\omega), \qquad s = \left(\frac{m^* a^2(eFa - \hbar\omega)}{2\hbar^2}\right)^2 \tag{15.32}$$

The spontaneous emission rate in GaN for the resonant case $eFa = \hbar\omega$ is illustrated in Fig. 15.12b. Note that, since for GaN $\omega\tau \approx 1$, the resonant condition barely satisfies the condition for Bloch oscillation. The rates of Fig. 15.12b are maximum.

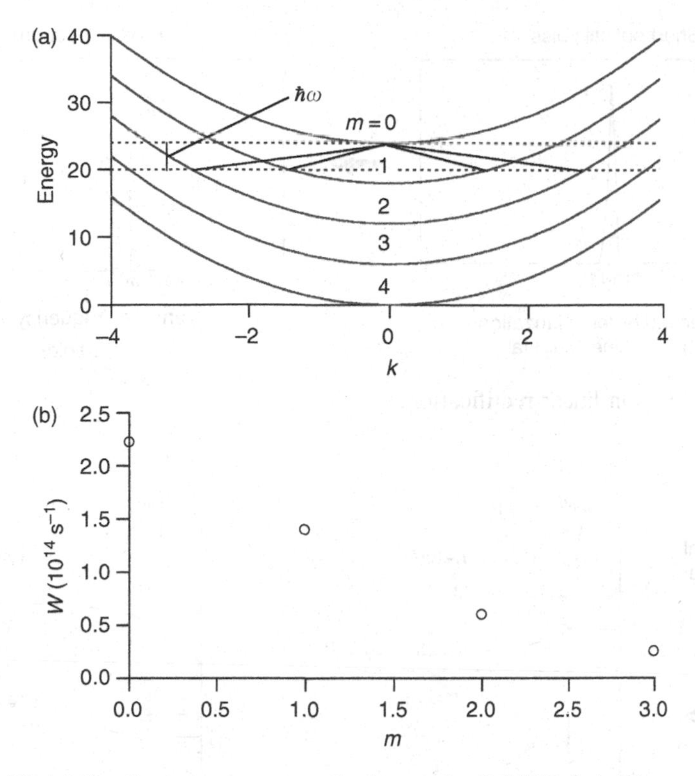

Fig. 15.12 (a) Band-structure perpendicular to the field ($Evk_\perp$). The case illustrated is for $\hbar\omega < eFa$. The transitions for m=1 and m=2 are indicated for emission; (b) W-S rates in GaN for the case $eFa = \hbar\omega$. For $m = 0$ the initial energy for the perpendicular motion is taken to be $E_\perp = \hbar\omega$, otherwise $E_\perp = 0$.

15.5 Femtosecond Generators

The femtosecond Ti-sapphire laser has been exploited to create transient electron-hole populations near the surface of a semiconductor, and this generates THz radiation. There is a range of processes that can contribute to the emission of electromagnetic emission in the THz region. We summarize these in what follows.

15.5.1 Optical Non-Linear Rectification.

Semiconductor crystals, like others, exhibit optical non-linearity, a measure of which is the non-linear susceptibility $\chi^{(2)}$. In the case of a monochromatic beam, the non-linearity produces rectification and a second-harmonic. In the case of a pulse, there is a broad spectrum that can be produced (Fig. 15.13) with the non-linear polarization proportional to the strength of the pulse.

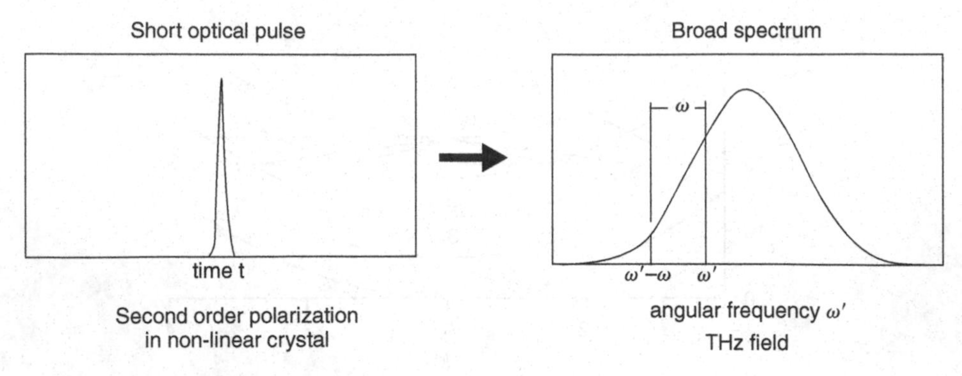

Fig. 15.13 Non-linear rectification.

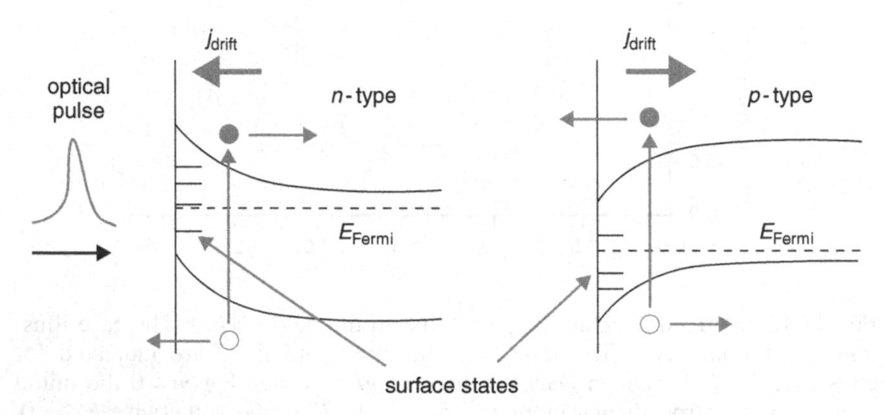

Fig. 15.14 Surface generation in the presence of a surface field and the consequent surge current.

15.5.2 Surge Current

Most semiconductors exhibit band bending at the surface. This occurs as a consequence of the Fermi level being pinned at the neutrality point of the surface states in an extrinsic semiconductor (see Mönch, 2004). The generated carriers therefore get accelerated by the surface field and the resultant transient current generates THz radiation (Fig. 15.14).

15.5.3 Dember Diffusion

In the absence of a surface field a quasi-neutrality is eventually established in which there is a bipolar diffusion at a rate determined by the lower mobility carrier. Neutrality is established by the flow of electron and hole diffusion currents, and this generates THz radiation (Fig. 15.15).

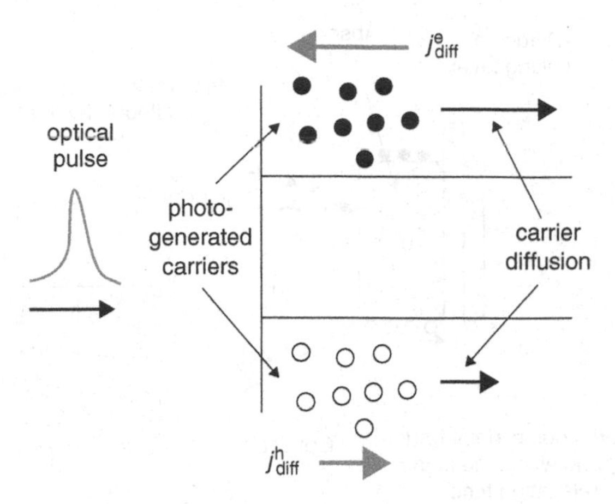

Fig. 15.15 Diffusion of carriers in the absence of strong band bending.

15.5.4 Coherent Phonons

The incident pulse can coherently excite LO/TO and polariton modes which have frequencies typically in the THz regime. Where the density of carriers is large, the excitation can be of coupled phonon–plasmon modes.

15.5.5 Photoconductive Switch

An Austin switch consists of a photoconductive material of high resistivity, short carrier lifetime, and photogenerated carriers of high mobility, in which a high electric field can be applied parallel to the surface between two closely spaced contacts. Excitation by a short light pulse generates currents that, in turn, generate THz radiation. Typical materials are radiation-damaged silicon and low-temperature-grown GaAs. Closely allied to the switching principle is the use of reversed-bias pin structures, in which the built-in high field is exploited.

15.6 CW Generators

15.6.1 Photomixing

The reverse-biased pin structure can also be used as a photomixer. In this device electrons and holes are generated in the p side of the pin by two lasers with frequencies that differ by THz. Since the generation rate is determined by intensity, which is quadratic in the amplitude, there is a component at the difference frequency. Generation in the p side of the junction means that the slower-moving holes are collected at the cathode and the faster electrons diffuse to the junction and provide the oscillating current that is the source of the THz radiation (Fig. 15.16).

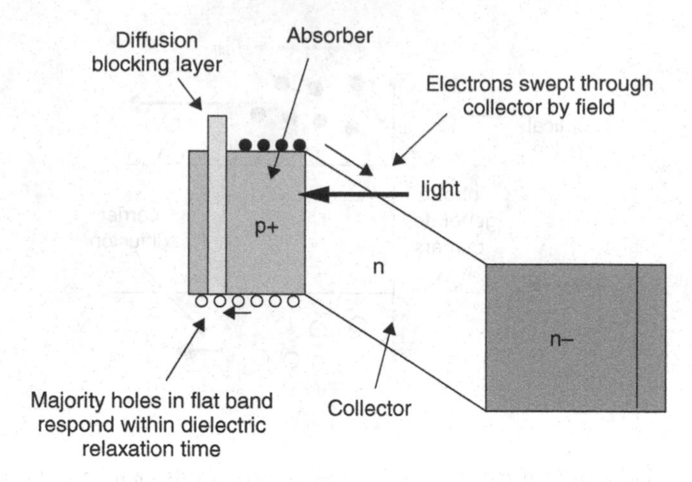

Fig. 15.16 Photoconductive mixing.

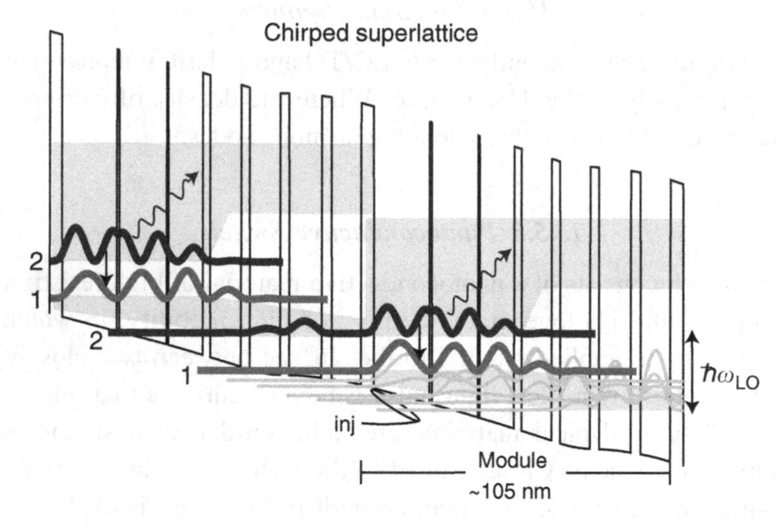

Fig. 15.17 Quantum-cascade laser: chirped superlattice (Williams, 2007).

15.6.2 Quantum-Cascade Lasers

As in all lasers the principle is to engineer an inverted distribution. In the cascade laser this is achieved by injecting electrons into subband 2 of a quantum well (or coupled wells) with, ideally, subband 1 empty. There are many ingenious designs; we consider just two (Williams, 2007). In the chirped superlattice the minibands are broad, ensuring efficient transport, and the optical transition is from the bottom of subband 2 to the top of subband 1. Electrons at the top of subband 1 are removed rapidly to the bottom by intrasubband scattering processes

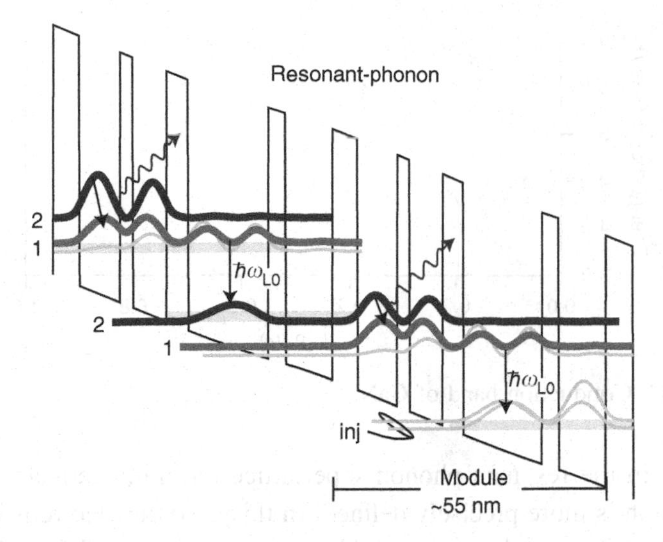

Fig. 15.18 Quantum-cascade laser: resonant phonon relaxation (Williams, 2007).

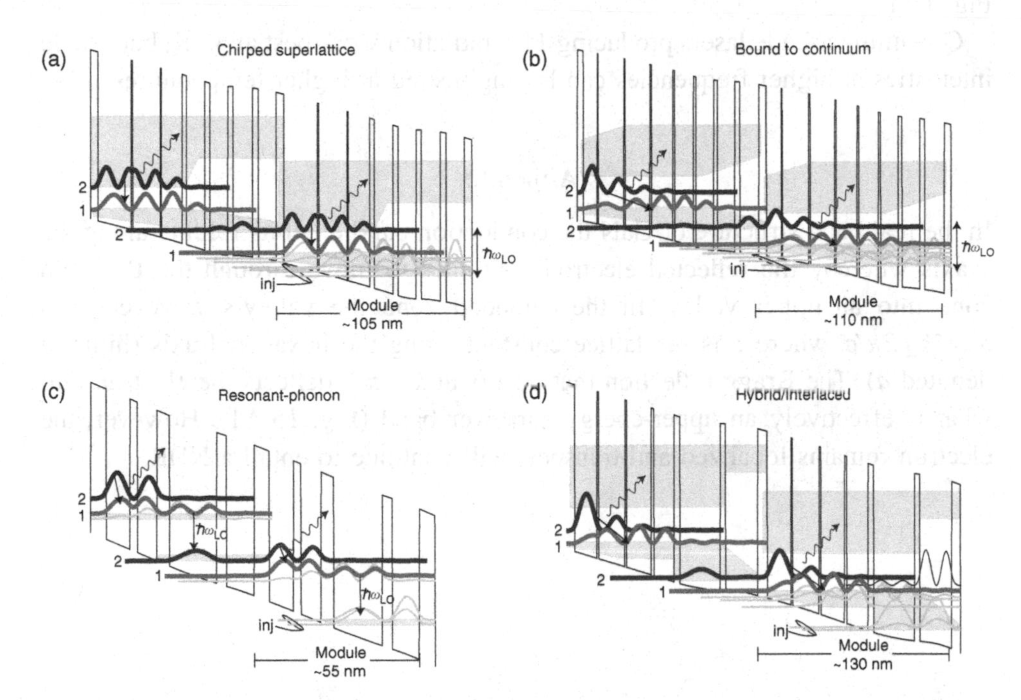

Fig. 15.19 Various designs of quantum-cascade lasers showing minibands and squared amplitude of wavefunctions (Williams, 2007).

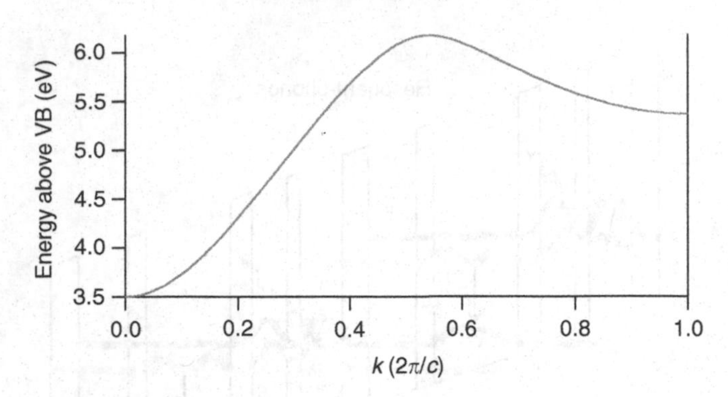

Fig. 15.A1 Conduction band of GaN.

(Fig. 15.17). In the resonant-phonon superlattice the minibands are narrower so the THz photon is more precisely defined. In this case the electrons in subband 1 are removed by the transition from subband 1 to the subband 2 of the following section, which is arranged to be exactly below subband 1 by the LO phonon energy giving the optimum emission rate (Fig. 15.18). Other designs are shown in Fig. 15.19.

Quantum-cascade lasers producing THz radiation work best at 4.2K, but usable intensities at higher frequencies can be engineered at higher temperatures.

Appendix

In the real band structure of GaN the conduction band exhibits folding along the c axis whereby the reflected electron continues to move through the Brillouin zone into an upper valley. In the extended zone the valley's wavevector is $\pi/c \leq k \leq 2\pi/c$, where c is the lattice constant along the hexagonal axis (hitherto denoted a). The Bragg reflection that occurs at $k = \pi/c$ deflects the electron into what is effectively an upper-energy narrower band (Fig. 15.A1). However, the electron remains localized and transport will continue to entail a NDR.

Appendix 1 The Polar-Optical Momentum-Relaxation Time in a 2D Degenerate Gas

Returning to Eq. (11.3) and retaining products of symmetric–antisymmetric occupation factors for the 2D case, we obtain for the antisymmetric components

$$
\begin{aligned}
eFv(E)\frac{\partial f_0(R)}{\partial E} &= W_0\left(\frac{\hbar}{2m*\omega}\right)^{1/2}\frac{\hbar\omega}{E}(I_1+I_2)\\
I_1 &= [f_1(E+\hbar\omega)H_a(a',E)--f_1(E)H_a(E)]n(\omega)\\
&\quad+[(1+f_0(E))f_1(E+\hbar\omega)H_a(a',E)-f_0(E+\hbar\omega)f_1(E)H_a(E)]\\
I_2 &= [f_1(E-\hbar\omega)H_e(a'',E)-f_1(E)H_e(E)]n(\omega)\\
&\quad+[f_0(E)f_1(E-\hbar\omega)H_e(a'',E)-(1-f_0(E-\hbar\omega))f_1(E)H_e(E)]
\end{aligned}
$$

These equations are a generalization of Eq. (11.37b) to the degenerate case. The ladder nature is brought out by substituting $E=\xi+j\hbar\,\omega$ where j is an integer and $0\le\xi\le\hbar\omega$. Using the relations between H_e and H_a we can express the sum as follows:

$$
eF\sum_{j=0}^{\infty}v(\xi+j\hbar\omega)\frac{\partial f_0(\xi+j\hbar\omega)}{\partial\xi}=-\sum_{j=0}^{\infty}\frac{f_1(\xi+j\hbar\omega)}{\tau(\xi+j\hbar\omega)}
$$

where

$$
\begin{aligned}
\frac{1}{\tau(\xi+j\hbar\omega)} &= W_0\left(\frac{\hbar}{2m*\omega}\right)^{1/2}[H_{me}\{n(\omega)+1-f_0(\xi+(j-1)\hbar\omega)\}\\
&\quad+H_{ma}\{n(\omega)+f_0(\xi+(j+1)\hbar\omega)\}]
\end{aligned}
$$

and

$$
\begin{aligned}
H_{me} &= H_e(\xi+j\hbar\omega)-H_e(a'',\xi+j\hbar\omega)\\
H_{ma} &= H_a(\xi+j\hbar\omega)-H_e(a',\xi+j\hbar\omega)
\end{aligned}
$$

We now exploit the fact that the differential on the left is approximately replaceable by the delta function $\delta(+j\hbar\omega-E_F)$, where E_F is the Fermi energy, an approximation that gets better the lower the temperature. This allows us to define a momentum-relaxation time

$$eFv(E_F)\delta(E - E_F) = -\frac{f_1(E_F)}{\tau(E_F)}$$

where

$$\frac{1}{\tau(E_F)} = W_0\left(\frac{\hbar}{2m*\omega}\right)^{1/2}\frac{\hbar\omega}{E_F}(H_{me} + H_{ma})(n(\omega) + f_0(E_F + \hbar\omega))$$

and it is understood, as usual, that Hme is zero unless $E_F\ \hbar\omega$. The mobility is then simply

$$\mu = \frac{e\tau(E_F)}{m*}$$

Appendix 2 Electron/Polar Optical Phonon Scattering Rates in a Spherical Cosine Band

In GaN along the hexagonal axis ($\Gamma - A$, the c axis) the band profile can be approximated with only small error by a cosine function (Fig.15.3):

$$E = (E_B/2)(1 - \cos ka) \tag{A2.1}$$

where E is the kinetic energy of the electron of wave-vector k, $-\pi/a < k < \pi/a$, E_B is the band width and a (=5.186Å) is the lattice constant along the c axis. The band profile along a principle direction perpendicular to the c axis ($\Gamma - M$) is virtually identical, and along the other principle perpendicular direction ($\Gamma - K$) it deviates from the form along the c axis only towards high energies. Thus, a reasonably valid approximation can be made that assumes spherical symmetry, which can be used to estimate density of states and scattering rates for energies not too close to the upper band edge. The density of states per unit energy interval is given by (Fig. A2.1):

$$N(E) = \frac{1}{2\pi^2 a^3 E_B} \frac{[\cos^{-1}\{1-(2E/E_B)\}]^2}{\sqrt{(E/E_B)-(E/E_B)^2}} \tag{A2.2}$$

We take the scattering to be predominantly associated with the interaction with polar-optical modes at 300K. There are three rates of importance – the scattering rate, the momentum-relaxation rate and the reciprocal of the energy-relaxation time. Analytical

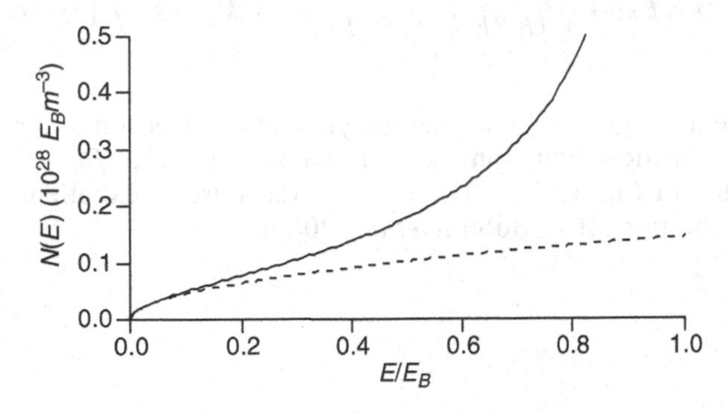

Fig.1 Density of states. The dashed curve is the parabolic approximation.

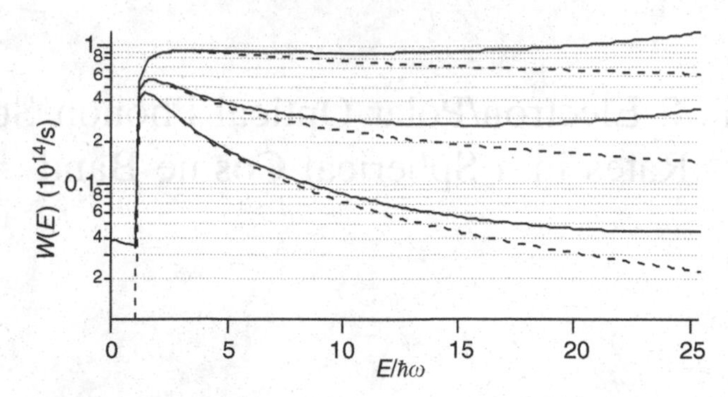

Fig.2 Scattering rate (top), momentum relaxation rate (middle) and energy relaxation rate (bottom). The dashed curves are for the parabolic approximation.

results for these for an electron with energy E is obtained by neglecting the anisotropy of the phonon frequency (which is small) and taking the phonon energy to be 91meV and independent of the wave-vector.

$$W(k) = \frac{e^2\omega}{4\pi\varepsilon_p E_B a}[n(\omega) + \frac{1}{2} \mp \frac{1}{2}]\frac{k'}{k}\frac{1}{\sqrt{(E'/E_B) - (E'/E_B)^2}}\ln\left(\frac{(k'+k)^2}{|k'^2 - k^2|}\right) \quad (A2.3)$$

$$\begin{aligned} W_m(k) = &\frac{e^2\omega}{4\pi\varepsilon_p E_B a}[n(\omega) + \frac{1}{2} \\ &\mp \frac{1}{2}]\frac{k'}{k}\frac{1}{\sqrt{(E'/E_B) - (E'/E_B)^2}}\left[\frac{k'}{k} + \frac{k^2 - k'^2}{2k^2}\right]\ln\left(\frac{(k'+k)^2}{|k'^2 - k^2|}\right) \end{aligned} \quad (A2.4)$$

where the dash denotes the final state and $E' = E \pm \hbar\omega$. The energy-relaxation rate is:

$$\frac{E - E_0}{\tau_E} = \frac{e^2\omega}{4\pi\varepsilon_p E_B a}\frac{k'}{k}\frac{\hbar\omega}{\sqrt{(E'/E_B) - (E'/E_B)^2}}\ln\left(\frac{(k'+k)^2}{|k'^2 - k^2|}\right)\begin{cases} n(\omega) + 1 \\ -n(\omega) \end{cases} \quad (A2.5)$$

where E_0 is the average energy at thermodynamic equilibrium and τ_E is the energy-relaxation time. In these equations: $k = (1/a)\cos^{-1}\{1 - (2E/E_B)\}$. $W(k)$, $W_m(k)$ and $1/\tau_E$ are depicted in Fig. (A2.2). These model rates are in excellent agreement with rates computed numerically (Bulutay *et al.*, 2000).

References

Abram R. A., Rees G. J., and Wilson B. L. H. (1978) *Adv. Phys.* **27** 799.
Akero H. and Ando T. (1989) *Phys. Rev.* **B40** 2914.
Al-Dossary O., Babiker M., and Constantinou N. C. (1992) *Semicond. Sci. Technol.* **7** B91.
Altarelli M. (1986) *Heterojunctions and Semiconductor Superlattices* (ed. G. Allan, G. Bastard, N. Boccara, M. Lannou, and M. Voos). Springer-Verlag, Berlin.
Altarelli M., Ekenburg U., and Fasolino A. (1985) *Phys. Rev.* **B32** 5738.
Ambacher A. (1998) *J. Phys D Appl. Phys.* **31** 2653.
Ambacher O., Smart, J., Shealy J. R. *et al.* (1999) *J. Appl. Phys.* **85** 3222.
Ambacher O., Majewski J., Miskys C. *et al.* (2002) *J. Phys. Condens. Matter* **14** 3399.
Ando T. (1977) *J. Phys. Soc. Jpn.* **43** 1616.
Ando T. (1982) *J. Phys. Soc. Jpn.* **51** 3900.
Ando T. and Akera H. (1989) *Phys. Rev.* **40** 11619.
Ando T., Fowler A. B., and Stern F. (1982) *Rev. Mod. Phys.* **54** 437.
Ando T., Wakahara S., and Akera H. (1989) *Phys. Rev.* **B40** 11609.
Andreani L. C., Pasquarello A., and Bassani F. (1987) *Phys. Rev.* **B36** 5887.
Avron J. E. (1982) *Annals of Phys.* **143** 33.
Babiker M. (1986) *J. Phys. C: Solid St. Phys.* **19** 683.
Babiker M. (1994) private communication.
Babiker M., Chamberlain M. P. and Ridley B. K. (1987) *Semicond. Sci. Technol.* **2** 582.
Babiker M., Constantinou N. C., and Ridley B. K. (1993) *Phys. Rev.* **B48** 2236.
Babiker M., Ghosal A. and Ridley B. K. (1989) *Superlatt. Microstruct.* **5** 133.
Barman S. and Srivastava G. P. (2004) *Phys. Rev. B* **69** 235208.
Baroni S., Giznozzi P., and Molinari E. (1990) *PR* **B41** 3870.
Bastard G. and Brum J. A. (1986) *IEEEJ. Quantum Electron.* **QE-22** 1625.
Bechstedt F. and Enderlein R. (1985) *Phys. Stat. Sol.* (b) **131** 53.
Bechstedt F. and Gerecke H. (1989) *Phys. Stat. Sol.* (b) **154** 565.
Beck M., Hofstetter D., Aellen T. *et al.* (2002) *Science* **295** 301.
BenDaniel D. J. and Duke C. B. (1966) *Phys. Rev.* **152** 683.
Bennett C. R., Constantinou N. C., and Tanatar B. (1995a) *J. Phys: Condens. Matt.* **7 L** 669.
Bennett C. R., Amato M.A., Zakhlenuik N. A., Ridley B. K. and Babiker M. (1998) *J. Appl. Phys.* **83** 1499.
Bennett C. R., Constantinou N. C., Babiker M., and Ridley B. K. (1995) *J. Phys. Condens. Matter* **7** 9819.
Bennett C. R., Tanatar B., Constantinou N. C., and Babiker M. (1994) *Solid St. Commun.* **92** 947.

Bernadini F. and Fiorentini V. (2001) *Phys. Rev. B* **64** 085207.
Bernadini F., Fiorentini V., and Vanderbilt D. (2001) *Phys. Rev. B* **63** 193201.
Bhatt A. R., Kim K. W., Stroscio M. A., Dutta M., Grubin H. L., Haque R., and Zhu X. T. (1993) *J. Appl. Phys.* **73** 2338.
Bir G. L., Aranov A. G., and Pikus G. E. (1976) *Sov. Phys. JETP* **42** 705.
Bloch F. (1928) *Z. Physik* **52** 555.
Blom P. W. M., Smit C., Haverkort J. E. M., and Wolter J. H. (1993) *Phys. Rev.* **B47** 2072.
Bockelmann U. and Bastard G. (1990) *Phys. Rev.* **B42** 8947.
Born M. and Huang K. (1954) *Dynamical Theory of Crystal Lattices*. Clarendon, Oxford.
Brockhouse B. N. (1959) *Phys. Rev. Lett.* **2** 256.
Brockhouse B. N. and Iyengar P. K. (1958) *Phys. Rev.* **111** 747.
Brooks H. and Herring C. (1951) *Phys. Rev.* **83** 879.
Bryksin V. V. and Firsov Yu. A. (1972) *Sov. Phys. JETP* **34** 1272.
Bryksin V. V. and Firsov Yu. A. (1972) *Sov. Phys.- Solid State* **13** 2729.
Buks E., Heiblum M., Levinson Y., and Shtrikman H. (1994) *Semicond. Sci. Technol.* **9** 2031.
Bulutay C., Ridley B. K., and Zakhlenuik N. A. (2000) *Phys. Rev. B* **62** 15754.
Bulutay C., Ridley B. K., and Zakhlenuik N. A. (2003) *Phys. Rev, B* **68** 115205.
Burt M. G. (1988) *Semicond. Sci. Technol.* **3** 739, 1224.
Burt M. G. (1992) *J. Phys. Condens. Matter* **4** 6651.
Bychkov Yu. A. and Dykhne A. M. (1965) *Sov. Phys. JETP* **21** 779 783.
Bychkov Yu. A. and Rashba E. I. (1984) *J. Phys. C.: Solid State Phys.* **17** 6039.
Callen H. B. (1949) *Phys. Rev.* **76** 1394.
Camley R. E. and Mills D. L. (1984) *Phys. Rev.* **B29** 1695.
Cao J. C. *et al.* (2001) *Appl. Phys. Lett.* **78** 2524.
Cardona M. (1989) *Superlatt. Microstr.* **5** 27; (1990) **7** 183.
Chamberlain M. and Cardona M. (1994) *Semicond. Sci. Technol.* **9** 749.
Chamberlain M. P., Cardona M., and Ridley B. K. (1993) *Phys. Rev.* **B48** 14356.
Chamberlain M. P., Trallero-Giner C., and Cardona M. (1994) *Phys. Rev.* **B50**.
Chang I. F. and Mitra S. S. (1971) Adv. *Phys.* **20** 359.
Chang Y.-C. (1982) *Phys. Rev.* **B25** 605.
Chazalviel J.-N. (1975) *Phys. Rev. B* **11** 1555.
Chen B. and Nelson D. F. (1993) *Phys. Rev.* **48** 15365.
Chu H., Ren S.-F., and Chang Y.-C. (1988) *Phys. Rev.* **B37** 10746.
Chuang S. L. and Chan C. S. (1996) *Phys. Rev. B* **54** 2491.
Churchill J. N. and Holmstrom F. E. (1982) *Am. J. Phys.* **50** 848.
Colvard C., Gant T. A., Klein M. V., Merlin R., Fischer R., Morkoc H., and Gossard A. C. (1985) *Phys. Rev.* **B31** 2080.
Constantinou N. C. (1991) *J. Phys. Condens. Matter* **3** 6859.
Constantinou N. C. (1993) *Phys. Rev.* **B48** 11931.
Constantinou N. C. (1995) private communication.
Constantinou N. C., Al-Dossary O., and Ridley B. K. (1993) *Solid St. Comm.* **86** 191; **87** 1087(E).
Constantinou N. C. and Ridley B. K. (1990) *Phys. Rev.* **B41** 10622.
Constantinou N. C. and Ridley B. K. (1994) *Phys. Rev.* **B49** 17065.
Conwell E. M. (1967) *High Field Transport in Semiconductors*. Academic Press, New York.
Conwell E. M. and Weisskopf V. F. (1950) *Phys. Rev.* **77** 388.
Cottam M. G. and Tilley D. R. (1976) *Introduction to Surface and Superlattice Excitations*. Cambridge University Press, Cambridge.
Csavinsky P. (1963) *Phys. Rev.* **131** 2033.

Csavinsky P. (1976) *Phys. Rev.* **B14** 1649.
D'yakonov M. I. and Perel V. I. (1972) *Fiz, Tverd. Tela* **13** 3581 [*Sov. Phys. Solid State* **13** 3023 (1972)].
Daniels M. E., Ridley B. K., and Emeny M. (1989) *Solid St. Electron* **32** 1207.
Das Sarma S., Jain J., and Jalabert R. (1988) *Phys. Rev.* **B37** 1228, 4560.
Das Sarma S. and Mason B. A. (1985) *Phys. Rev.* **B31** 5536.
Davison S. G., English R. A., Miskovic Z. L. *et al.* (1997) *J. Phys. Condens,: Matter* **9** 6371.
Dyson A. and Ridley B. K. (2004) *Phys. Rev. B* **69** 125211; (2005) *Phys. Rev. B* 72 045326.
Dyson A. and Ridley B. K. (2005) *Phys. Rev. B* **72** 193301.
Dyson A. and Ridley B. K. (2006) *Semicond. Sci. Technol.* **21** 210.
Dyson A. and Ridley B. K. (2008) *J. Appl. Phys.* **103** 114507.
Dyson A., Ridley B. K., Aslan B. *et al.* (2006) *Physica (c).*
Ehrenreich H. and Cohen M. H. (1959) *Phys. Rev.* **115** 786.
Ekenburg U. (1989) *Phys. Rev.* **B40** 7714.
El-Ghanem H. M. A. and Ridley B. K. (1980)7. *Phys. C: Solid St. Phys.* **13** 2041.
Elliot R. J. (1954) *Phys. Rev.* **96** 266; **96** 280.
Emyura S., Nakagawa T., Gonda S., and Shimizu S. (1987) *J. Appl. Phys.* **62** 4632.
Esaki L. and Tsu R. (1970) *IBM J. Res Develop.* Jan. p. 61.
Esipov S. E. and Levinson Y. B. (1987) *Adv. Phys.* **36** 331.
Faist J., Capasso F., Sivco D. L. *et al.* (1994) *Science* **264** 553.
Falicov L. M. and Cuevas M. (1967) *Phys. Rev.* **164** 1025.
Fang F. F. and Howard W. E. (1966) *Phys. Rev. Lett.* **16** 797.
Fang F. F. and Howard W. E. (1967) *Phys. Rev.* **163** 816.
Fischetti M. V. and Laux S. E. (1993) *Phys. Rev.* **B48** 2244.
Fishman G. and Lampel G. (1977) *Phys. Rev. B* **16** 820
Foreman B. A. (1993) *Phys. Rev.* **B48** 4964; (1994) **B49** 1757.
Foreman B. A. (1994), (1995) private communication.
Foreman B. A. (1995) Ph.D. Thesis, School of Engineering, Cornell University; *Phys. Rev.* (1995) **B52** 12241.
Foreman B. A. (1995a) *Phys. Rev.* **B52** 12260.
Foreman B. A. (1998) *Phys. Rev. Lett.* **80** 3823.
Foreman B. A. and Ridley B. K. (1999) *Proc. ICPS24 CDROM Section V-E3.*
Foreman B. A. (2000) *J. Phys: Condens. Matter* **12** R435.
Friedman B. (1956) *Principles and Techniques of Applied Mathematics.* Wiley, New York.
Frohlich H. (1937) *Proc. Roy. Soc. A* **160** 230.
Fuchs R. and Kliewer K. L. (1965) *Phys. Rev.* **140A** 2076.
Fukuyama H., Bari R. A., and Fogedby H. C. (1973) *Phys. Rev. B* **8** 5579.
Gerecke H. and Bechstedt F. (1991) *Phys. Rev.* **B43** 7053.
Goepfert I. D., Schubert E. F., Osinsky A., Norris P. E., and Faleev N. N. (2000) *J. Appl. Phys.* **88** 2030.
Gold A. and Ghazali A. (1990) *Phys. Rev.* **B41** 7626.
Goodnick S. M., Ferry D. K., Wilmsen C. W., Liliental Z., Fathy D., and Krivanek (1985) *Phys. Rev.* **B32** 8171.
Goodnick S. M. and Lugli P. (1988) *Solid St. Electron.* **31** 463.
Goodnick S. M. and Lugli P. (1992) *Hot Carriers in Semiconductor Nanostructures* (ed. J. Shah). Academic Press, London.
Grahn H. T. (ed.) (1995) *Semiconductor Superlattices, Growth and Electronic Properties.* World Scientific.

Grubnikov Z. S., Bashirov R. R., and Mitin V. V. (2001) *IEEE J. Selected Topics in Quantum Electronics* **7** 630.
Guillemot C. and Clerot F. (1991) *Phys. Rev.* **B44** 6249.
Gupta R., Balkan N., Ridley B. K., and Emeny M. (1991) *SPIE* **1362** 798.
Gupta R., Balkan N., and Ridley B. K. (1992) *Phys. Rev.* **B46** 7745.
Gupta R. and Ridley B. K. (1990) *SPIE* **1362** 790.
Gupta R. and Ridley B. K. (1993) *Phonons in Semiconductor Nanostructures* (ed. J. P. Leburton *et al.*). Kluwer Academic, London.
Hackenberg W. and Fasol G. (1989) *Solid St. Electron.* **32** 1247.
Hacker K. (1969) *Phys. Stat. Sol.* **33** 607.
Haines M. and Scarmarcio G. (1992) *Phonons in Nanostructures*. NATO ARW St. Filiu, Spain. Kluwer Academic, London.
Harris J. J., Pals J. A., and Woltjer R. (1989) *Rep. Prog. Phys.* **52** 1217.
Harris J. J. and Ridley B. K. (1973) *J. Phys. Chem. Solids* **34** 197.
Harrison W. A. (1970) *Solid State Theory*. McGraw-Hill, New York.
Haupt R. and Wendler L. (1991) *Phys. Rev.* **B44** 1850.
Herbert D. C. (1973) *J. Phys. C: Solid St. Phys.* **6** 2788.
Hirakawa K. and Sakaki H. (1986) *Appl. Phys. Lett.* **49** 889.
Hofstetter D., Schad S.-S., Wu H., Schaff W. J., and Eastman L. F. (2007) *Appl. Phys. Lett.* **91** 131115.
Houston W. V. (1940) *Phys. Rev.* **57** 184.
Huang D., Gumbs G., Zhao Y., and Auner G. W. (1995) *Phys. Lett.* **A200** 459.
Huang K. and Zhu B. (1988) *Phys. Rev.* **B38** 2183, 13377.
Ichimaru S. (1980) *Basic Principles of Plasma Physics*. Addison-Wesley, Reading, Massachusetts.
Jaros M. and Wong K. B. (1984) *J. Phys.* **C17** L765.
Jaros M., Wong K. B., and Gell M. A. (1985) *Phys. Rev.* **B31** 1205.
Jones W. E. and Fuchs R. (1971) *Phys. Rev.* **B4** 3581.
Kanallis G., Morhange J. F., and Balkanski M. (1983) *Phys. Rev.* **B28** 3390, 3398, 3406.
Kane E. O. (1959) *J. Phys. Chem. Solids* **12** 181.
Kash J. A., Tsang J. C., and Huam J. M. (1985) *Phys. Rev. Lett.* **54** 2151.
Keating P. N. (1966) *Phys. Rev.* **145** 637.
Kim K., Laambrecht W. R.L., Segall B., and van Schilfgaarde M. (1997) *Phys. Rev. B* **56** 7363.
Kim M. E., Das A., and Senturia S. D. (1978) *Phys. Rev.* **B18** 6890.
Kim O. K. and Spitzer W. G. J. (1979) *J. Appl. Phys.* **50** 4362.
King P. D.C., Veal T. D., McConville C. F. *et al.* (2008) *Phys. Rev. B* **77** 045326.
Kishino K., Kikuchi A., Kanazawa H., and Tachibana T. (2002) *Appl. Phys. Lett.* **81** 1234.
Kittel C. (1963) *Quantum Theory of Solids*. Wiley, New York.
Klein M. V. (1986) *IEEE J. Quant. Electron.* **QE-22** 1760.
Klemens P. G. (1966) *Phys. Rev.* **148** 845.
Knipp P. A. and Reinecke T. L. (1992) *Phys. Rev.* **B45** 9091.
Knipp P. A. and Reinecke T. L. (1994) *Solid St.. Electron.* **37** 1105.
Knox W. H., Hirlimann C., Miller D. A. B., Shah J., Chemla D. S., and Shank D. V. (1986) *Phys. Rev. Lett.* **56** 1191.
Kocevar P. (1987) *Festkorperprobleme* **27** 197.
Kogan S. M. (1963) *Sov. Phys.-Solid St.* **4** 1813 (F.T.T. **4** 2474).
Köhler R., Tredicucci A., Beltram F. *et al.* (2002) *Nature* **417** 156.
Kozlov V. A., Nokolaev A. V., and Samokhvalov A. V. (2004) *Semicond. Sci. Technol.* **19** S99.

Kozody P., Hansen M., DenBaars S. P., and Mishra U. K. (1999) *Appl. Phys. Lett.* **74** 3681.
Krieger J. B. and Iafrate G. J. (1986) *Phys. Rev. B* **33** 5494.
Krishnamurthy S., van Schlifgaarde M., Sher A., and Chen A.-B. (1997) *Appl. Phys. Lett.* **71** 1999.
Krömer H. (1958) *Phys. Rev.* **109** 1856.
Krumhansl J. A. (1965) *Lattice Dynamics* (ed. R. F. Wallis). Pergamon Press, Oxford, p. 298.
Kunin I. A. (1982) *Elastic Media with Structure*. Springer-Verlag, Berlin.
Lancefield D., Adams A. R., and Fisher M. A. (1987) *J. Appl. Phys.* **62** 2342.
Landau L. D. and Lifshitz E. M. (1977) *Quantum Mechanics*. Pergamon Press, Oxford.
Landau L. D. and Lifshitz E. M. (1986) *Theory of Elasticity*, 3rd edn. Pergamon Press, Oxford.
Landheer D., Liu H. C., Buchanan M., and Stoner R. (1989) *Appl. Phys. Lett.* **54** 1784.
Landolt-Börnstein (1987) **Vol.111/22a** Springer-Verlag, Berlin.
Landsberg P. T. (1986) *Phys. Rev.* **B33** 8321.
Larmor J. (1894) *Phil. Trans. Roy. Soc.* (ser. A) **185**.
Lassnig R. (1984) *Phys. Rev.* **B30** 7132.
Lassnig R. (1988) *Solid St. Commun.* **65** 765.
Leburton J. P. (1992) *Phys. Rev.* **B45** 11022.
Lee I., Goodnick S. M., Gulia M., Molinari E., and Lugli P. (1995) *Phys. Rev.* **B51** 7046.
Lee J. and Spector H. N. (1983) *J. Appl. Phys.* **54** 6989; **57** 366.
Lei X. L. (1985) *J. Phys. C: Solid St. Phys.* **18** L593.
Lei X. L., Horing N. J. M., and Cui H. L. (1995) *J. Phys.: Condens. Matter* **7** 9811.
Leo J. and MacKinnon A. (1989) *J. Phys.: Condens. Matter* **1** 1469.
Levinson I. B. and Vasevichyute Ya. (1972) *Sov. Phys. JETP* **35** 991.
Li Q. P. and Das Sarma S. (1951) *Phys. Rev.* **B43** 11768; (1991) **B44** 6277.
Lindhard J. (1954) *Kgl. Danske Videnskab. Selskab. Mat. Fys. Medd.* **28**.
Llewellyn F. B. (1941) *Electron-Inertia Effects*. Cambridge University Press, Cambridge.
Loudon R. (1964) *Proc. Phys. Soc.* **84** 379.
Love A. E. H. (1927) *A Treatise on the Mathematical Theory of Elasticity*. Cambridge University Press, Cambridge.
Lowe D. and Barker J. R. (1985) *J. Phys. C: Solid St. Phys.* **18** 2507.
Lu H., Schaff W. J., Eastman L. F., and Stutz C.E. (2003) *Appl. Phys. Lett.* **82** 1736.
Lucovsky G. and Chen M. F. (1970) *Solid St. Commun.* **8** 1397.
Lugli P., Bordone P., Reggiani L., Rieger M., Kocevar P., and Goodwich S. M. (1989) *Phys. Rev.* **B 39** 7834.
Lyon S. A. (1986) *J. Lumin.* **35** 121.
MacCullagh J. R. and Dublin, R. (1839) *Irish Acad. Trans.* **21**.
MacCullagh J. R. and Dublin, R. (1880) *Collected Works*. Dublin.
Mannion S. J., Artaki M., Emanuel M. A., Coleman J. J., and Hess K. (1987) *Phys. Rev.* **B35** 9203.
Maradudin A. A. and Stegemann G. I. (1991) *Surface Phonons* (ed. W. Kress and F. W. de Wette). Springer-Verlag, New York.
Martin R. M. (1970) *Phys. Rev. B* **1** 4005.
Matthews J. W. and Blakeslee A. E. (1974) *J. Crystal Growth* **32** 265.
McLachlan N. W. (1947) *Theory and Applications of Mathieu Functions*. Clarendon, Oxford.
Mendez E. E., Agullo-Rueda F., and Hong J. M. (1988) *Phys. Rev. Lett* **60** 2426.
Menendez J. (1989) *J. Lumin* **44** 285.

Messiah A. (1966) *Quantum Mechanics*. North-Holland, Amsterdam.
Meyer J. R. and Bartoli F. (1983) *J. Phys. Rev.* **B28** 915.
Milsom P. K. and Butcher P. N. (1986) *Semicond. Sci. Technol.* **1** 58.
Mirlin D. N., Karlick I. Ya., Nikitin N. P., Reshina I. I., and Sapega (1980) *Solid St. Comtn.* **34** 757.
Molinari E., Fasilino A., and Kunc K. (1986) *Superlatt. Microstr.* **2** 397.
Molinari E., Fasolino A., and Kunc K. (1987) *Proc. Physics of Semiconductors*. World Scientific, London, p. 663.
Monch W. (1996) *Electronic Properties of Semiconductor Interfaces*. Springer, Berlin.
Moore E. J. (1967) *Phys. Rev.* **160** 607, 618.
Mori N. and Ando T. (1989) *Phys. Rev.* **B40** 6175.
Mosko M. and Moskova A. (1994) *Semicond. Sci. Technol.* **9** 478.
Mosko M., Moskova A., and Cambel V. (1995) *Phys. Rev.* **B51** 16860.
Mott N. F. (1936) *Proc. Camb. Philos. Soc.* **32** 281.
Mott N. F. (1993) *Phys. Rev.* **B47** 2072.
Mowbray D. J., Cardona M., and Ploog K. (1991) *Phys. Rev.* **B43** 1598.
Nag B. R. (1972) *Theory of Electrical Transport in Semiconductors*. Pergamon, Oxford.
Nag B. R. (1980) *Electron Transport in Compound Semiconductors*. Springer-Verlag, Berlin.
Nash K. J. (1992) *Phys. Rev.* **B46** 7723.
Nash K. J., Skolnick M. S., and Bass S. J. (1987) *Semicond. Sci. Technol.* **2** 329.
O'Reilly E. P. (1989) *Semicond. Sci. Technol.* **4** 121.
Onipko A. and Malysheva L. (2001) *Phys. Rev. B* **63** 234510.
Ozturk E., Constantinou N. C., Straw A., Balkan N., Ridley B. K., Ritchie D. A., Linfield E. H., Churchill A. C., and Jones G. A. C. (1994) *Semicond. Sci. Technol.* **9** 782.
Perez-Alvarez R., Garcia-Moliner F., Velasco V. R., and Trallero-Giner C. J. (1993) *Phys. Condens. Matter* **5** 5389.
Pikus G. E. and Bir G. L. (1959) *Sov. Phys.-Solid St.* **1** 1502.
Pikus G. E. and Titkov A. N. (1984) *Optical Orientation*. Elsevier, Amsterdam, p.73.
Piprek J. (ed.) (2007) *Nitride Semiconductor Devices: Principles and Simulation*. Wiley-VCH Verlag GmbH & Co. KgaA, Weinheim.
Polonowski J.-P. and Tomazawa K. (1985) *Jpn. J. Appl. Phys.* **24** 1611.
Price P. J. (1981) *J. Vac. Sci. Technol.* **19** 599.
Price P. J. (1982) *J. Appl. Phys.* **53** 6863.
Price P. J. (1984) *Surf. Sci.* **143** 1456.
Rabinovitch A. (1971) *Phys. Rev. B* **4** 1017.
Rabinovitch A. and Zak J. (1971) *Phys. Rev. B* **4** 2358.
Ralph H., Simpson G., and Elliot R. J. (1975) *Phys. Rev.* **B11** 2948.
Register L. F. (1992) *Phys. Rev.* **B45** 8756.
Ren S. F., Chu H., and Chang Y.-C. (1987) *Phys. Rev. Lett.* **59** 1841.
Ren S. F., Chu H., and Chang Y.-C. (1988) *Phys. Rev.* **B37** 8899.
Ren S.-F., Hanyou C., and Chang Y.-C. (1989) *Phys. Rev.* **B40** 3060.
Richter E. (1986) *Diplomarbeit Universitat Regensburg*.
Richter E. and Strauch D. (1987) *Solid St. Comm.* **64** 867.
Riddoch F. A. and Ridley B. K. (1983) *J. Phys. C: Solid St. Phys.* **16** 6971.
Riddoch F. A. and Ridley B. K. (1984) *Surf. Sci.* **142** 260.
Riddoch F. A. and Ridley B. K. (1985) *Physica* **134B** 342.
Ridley B. K. (1963) *Proc. Phys. Soc.* **82** 954.
Ridley B. K. (1977) *J. Phys. C: Solid St. Phys.* **10** 1589.
Ridley B. K. (1983) *J. Phys. C.: Solid St. Phys.* **16** 3373.

Ridley B. K. (1982) *J. Phys. C.: Solid State Phys.* **15** 5899; (1984) **17** 5357.
Ridley B. K. (1988) *Semicond. Sci. Technol.* **3** 111.
Ridley B. K. (1989) *Phys. Rev.* **B39** 5282.
Ridley B. K. (1991) *Phys. Rev.* **B44** 9002.
Ridley B. K. (1991) *Rep. Progr. Phys.* **54** 169.
Ridley B. K. (1992) *Proc. SPIE* **1675** 492.
Ridley B. K. (1993) *Phys. Rev.* **B47** 4592.
Ridley B. K. (1993) *Quantum Processes in Semiconductors*, 3rd edn. Oxford University Press, Oxford.
Ridley B. K. (1994) *Phys. Rev.* **B49** 17253.
Ridley B. K. (1996a) *Semicond. Sci. Technol.* (to be published).
Ridley B. K. (1996b) *J. Phys.: Condens. Matter* **8** L511.
Ridley B. K. (1998) *J. Phys. Condens. Matter* **19** L607.
Ridley B. K. (1999) *Quantum Processes in Semiconductors* (4th edn). Oxford University Press.
Ridley B. K. (2001) *J. Phys. Condens. Matter* **13** 2799.
Ridley B. K., Al-Dossary O., Constantinou N. C., and Babiker M. (1994) *Phys. Rev.* **B50** 11701.
Ridley B. K. and Babiker M. (1991) *Phys. Rev.* **B43** 9096.
Ridley B. K. and Gupta R. (1991) *Phys. Rev.* **B43** 4939.
Ridley B. K. and Zakhleniuk N. A. (1996) submitted.
Ridley B. K., Schaff W. J., and Eastmann L. F. (2003) *J. Appl. Phys.* **94** 3972.
Ridley B. K., Schaff W. J., and Eastman L. F. (2005) *J. Appl. Phys.* **97** 094503.
Ridley B. K., Zakhleniuk N. A., and Bennet C. R. (2000) *Proc. Modeling and Simulation of Microsystems 2000* 412.
Rorison J. M. and Herbert D. C. (1986) *J. Phys. C: Solid St. Phys.* **19** 3991.
Rossi F., Bungaro C., Rota L., Lugli P., and Molinari E. (1994) *Solid St. Electron.* **37** 761.
Rotvig J., Jauho A.-P., and Smith H. (1996) *Phys. Rev. B* **54** 17691.
Rücker H., Molinari E., and Lugli P. (1991) *Phys. Rev.* **B44** 3463.
Rücker H., Molinari E., and Lugli P. (1992) **B45** 6747.
Rudin S. and Reinecke T. L. (1990) *Phys. Rev.* **B41** 7713.
Rudin S. and Reinecke T. L. (1991) **B43** 9298.
Rytov S. M. (1956) *Sov. Phys. Acoustics* **2** 67.
Sakaki H., Noda T., Hirakawa K., Tanaka M., and Matsuse T. (1987) *Appl. Phys. Lett.* **51** 1934.
Sawaki N. (1986) *J. Phys. C: Solid St. Phys.* **19** 4965.
Sawaki N. and Nishinaga T. (1977) *J. Phys. C.: Solid State Phys.* **10** 5003.
Schubert E. E., Pfeiffer L., West K. W., and Izabelle A. (1989) *Appl. Phys. Lett.* **54** 1350.
Seaford M. L., Martin G., Hartzell D., Massie S., and Eastman L. F. (1995) *Proc. Materials Res. Soc.* Spring.
Shah J. (1986) *IEEEJ. Quant. Electron.* **QE-22** 1728.
Shah J., Pinczuk A., Stormer H. L., Gossard A. C., and Wiegmann W. (1983) *Appl. Phys. Lett.* **42** 55.
Shanabrook B. V., Bennett B. R., and Wagner R. J. (1993) *Phys. Rev.* **B48** 17172.
Shannon C. E. (1949) *Proc. IRE* **37** 10.
Shealy J. R., Prunty T. R., Chumbes E. M., and Ridley B. K. (2003) *J. Crystal Growth* **250** 7.
Shi J.-J. (2003) *Phys. Rev. B* **68** 165335.
Shulman J. N. and Chang Y.-C. (1981) *Phys. Rev.* **B24** 4445.
Shulman J. N. and Chang Y.-C. (1985) **B31** 2056.

Sirenko Yu. M., Jeon J.-B., Kim K. W., Littlejohn M. A., and Stroscio M. A. (1997) *Phys. Rev. B* **53** 1997 (1996); *Phys. Rev. B* **55** 4360.
Sokolov V. N., Kim K. W., Kochelap V. A., and Woolard D. L. (2004) *Appl. Phys. Lett.* **84** 3630; (2005) *J. Appl. Phys.* **98** 064507.
Sood A. K., Menendez J., Cardona M., and Ploog K. (1985) *Phys. Rev. Lett.* **54** 2111, 2115.
Starikov E., Shiktorov P., Gruzinskis V. *et al.* (2001) *J. Appl. Phys.* **89** 1161; (2002) *Physica B* **314** 171.
Stern F. (1967) *Phys. Rev. Lett.* **18** 546.
Stern F. and Howard W. E. (1967) *Phys. Rev.* **163** 816.
Stratton R. (1962) *J. Phys. Chem. Solids* **23** 1011.
Strauch D. and Dorner B. J. (1990) *Phys. Condens. Matter* **2** 1457.
Stroscio M. A., Kim K. W., Iafrate G. J., Dutta M., and Grubin H. L. (1992) *Phil. Mag. Lett.* **65** 173.
Susuki M., Uenoyama T., and Yanase A. (1995) *Phys. Rev. B* **52** 8132.
Suzuki N. and Iizuka N. (1999) *Jpn. J. Appl. Phys.* **38** L363.
Takada Y. and Uemura Y. (1977) *J. Phys. Soc. Jpn.* **43** 139.
Takimoto N. (1959) *J. Phys. Soc. Jpn.* **14** 1142.
Trallero-Giner C. and Comas F. (1988) *Phys. Rev.* **B37** 4583.
Trallero-Giner C., Garcia-Moliner F., Velasco V. R., and Cardona M. (1992) *Phys. Rev.* **B45** 11944.
Tripathi P. and Ridley B. K. (2002) *Phys. Res.* **66** 195301.
Tripathi P. and Ridley B. K. (2003) *J. Phys. Condens.* Matter **15** 1057.
Tsang J. C. and Kash J. A. (1986) *Phys. Rev.* **B34** 6003.
Tsang J. C. and Kash J. A. (1991) in *Light Scattering in Solids VI* (ed. M. Cardona and G. Gunterodt). Springer, Berlin.
Tselis A. C. and Quinn J. J. (1984) *Phys. Rev.* **B29** 3318.
Tsen K. J., Joshi R. P., Ferry D. K., and Morkoc H. (1989) *Phys. Rev.* **B39** 1446.
Tsen K. J., Tsen S. Y., and Morkoc H. (1988) *SPIE* **942** 114.
Tsen K. T., Wald K. R., Ruf T., Yu P.-Y., and Morkoc H. (1991) *Phys. Rev. Lett.* **67** 2557.
Tsuchiya T., Akera H., and Ando T. (1989) *Phys. Rev.* **B39** 6025.
Ushioda S. and Loudon R. (1982) in *Surface Polaritons* (ed. V. M. Agranovitch and D. Mills). North-Holland, Amsterdam.
Vallée F. and Bogani F. (1991) *Phys. Rev. B* **43** 12049.
van Hall P. J. (1989) *Superlat. Microstruct.* **6** 213.
van Hall P. J., Klaver T., and Wolter J. H. (1988) *Semicond. Sci. Technol.* **3** 120.
Vassell M. O., Ganguly A. K., and Conwell E. M. (1970) *Phys. Rev.* **B2** 948.
Vickers A. J. (1992) *Phys. Rev.* **B46** 13313.
Voisin P., Bleuse J. Bouche C. *et al.* (1988) *Phys. Rev. Lett.* **61** 1639.
von Hippel A. (1937) *J. Appl. Phys.* **8** 815.
von der Linde D., Kuhle J., and Klingenberger H. (1980) *Phys. Rev. Lett.* **44** 1505.
Wacker A. (1998) *Theory of Transport Properties of Semiconductor Nanostructures* (ed. E. Schöll E.). Chapman and Hall, London, p. 321.
Wang X. F. and Lei X. L. (1994) *Solid St. Commun.* **91** 513.
Wannier G. H. (1960) *Phys. Rev.* **117** 432.
Wannier G. H. (1962) *Rev. Mod. Phys.* **34** 645.
Wannier G. H. (1969) *Phys. Rev.* **181** 1364.
Watson G. N. (1944) *A Treatise on the Theory of Bessel Functions*. Cambridge University Press, Cambridge.
Weber G. and Ryan J. F. (1992) *Phys. Rev.* **B45** 11202.

Weimann N. G. and Eastman L. F. (1998) *J. Appl. Phys.* **83** 3656.
Welch D. F., Wicks G. W., and Eastman L. F. (1984) *J. Appl. Phys.* **55** 3176.
Wendler L. and Haupt R. (1987) *Phys. Stat. Sol.* (b) **141** 493.
Wendler L., Haupt R., Bechstedt F., Rucker H., and Enderlein R. (1988) *Superlatt. Microstr.* **4** 577.
Wendler L. and Pechstedt R. (1986) *Phys. Stat. Sol.* (b) **138** 197.
Wendler L. and Pechstedt R. (1987) *Phys. Stat Sol.* **141** 129.
Whittaker E. T. (1915) *Proc. Roy. Soc. (Edin.)* **35** 181.
Wiley J. D. (1971) *Phys. Rev.* **B4** 2485.
Williams B. S. (2007) *Nature Photonics* **1** 517.
Wu J., Walukiewicz W., Shan W. *et al.* (2002) *Phys. Rev. B* **66** 201403.
Yafet Y. (1963) *Solid State Phys.* **14** 1.
Yanchev I. Y., Arnovdov B. G., and Evtimova S. K (1979) *J. Phys. C: Solid St. Phys.* **12** L765.
Yeo Y. C., Chong T. C., and Li M. F. (1998) *J. Appl. Phys.* **83** 1429.
Yip S.-K. and Chang Y.-C. (1984) *Phys. Rev.* **B30** 7073.
Yu-Kuang Hu B. and Das Sarma S. (1993) *Phys. Rev.* **B48** 5469.
Yussouf M. and Zittartz J. (1973) *Solid St. Commun.* **12** 959.
Zak J. (1968) *Phys. Rev. Lett.* **20** 1477.
Zak J. (1969) *Phys. Rev.* **81** 1366.
Zak J. (1996) *J. Phys.: Condens. Matter* **8** 8295.
Zakhleniuk N. A., Bennett C. R., Constantinou N. C., Ridley B. K., and Babiker M. (1996) *Phys. Rev.* B **54** 17838.
Zawadski W. and Szymanska W. (1971) *Phys. Stat. Sol.* (b) **45** 415.
Zener C. (1934) *Proc. Roy. Soc.* **145** 523.
Zianni X., Butcher P., and Dharssi J. (1992) *J. Phys. Condens. Matter* **4** L77.
Zucker J. E., Pinczuk A., Chemla D. S., Gossard A., and Wiegman W. (1984) *Phys. Rev. Lett.* **53** 12280.

Index

2D electrons 2
acoustic boundary conditions 67, 84, 92
acoustic mode spectrum 10
acoustic modes 3
acoustic-phonon scattering 290–6
Airy functions 360, 366, 368
alloy 205–6
  scattering 6, 230–1
AlN/GaN superlattice 360–6
angular momentum 62
anisotropy 61
anticrossings 61
asymmetric modes 18

ballistic transport 378–87
band-mixing effects 64
barrier modes 182, 195
biaxial strain 56
Bir–Aranov–Pikus mechanism 326–9
Bloch oscillations 370, 383–7
Bohr energy 220, 311
Boltzmann equation 275–6, 301, 314
Born approximation 219, 225, 231
Bose–Einstein number 276
boundary conditions 3, 6, 9, 10, 44, 58, 60, 68, 69, 85, 92
  effective mass 48, 49, 50
  EM 18
  mechanical 17, 18, 68, 119
Bragg reflections 370
branch point energy 350
Brillouin zone boundary 370
bulklike phonons 6
  spectrum 22

Camley–Mills dispersion 154
capture processes 2
cascade lasers 369
charged-impurity scattering 6, 217, 223, 236, 345
  phonons 239–40
classic elasticity 97
coherence length 2
coherent scattering 219
complex band structure 65, 66
conducting layers 160
confinement mass 65
connection rules 71, 72
continuity of current 42
continuity of probability 42
continuum model 21
continuum theory 67
Coulomb potential 217
crystal momentum 24, 27
crystals 114, 359
current continuity 3
CW generators 389

DC model *see* dielectric-continuum (DC) model
Debye spectrum 26
deformation potential 24
  interaction 17
density matrix 244
density-of-states 2, 29
dielectric-continuum (DC) model 6, 11–16, 35, 38, 41, 112, 126, 137, 139, 154, 173, 175, 176, 179, 199, 200, 201, 204, 212
dielectric response function 260, 268
dipole fluctuation model 225
Dirac delta function 77
dislocation lines 345
dispersion 22, 70
double hybrids 22, 142, 144
DX centre 160, 225
D'yakonov–Perel process 317–22

effective-mass approximation 3, 43
effective-mass equation 42–9, 43, 44, 48
effective-mass tensor 43
effective-mass theory 9, 10, 43, 51
eigenfunctions 55, 340–2
Einstein spectrum 26
electromagnetic (EM)
  boundary conditions 12, 68
electron
  affinity 42
  confinement 9, 24, 49–53

electron (*cont.*)
  scattering rate 17
  states
    symmetry 24
  temperature 306
electron–electron scattering 6, 231–6
electron–hole scattering 236
electron–hybridon interaction 41
electron–phonon interaction 6, 10, 11, 134
electron–phonon scattering 22
electron transfer mechanism 373
Elliot–Yafet process 313–17, 340
elliptical cross-section 176
EM *see* electromagnetic
energy relaxation 24
envelope function
  equation 47, 48, 77
  theory 4, 5, 9, 43, 45
equations of motion 81, 82
Esaki–Tsu approach 375

Fang–Howard 51, 353
Fermi–Dirac 290, 300, 301
Fermi's golden rule 23
ferroelectrics 337
field plate 354
field quantization 113
force-constant approximation 72
form factor 31
Franz–Keldysh effect 353
Fresnel equation 344
Fröhlich interaction 13, 26, 135
Fuchs–Kliewer surface modes 13, 175

GaAs quantum wells 125
GaAs/AlAs superlattice 148
Gaussian 220, 224
Ge/Si 119
Gibbs phenomenon 47
Gruneisen constant 237
Gruneisen–Bloch 304
Gunn effect 373

Hall mobilities 345
Hamiltonian density 107, 113
Hartree approximation 248
HD model *see* hydrodynamic (HD) model
HEMT *see* high electron mobility transistor
hermiticity 44
HFET 354
high electron mobility transistor (HEMT) 51
high-temperature approximation 306
holes 53
  confinement 53–61
  heavy 53, 55, 312
  light 53, 55, 312
hot-electron luminescence 280
Hooke's law 97
hot-electron theory 24
hot-phonon effects 7, 236
hybrid model 41
hybrid modes 21–2, 127, 138, 143
hybrid theory 6
hybrids
  scattering 184–5, 187–9
hydrodynamic (HD) model 16–18, 35, 37, 38, 73, 126
hyperfine coupling 329–32

InAs/GaSb system 86, 91, 93
in-plane effective mass 60, 61
interband coupling 48
interband transition 17
interface effects 86, 91
interface modes 10, 14, 16, 38
interface phonons 13
interface polaritons 13
  scattering 194–7
interface-roughness scattering 6, 227–30, 241–3
interface terms 85
intersubband 2, 198–9, 364
intrasubband 2, 182, 197–8
IP
  dielectric response 257
  dispersion 127
  modes 133, 154
    field factor 132

Jacobi polynomials 63

Kane 338
k.p.
  interaction 53, 55, 340, 342
  theory 43, 313, 317, 338, 360

ladder technique 320
Lagrangian 78, 79, 80, 104
Landau damping 257, 259, 271, 272, 274
lattice
  dynamics 148
  mismatch 53
  screening 265–6
Legendre polynomials 282, 283
Lindhard formula 250, 273
linear-chain model 68, 69, 69–76
Liouville equation 246
LO *see* longitudinally polarized (LO)
local approximation 47, 82
local potential 46
localized mode 74
longitudinally polarized (LO)
  hybridization 68, 183
  modes 11
  rates 199
  scattering 374–5
  vibration patterns 12
LS coupling 331
Luttinger parameters 55
Lyddane–Sachs–Teller formula 14, 105, 113, 115, 118

magnetic dipole moment 311
mass approximation 72

mass reversal 60
Matthews–Blakeslee limit 353
Maxwell–Boltzmann statistics 279, 281, 290, 292
MCA *see* momentum conservation approximation
metacontinuum 92
metal–semiconductor structures 170–3
micro-Raman scattering 22
microscopic calculations 18
microscopic potentials
  local 46
  non-local 46
microscopic theory 6
mode hybridization 22
mode patterns 17
modulation doping 217
momentum conservation approximation (MCA) 29, 30, 31, 34
momentum mass 65
momentum relaxation 24, 303
monolayer 53, 160, 161
  single 162–6
  double 166–8
Monte Carlo simulations 235
Mott formula 231

NaCi slab 173
negative differential resistance (NDR) 272, 372, 373, 373–4, 380–3
non-local operator 78, 79, 82
non-polar interaction 302
non-parabolicity 61, 64, 65
nuclear dipole 329

off-diagonal elements 58
optical excitation 278–81
optical mode 3
  confinement 10
  hybridization 6, 41
optical phonon, 343
  confinement of 3, 6
  engineering 160
  tunnelling 127
optical stress
  boundary conditions 85
overlap integral 23, 27, 62, 63

pair correlation 220
parallel spins 235
particle in a box 10
Peierls transition 261
permittivity 14
phase-shift analysis 219, 231
phonon
  cascade 280
  confinement 9, 24
    continuum model 11
  lifetime 236
  resonances 206–8
phonon frequency
  quantization 10
photoluminescence spectrum 279
piezoelectric interaction 26, 302
piezoelectric polarization 337, 349, 351
piezoelectric rate 31
piezoelectric scattering 296, 309
plasma frequency 259
plasma oscillations 251
plasma thermalization 235
plasmons 251, 265, 272
Poisson statistics 220, 225
Poisson's equation 248
polar interaction 17, 184
polariton dispersion 147
polarization selection rules 17
polarization superlattice 358–60
pseudopotential techniques 66
pyroelectric 337

quantum-cascade laser 366
quantum dots 51, 160, 181
quantum well 17
quantum wires 51, 160, 173, 176–81, 296, 297, 309, 324–6
quasi-1D dielectric function 259–65
quasi-2D dielectric function 252–9
quasi-continuum 9, 46, 68, 76

Raman scattering 10, 16, 17, 154, 175, 280
Rashba mechanism 326
Rayleigh waves 175
rectangular wires 34
reformulated-mode (RM) model 18–21
reformulated modes 19, 126
RM *see* reformulated-mode (RM) model
resonant scattering 220
rotation matrix 62

scalar potential 12, 16, 182, 212–16
scattering 2
  single charges 220–3
scattering rate 6, 24
Schottky barrier 356
Schrödinger equation 54
screening 272, 302, 306
secular determinant 57
semiconductor multilayer 2
Shubnikov–de Haas oscillations 300, 306, 309
single-particle excitation 251, 257
single quantum well 49, 50
slab modes 173–5
Slater determinant 245
Sommerfield factor 328
spin-polarized population 312
spin relaxation 314, 318
spin states 64
spin-density matrix 318
spin-orbit coupling 53
spin-orbit splitting 55
split-off bands 53, 55, 61, 63
spintronics 311
spontaneous electric polarization 337, 346, 346–7, 349, 353

static-screening approximation 225
strain splitting 60
strained layer 56
stress–strain energy density 98
subbands 2, 52
sum rule 22, 168, 209–12
superlattices 50, 141, 160, 205, 373
surface barrier heights 350
surface plasma 172
symmetric modes 17

threshold rates 189–92
tight-binding techniques 66
triple hybrids 22, 141, 182
two-mode case 116

umklapp processes 27, 222

valence band 56, 359
vector potential 16–354, 182, 212–16

Wannier–Stark states 370, 371, 373
wavevector-space method 9
well modes 182, 195, 200
Wurtzite lattice 336–7

Printed in the United States
By Bookmasters